D1767112

51 Springer Series in Solid-State Sciences

Edited by Manuel Cardona

Springer Series in Solid-State Sciences

Editors: M. Cardona P. Fulde H.-J. Queisser

Volume 40 **Semiconductor Physics** - An Introduction By K. Seeger

Volume 41 **The LMTO Method** Muffin-Tin Orbitals and Electronic Structure
By H.L. Skriver

Volume 42 **Crystal Optics with Spatial Dispersion and the Theory of Excitations**
By V.M. Agranovich and V.L. Ginzburg

Volume 43 **Resonant Nonlinear Interactions of Light with Matter**
By V.S. Butylkin, A.E. Kaplan, Yu.G. Khronopulo, and E.I. Yakubovich

Volume 44 **Elastic Media with Microstructure II** Three-Dimensional Models
By I.A. Kunin

Volume 45 **Electronic Properties of Doped Semiconductors**
By B. I. Shklovskii and A. L. Efros

Volume 46 **Topological Disorder in Condensed Matter**
Editors: F. Yonezawa and T. Ninomiya

Volume 47 **Statics and Dynamics of Nonlinear Systems**
Editors: G. Benedek, H. Bilz, and R. Zeyher

Volume 48 **Magnetic Phase Transitions**
Editors: M. Ausloos and R. J. Elliott

Volume 49 **Organic Molecular Aggregates,** Electronic Excitation and Interaction Processes
Editors: P. Reineker, H. Haken, and H. C. Wolf

Volume 50 **Multiple Diffraction of X-Rays in Crystals**
By Shih-Lin Chang

Volume 51 **Phonon Scattering in Condensed Matter**
Editor: W. Eisenmenger

Volumes 1 - 39 are listed on the back inside cover

Phonon Scattering in Condensed Matter

Proceedings of the
Fourth International Conference
University of Stuttgart, Fed. Rep. of Germany
August 22–26, 1983

Editors:
W. Eisenmenger, K. Laßmann, and S. Döttinger

With 321 Figures

Springer-Verlag
Berlin Heidelberg New York Tokyo 1984

Professor Dr. Wolfgang Eisenmenger
Dr. Kurt Laßmann
Dr. Siegfried Döttinger
Physikalisches Institut der Universität Stuttgart, Pfaffenwaldring 57
D-7000 Stuttgart 80, Fed. Rep. of Germany

Series Editors:
Professor Dr. Manuel Cardona
Professor Dr. Peter Fulde
Professor Dr. Hans-Joachim Queisser
Max-Planck-Institut für Festkörperforschung, Heisenbergstrasse 1
D-7000 Stuttgart 80, Fed. Rep. of Germany

ISBN 3-540-12954-5 Springer-Verlag Berlin Heidelberg New York Tokyo
ISBN 0-387-12954-5 Springer-Verlag New York Heidelberg Berlin Tokyo

Offset printing: Beltz Offsetdruck, 6944 Hemsbach/Bergstr. Bookbinding: J. Schäffer OHG, 6718 Grünstadt.
2153/3130-543210

Preface

This volume contains the proceedings of the Fourth International Conference on Phonon Scattering in Condensed Matter held from August 22-26, 1983 at the University of Stuttgart. The preceding conferences were organized at Saint Maxime and Paris in 1972, at the University of Nottingham in 1975, and at the Brown University Providence/Rhode Island in 1979.

The Stuttgart conference, like the preceding conferences, was mainly concerned with "propagating" high-frequency acoustic phonons, mechanical waves and heat up to the lattice limiting frequency. Lattice dynamics, optical phonons, phase transitions, etc., were included as far as they are involved in acoustical phonon scattering, propagation and generation. In this context the conference covered all aspects of acoustical phonon physics, especially generation of phonons, propagation, scattering and detection. Since acoustic phonons participate in most energy-transfer processes in solids and liquids, the field of interest is growing rapidly. Therefore exciting new developments of acoustic phonon physics could be presented at the Stuttgart conference as well as important progress with respect to well-known problems, as, for example, the Kapitza resistance.

Two hundred and six scientists from 21 countries attended the conference. Thirteen invited papers and 105 contributed papers, with 34 as posters, were presented. The discussions are included in this volume.

A discussion session on large wave vector phonons was organized and chaired by V. Narayanamurti. A discussion session on phonon scattering at interfaces was organized and chaired by R.O. Pohl.

The conference was supported and sponsored by the Deutsche Forschungsgemeinschaft, the International Union of Pure and Applied Physics, the Land Baden Württemberg, the University of Stuttgart, the Deutsche Physikalische Gesellschaft and the E. Leitz Company, Wetzlar. The support and sponsorship of these organizations is gratefully acknowledged.

We extend our thanks to the Rektor of the University of Stuttgart, Prof.Dr.rer.nat. H. Zwicker, who opened the conference and gave a reception for the delegates.

We should like to thank all members of the international advisory committee for their extremely helpful suggestions and recommendations.

We thank our colleagues of the organization committee K. Dransfeld, E. Sigmund and M. Wagner for close and fruitful cooperation. Special thanks are due to our colleague H.-J. Bauer for his advice and assistance in adapting our data handling system to the needs of the conference. In particular we thank Mrs. R. Mann, Mrs. I. Poljak and Mrs. G. Untereiner as well as all our co-workers and students for their invaluable assistance and help during all stages of the conference.

Stuttgart, December 1983 *W. Eisenmenger · K. Laßmann · S. Döttinger*

Contents

Part I Acoustic Phonon Spectroscopy

*Crossing Effects in Phonon Scattering
By L.J. Challis (With 10 Figures) ... 2

*Far Infrared Volume Generation and Detection of Phonons
By K.F. Renk (With 5 Figures) ... 10

*The Josephson Junction – A New Tunable Phonon Source with High-Frequency Resolution. By P. Berberich and H. Kinder (With 5 Figures) ... 18

*Two Applications of Microwave Acoustics in Liquid Helium: High Resolution Microscopy and Direct Measurement of Phonon Dispersion
By D. Rugar (With 5 Figures) ... 26

Heater Film Dynamics, Phonon Diffusion and Phonon Decay. By T.E. Wilson, W.E. Bron, F.M. Lurie, and W.L. Schaich (With 2 Figures) ... 34

Infrared Excitation of High-Frequency Phonons by Multiphonon Absorption
By U. Happek, R. Baumgartner, and K.F. Renk (With 2 Figures) ... 37

Zero-Field Hyperfine Splitting of $Al_2O_3:V^{3+}$ by Josephson Phonon Spectroscopy. By A. Schick, P. Berberich, W. Dietsche, and H. Kinder (With 2 Figures) ... 40

Subnanosecond Bolometry Using Niobium Films
By J.P. Maneval, J. Desailly, and B. Pannetier (With 2 Figures) ... 43

Generation of High-Frequency Phonons by Metallic Point Contacts
By R.J.G. Goossens, A.G.M. Jansen, J.I. Dijkhuis, P. Wyder, and H.W. de Wijn (With 4 Figures) ... 46

Nonequilibrium Phonon Distribution in a Quantizing Magnetic Field: A Tunable GHz to THz Phonon Generator?
By G.W. Slater and A.-M.S. Tremblay (With 3 Figures) ... 49

Relaxation Processes and Response Time of Superconductors to a Periodic Excitation. By C. Vanneste and N. Perrin (With 3 Figures) ... 52

Amplitude Modulated Heat Pulses
By L. Hirschbiegel, M. Siemon, and W. Grill (With 2 Figures) ... 55

*invited paper

Thermal Conductivity Measurement Under Applied Uniaxial Stress
By B. Salce (With 2 Figures) ... 58

Thin Piezoelectric PVDF-Layers as Ultrasonic Transducers. By A. Ambrosy, K. Holdik, W. Scheitler, and H. Schulze (With 7 Figures) ... 61

Discussions ... 64

Part II Phonon Focusing

Phonon Focusing in Germanium Imaged by Electron-Beam Scanning
By W. Metzger, R. Eichele, H. Seifert, and R.P. Huebener (With 3 Figures) ... 72

Nonlinear Phonon Focusing. By D. Armbruster, G. Dangelmayr, and W. Güttinger (With 7 Figures) ... 75

Phonon Focusing in Highly Dispersive and Isotopically Impure Crystals
By S. Tamura (With 4 Figures) ... 78

Spatial Variation of Phonon Distribution in Thermal Conduction
By F.W. Sheard, G.A. Toombs, and S.R. Williams (With 2 Figures) ... 81

Discussions ... 84

Part III Large Wave-Vector Phonons

Observation of a Quasi-Diffusive Phonon Propagation Mode. By W.E. Bron, J.M. O'Connor, and Y.B. Levinson (With 2 Figures) ... 88

Propagation of Near Zone Boundary Acoustic Phonons in Solid (hcp) ^{4}He
By T. Haavasoja, V. Narayanamurti, and M.A. Chin (With 4 Figures) . 91

Scattering of Debye Phonons by Substitutional Defects in Solids
By V.P. Srivastava ... 94

Direct Observation of Ballistic Large-Wavevector Phonon Propagation in Gallium Arsenide. By B. Stock, R.G. Ulbrich, and M. Fieseler (With 2 Figures) ... 97

Search for Large k-Vector Phonons in GaAs
By J.P. Wolfe and G.A. Northrop (With 3 Figures) ... 100

On the Propagation of Long-Lived Short Wavelength TA Phonons in GaAs
By M. Lax, V. Narayanamurti, R. Ulbrich, and N. Holzwarth (With 4 Figures) ... 103

Temperature Dependence of Optical Phonon Lifetimes. By J. Kuhl, B.K. Rhee, and W.E. Bron (With 2 Figures) ... 106

Anharmonic Decay of High-Energy LA Phonons
By S. Tamura and K. Okubo (With 3 Figures) ... 109

Lifetime and Linewidth of Resonant 40-cm^{-1} Phonons in Alexandrite
By R.J.G. Goossens, J.I. Dijkhuis, and H.W. de Wijn (With 3 Figures) 112

High-Frequency Phonon Dynamics in LaF_3 Using Monoenergetic Optical Detection Methods. By R.S. Meltzer, J.E. Rives, D.J. Sox, and G.S. Dixon (With 3 Figures) 115

Raman Spin-Lattice Relaxation Induced by Optically Generated Zone-Boundary Phonons in Ruby. By J.G.M. van Miltenburg, J.I. Dijkhuis, and H.W. de Wijn (With 2 Figures) 118

Raman Probe of the Brillouin Zone for Nonequilibrium Phonons in GaAs By R. Bray, K.T. Tsen, and K. Wan (With 1 Figure) 121

Phonon Decay in X-Ray Irradiated Ruby Crystals By M. Engelhardt and K.F. Renk (With 4 Figures) 124

Decay of a Highly Excited Phonon Mode By P. Ullersma (With 1 Figure) 127

Stimulated Emission and Decay of Phonons Resonant Between the Zeeman States of $\bar{E}(^2E)$ in Ruby. By J.G.M. van Miltenburg, G.J. Jongerden, J.I. Dijkhuis, and H.W. de Wijn (With 3 Figures) 130

Raman Scattering and the Two-Phonon Density of States in GaAs By M. Lax, V. Narayanamurti, R.C. Fulton, R. Bray, K.T. Tsen, and K. Wan (With 2 Figures) 133

Propagation of Phonons Generated by Nonradiative Transitions in KCl-NO_2^- By I. Sildos, G. Zavt, and I. Dolindo (With 2 Figures) 136

Discussions 139

Part IV Surfaces, Interfaces, Kapitza Resistance

*Phonon-Induced Desorption of Helium By P. Taborek (With 4 Figures) 148

*Thermal Boundary Resistance Between Small Particles and Liquid He-3 By T. Nakayama (With 5 Figures) 155

*Scattering and Absorption of Ballistic Phonons by the Electron Inversion Layer in Silicon: Theory and Experiment. By J.C. Hensel, R.C. Dynes, B.I. Halperin, and D.C. Tsui (With 6 Figures) 163

*Low Wavevector Phonons in the 2-Dimensional Electron Solid on Liquid Helium. By F.I.B. Williams (With 5 Figures) 171

Reciprocity Theorem for Phonon Transitions at Ideal Interfaces Within the Acoustic Mismatch Model. By O. Weis (With 3 Figures) 179

Phonon Scattering by Twin Planes By J.W. Vandersande, P.N. Chopra, and R.O. Pohl (With 3 Figures) .. 182

Boundary and Dislocation Scattering of Phonons in Lead Single Crystals By W. Odoni, P. Fuchs, and H.R. Ott (With 2 Figures) 185

Diffuse Scattering of Thermal Phonons at Crystal Surfaces By T. Klitsner and R.O. Pohl (With 2 Figures) 188

*invited paper

Imaging of Specularly Reflected Phonons from a Crystal Boundary
By G.A. Northrop (With 2 Figures) 191

Critical-Cone Channeling of Thermal Phonons from Solid/Solid Interfaces
By A.G. Every, G.L. Koos, G.A. Northrop, and J.P. Wolfe
(With 2 Figures) 194

Anomalous Low Temperature Kapitza Resistance of a Paramagnetic Salt
By G.J. Batey and P.C. Main (With 3 Figures) 197

A Size Effect in the Kapitza Resistance to Dilute ^{3}He-^{4}He Mixtures
By F. Guillon, J.P. Harrison, and A. Sachrajda (With 2 Figures) ... 200

Kapitza Resistance Near 1 mK — The Shaking Box Model. By A.R. Rutherford,
J.P. Harrison, and M.J. Stott (With 2 Figures) 203

Phonon and Roton-Induced Evaporation. By A.F.G. Wyatt, M.J. Baird, and
F.R. Hope (With 4 Figures) 206

Spectral Dependence of the Kapitza Resistance Between 0.5 K and 2.3 K
By O. Koblinger, E. Dittrich, U. Heim, M. Welte, and W. Eisenmenger
(With 3 Figures) 209

Kapitza Resistance of Laser-Annealed Surfaces. By H.C. Basso,
W. Dietsche, H. Kinder, and P. Leiderer (With 2 Figures) 212

Discussions 215

Part V Quantum Liquids and Crystals

*Crystallization Waves in Helium. By A.Ya. Parshin (With 6 Figures) 226

*Phonon Transmission and the Kapitza Resistance Between Liquid and
Solid Helium. By H.J. Maris (With 7 Figures) 234

*Transmission of Sound at the Solid-Liquid Interface of ^{4}He
By B. Castaing, L. Puech, and G. Bonfait (With 3 Figures) 241

Propagation of High Frequency Phonons in Liquid He II. By T. Haavasoja,
V. Narayanamurti, and M.A. Chin (With 4 Figures) 249

Ultrasonic Attenuation in KCl:OH$^-$ with High OH Concentration
By M. Saint-Paul, R. Nava, and J. Joffrin (With 2 Figures) 252

Localized Phonon Mode Associated with Dislocations in Alkali Halide
Crystals. By Y. Hiki, Y. Kogure, and F. Tsuruoka (With 3 Figures) . 254

^{3}He-^{3}He Interaction in ^{3}He-^{4}He Liquid Mixtures Determined from Sound
Attenuation. By A.J. Ikushima, I. Fujii, M. Fukuhara, and
K. Kaneko (With 2 Figures) 257

Ultrasonic-Attenuation and Pressure Measurements in Phase-Separated
Solid ^{3}He-^{4}He Mixtures
By I. Iwasa and H. Suzuki (With 2 Figures) 260

*invited paper

Effects of Anharmonicities and Broken Time-Reversal Invariance on Static and Dynamic Properties of Two-Dimensional Electron Solids
By G. Meissner (With 1 Figure) 263

Charge-Induced Deformation of the ^4He Solid-Superfluid Interface
By J. Bodensohn, P. Leiderer, and D. Savignac (With 2 Figures) 266

Pinning of Dislocations in Helium by Large Ultrasonic Stresses
By J.R. Beamish and J.P. Franck (With 1 Figure) 269

Discussions 272

Part VI Cooperative Phenomena

Ultrasonic Velocity and Attenuation Near the Cooperative Jahn-Teller Dilation in Cerium Ethyl Sulphate
By J.T. Graham and J.H. Page (With 2 Figures) 278

Ultrasonic Attenuation at Megahertz Frequencies in the Cooperative Jahn-Teller System $TmVO_4$. By J.H. Page and S.R.P. Smith (With 3 Figures) 281

Thermoelectric Power of $TiSe_{2-x}S_x$ Mixed Crystals. By A.A. Lakhani, S. Jandl, J.P. Jay-Gerin, and C. Ayache 284

Ultrasonic Study of Melting of Crystalline Solids
By Y. Hiki and J. Tamura (With 3 Figures) 285

Piezoacoustic Observation of Acoustic Soft Mode in KH_2PO_4 Crystal
By J.Y. Koo, T.W. Yoo, and J.J. Kim (With 3 Figures) 288

Brillouin-Scattering Study of Sound Velocity of Quartz at α-β Transition. By H. Unoki, H. Tokumoto, and T. Ishiguro (With 2 Figures) 292

Phonon-Soliton Interaction in K_2SeO_4
By W. Rehwald (With 3 Figures) 295

Thermal Conductivity of Cooperative Jahn-Teller E-b_1,b_2 Systems
By W. Mutscheller and M. Wagner (With 2 Figures) 298

Part VII Free Carriers

Magnetic Field Dependence of Ultrasonic Attenuation in Heavily Doped Ge:Sb. By H. Sakurai, K. Suzuki, and T. Miyasato (With 2 Figures)... 302

Phonon Attenuation in Heavily Doped p-Type Semiconductors
By T. Sota, K. Suzuki, and D. Fortier (With 2 Figures) 304

Ballistic Phonon Transport in Ge:P Under Magnetic Field
By T. Miyasato, M. Tokumura, and K. Suzuki (With 4 Figures) 307

Hole-Phonon Interaction in Wurtzite-Type Semiconductors
By M. Singh and J. Leotin (With 2 Figures) 310

The Observation of Strongly Coupled Magnetic Ions in Al_2O_3 by Low Temperature Thermal Expansion Measurements
By I.J. Brown and M.A. Brown (With 2 Figures) 313

Sound Velocity Measurements in Highly Oriented and Intercalated Graphite Specimen by Direct Electromagnetic Excitation of Ultrasound. By K. de Groot, V. Müller, D. Maurer, V. Geiser, and H.-J. Güntherodt (With 3 Figures) 316

The Effect of Electron Relaxation on Damping of Long-Wavelength Phonons in Metals and Heavily Doped Semiconductors. By I.P. Ipatova, A.V. Subashiev, and V.A. Shchukin (With 1 Figure) 319

Nuclear Acoustic Resonance Measurements of the Electron-Phonon Interaction in bcc Transition Metals. By V. Müller, E.-J. Unterhorst, and W. Neumann (With 3 Figures) 322

Revision of the Statistical Mechanics of Phonons to Include Phonon Linewidths. By W.C. Overton Jr. (With 1 Figure) 325

Phonon Emission and Electron Heating in a Two-Dimensional Electron Gas
By M.A. Chin, V. Narayanamurti, H.L. Stormer, and J.C.M. Hwang (With 4 Figures) 328

Density and Field Dependence of the Phonon-Limited Hot-Electron Temperature in n-Si Inversion Layers. By R.A. Höpfel, E. Vass, and E. Gornik (With 2 Figures) 331

Amplification of Total-Reflection-Mode Surface Phonons in n-Type InSb Films. By C.-C. Wu (With 3 Figures) 335

Time-Resolved Photoluminescence and Phonon Transport in Amorphous Si:H Films. By U. Strom, J.C. Culbertson, P.B. Klein, and S.A. Wolf (With 3 Figures) 338

Surface Acoustic Waves in Metals. By J. Heil, I. Kouroudis, C. Lingner, and B. Lüthi (With 2 Figures) 341

Discussions 344

Part VIII Defects

*Phonon Scattering by Dislocations. By A.C. Anderson (With 4 Figures) . 348

The Acoustic Paramagnetic Resonance of Cr^{2+} in n-Type GaAs
By A.S. Abhvani, C.A. Bates, P.J. King, D.R. Pooler, V.W. Rampton, P.C. Wiscombe, and P. Bury (With 1 Figure) 355

Magnetic-Field-Dependent Phonon Scattering by Li Ions in Si
By L.J. Challis, A.P. Heraud, V.W. Rampton, M.K. Saker, and M.N. Wybourne (With 2 Figures) 358

Influence of Defects on the Splitting of the Acceptor Ground State in Silicon. By A. Ambrosy, A.M. de Goër, K. Laßmann, B. Salce, and H. Zeile (With 4 Figures) 361

Phonon Scattering at Electronically Degenerative Systems: An Application to the Defect Systems Si(In), Si(B), and GaAs (Mn)
By J. Maier and E. Sigmund (With 6 Figures) 364

*invited paper

Magnetothermal Conductivity of Boron-Doped-Silicon
By L.J. Challis and A.P. Heraud (With 2 Figures) 368

An Additive Conservation Law for Phonon Collision Operator in Molecular Crystals. By B. Perrin (With 2 Figures) 371

Discussions 374

Part IX Two-Level Systems

*Low-Energy Excitations in Disordered Solids: New Aspects
By S. Hunklinger (With 3 Figures) 378

Anomalous Low-Temperature Ultrasonic Behaviour in a Fluoride Glass Containing Mn. By P. Doussineau, A. Levelut, M. Matecki, and W.D. Wallace (With 2 Figures) 386

Vibrational Dynamics of Lithium Ions in β-Alumina Crystals
By R. Di Valerio, A. Fontana, G. Mariotto, and M. Montagna (With 4 Figures) 389

Low-Energy Excitations in $(KBr)_{1-x}(KCN)_x$ in the Orientational Glass State. By M. Meissner, J.J. de Yoreo, R.O. Pohl, and S. Susman (With 2 Figures) 392

Spectral Hole Burning of Dyes, Probing Phonon Processes at Surfaces and in Amorphous Systems. By U. Bogner and G. Röska (With 2 Figures) .. 395

Evidence of Two-Level Systems in Electrolyte Glass by Brillouin Scattering. By J. Pelous, R. Vacher, A. Essabouri, U. Reichert, and M. Schmidt (With 2 Figures) 398

Acoustic Absorption Due to Hydrogen Tunneling in $NbN_{0.0015}H_{0.0025}$
By J.L. Wang, G. Weiss, H. Wipf, and A. Magerl (With 1 Figure) 401

Phonon Scattering in Phosphorous-Implanted Silicon
By M. Grimshaw and G. Feuillet (With 1 Figure) 404

Relaxation Ultrasonic Attenuation Measurements in Quartz Slightly Disordered by Neutron Irradiation. By C. Laermans and V. Esteves (With 2 Figures) 407

Low-Energy Excitations in Zr-Based Amorphous Alloys Studied by Thermal Conductivity and Specific Heat. By J.C. Lasjaunias, A. Ravex, and O. Béthoux (With 3 Figures) 410

Low-Temperature Thermal Properties of Amorphous Zr_xCu_{1-x} After Structural Relaxation. By H.J. Schink, S. Grondey, H. v. Löhneysen, and K. Samwer (With 1 Figure) 413

Time-Dependent Specific Heat of Vitreous Silica Between 0.1 and 1 K
By W. Knaak and M. Meissner (With 3 Figures) 416

Phonon-Dispersion Measurements in Glasses. By M. Rothenfusser, W. Dietsche, and H. Kinder (With 3 Figures) 419

*invited paper

The Influence of Two-Level States on the Thermal Conductivity of Amorphous Materials. By D.E. Farrell, J.E. de Oliveira, and H.M. Rosenberg (With 2 Figures) 422

Low-Frequency Elastic Loss in Dielectric and Metallic Glasses at Low Temperature. By H. Tietje, M. v. Schickfus, E. Gmelin, and H.-J. Güntherodt (With 3 Figures) 425

Phonon Absorption Due to Two-Level Systems in Metallic Glasses By N. Thomas (With 1 Figure) 428

Resonant Interaction of Acoustic Waves with Two-Level Systems in a Fluorozirconate Glass. By R. Vacher, J. Pelous, M. Schmidt, P. Doussineau, and A. Levelut (With 2 Figures) 431

Heat Treatment Effects on the Phonon-Electron Contribution to Thermal Conduction in Amorphous $Zr_{70}Cu_{30}$. By P. Esquinazi and F. de la Cruz (With 2 Figures) 434

Disorder-Induced Light Scattering in α-AgI. By E. Cazzanelli, A. Fontana, G. Mariotto, V. Mazzacurati, G. Ruocco, and G. Signorelli (With 3 Figures) 437

Discussions 440

Part X Phonon Echoes

Rotary Phonon Echoes in Silica Glass. By B. Golding, D.L. Fox, and W.H. Haemmerle (With 2 Figures) 446

Pseudospin Echoes in Borate Glasses. By M. Devaud, J.-Y. Prieur, and W.D. Wallace (With 2 Figures) 449

The Enhancement of Phonon Echo Generation by Defects in Crystals By D.J. Meredith, H. Mkhwanazi, J.K. Wigmore, and T. Miyasato (With 2 Figures) 452

On Theory of Echo Phenomena in Amorphous Materials By V.S. Kuz'min and A.P. Sayko 455

Discussions 457

Part XI Spin-Phonon Interaction

Evidence for Phonon Scattering by Magnetic Two-Level Systems in Crystalline Spin-Glass $Eu_xSr_{1-x}S$. By C. Arzoumanian, B. Salce, A.M. de Goër, and F. Holtzberg (With 2 Figures) 460

Phonon Spectroscopy of MnF_2/ZnF_2 Mixed Crystals. By P.J. King, D.T. Murphy, and V.W. Rampton (With 2 Figures) 463

Heat Transport by Phonons and Magnons in Ferromagnetic EuS By G.V. Lecomte, H. v. Löhneysen, and W. Zinn (With 2 Figures) 466

Discussions 469

Index of Contributors 471

Part I

Acoustic Phonon Spectroscopy

Chairmen:
W. Dietsche K. Dransfeld T. Ishiguro G. A. Northrop
J. P. Wolfe A. F. G. Wyatt

Crossing Effects in Phonon Scattering

L.J. Challis

Department of Physics, University of Nottingham, University Park
Nottingham NG7 2RD, England

Resonant scattering from a phonon current can occur at the transition frequencies ν_i associated with the electronic or motional energy levels of impurity ions or centres present in the crystal. If the concentration of centres is 'small' the scattering will only be significant within a narrow frequency bandwidth Δ [1] which may be much less ($\lesssim$ 1 GHz) than that of the phonon current ($\sim$ 2kT for a thermal current or $\sim$ 40 GHz at 1K). Under these conditions the frequency spectrum of the phonon current has sharp holes burned in it as shown in fig. 1(a). In many cases these can be moved to and fro by applying an external perturbation such as a magnetic field. Sharp features in the total scattering occur when two of these holes cross ($\nu_i = \nu_j$) or in some cases anticross and these provide spectroscopic information which can be of quite high precision and resolution.

There are three main effects [2], frequency, level and anticrossing.

1. Frequency Crossing

This effect only requires that 2 transition frequencies become equal. It was first observed and accounted for qualitatively by BERMAN et al. [3] using the heuristic model illustrated in fig. 1(b) which assumes that when resolved, the 2 resonant processes completely block the 2 rectangular conduction channels shown shaded. However, when they are crossed only one channel is blocked so the total conduction rises giving a signal $\Delta K/K_0 \sim$ channel conduction/total conduction and width $\sim$ channel width. In practice of course a channel is never entirely blocked. Its conduction depends on $\tau(\nu)$, the phonon relaxation time and $\tau(\nu) = (\tau_B^{-1} + \tau_1^{-1} + \tau_2^{-1})^{-1}$, where τ_B^{-1}, τ_1^{-1} and τ_2^{-1} are respectively the background and two resonant scattering rates. Expansion of $\tau(\nu)$ contains functions of the product $\tau_1^{-1}\tau_2^{-1}$ which is essentially zero at all frequencies when the two processes are resolved, but becomes non-zero for $\nu \sim \nu_1 = \nu_2$ when the processes are crossed [Fig 2]. So the crossing signal is a consequence of the non-linear dependence of $\tau(\nu)$ on the scattering rates.

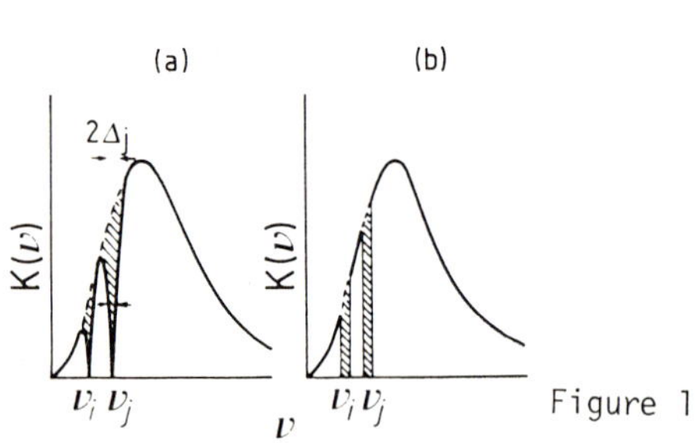

Figure 1

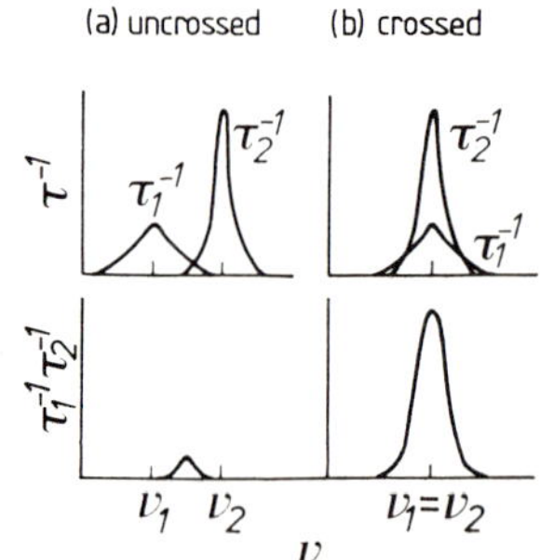

Figure 2

At low temperatures and in a Debye model the conductivity differs from that of the pure crystal by

$$\Delta K = K_0 - K = CT^3 \int_0^\infty \frac{x^4 e^x}{(e^x-1)^2} \left[\tau_B - \tau(\nu) \right] d\nu \text{ where } x = h\nu/kT.$$

In general the crossing signal $\Delta W/W_0 \simeq -\Delta K/K_0$ ($W = 1/K$) can be computed from τ_1^{-1} and τ_2^{-1} by calculating ΔK as a function of the separation of the 2 frequencies in the region of the crossing frequency ν_0. However, when the holes have widths <<kT, we can simplify the calculation since

$$\Delta K \simeq CT^3 \left[\frac{x^4 e^x}{(e^x-1)^2}\right]_{\nu_0} \int_0^\infty \left[\tau_B - \tau(\nu)\right] d\nu.$$ Indeed in some limits the integral can be solved analytically for both Lorentzian and Gaussian line shapes giving simple expressions for the form of the signal [4]. So the size and shape of frequency crossing signals can provide information on the resonant scattering rates and hence the spin-phonon coupling constants or, if these are known, on the ionic concentrations. Their positions can provide spectroscopic information such as the spin-Hamiltonian parameters. Although this discussion refers to steady heat currents, similar analysis of course applies to heat pulses.

An example of a frequency crossing signal is shown in fig. 4 at low resolution. The signal shows the change in temperature difference along a sample as the field is swept and is usually displayed at $\Delta W/W_0$. The system is V^{3+} in Al_2O_3 whose levels for B||c axis correspond to those of fig. 3 and give rise to a crossing at a field $B = D/3g_{11}\beta$. The crossing moves to higher fields when these are moved away from the c axis, and D and $g_\perp$ can be obtained rather accurately from the straight line plot of B^{-2} against $\cos^2\theta$ (θ is the angle to the c axis) if g_{11} is known from EPR which is very often the case in such systems [5].

The upper 2 levels of fig.3 are in fact each split into 8 by hyperfine interaction, as shown in fig. [3]. This gives rise to the structure shown in fig. [5], [7]. Figure 6 shows the differential of this signal obtained by applying a small modulation field to the sample [8].

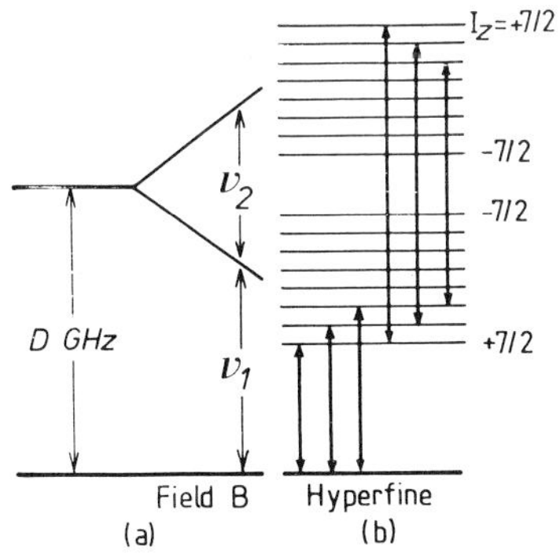

Fig.3: the energy levels of V^{3+} in Al_2O_3. If we neglect hyperfine splitting ($\sim 10^{-3}$D), $\nu_1 = D - g_{11}\beta B$ and $\nu_2 = 2g_{11}\beta B$ so $\nu_1 = \nu_2$ when $B = D/3g_{11}\beta$

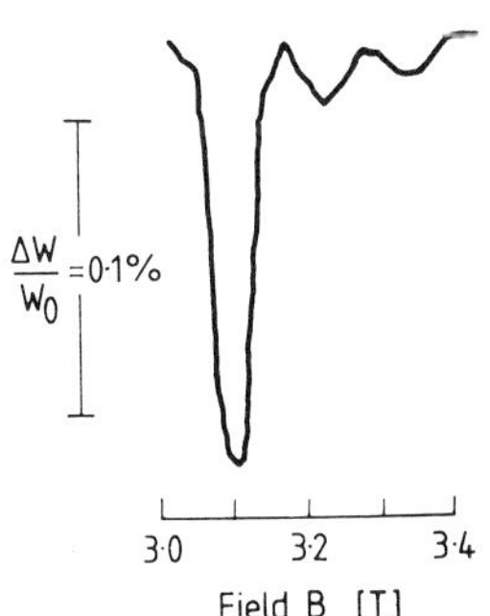

Fig 4: shows the frequency crossing signal of fig 3 displayed at low resolution. The 2 signals at higher fields are due to V^{3+} pairs present at a concentration $\sim 10^{-2}$ppm [6]

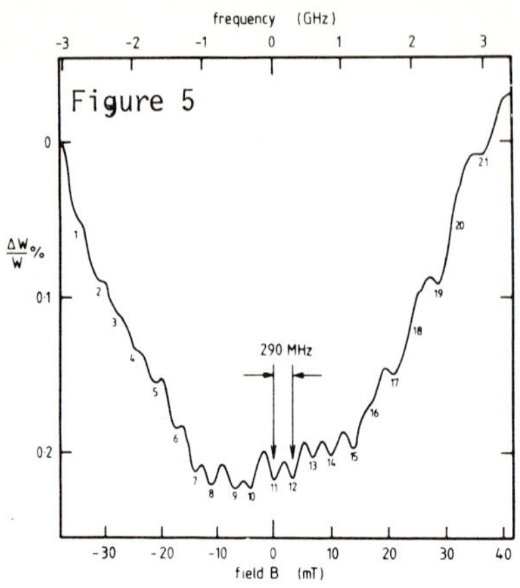

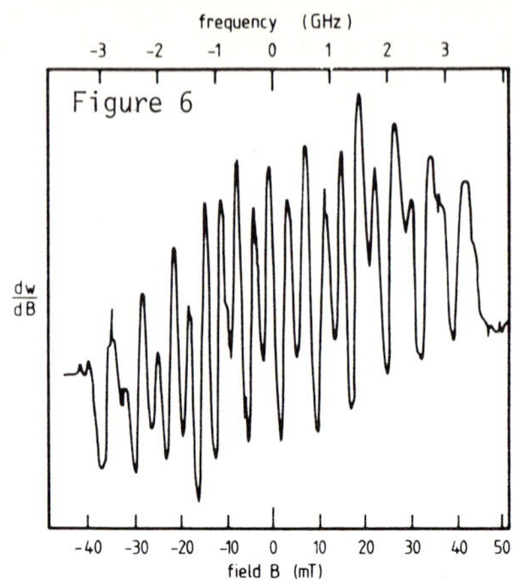

In this case and many others that have been reported the 2 scattering processes were both from the same type of ion but crossing signals have also been seen when the processes are from 2 ions in inequivalent sites [3], an ion and an ion pair [6] and 2 entirely different ionic species [9], [10]. So, as WALTON first stressed [9] one can use a known ion as a probe to investigate another and the fact recently demonstrated that they do not have to be in the same part of the sample opens up the scope of the technique.

2. Frequency Crossing between spatially separated ions

This has been demonstrated [11] using an Al_2O_3 bicrystal doped with Fe^{2+} ions in one half and V^{3+} in the other [Fig.7]. Several crossings occur between the transition frequencies of the two ions and fig. 7 shows two of these(lines C and D). The increase in their size when heat injection is switched from H_V to H_{Fe} shows that the Fe^{2+} holes in the phonon current are carried across the interface; the V^{3+} ions have little effect on this current when tuned to a hole. This technique seems capable of being developed in various ways and we hope to use it to study ions in epitaxial and diffused layers, electrons in inversion layers,etc. by crossing their frequencies with those of probe ions in the bulk crystal [12].

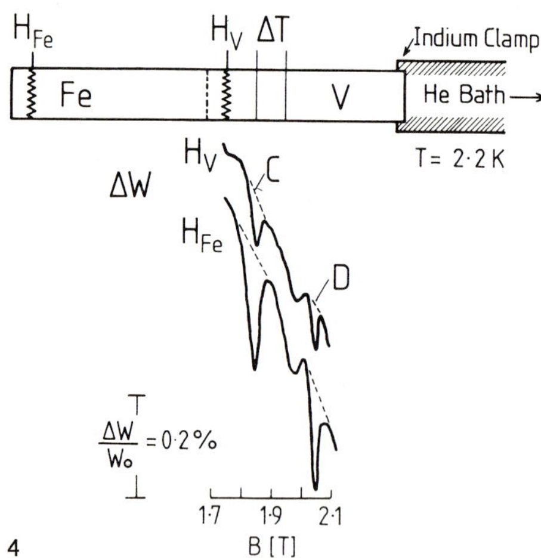

Figure 7: shows frequency crossing signals from the Fe/V Al_2O_3 bicrystal shown in the upper part of the figure. The signals ΔW are minima in the temperature difference on the V-doped half. The signals are $\sim$ 3 times larger when heat is injected at H_{Fe} rather than at H_V in the control experiment when the signals are due to traces of Fe^{2+} in the V half

3. Level Crossing

The effects that should be observable here using phonons are analogous to the HANLE effect seen in light scattering in 1924 [13] and first explained by BREIT in 1933 [14]. In recent years it has been used as a technique for high resolution spectroscopy [15]. Suppose an ion has 2 crossing levels with energies $h\nu_1$ and $h\nu_2$ above a singlet ground state and that the upper 2 levels can be made to cross by a magnetic field (fig.8(a)) making $\nu_1 = \nu_2$. If white light is incident on the ion, a photomultiplier can be used to measure the total intensity I_T of the components at ν_1 and ν_2 scattered into a particular small solid angle. Now if $|\nu_1 - \nu_2| >> \gamma$ (γ is their combined line width), $I_T = I(\nu_1) + I(\nu_2)$ but if $|\nu_1 - \nu_2| \lessapprox \gamma$, interference takes place in the emission process and $I_T = |A(\nu_1) + A(\nu_2)|^2 = I_1 + I_2$ plus an interference term. So the photomultiplier records a signal of width γ at the crossing point. There seems no reason to suppose that similar effects should not occur with phonons although so far as I know, no experiments have been reported. It might be possible to see them in thermal conduction which would certainly be affected by changes in the angular distribution of the scattered phonons that occur at a level crossing. However, at the concentrations used in phonon experiments so far, these changes would be accompanied by frequency crossing effects and while the resulting signal would certainly differ from a pure frequency crossing signal [16], it is not clear whether these differences could be identified. So although it would seem that it would be interesting to explore these effects in this way, experiments looking more directly at the angular distribution would seem to be more promising.

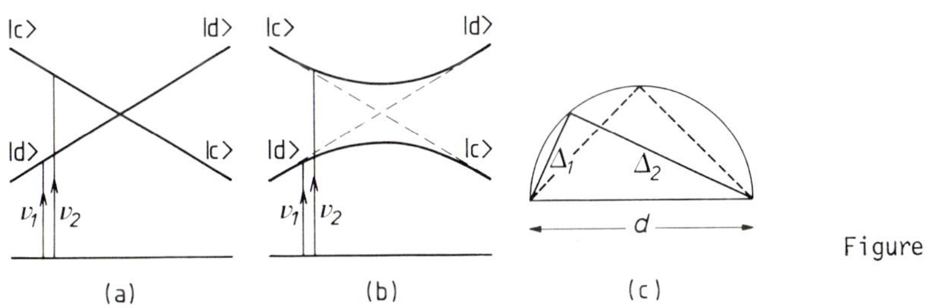

Figure 8

A complication in level crossing experiments comes from the effect of random strain which may prevent levels from approaching to within their linewidths (anticrossing)

4. Anticrossing

If there is coupling between the 2 approaching levels of fig.8(a), they will repel each other or anticross as shown in fig.8(b). So if the holes in the heat current at ν_1 and ν_2 are narrower than their nearest separation, they can never overlap and no frequency crossing signal can occur. However, there can still be another sort of signal because of the state mixing that occurs at the anticrossing. This changes the widths Δ_1, Δ_2 of each of the holes while keeping $\Delta_1^2 + \Delta_2^2$ constant. The additional thermal resistance caused by the existence of the holes α $(\Delta_1 + \Delta_2)$ which has a maximum at the centre of the anticrossing when the 2 hole widths become equal $(\Delta_1 = \Delta_2)$. This can be seen in Fig 8(c) where $\Delta_1^2 + \Delta_2^2 = d^2$ (d = diameter of semicircle)

: $(\Delta_1 + \Delta_2)$ increases from d when $\Delta_1 << \Delta_2$ (or vice versa) to a maximum of $\sqrt{2}d$ when $\Delta_1 = \Delta_2$. This anticrossing signal has a Lorentzian lineshape whose width is equal to the closest separation which is twice the matrix element coupling the two levels [16].

Fig.9 shows an anticrossing signal for Fe^{2+} in Al_2O_3 [17]. The coupling term has 2 components : one is fixed and due to random strain terms while the other is due to fields normal to the c axis of quantization and can be made nearly zero by careful alignment ($\theta = 0^o$). In the experiment the separation of the 2 holes at the centre could be made sufficiently small that they started to overlap, giving rise to a frequency crossing minimum at the centre of the anticrossing. Anticrossing effects have since been seen in Cr^{2+} in MgO [18] and V^{3+} in Al_2O_3 [19]. Signals observed in this way seem likely to be dominated by the incoherent effects described here, although it may again be possible to see the coherent effects which should also be present [16].

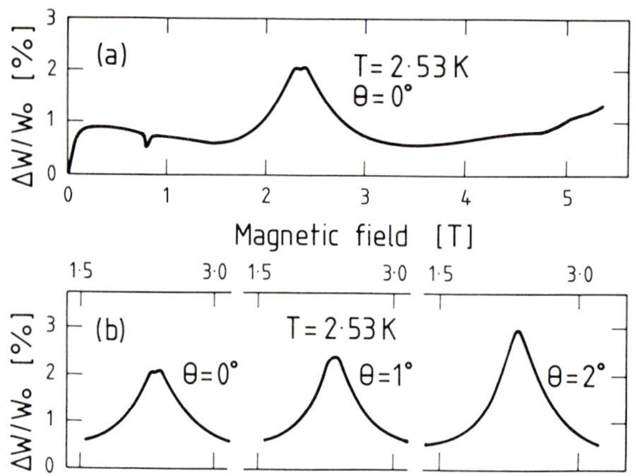

Figure 9: The Fe^{2+} anticrossing occurs at 2.4T (the minimum at 0.8T in the upper plot is a frequency crossing signal similar to that for V^{3+} in fig.4)

5. Observations of inelastic scattering at surfaces by frequency crossing

This can be observed using the Al_2O_3 bicrystal described earlier. In the arrangement shown in fig.10, the phonon current injected at H_{Fe} leaves the Fe^{2+} half with holes at the transition frequencies of the Fe^{2+} ions. As the heat current passes down the other half (which we shall for the moment suppose to be pure), these holes must gradually fill up as the phonon spectrum adjusts itself to the non-resonant boundary scattering present. In the bulk this adjustment or thermalization can only take place through 3 (or more) phonon processes but these are very weak at these temperatures and frequencies so in the absence of inelastic surface scattering this would take distances ∿ metres. Now since the size of a Fe^{2+}/V^{3+} crossing signal depends on the size of the Fe^{2+} hole we can use the signal to monitor hole decay. Fig.10 shows that the signals near to the interface (ΔT_n) are 3 times bigger than those further away (ΔT_f) which in fact have fallen approximately to the background level due to Fe^{2+} trace impurities (fig.7) [11]. This experiment shows that at these surfaces the holes are effectively filled during 1 collision with the surface and we attribute

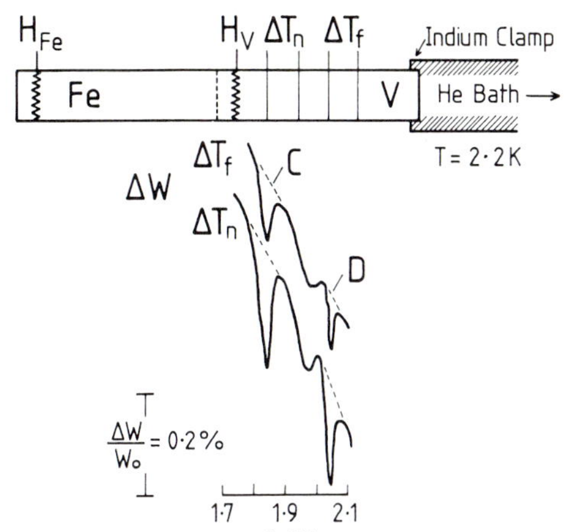

Figure 10: shows the temperature differences ΔT_n, ΔT_f at different points along the V half of the bicrystal (n = near, f = far from the interface). The heat is injected at H_{Fe}. The signals on ΔT_f are $\sim$ 3 times smaller than those on ΔT_n. This shows that the Fe^{2+} holes in the phonon current leaving the Fe half have effectively decayed by the time they reach ΔT_f

this to the presence of resonant defects. (Thermalization could occur by one-phonon processes of absorption and emission if the defects interact strongly amongst themselves or by two-phonon processes of the type $\nu \to \nu - \delta$ in which the defect is excited by δ). These experiments were carried out on samples with fine ground surfaces which had been annealed to 1200°C [11] and are being repeated after annealing the samples to 1775°C. This causes structural changes to the surface and electron microscopy shows rounding of scratch edges, etc. These new experiments are at an early stage [20] but do suggest that thermalization is still taking place. Similar experiments are in preparation using a Si bicrystal doped with B and Li [12]. This would have several advantages. Si surfaces are more easily treated than Al_2O_3 and the trace level of impurities should be much lower.

Thermalization at a surface has of course also been seen in other experiments [21 - 24]. In the reverberation experiments of TRUMPP and EISENMENGER [23] the decay of 2Δ phonons (280GHz) in a Si sample was observed as a function of time. For carefully prepared surfaces, inelastic scattering only occurred on average after 10 collisions with an uncovered surface but became much more probable when they were coated with tin.

TABOREK and GOODSTEIN [24] looked at a diffusely reflected heat pulse and observed, as others have, that there is loss in detected signal when a surface film is added. However, this lost signal did not appear later and they concluded that it must have been reduced in frequency, by processes associated with the film, to below some cut-off in their bolometer. Evidence that the proportion of diffuse scattering can be increased by a surface coating and also by surface damage comes from recent low temperature thermal conductivity measurements on Al_2O_3 by POHL and STRIZKER [25]. In one treatment causing diffuse scattering, a 50nm damaged layer was produced by ion bombardment. This had no observable effect on the surface structure, implying that the additional diffuse scattering was not due to an increase in roughness.

Since diffuse scattering and thermalization can be caused by similar treatments it is tempting to suppose that they are due to the same defects. This does not necessarily require that they are caused by the same processes since those causing diffuse scattering can be either elastic or inelastic while those resulting in thermalization must of course be inelastic. At low frequencies, for a two-level system, the inelastic one-phonon process normally dominates so most diffuse scattering caused by these defects will probably be inelastic. However at high frequencies, the elastic (resonance fluorescence) process dominates and most diffuse scattering will now become elastic. Evidence consistent with this comes from the reflection experiments of MARX and EISENMENGER [26]. They conclude that at ≳ 280GHz, chemically polished Si surfaces are very nearly wholly diffuse while as we have noted similar surfaces have been shown to be very nearly elastic [23]. It is always possible though that for these surfaces, the diffuse scattering is mainly by surface roughness.

I have discussed these results mainly in terms of resonant defects - two level systems - since recent work suggests these are the most likely source of inelastic scattering in surfaces. SCHUBERT et al. [27] observed saturation in the phonon absorption at 25GHz caused by a paraffin layer on the surface of a quartz crystal indicating that the absorption is due to defects with a finite number of levels. This work and many of the other experiments are very relevant to discussions of the anomalous Kapitza conductance but I must restrain myself from digressing as far as that. I will, however, mention that one experiment that may have been influenced by surface defects is that by HENSEL et al. [28] who observed anomalously strong electron-phonon interaction between a heat pulse and the electrons in an inversion layer. If defects were present the phonons might have rather poorly defined wave vectors in this region and might also have a somewhat lower effective temperature.

From this discussion we have seen that it is possible by careful treatment to produce surfaces that are largely elastic. These will be needed if we are to exploit fully the fact that frequency crossing spectroscopy can be carried out between spatially separated ions. Apart from experiments with films and layers that I have already mentioned it may be possible, if both the surfaces are elastic, to do frequency crossing spectroscopy by bonding a crystal containing a probe ion onto a sample to be investigated. It may even be possible to put a pure crystal between the two to act as a phonon guide so that the sample could be in zero magnetic field. However, so far the only example I know of largely elastic transmission across an interface at high frequencies (200 GHz) occurred when there was a good lattice match [29] and it may be that inelastic defects are an inevitable feature of a change in lattice parameter or structure.

I should like to thank all my collaborators with whom I have had many stimulating discussions and the Science and Engineering Research Council for their support.

References

[1] Δ is usually greater than the resonance linewidth and is defined by the frequencies at which the resonance scattering rate equals the boundary scattering rate.

[2] A more detailed description of these will be given by the author in 'Non-Equilibrium Phonons in Non-metallic crystals'. (North-Holland, Amsterdam) eds. W.Eisenmenger and A.A. Kaplyanskii.

[3] R.Berman, J.C.F. Brock and D.J. Huntley : Phys.Lett. 3, 310 (1963).

[4] B.R. Anderson and L.J. Challis : J. Phys C : Sol. State Phys. 8, 1475 (1975), L.J. Challis and M.N. Wybourne : J. Phys. C : Sol. State Phys. 12, L711 (1979).

[5] L.J. Challis and D.L. Williams : J Phys C : Sol. State Phys. 10, L621 (1977).
[6] L.J. Challis, A.A. Ghazi, D.J. Jefferies, D.L. Williams and M.N. Wybourne : Phys. Rev. Letts. 40, 519 (1978).
[7] A.P. Heraud, A.A. Ghazi and L.J. Challis (unpublished).
[8] M.N. Wybourne, L.J. Challis and A.A. Ghazi : J. Phys. C : Sol. State Phys. 13, 6495 (1980).
[9] D. Walton : Phys. Rev. 151, 627 (1966) and Phys. Rev. Letts. 19, 305 (1967) ; M.C. Heltzer Jr. and D. Walton : Phys. Rev. B8, 4801 (1973).
[10] A.A. Ghazi, L.J. Challis, D.L. Williams and J.R. Fletcher in 'Phonon Scattering in Condensed Matter' (Plenum, New York) ed. H.J. Maris p 113.
[11] L.J. Challis, A.A. Ghazi and M.N. Wybourne : Phys. Rev. Letts. 48, 759 (1982), and J. Phys. C : Sol. State Phys. 15, 77 (1982).
[12] This work is being carried out in collaboration with Dr M. N. Wybourne of the GEC Hirst Research Centre.
[13] W. Hanle : Z. Phys. 30, 93 (1924).
[14] G. Breit : Rev. Mod. Phys. 5, 91 (1933).
[15] e.g. A. Corney : Atomic and Laser Spectroscopy (Oxford University Press, Oxford 1977).
[16] B.R. Anderson and L.J. Challis : J. Phys. C : Sol. State Phys. 8, 1495, (1975).
[17] B.R. Anderson and L.J. Challis : J. Phys. C : Sol. State Phys. 7, L440, (1974).
[18] J.L. Patel and J.K. Wigmore : J. Phys. C : Sol. State Phys. 10, 1829 (1977).
[19] M.N. Wybourne, (private communication).
[20] S.V.J. Kenmuir, A.P. Heraud and L.J. Challis - to be published.
[21] J.K. Wigmore : Phys. Letts. 37A, 293 (1971).
[22] K. Böhm, Diplomarbeit, Stuttgart University, 1975 (unpublished).
[23] H.J. Trumpp and W. Eisenmenger, Z. Phys. 28, 159 (1977).
[24] P. Taborek and D.L. Goodstein : Sol. State Comm. 38, 215 (1981).
[25] R.O. Pohl and B. Strizker : Phys. Rev. B25, 3608 (1982).
[26] D. Marx and W. Eisenmenger : Phys. Letts. 93A, 152 (1983).
[27] H. Schubert, P. Leiderer and H. Kinder : J. Low Temp. Phys. 39, 363 (1980) and Phys. Rev. B26, 2317 (1982).
[28] J.C. Hensel, R.C. Dynes and D.C. Tsui : J. de Physique, C6, 308, (1981); Surf. Sci. 113, 249 (1982).
[29] V. Narayanamurti, H.L. Störmer, M.A. Chin, A.C. Gossard and W. Wiegmann: Phys. Rev. Letts. 43, 2012 (1979)

Far Infrared Volume Generation and Detection of Phonons

K.F. Renk

Institut für Angewandte Physik, Universität Regensburg
D-8400 Regensburg, Fed. Rep. of Germany

Recently developed far infrared techniques of phonon generation and detection are discussed. Detection was performed by far infrared phonon-difference absorption; the method was applied to determine lifetimes of zone-boundary phonons in TlCl and TlBr crystals at different crystal temperatures. Phonon generation was performed by multiphonon infrared absorption; it was found that an originally broadband phonon spectrum narrowed with increasing time after pulsed excitation. Extremely monochromatic phonons were generated by far infrared excitation and relaxation of electronic impurity states; the far infrared generation was applied, together with monochromatic optical detection, to study bottleneck effects for phonons in Al_2O_3 crystals that contained either excited Cr^{3+} ions or V^{4+} ions.

1. Introduction

Evidence for far infrared surface generation of phonons was reported by GRILL and WEIS [1]. Far infrared volume generation was studied by use of different techniques of phonon detection: BRON and GRILL [2] generated phonons, in $Al_2O_3:V^{4+}$, by pulsed far infrared excitation and relaxation of impurity ions and investigated the phonons by time-of-flight techniques using bolometric detection. LENGFELLNER and RENK [3] generated phonons, in Al_2O_3 containing excited Cr^{3+} ions, also by far infrared excitation and relaxation of impurity ions, but detected the phonons monochromatically by observation of phonon-induced fluorescence; generation and detection frequencies were equal. In a modified experiment by ENGELHARDT and RENK [4] phonons in Al_2O_3 were generated at one frequency and detected at another frequency. By simultaneous volume generation and detection, surface effects were eliminated and phonons in bulk crystals were studied. Because of the monochromasy of generation and detection, the techniques [3,4] are very suitable to study elastic and inelastic phonon scattering within crystals (section 4). Defect-induced one-phonon absorption, as an alternative method for monochromatic phonon generation [3] will be described in section 5.

A new technique for generation of high-frequency phonons was recently developed by HAPPEK et al. [5]; phonons were generated by multiphonon absorption of infrared radiation (section 3). A first far infrared technique of phonon detection was developed by LENGFELLNER and RENK [6]; phonons were detected by observation of phonon-difference absorption (section 2).

2. Detection of Zone-Boundary Phonons by Phonon-Difference Absorption

High-frequency transverse acoustic phonons are expected to have, in pure crystals at low temperature, extremely long lifetimes [7,8]. From time-of-flight experiments long lifetimes were concluded for large-wave-vector phonons in GaAs [9] and from a far infrared experiment long lifetimes were suggested for zone-boundary phonons in TlCl and TlBr [6]. This experiment will be discussed here. Phonons were detected by phonon-difference absorption (Fig. 1a) in which transverse acoustic phonons are involved. The difference absorption occurs only if the transverse acoustic phonon branch is populated. Therefore nonequilibrium phonon populations can be probed by transient transmission measurements. Phonons were generated by pulsed optical pumping with radiation (at 532 nm) of a Nd:YAG laser; a small portion of the radiation was absorbed by impurities in TlCl and by nonradiative transitions high-frequency phonons, including zone-boundary phonons, were generated. For detection far infrared radiation of a HCN laser (emission frequency 891 GHz) was transmitted through the sample and detected with a fast detector (inset of Fig. 1b). After pulsed excitation the far infrared transmissivity of a 4 mm thick sample was reduced by typically 10%. A transient-transmission signal is shown in Fig. 1b. The first fast signal is attributed to nonthermal phonons, the signal decrease at late times is due to heating of the sample by the absorbed laser power. In case of fast thermalisation the dashed curve (Fig. 1b) was expected.

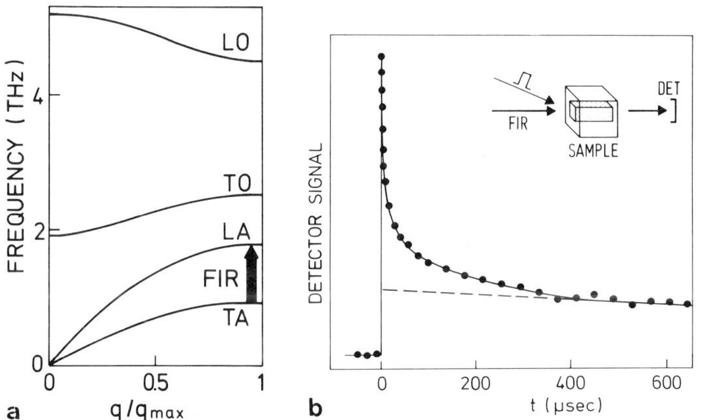

Fig. 1 Detection of zone-boundary phonons by phonon-difference absorption; (a) principle; (b) transient far infrared transmission

The fast signal is shown in Fig. 2a for different crystal temperatures. At low temperature (lower curve) a fast decay time of $5\cdot10^{-6}$s is found; this time is attributed to an effective lifetime of transverse acoustic phonons at the zone boundary. The fast decay time becomes shorter with increasing crystal temperature (upper curves). Experimental decay times are summarized in Fig.2b for TlCl and TlBr crystals. For a discussion we take into account that mode mixing occurs due to fast elastic scattering at impurities and that therefore the transverse acoustic phonons are in an equilibrium with the longitudinal phonons of the same frequency. At low temperature phonon decay is possible via anharmonic splitting processes of longitudinal acoustic phonons, the effective decay time of the phonon mixture is $\tau = \tau_L(\nu)D_T(\nu)/D_L(\nu)$, where $\tau_L(\nu)$ is the decay time of the longitudinal phonons and $D_T(\nu)$ and $D_L(\nu)$ are the density of states of transverse and longitudinal phonons. With $D_T(\nu)/D_L(\nu) \simeq 10^2$ a value $\tau_L \simeq 10^{-8}$s (for phonons at 0.9 THz) is obtained, the value is in reasonable agreement with anharmonic phonon decay in other crystals [10]. The nonexponential behavior of the phonon decay (lower curve in Fig. 2a) is most

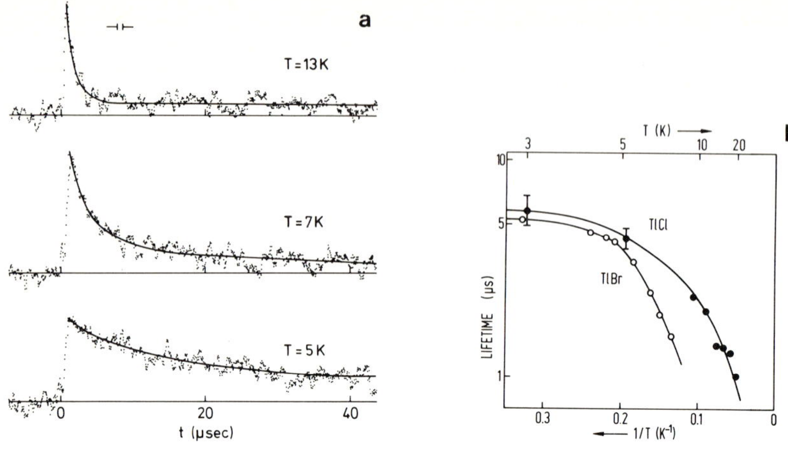

Fig. 2 (a) Transient far infrared absorption of TlCl at different crystal temperatures and (b) decay times of phonons in TlCl and TlBr

likely due to bottleneck effects in the phonon cascade decay towards lower frequencies. At high temperatures, transverse acoustic phonons can directly decay anharmonically by interaction with thermal phonons. The phonon decay can well be described by anharmonicity theory [8] which predicts that the decay rate increases with the population of the interacting thermal phonons. A detailed analysis showed [6] that the detected phonons have frequencies around 0.9 THz and wave vectors near the $[\frac{1}{2},\frac{1}{4},\frac{1}{4}]$ point of the zone boundary while thermal phonons which collide with these phonons are most likely phonons near $[\frac{1}{2},\frac{1}{2},0]$ and have frequencies of about 0.3 THz.

By using far infrared lasers with other frequencies it seems possible to study phonons in TlCl and TlBr that have different wave vectors at the zone boundary. Phonon detection by phonon-difference absorption should be applicable also for other crystals.

3. Phonon Generation by Multiphonon Infrared Absorption

Phonons at frequencies above 1 THz were generated by heat pulse techniques or by electronic excitation techniques; electronic excitation was either performed by interband excitation of semiconductors with phonon generation by electron-hole recombination [9] or by excitation of impurities with phonon generation by multiphonon nonradiative transitions [6,10] or with phonon generation by one-phonon relaxation (see sections 2 and 4). Here, the new method for generation of high-frequency phonons by multiphonon absorption shall be described shortly, an extended description is given in [5].

Pulsed CO_2 laser radiation was used to excite phonons by three-phonon absorption in CaF_2. In an absorption process a photon is transformed into three phonons; in the process energy and momentum conservation laws are fulfilled. The absorption leads to generation of optical phonons which decay anharmonically into acoustic phonons. The spectrum of these acoustic phonons was studied by vibronic sideband spectroscopy. In a first experiment [5] phonons were excited homogeneously in a large crystal. It was found that a broadband phonon spectrum narrowed with increasing time after pulsed excitation (Fig.3).

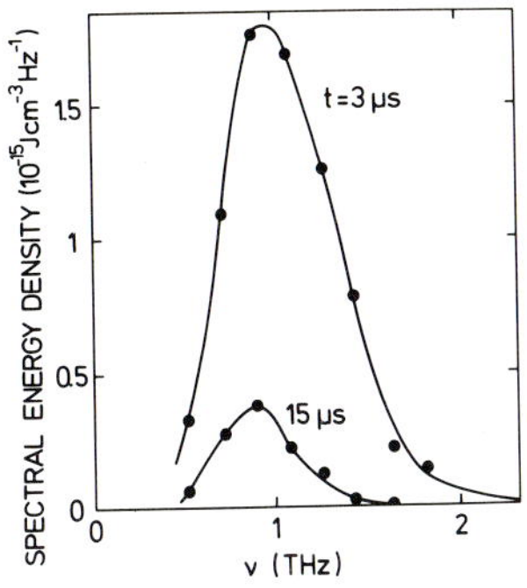

Fig. 3 Transient spectra of phonons in a CaF_2 crystal (size 8 x 8 x 10 mm^3, 0.05% Eu^{2+}) after pulsed excitation with radiation of a CO_2 laser

Monochromatization occurred since the anharmonic phonon decay rate increased strongly with frequency (as ν^5) and since the spatial escape rate by diffusion decreased strongly (as ν^{-4}). The maximum of the distribution, at a frequency $\nu_m \simeq 0.9$ THz, occurs for phonons which have a minimum total decay rate. Phonons at $\nu > \nu_m$ decay mainly anharmonically by splitting processes and phonons at $\nu < \nu_m$ escape from the crystal by spatial diffusion.

Phonon generation by multiphonon absorption has, compared to other optical techniques, the advantage that the infrared radiation is directly transformed into vibrational energy and that no electronic excitations occur and that no impurities are needed. In comparison with heat pulse techniques that lead to spatially inhomogeneous phonon distributions at crystal surfaces, multiphonon excitation allows for generation of high-frequency phonons with spatially homogeneous distributions. The technique is generally applicable for the generation of high-frequency phonons in dielectric crystals since phonon absorption by combination processes occurs in each dielectric crystal either due to crystal anharmonicity or due to nonlinear dipole moments.

4. Resonance Trapping of Phonons in $Al_2O_3:Cr^{3+}$ and $Al_2O_3:V^{4+}$

Resonance interaction with excited Cr^{3+} ions can lead to spatial trapping of 29 cm^{-1} phonons in ruby [11-22]. Besides spatial diffusion caused by the elastic resonance scattering, different mechanisms for the phonon escape from a resonance volume were discussed, as anharmonic phonon decay [11-14,16] and processes of spectral-spatial diffusion due to inelastic scattering of phonons at excited Cr^{3+} ions coupled to other Cr^{3+} ions [18,20,21]. In a recent study Retzer et al. [23] proposed inelastic scattering at impurities, other than Cr^{3+} ions, as a further mechanism. The results of this study will be presented here.

Resonance trapping of phonons in $Al_2O_3:V^{4+}$ [24,4] will be discussed in comparison with the trapping in $Al_2O_3:Cr^{3+}$. Fig. 4 shows resonance lines of excited Cr^{3+} ions and of V^{4+} ions in Al_2O_3 crystals at low temperature. The $\overline{E} \rightarrow 2\overline{A}$ line of excited Cr^{3+} is, in ruby crystals of high quality, homogeneously broadened due to the finite lifetime of the $2\overline{A}$ level against phonon emission as was recently shown by a far infrared transmission experiment [25]. The $E_{3/2} \rightarrow {}_1E_{1/2}$ line of V^{4+} is most likely also homogeneously broadened [4]. Each of the lines can therefore be described by a Lorentzian curve $g(\nu) = [1 + 4(\nu - \nu_0)^2/\Gamma^2]^{-1}$ where ν_0 is the resonance frequency and Γ the halfwidth. Accordingly, the cross section for resonance scattering of phonons of a fre-

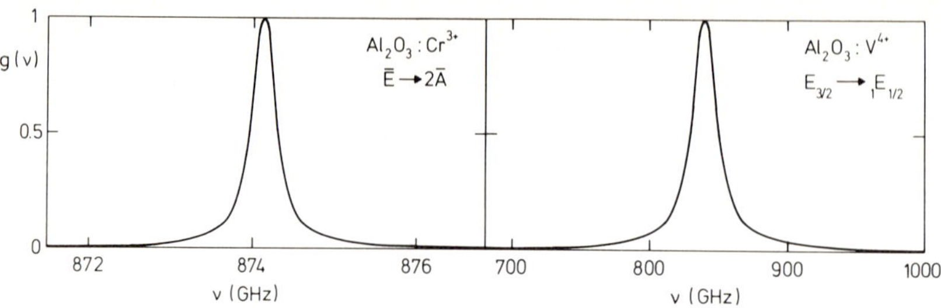

Fig. 4 Resonance lines of excited Cr^{3+} ions (left) and of V^{4+} (right) in Al_2O_3 crystals

quency ν is given by $\sigma(\nu) = \sigma_0 g(\nu)$, where $\sigma_0 = (8\pi/3)k^{-2}$ is the cross section, $k = 2\pi\nu_0/v$ an average wave vector of phonons at ν_0, and v an average velocity of sound for Al_2O_3. Since the values of ν_0 differ only slightly for the two lines (Fig. 4), the values of σ_0 are nearly equal, but, because of a much larger homogeneous width, $\Gamma(V^{4+}) \simeq 60\,\Gamma\,(Cr^{3+})$, phonon trapping is, at equal concentrations of resonant-scattering centers, efficient in $Al_2O_3:V^{4+}$ over a much larger frequency range than in Al_2O_3 containing excited Cr^{3+} ions.

For a discussion of experiments with monochromatic phonons in ruby we imagine that the frequency ν of far infrared radiation is tuned over the resonance line of excited Cr^{3+} (Fig. 4) and that, by far infrared excitation and relaxation of excited Cr^{3+} ions, phonons at exactly the frequency ν are generated. In case of continuous far infrared excitation we expect a stationary population N of the $2\bar{A}$ level given by

$$N = Q\, T_1\, N^*\, g(\nu) \cdot [1 + v\, N^*\, \sigma_0\, g(\nu)\, \tau(d,N^*,\nu)] \tag{1}$$

where Q is the $2\bar{A}$ population rate by the far infrared absorption per excited Cr^{3+} ion at the resonance frequency ν_0, T_1 ($5\cdot10^{-10}$s) the lifetime of the $2\bar{A}$ level for phonon emission, N^* the concentration of excited Cr^{3+} ions and $\tau(d,N^*,\nu)$ the lifetime of a phonon for phonon escape, and where d is a typical dimension of the resonance volume. The second term in the bracket describes the number of reabsorption processes of a phonon during its lifetime τ. At sufficiently large N^* this term is large compared to 1 and N is mainly determined by the trapped phonons. According to eq.(1) an experiment with monochromatic far infrared radiation can lead to a decision about the escape mechanism if either the volume , or the N^* , or the frequency dependence of N is determined. It is expected that in case of diffusive escape τ increases quadratically with d, linearly with N^* and is proportional to $g(\nu)$ and that therefore $N \sim N^{*3}\, d^2\, g^3(\nu)$. In case of spectral-spatial diffusion τ is expected to depend on d, on N^* (e.g., as N^{*-1}) and possibly on ν. In case of total spatial trapping and inelastic phonon decay τ is independent of d,N^* and ν supposing that in an inelastic scattering process the frequency shift is much larger than τ and that therefore an inelastically scattered phonon is not reabsorbed.

Instead of tuning the far infrared frequency, the resonance frequency of the line was magnetically tuned over the frequency ν_{FIR} of a far infrared laser [23]. By applying a magnetic field B parallel to the crystal c axis the $\bar{E}$ level was split into $\bar{E}_\pm$ (g value 2.4), and the $2\bar{A}$ level into $2\bar{A}_\pm$ (g

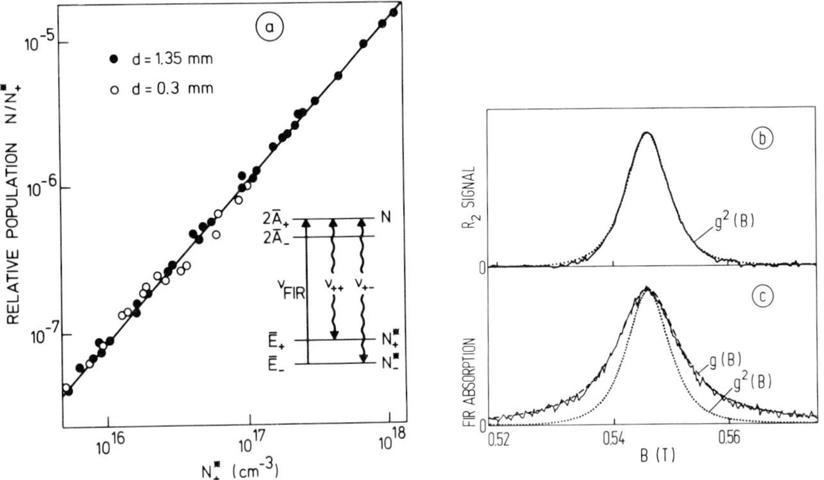

Fig. 5 Resonance scattering of phonons in ruby; (a) N^* and volume dependence of the $2\bar{A}$ population; (b) magnetic-field dependence of an R_2 signal; (c) far infrared absorption; g(B) = Lorentzian curve

value 1.6). The crystal was optically pumped with radiation of an argon-ion laser and contained $N_+^* \approx N_-^*$ excited Cr^{3+} ions in the $E_\pm$ levels. The relative population N/N_+^* was determined from the intensity ratio of R_1 and R_2 fluorescence that is due to transitions from $\bar{E}_\pm$ and $2\bar{A}_+$, respectively, to the Cr^{3+} ground state sublevels. Far infrared absorption by $\bar{E}_- \to 2\bar{A}_+$ transitions at frequency ν_{FIR} (inset of Fig. 5a) resulted in generation of phonons at ν_{++} by non-spinflip and at $\nu_{+-} = \nu_{FIR}$ by spinflip relaxation processes. By solving rate equations for the system in the magnetic field, it is found that eq.(1) is still approximately valid if N^* is replaced by N_+^* and ν by a corresponding value of the magnetic field strength B. For a decision about the phonon escape mechanism two samples of different thicknesses d were studied; the samples were cut from the same large crystal (containing 0.05 mol% Cr_2O_3). It was found (Fig. 5a) that at constant N_+^* the relative population N/N_+^* was independent of d. This shows that the phonons were totally spatially trapped and escaped by inelastic scattering. Total trapping and phonon escape with a decay time τ independent of N^* was confirmed by the observation of a linear N_+^* dependence of N/N_+^* (Fig. 5a). Finally, it was found that the R_2 signal (Fig. 5b) was proportional to $g^2(B)$, where g(B) is the Lorentzian shape of the resonance line determined by a far infrared transmission experiment (Fig. 5c). This shows that τ was independent of B and therefore of frequency ν. The results indicate that in the present experiment, spatial escape and spectral-spatial escape were not the main mechanisms for the phonon escape, but that the phonons were totally spatially trapped and decayed inelastically. Using eq.(1), from an estimate for Q and from the values N/N_+^* in Fig. 5a an inelastic decay time $\tau \simeq 5\cdot10^{-7}$s was estimated. It is suggested [23] that the decay is due to Cr^{2+} ions which may be present in $Al_2O_3{:}Cr^{3+}$ crystals. With an inelastic cross section $\sigma_{in} \simeq 10^{-17}$ cm^2, from the decay time a concentration of about 10 ppm Cr^{2+} ions (about 2% of the chromium ions) is estimated. Cr^{2+} ions as possible inelastic scattering centers have been identified by Engelhardt and Renk [26]: X-ray irradiation of a ruby crystal has been found to lead to a decrease of the decay time τ of trapped phonons. The effect has been attributed to Cr^{2+} ions produced by the X irradiation [27]; from an estimated Cr^{2+} concentration (10^{18} cm^{-3}) and a lifetime τ ($2\cdot10^{-7}$s) in the X-irradiated crystal σ_{in} was obtained.

Direct information about an average Raman shift ν_R occurring in an inelastic scattering process was obtained from a trapping experiment with phonons in an X-ray irradiated Al_2O_3 crystal that contained both V^{4+} and Cr^{3+} ions [4,26]. Using V^{4+} ions for phonon generation and Cr^{3+} ions for detection it was found that phonons generated at one frequency lead to a strong detector signal at a detection frequency that was different from the generation frequency. From an analysis of the experiment $\nu_R \simeq 4$ GHz was concluded. It was suggested [4,25] that the phonon propagation was governed by both multiple resonance scattering at V^{4+} ions and multiple inelastic scattering at impurities (other than V^{4+} ions) and that phonon escape can occur via the far wings of the V^{4+} line (Fig. 4). Accordingly, the spectrum of phonons evanescent from a resonance volume is expected to show self-reversal [26]. The experiment suggests also that the anharmonic lifetime of a mixture of longitudinal and transverse phonons with frequencies near 1 THz is larger than $5\cdot10^{-4}$s [4,26]. Propagation of phonons [24] in X-irradiated Al_2O_3 crystals that contained V_2O_3, but no Cr_2O_3 doping, may also be influenced by inelastic scattering at impurities, such as V^{3+} or Mn^{3+} or other ions.

A support for inelastic scattering in ruby is obtained from a time-of-flight experiment: Basun and Kaplyanskii [28] found that longitudinal and transverse 29 cm^{-1} phonons in Al_2O_3:Cr^{3+} (0.02%) had lifetimes of the order of 10^{-6}s and that the lifetimes depended on the propagation direction. This behavior is consistent with inelastic scattering at the Cr^{2+} Jahn–Teller ions.

It is suggested [23] that in experiments at $N^* > D(\nu_0)\Gamma$ where $D(\nu_0)$ is the density of states of phonons at the resonance frequency ν_0 in Al_2O_3, with phonon generation by optical pumping, trapping of broadband phonons may have been responsible for several effects as volume , N^* and magnetic-field dependences which were related either to spatial diffusion or to spectral-spatial diffusion. Broadband phonons around 29 cm^{-1} may be generated, in addition to monochromatic phonons obtained by $2\bar{A} \rightarrow \bar{E}$ relaxation, by anharmonic decay or by decay due to inelastic scattering at impurities, of high-frequency phonons produced by nonradiative transitions.

In conclusion, the far infrared experiments give strong evidence that inelastic scattering at impurity ions, as Cr^{2+} ions, is an important decay mechanism for high-frequency phonons in Al_2O_3. It is suggested that this process should be included in the analysis of phonon-trapping experiments which were performed with other techniques of phonon generation.

5. Phonon Generation by Impurity-Induced One-Phonon Absorption

In crystals containing impurities or other defects acoustic phonons can be excited directly by defect-induced one-phonon absorption which is possible because of electronic dipole moments of perturbed phonon modes. First evidence for one-phonon absorption was reported by LENGFELLNER and RENK [3]. Phonon absorption was attributed to Cr^{3+} impurities in Al_2O_3. The technique, yet not well developed, seems to be very promising for the generation of acoustic phonons, since defect-induced absorption occurs in all crystals that contain impurities and also in amorphous materials.

6. Conclusion

Recent far infrared experiments with high-frequency phonons lead to experimental information on the decay mechanisms of zone-boundary phonons in TlCl and TlBr, on the decay of a broadband nonequilibrium phonon spectrum indicating a narrowing effect of the transient spectrum of phonons in CaF_2. Results of experiments with monochromatic phonon generation and detection suggest a new description of the resonance trapping of acoustic phonons in Al_2O_3 crystals that contain excited Cr^{3+} ions or V^{4+} ions. It should be pointed out that there is a wide range of further applications of the far infrared techniques and that the observed phonon scattering effects are important also in many other crystals and also in amorphous materials.

1. W. Grill, O. Weis: Phys.Rev.Lett. 35, 588 (1975)
2. W.E. Bron, W. Grill: Phys. Rev. Lett. 40, 1459 (1978)
3. H. Lengfellner, K.F. Renk: Semiconductors and Insulators 3, 113 (1978)
4. M. Engelhardt, U. Happek, K.F. Renk: Phys. Rev. Lett. 50, 116 (1983)
5. U. Happek, R. Baumgartner, K.F. Renk: "Infrared Excitation of High-Frequency Phonons by Multiphonon Absorption"; this volume
6. H. Lengfellner, K.F. Renk: Phys. Rev. Lett. 46, 1210 (1981)
7. G.L. Slonimskii: JETP 7, 1457 (1937)
8. R. Orbach, L.A. Vredevoe: Physics 1, 90 (1964)
9. R.G. Ulbrich, V. Narayanamurti, M.C. Chin: Phys.Rev.Lett. 45, 1432 (1980)
10. R. Baumgartner, M. Engelhardt, K.F. Renk: Phys.Rev. Lett. 47, 1403 (1981)
11. S. Geschwind, G.E. Devlin, R.L. Cohen, S.R. Chinn: Phys. Rev. 137, A1087 (1965)
12. R. Adde, S. Geschwind, L.R. Walker: "The Observation of a Phonon Bottleneck in the Orbach Relaxation of the $\bar{E}(^2E)$ State in Ruby", in North Holland, Amsterdam 1969 Proc. Colloque Ampère XV, p. 460
13. K.F. Renk, J. Deisenhofer: Phys. Rev. Lett. 26, 764 (1971)
14. K.F. Renk, J. Peckenzell: J. Physique C4, 103 (1972)
15. A.A. Kaplyanskii, S.A. Basun, V.A. Rachin, R.A. Titov: Sov. Techn. Phys. Lett. 1, 281 (1975)
16. J.I. Dijkhuis, A. van der Pal, H.W. de Wijn: Phys. Rev. Lett. 37, 1554 (1976)
17. R.S. Meltzer, J.E. Rives: Phys. Rev. Lett. 38, 421 (1977)
18. G. Pauli, G. Klimke, H.J. Kreuzer, K.F. Renk: phys. stat. sol. (b) 95, 503 (1979)
19. J.I. Dijkhuis, H.W. de Wijn: Phys. Rev. B 20, 1844 (1979)
20. W.C. Egbert, R.S. Meltzer, J.E. Rives: "Dynamics of 29 cm^{-1} Phonons in Ruby using Optical Generation and Detection", in Plenum Publ. 1980 Proc. Conf. Phonon Scattering in Condensed Matter, p. 365
21. R.S. Meltzer, J.E. Rives: Phys. Rev. B 25, 3026 (1982)
22. S.A. Basun, A.A. Kaplyanskii, V.L. Shekthman: Sov. Phys. Sol. St. 24, 1093 (1982)
23. N. Retzer, U. Werling, H. Lengfellner, K.F. Renk: to be published
24. W.E. Bron, Y.B. Levinson, J.M. O'Connor: Phys. Rev. Lett. 49, 209 (1982)
25. N. Retzer, H. Lengfellner, K.F. Renk: Phys. Lett. 96A, 487 (1983)
26. M. Engelhardt, K.F. Renk: "Phonon Decay in X-Ray Irradiated Ruby Crystals"; this volume
27. I.J. Brown, M.A. Brown: Phys. Rev. Lett. 46, 835 (1981)
28. S.A. Basun, A.A. Kaplyanskii: Sov. Phys. Sol. St. 22, 2055 (1980)

The Josephson Junction – A New Tunable Phonon Source with High-Frequency Resolution

P. Berberich and H. Kinder

Physik Department der Technischen Universität München
D-8046 Garching, Fed. Rep. of Germany

1. Introduction

Several methods of monochromatic phonon generation have been presented in the past, like piezoelectric surface excitation,[1] microwave [2] or optical [3] pumping of impurity ions, and single-particle tunneling in superconducting junctions [4,5]. Of these methods, tunneling junctions have the advantage of being independent of the sample under study. On single-particle tunneling, "bremsstrahlung" phonons are generated which can be made quasi monochromatic by a modulation technique [5]. Their frequency, $f = (eV-2\Delta)/h$, is tunable by the applied voltage V. Their frequency resolution depends on the sharpness of the energy gap Δ, usually a few percent. The energy gap comes into play because Cooper pairs are being broken during the single particle tunneling process. If the pairs could be left intact, i.e. if Josephson tunneling [6] could be used for phonon generation, one should obtain a much higher frequency resolution. We have recently observed that it is indeed possible to generate monochromatic phonons by the Josephson effect, so that very high resolution spectroscopy is now feasible. The present paper accounts for the results now available on the generation process. The first application to high-resolution spectroscopy is presented in a following paper by Schick et al.[8].

2. Theory

All Josephson phenomena are based on the equation [6]

$$j = j_c \sin\varphi \tag{1}$$

which relates the actual current density j to the critical current density j_c and the phase φ of the coupled ground states of the two superconductors. In the presence of electromagnetic fields, the latter is determined by

$$\partial\varphi/\partial t = 2eV/\hbar \tag{2}$$
$$\partial\varphi/\partial\vec{r} = 2e\vec{A}/\hbar \tag{3}$$

where V and $\vec{A}$ are the electric and vector potentials. If V_o is constant and $\vec{B}_o = (\partial/\partial\vec{r})\times\vec{A}$ is parallel to the barrier and homogenous, the integration of (2) and (3) yields a current wave

$$j(r,t) = j_c \sin(\omega_o t - \vec{k}_o\vec{r}) \tag{4}$$

with the Josephson frequency $\omega_o = 2eV/\hbar$ and with the wave vector $k_o = 2eB_o d_m/\hbar$ parallel to the barrier and perpendicular to the magnetic field. d_m is the "magnetic" barrier thickness including the penetration depths of the adjacent superconductors.

If the phase velocity $v = \omega_o/k_o = V_o/B_o d_m$ of this Josephson current wave matches that of the electromagnetic waves,[9] $\bar{c}$, in the junction, the cavity modes of the junction can be resonantly excited. This means that an ac voltage is induced which adds to the dc voltage V_0. The integration of (2) then leads to a series of sidebands at a distance $n\omega_0$ (with n integer) away from the "carrier" frequency ω_0. The sidebands with $n \neq -1$ are identical to harmonics in this case of self-modulation, and the sideband with $n = -1$ is always at zero frequency, that is, a dc current. This dc current shows up in the current-voltage characteristic of the junction which therefore reflects both the matching condition by a current peak around the voltage $V = B_0 d_m \bar{c}$, the so-called Eck bump, and the cavity mode frequencies of the junction, by superimposed little peaks or shoulders, the so-called Fiske steps [11]. Actually measured dc characteristics are shown by inset (b) in Fig. 1 (see further below). Only the regions of the characteristics with positive resistance were measured because of constant current conditions. The position of the Eck bump was varied by the external magnetic field ($\approx 10^{-3}$T). The Fiske steps are most clearly seen at the lower voltages. Because the Eck bump can be tuned to any voltage, strong electromagnetic fields can be generated at any frequency inside the junction. Coupling the radiation out to free space, however, is very inefficient, because the junction is an extremely short antenna. Rather, almost all of the energy is eventually dissipated.

Two basic processes were proposed by Werthamer [12] as being responsible for the dissipation at zero temperature, namely (i) photon-assisted single-particle tunneling and (ii) direct pair breaking in the superconductors.

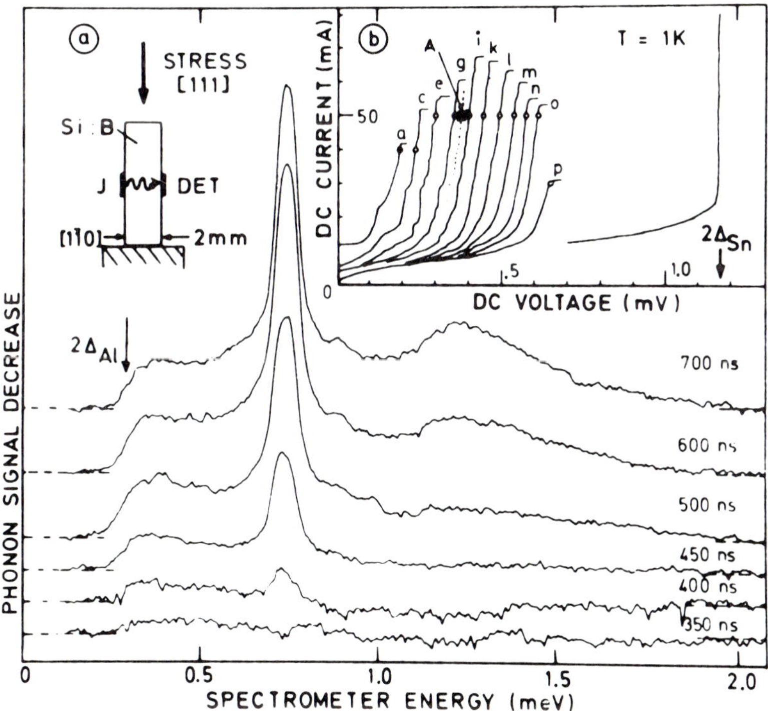

Fig. 1. Time evolution of Josephson phonon spectrum at a fixed operating point indicated by A on the junction characteristic, inset (b). The experimental arrangement is shown in inset (a)

Both processes require a threshold photon energy, $\hbar\omega > 2\Delta - eV$ for case (i), and $\hbar\omega > 2\Delta$ for case (ii), but higher-order processes appear at lower voltages, via multiple photon-assisted tunneling, and via harmonic generation by the self-modulation process described above. Eventually, all these dissipation processes lead to the excitation of single particles and hence to phonons of the relaxation ("bremsstrahlung") and recombination type.

The resulting recombination phonons of a Sn junction were studied by Kinder [13] some time ago, and the thresholds corresponding to the various processes were observed. Phonons of frequencies lower than the gap could not be detected in that experiment, because the detector was also a Sn junction. The full phonon spectrum was studied only recently [7] by using a stress tuned phonon spectrometer [14,15].

3. Spectrometer

As a spectrometer for the emitted phonons, we used a silicon substrate crystal with boron acceptors [15] which exhibit a stress tunable absorption line. The arrangement is shown by inset (a) in Fig. 1. If the line is positioned to a given frequency by stress, the total phonon signal, as monitored by an Al/Al detector junction, is diminished by an amount which is proportional to the intensity of emitted phonons at that frequency. The full phonon spectrum is thus obtained by sweeping the line position. Because the line increases in strength with increasing frequency [15], the raw spectra still contain a weight factor which can be easily corrected for, however. The line position versus stress was calibrated using "bremsstrahlung" phonons.

4. Josephson Phonon Spectra

Although the electromagnetic radiation of the Josephson junction is expected to be undetectably weak, one should make sure that the detector signal is due to phonons only, by observing their time of flight. If usual pulse techniques are used, however, the ringing of the pulse top would modulate the Josephson frequency. Therefore, square waves with sharp fall times were applied to the junction, so that there was ample time for the voltage to settle at the top of the square wave, but the time delay of the detector signal could still be measured by a boxcar averager whose 100ns-gate was set to some time t after the fall (t=0) of the square wave. At this gate time, the fall of the detector signal represented the time integral of the intensity from 0 to t. The frequency spectra of these integrated intensities were then obtained by sweeping the absorption line and monitoring the signal change as a function of stress. These are plotted in Fig. 1 for a set of gate times.

There was no intensity before 350 ns, indicating that electromagnetic radiation was negligible indeed. After 400 ns, the time of flight of fast transverse phonons which are focused into the (110) direction [16], the spectrum consisted of a sharp peak at the Josephson frequency and a broad background which was cut off at the energy gap of the detector junction. At later gate times, a second broader peak evolved at 1.2 meV due to recombination phonons, which were delayed by trapping in the superconductors.

Fig. 2 represents similar spectra at a fixed delay time of 700 ns, but with varying bias voltage. The operating points corresponding to the different traces are indicated in Fig. 1, inset (b). At low bias voltage, (traces a-e) a second harmonic peak is nicely seen. At $eV > 2\Delta/4$, trace f, the second harmonic frequency exceeds the energy gap. Therefore, harmonic generation is damped by direct pair breaking, in favour of the peak at the Josephson frequency (trace f). The further increase of that peak (traces f to i) is mostly due to the spectrometer weight factor mentioned above. If the spectra are quantitatively evaluated taking the weight factor into account, we find a maximum efficiency of 30% Josephson power for $eV < 2\Delta/3$. At $eV >$

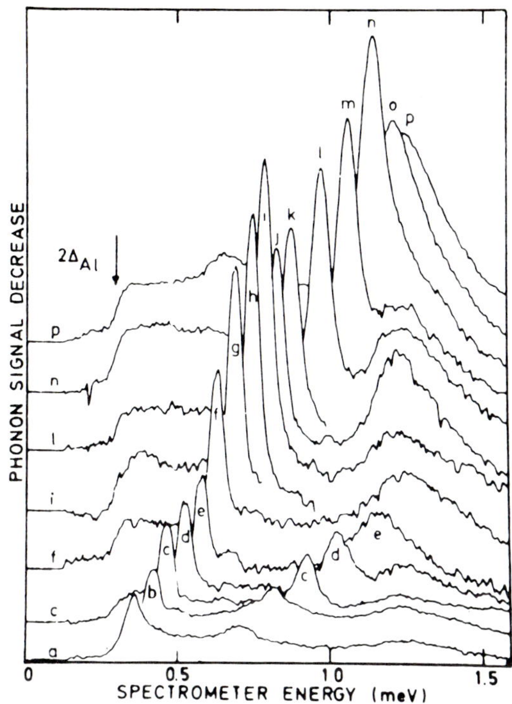

Fig. 2. Spectra for various bias voltages, correspondingly displaced in the vertical direction. The operating points are indicated in Fig. 1, inset (b)

$2\Delta/3$, trace j, direct photon assisted tunneling becomes possible which now favours the recombination phonon output. Accordingly, the Josephson efficiency drops to 15%. At eV > $2\Delta/2$ eventually, the Josephson frequency itself reaches the energy gap so that pairs are directly broken, the Eck bump fades away, and the recombination phonons take over completely. All this behaviour demonstrates the interplay between the Werthamer loss processes and an unexpected new mechanism which directly converts photons into phonons of the same frequency. This new mechanism makes it possible indeed to use Josephson junctions as monochromatic phonon sources.

Similar results were also obtained in lead junctions [7], where the efficiency of Josephson power was 35 % at eV < $2\Delta/3$ and 20 % at eV > $2\Delta/3$. The maximum efficiency was observed here at eV > $2\Delta/4$ to be 40 %.

5. Angular Distribution

To gain more insight into the generation mechanism, we have studied the angular distribution of the Josephson phonons. A detector array[17] was used for this purpose as indicated in the inset of Fig. 3. The effect of phonon focusing is eliminated, if one compares the Josephson phonon intensity with that of the recombination phonons for reference, because the latter are known to have a broad angular distribution [18]. For a Sn junction as the generator, Fig. 3 shows the spectra obtained for two angles, 0° and 40°, respectively. The vertical scales were adjusted for equal heights of the reference peaks. The Josephson peak in the 40° direction is somewhat larger, but essentially these phonons have the same broad distribution as the recombination phonons.

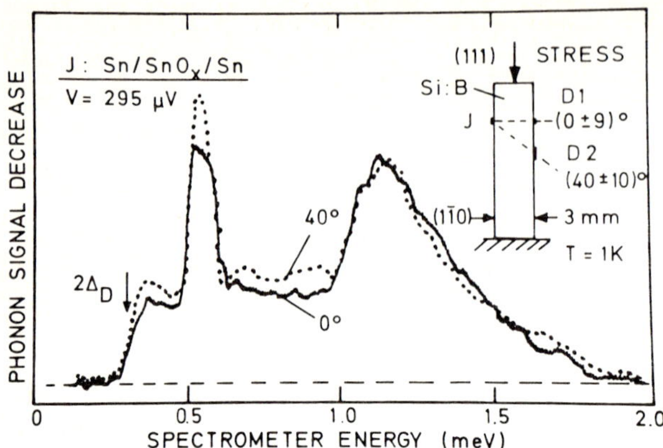

Fig. 3. Angular distribution of the phonon emission by a Sn Josephson junction J. The inset shows the detector array used. Solid line: detector D1, dotted line: detector D2. The vertical scales were adjusted for equal 2Δ-phonon peaks

For Sn junctions, the above result was not unexpected because they have a milky appearance and therefore a ragged oxide plane. Smooth junctions can be made of Al films as is manifested by standing wave patterns [19]. Therefore, we tried to use granular Al/Al junctions as generators and pure Al junctions with smaller gaps as detectors. Unfortunately, we could not observe any Josephson phonons in the frequency span between detector and generator gaps. We believe that the cause was twofold, namely the competition by photon

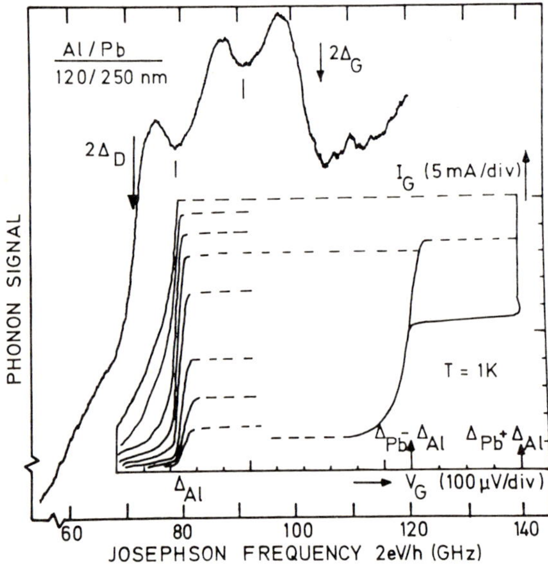

Fig. 4. Phonon standing wave pattern vs Josephson frequency for Al/Pb junction at 25mA bias current. Inset: I-V characteristic of the junction in various magnetic fields

assisted tunneling in this frequency range, and the large thermal tunneling current at 1 K. Both causes were eliminated by using Al/Pb heterojunctions which showed good Josephson efficiency indeed.

The I-V characteristic of an Al/Pb junction is shown in the inset of Fig. 4. The Eck bumps vanished at $V = \Delta_{Al}/e$, i.e.at $\hbar\omega_0 = 2\Delta_{Al}$ where direct pair breaking set in. If the junctions were made with Al as the bottom and Pb as the top film on substrate surfaces finished by a mechanical/chemical polish, standing waves could be observed indeed as shown in Fig. 4. For this measurement, the bias current was held fixed while sweeping the "Eck bump", and hence the Josephson frequency, by the external magnetic field.

The spacing of the standing wave pattern, Δf = 13 GHz, together with the thickness of the Al film, a = 120 nm ,yields a sound velocity of $v = 2a/\Delta f$ = 3.1 km/s which is that of transverse phonons in granular Al [19]. The depth of the pattern is consistent with that estimated from the reflection coefficients of the Al/Si and the Al/Pb interfaces. Therefore, the oxide must have been reasonably smooth. For these junctions, we have again measured the angular distribution in the same way as before, except that the reference phonons were generated at a separate operating point of the generator. The resulting Josephson and reference spectra were again plotted, see Fig. 5, for two angles with the scales adjusted to yield the same height of the reference phonons. In fact the Josephson phonons are now somewhat stronger in the 0° direction, but less than a factor of two.

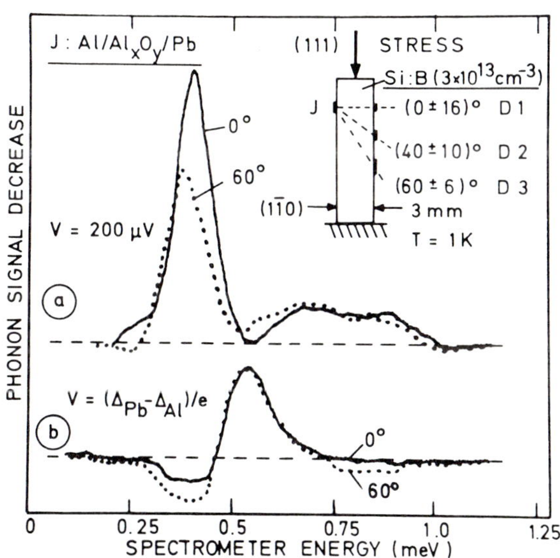

Fig. 5. Angular distribution of an Al/Pb junction; (a) Josephson phonons, (b) reference phonons generated at a different operating point. Solid lines: detector D1, dotted lines: detector D3. Scale of dotted traces was adjusted for equal reference phonons

6. Discussion

Because we have not observed the pronounced forward peaking of the angular distribution expected for plane waves, we can rule out the possibility of a preferentially oriented oxide barrier with piezoelectric properties. Rather, the oxide should have a disordered structure with the properties of a glass. The interaction of glasses with electromagnetic waves at low temperatures was extensively studied by far infrared absorption measurements [20,21]. In terms of this absorption α, the ac ad-mittance Y of the junction [7] can be expressed by

$$Y = c\, n\alpha\, \omega_0\, A/2d\,, \tag{5}$$

where c is the vacuum speed of light, n the refractive index of the glass, A the junction area, and d the oxide thickness of the junction. From the spectra of Fig. 2, we have estimated [7] an admittance $Y \simeq 200\Omega^{-1}$ for the 1:1 conversion process. If we assume that this conversion process is actually identical with the infrared absorption process, (5) would require $n\alpha \simeq 2$ cm^{-1}. This is well within the range of the available data for various glasses in our frequency range [20]. The infrared absorption at low temperatures was attributed to two processes [21]. One is the coupling of electromagnetic waves by charge fluctuations to phonons of the same frequency as proposed by Schlömann [22]. The other one is the coupling to localized "two-level systems" (TLS) [23]. The excited TLS will relax by emission of phonons of the same frequency [7] if the interaction between TLS is negligible. Both of these processes should also be responsible for the Josephson phonon generation, although modifications due to the restricted thickness of the barrier should be considered in this case.

7. Conclusion

At least the part of the phonons generated by the Schlömann mechanism [22] will have exactly the same frequency as the elctromagnetic waves, and will be coherent in time. The frequency resolution will therefore be determined only by that of the photons, that is, by the stability of the voltage applied to the junction. Experimentally, a frequency resolution of 1:5000 was so far reached [8]. The spatial distribution is definitely not that of plane waves. Therefore, the Josephson junctions will probably not be useful as coherent detectors by reversing the generation processes.

We are indebted to T. v.d. Ropp for carrying out some of the experiments and to Dr. H. Föll (Siemens München) for advice with the polishing procedure.

References

1. H. E. Bömmel, K. Dransfeld: Phys. Rev. Lett. 2, 298 (1959)
W. Grill, O. Weis: Phys. Rev. Lett. 35, 588 (1975)
2. C. H. Anderson, E. S. Sabisky: Phys. Rev. Lett. 18, 236 (1967)
3. K. F. Renk, J. Deisenhofer: Phys. Rev. Lett. 26, 764 (1971)
R. S. Meltzer, J. E. Rives: Phys. Rev. Lett. 38, 421, (1977)
4. W. Eisenmenger, A. H. Dayem: Phys. Rev. Lett. 18, 125 (1967)
5. H. Kinder: Phys. Rev. Lett. 28, 1564 (1972)
6. B. D. Josephson: Phys. Lett. 1, 251 (1962); for a review see: D.N. Langenberg, D.J.Scalapino, B. N. Taylor: Proc. IEEE 54, 560 (1966)
7. P. Berberich, R. Buëmann, H. Kinder: Phys. Rev. Lett. 49, 1500 (1982)
8. A. Schick, P. Berberich, W. Dietsche, and H. Kinder, following article
9. J. C. Swihart: J. Appl. Phys. 32, 461 (1961)
10. R. E. Eck, D.J. Scalapino, B.N. Taylor: Phys. Rev. Lett. 13, 15 (1964)
11. M. D. Fiske: Rev. Mod. Phys. 36, 221 (1963)
12. N. R. Werthamer: Phys. Rev. Lett. 147, 255 (1966)
13. H. Kinder: Phys. Lett. 36A, 379 (1971)
14. R. C. Dynes, V. Narayanamurti, M. Chin: Phys. Rev. Lett. 26, 181 (1971)
15. P. Berberich, H. Kinder: J. Phys. (Paris) 42, C6-374 (1981)
16. C. Elbaum: In Proc. Int. Conf. on Phonon Scattering in Solids (Paris 1972), ed. by H. J. Albany, p.1

17. H. Kinder, J. Weber, W. Dietsche: In Proc. Int. Conf. on Phonon Scattering in Condensed Matter (Plenum, New York (1980) ed. by H. J. Maris, p 173
18. W. Eisenmenger: In Physical Acoustics Vol 12, ed. by W. P. Mason and R. N. Thurston (Academic Press, New York 1976), p. 79
19. H. Kinder: J. Phys. (Paris) 33, C 4-21 (1972)
20. U. Strom, P. C. Taylor: Phys. Rev., B16, 5512 (1977)
21. M. A. Bösch: Phys. Rev. Lett. 40, 879 (1978)
22. E. Schlömann: Phys. Rev. 135, A 413 (1964)
23. S. Hunklinger, M. v. Schickfus: In Topics in Current Physics Vol. 24, ed. by W. A. Phillips (Springer, Berlin 1981) p. 811

Two Applications of Microwave Acoustics in Liquid Helium: High Resolution Microscopy and Direct Measurement of Phonon Dispersion

D. Rugar

Department of Applied Physics, Edward L. Ginzton Laboratory
Stanford University, Stanford, CA 94395, USA

1. Introduction

During the past three years, ultrasonic techniques have been developed at Stanford University for efficient generation and detection of multi-gigahertz sound waves in liquid helium. This paper discusses two applications of these techniques: high resolution acoustic microscopy and direct measurement of low energy phonon dispersion. We begin with acoustic microscopy.

2. Liquid Helium Acoustic Microscopy

Fundamentals of acoustic microscopy have been reviewed elsewhere [1-4]. Briefly, the heart of the microscope is the transducer/lens element. This consists of a 1-3 mm long sapphire rod which has a planar thin-film ZnO transducer fabricated on one end. At the other end of the rod is a carefully polished spherical depression of radius 20-200 μm. When in contact with an acoustic coupling fluid, such as liquid helium, the spherical surface acts as a lens and focuses the plane waves generated by transducer to a diffraction-limited spot in the coupling medium. A quarter wavelength thick layer of carbon serves as an anti-reflection coating and improves transmission across the sapphire-helium interface.

An object of interest is placed at the focal plane of the lens and it reflects a portion of the incident acoustic energy back to the transducer/lens element. Images are formed by mechanically scanning the acoustic lens with respect to the object. Pixel intensity is assigned in proportion to the received acoustic power.

Liquid helium at temperatures less than about 0.2K is of interest as a coupling fluid since very short wavelengths may propagate with low loss over macroscopic distances. Liquid helium is unique among liquids in this regard. For example, the attenuation of 50 nm waves at 0.1K is less than 1 dB/mm. DYNES and NARAYANAMURTI [5], among others, have demonstrated that phonons with less than 1 nm wavelength can propagate millimeter distances with low loss. If such short wavelengths could be harnessed for imaging, resolution comparable to that of scanning electron microscopy would be attainable.

Presently we have a microscope operating at 4.2 GHz in SVP helium. The corresponding acoustic wavelength is 57 nm. A top-loading dilution refrigerator is used to cool the microscope to its operating temperature near 0.1K. The top-loading feature allows samples to be changed without warming the refrigerator. A mixture of 3% ^{3}He is used as the coupling fluid instead of pure ^{4}He in order to reduce reflux heating due to superfluid film flow up the access tube. In the low gigahertz frequency regime and for propagation distances less than a millimeter, the increased attenuation due to phonon-^{3}He quasi-particle scattering is acceptable.

The lens of the microscope has an f/1 aperture which results in a rather narrow depth of focus, approximately 150 nm. Because of the phase sensitive nature of the transducer, features which are in focus appear bright while out of focus features are dark. This behavior is illustrated in Fig. 1 where the object is a silicon bipolar integrated circuit [6].

Fig. 1 Images of a bipolar silicon integrated circuit at two different focuses. Arrow points to border of ion-implanted region

In Fig. 1(a) the silicon substrate is in focus and appears bright. The aluminum lines, which are 2.5 μm wide and 0.5 μm thick, are out of focus and appear dark. The arrow in Fig. 1(a) indicates the border of the ion-implanted region for the base of the transistor. This is a feature which had not been seen previously by the manufacturer using other forms of imaging. Visualization of the ion implantation is probably due to the great sensitivity of the microscope to surface topography and due to a slight surface displacement in the implanted region.

Figure 1(b) is an image taken with the microscope focused on the top of the aluminum lines, which now appear bright. Some evidence of grain structure in the aluminum is visible. The grain structure can be seen more clearly at higher magnification, as is shown in Fig. 2. Many grain boundaries are visible and these may be used to provide an estimate of resolution. Numerous features can be found with 3-dB full widths of less than 50 nm.

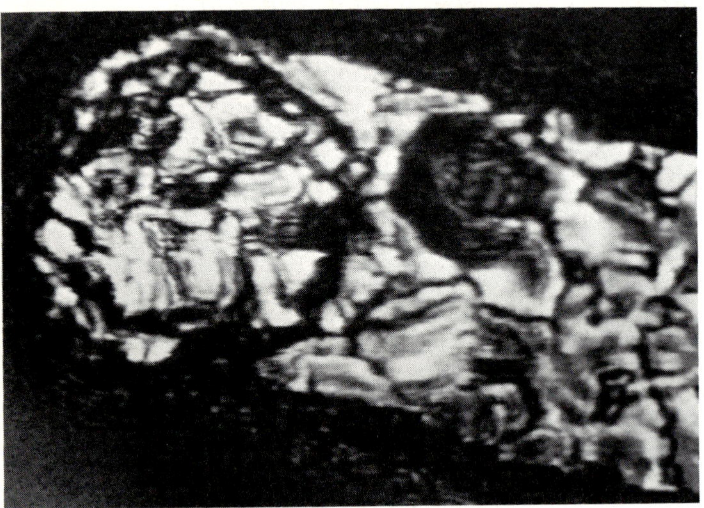

Fig. 2 Portion of an aluminum line shows high resolution and many grain structures. Line width is 2.5 μm

3. The Acoustic Microscope as a Phonon Detector

Since phonons in helium act to scatter acoustic waves, the microscope can be used as a phonon detector. Recent experiments have demonstrated detection capability with sub-micron spatial resolution, sub-50 nsec temporal resolution and high sensitivity. Figure 3 illustrates the technique.

The object shown in Fig. 3 is a thin film chromium heater on a sapphire substrate. The heater appears on the right side of the images and the bare sapphire on the left. Some dirt particles are visible and they appear dark. Figure 3(a) was taken with the heater off. The heater and substrate exhibit approximately equal brightness in the image. In Fig. 3(b), the heater was repeatedly pulsed so that phonons were emitted into the helium bath at a time coincident with arrival of acoustic pulses from the microscope. The heater appears relatively darker due to scattering of the focused acoustic beam by the emitted thermal phonons.

Intrinsic acoustic contrast may be separated from contrast due to phonon emission by dividing the recorded data in Fig. 3(b) by the data in Fig. 3(a). The resulting information, which is somewhat noisy, is displayed in Fig. 4. This image may be considered to be a map of phonon emission as a function of position, with regions of high phonon emission appearing dark. Thus, the sapphire appears bright and the heater dark. Note that dirt on the sapphire has been normalized out while dirt on the heater appears bright, indicating reduced phonon emission. Further details of this imaging technique will be published [7].

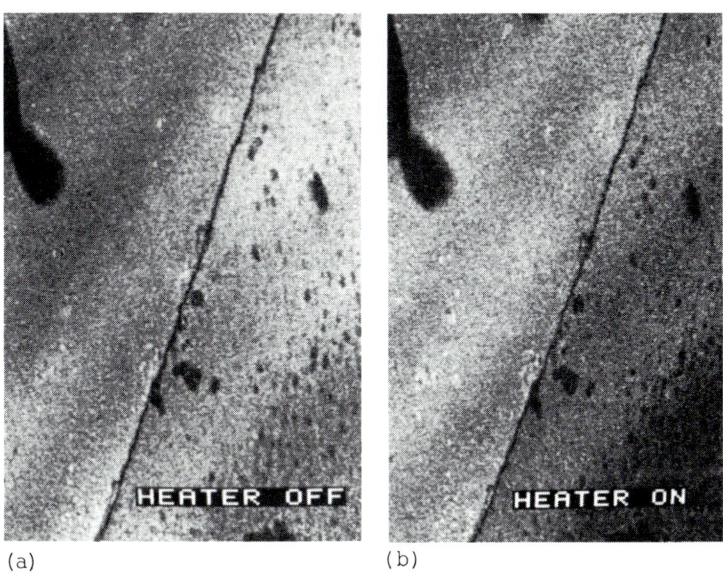

(a) (b)

Fig. 3 Images of chrome heater on sapphire. The heater is located on the right in both micrographs. Field width is 10 μm.

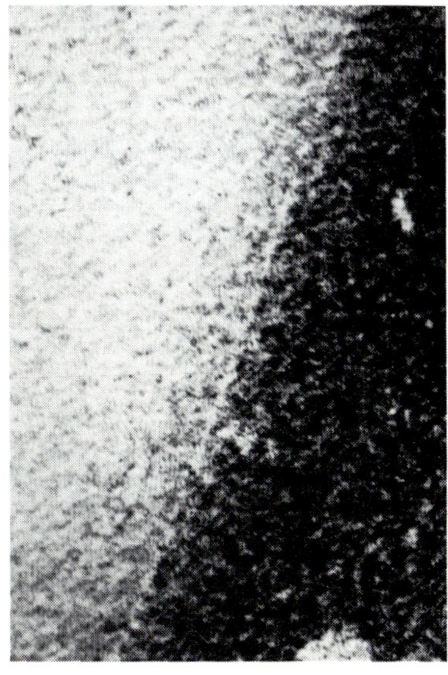

Fig. 4 Thermal phonon emission image obtained by dividing Fig. 3(b) by Fig. 3(a). Phonon emission appears dark

4. Direct Measurement of Low Energy Phonon Dispersion

By using the gigahertz acoustic wave technology developed for the acoustic microscope, we have succeeded in making a direct measurement of low energy phonon dispersion in liquid helium. Before describing the experimental details, we review the theory of weak second harmonic generation in a medium with dispersion.

We start with a lossless nonlinear wave equation similar to the one derived by TJØTTA and TJØTTA [8]:

$$\left(\nabla^2 - \frac{1}{[c(\omega)]^2}\frac{\partial^2}{\partial t^2}\right)\rho = A\,\frac{\partial^2}{\partial t^2}(\rho-\rho_o)^2 , \tag{1}$$

where ρ is the instantaneous density of the fluid, ρ_o is ambient density, $c(\omega)$ is the phonon phase velocity which we allow to be frequency dependent, and A is a constant.

We consider a general Fourier analyzed solution to (1)

$$\rho = \mathrm{Re}\,[\rho_o + \rho_1(x,y,z)\exp(-i\,\omega_1 t) + \rho_2(x,y,z)\exp(-i\,\omega_2 t) + \ldots] , \tag{2}$$

where $\rho_1(x,y,z)$, $\rho_2(x,y,z)$, etc., are complex-valued functions describing the amplitude and phase of the wave components. The angular frequencies are related by $\omega_n = n\omega_1$, where ω_1 is the angular frequency of the fundamental.

Substituting (2) into (1) and equating terms of like frequency, an infinite set of coupled equations is obtained. We now assume that depletion of the fundamental and generation of harmonics higher than the second can be ignored. In this limit of weak nonlinearity, we obtain two equations

$$(\nabla^2 + k_1^2)\,\rho_1 = 0 \tag{3}$$

$$(\nabla^2 + k_2^2)\,\rho_2 = B\,\rho_1^2 , \tag{4}$$

where $k_1 = \omega_1/c_1, k_2 = \omega_2/c_2, c_1$ is the phase velocity at ω_1, c_2 is the phase velocity at ω_2 and B is a constant. Equation (3) is the wave equation for the fundamental beam and it is linear since we are ignoring depletion. Equation (4) is for the generated second harmonic beam and it contains a nonlinear source term.

Specializing to the case of a plane wave fundamental beam, we write

$$\rho_1(z) = \rho_{10}\exp(ik_1 z) \tag{5}$$

$$\rho_2(z) = \rho_2'(z)\exp(ik_2 z) , \tag{6}$$

where ρ_{10} is a constant and $\rho_2'(z)$ is an envelope function for the second harmonic. Substituting these into (4) yields

$$\frac{d\rho_2'}{dz} = C\,\rho_{10}^2\exp[i(2k_1-k_2)z] , \tag{7}$$

where C is a constant. In deriving (7), we used the slowly varying envelope approximation

$$\left| k_2 \frac{d\rho_2'}{dz} \right| >> \left| \frac{d^2\rho_2'}{dz^2} \right| .$$

Equation (7) may be solved by simple integration.

The intensity of the second harmonic is proportional to $|\rho_2'(z)|^2$. Solving (7) with the boundary condition that $\rho_2'(z = 0) = 0$, the intensity of the second harmonic is found to be

$$I_2(z) = D \sin^2 (\frac{\Delta k}{2} z) , \tag{8}$$

where $\Delta k = 2k_1 - k_2$ and D is a constant. Thus the power in the second harmonic oscillates as a function of distance with period $L = 2\pi/\Delta k$. The same result is found for the analogous case in nonlinear optics [9].

An experiment was performed in SVP liquid ^{4}He at temperatures less than 80 mK to observe the dependence predicted by (8). A ZnO transducer was used to generate a pulsed fundamental beam at 2078 MHz. The transducer was coupled to the helium through a sapphire buffer rod with a carbon anti-reflection coating. Near field effects were avoided by making the buffer rod longer than a Fresnel length. A second transducer designed to receive 4156 MHz was located collinear with the transmitting transducer. The relative spacing and parallelism of the buffer rods were precisely controlled during the experiment by adjusting liquid pressure acting on a system of flexible bellows in the low temperature cell. To measure accurately the relative second harmonic power, the faces of the buffer rods needed to be parallel to better than 0.01 degrees at all separations. The acoustic path length through the helium bath was determined from the timing of the second harmonic pulses.

Figure 5 shows the results of the experiment. The squares indicate measured values of second harmonic power. Two successive nulls are observed with a spacing of 426 ± 10 μm. The solid line shows a plot of (8) for $|\Delta k| = 1.47 \times 10^4 \text{ m}^{-1}$. The fit between experiment and theory is excellent.

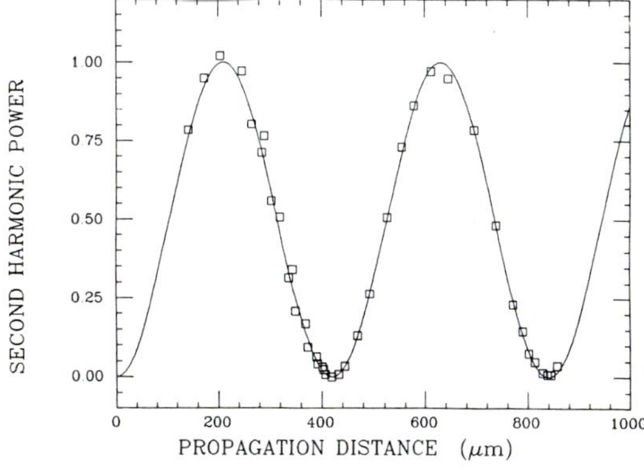

Fig. 5 Second harmonic power as a function of propagation distance. Solid line is computed from (8) and squares show measured values

By measuring the phase of the second harmonic as a function of distance, Δk was determined to be positive, indicating that phase velocity increases with frequency, as expected.

The dispersion of phonon phase velocity in helium is often expressed in power series form:

$$c(k) = c_0(1 + \alpha_1 k + \alpha_2 k^2 + \alpha_3 k^3 + \ldots) , \qquad (9)$$

where $c_0 = 238.3$ m/sec. ROACH et al. [10] in their experiments at 30 and 90 MHz concluded that $|\alpha_1| < 0.01$ Å. Other authors [11] have proposed that α_1 is identically zero. With this assumption and the reasonable assumption that cubic and higher order terms in (9) may be ignored for low gigahertz frequencies, the phase velocity can be written as

$$c(k) = c_0(1 + \gamma k^2) , \qquad (10)$$

where γ has been written in place of α_2 .

The parameter γ can be determined from our measured value of Δk according to

$$\gamma \approx \frac{c_0^3 \Delta k}{6 \omega_1^3} . \qquad (11)$$

Using experimental values for ω_1 and Δk, we find $\gamma = 1.49 \pm 0.04$ Å^2 .

Previous determinations of γ have yielded widely varying values, some of which are shown in Table I. Previous experimental methods involved using γ as an adjustable parameter in theories to fit data for specific heat, thermal expansion, ultrasonic attenuation, etc. Thus the obtained values depended on the theoretical framework of the fitting procedure.

The principal advantage of our method is its high accuracy and directness. The difference in phase velocity for phonons with a factor of two difference in energy is measured without theoretical complications. The accuracy is limited primarily by the uncertainty in the propagation path length. Since the experiment is performed at high frequencies and low temperatures, we

Table I. Some previous determinations of γ

Author	γ(Å^2)	Method	Reference
Present work	1.49 ± 0.04	Second harmonic coherence	-
Maris	1.1 ± 0.2	Specific heat and neutron scattering	12
Dynes & Narayanamurti	0.12	Phonon critical energies and group velocity	5
Berthold et al.	1.70 ± 0.33	Thermal expansion	13
Junker & Elbaum	1.78 ± 0.11	Ultrasonic wave interaction with thermal phonons	14
Aldrich et al.	1.5	Theory	15

have avoided complicating factors such as resonant interaction of the acoustic wave with thermal phonons in the bath.

Further experiments are planned at higher bath pressures and at different frequencies. By measuring the dependence of Δk on frequency, we can check the validity of the assumption that the other terms in (9) can be ignored.

5. Conclusion

We have described two recent applications which use gigahertz coherent phonons in liquid helium: acoustic microscopy and measurement of phonon dispersion.

The helium acoustic microscope is both a general purpose high resolution imaging instrument and a new tool for studying phonon physics. The microscope currently operates with 20 dB signal-to-noise ratio and exhibits 50 nm resolution. Future increases in operating frequency are planned, the primary obstacle being nonlinear excess attenuation which will increase 6 dB per doubling of frequency [3]. By using the microscope as a phonon detector, phonon emission from surfaces can be studied on a microscopic scale with good time resolution. This method should prove useful for imaging sub-surface structures, studying phonon emissivity at helium-solid interfaces (e.g., Kapitza resistance) and imaging samples which exhibit spatially variable phonon interaction (e.g., superconductors in intermediate or mixed states).

We have also described a new and highly accurate method of measuring low energy phonon dispersion in liquid helium. It is hoped that this method will help alleviate the confusion which currently exists in the literature concerning the true magnitude of low energy dispersion.

Acknowledgements

It is a pleasure to acknowledge the close collaboration of J. S. Foster and C. F. Quate in all of the described experiments. Supported by the U.S. Office of Naval Research and partially by the Air Force Office of Scientific Research.

References

1. C.F. Quate, A. Atalar, H.K. Wickramasinghe: Proc. IEEE 67, 1092 (1979)
2. J. Heiserman, D. Rugar, C.F. Quate: J. Acoust. Soc. Am. 67, 1029 (1980)
3. D. Rugar, J.S. Foster, J. Heiserman: in Acoustical Imaging, Vol. 12, ed. by E.A. Ash and C.R. Hill (Plenum, New York, 1982), pp. 13-25
4. J.S. Foster, D. Rugar: Appl. Phys. Lett. 42, 869 (1983)
5. R.C. Dynes, V. Narayanamurti: Phys. Rev. B 12, 1720 (1975)
6. Courtesy of TRW Corporation, Redondo Beach, CA.
7. J.S. Foster, D. Rugar, C.F. Quate: to be published
8. J.N. Tjøtta, S. Tjøtta: J. Acoust. Soc. Am. 69, 1644 (1981)
9. J.A. Armstrong, N. Bloembergen, J. Ducuing, P.S. Pershan: Phys. Rev. 127, 1918 (1962)
10. P.R. Roach, B.M. Abraham, J.B. Ketterson, M. Kuchnir: Phys. Rev. Lett. 29, 32 (1972)
11. C. Zasada, R.K. Pathria: Phys. Rev. Lett. 29, 988 (1972)
12. H.J. Maris: Phys. Rev. A 8, 1980 (1973)
13. J.E. Berthold, H.N. Hanson, H.J. Maris, G.M. Seidel: Phys. Rev. B 14, 1902 (1976)
14. W.R. Junker, C. Elbaum: Phys. Rev. B, 15, 162 (1977)
15. C.H. Aldrich, C.J. Pethick, D. Pines: J. Low Temp. Phys. 25, 691 (1976).

Heater Film Dynamics, Phonon Diffusion and Phonon Decay

T.E. Wilson, W.E. Bron, F.M. Lurie, and W.L. Schaich

Physics Department, Indiana University
Bloomington, IN 47405, USA

1 INTRODUCTION

A phonon spectrometer [1] has been used to measure the spectral, spatial and temporal distribution of phonons propagating in a SrF_2 crystal. The spectrometer uses as the detecting element the vibronic sidebands associated with certain luminescent transitions at Eu^{2+} probe ions placed in dilute concentration in the crystal. The phonons are generated in a constantan thin film through Joule heating and cross the interface between the film and the crystal.

2 EXPERIMENT

The primary experimental observation is the luminescence associated with the $4f^6 5d(\Gamma_8) \rightarrow 4f^7(^8S_{7/2})$ electronic transition of the Eu^{2+} ion. The crystal, containing 0.1 mole % of Eu^{2+}, is cut to the dimensions 7x7x10 mm, and polished in stages down to 0.25 μm sized diamond grit. During the experiment, all but one face of the sample is in a final vacuum of less than 10^{-8} Torr. The remaining face is attached to a 'cold finger' at the bottom of a helium reservoir in an optical Dewar. The ambient crystal temperature is 4.2 K between heat pulses. A 30 Ω constantan film, approximately 3000 Å thick, is evaporated onto a (100) end surface of the crystal. The film has the dimensions 5.0x.2 mm.

Luminescence is excited with a N_2 laser having a 10 ns pulse width, and 20 KW peak power. The laser beam is focused inside the crystal to a column approximately 0.15 mm in diameter, parallel to the long dimension of the heater film. The resulting luminescence is collected and focused on the entrance slit of a 1 m Czerny-Turner spectrometer, adjusted for a spectral resolution of 0.8 Å (5 cm^{-1}). The PMT output pulses are counted by an 850 MHz gated event counter. An LSI-11 minicomputer reads the counts, reads an A/D converted photodiode signal monitoring the laser intensity, reads the wavelength setting, strobes the wavelength drive, and performs the necessary calculations described below. The laser is pulsed at twice the repetition rate of the applied heat pulses (40 Hz), such that every other luminescent pulse is in the presence of injected phonons. The Eu^{2+} luminescence has a lifetime of 1 μs. Gated portions of the luminescence, ranging from 0.2 to 4.0 μs, were used as the signal to the counter. It was observed that varying the gate width over this range produced only small effects in the resulting phonon distributions.

The anti-Stokes signal, AS, was calculated according to the algorithm

$$AS = [\Sigma_i'(\text{counts}_i) - \Sigma_i(\text{counts}_i)/\Sigma_i(N_{2_i})\Sigma_i'(N_{2_i})]/\Sigma_i'(N_{2_i}) \,, \tag{1}$$

where the unprimed sums run over the number of gates without an applied heat pulse, and the primed sums run over the number of gates with a heater pulse. The number of gates used was nominally 20,000. An estimate for the standard deviation for an AS data point is given by

$$\sigma = [\mathrm{sqrt}(\Sigma_i'(\mathrm{counts}_i)) + \mathrm{sqrt}(\Sigma_i(\mathrm{counts}_i))]/|\Sigma_i'(\mathrm{counts}_i) - \Sigma_i(\mathrm{counts}_i)|. \quad (2)$$

The algorithm for AS eliminates any effects due to ambient concentrations of phonons generated at the Eu^{2+} sites from nonradiative decay from states excited by N_2 laser absorption to the $4f^65d(\Gamma_8)$ emitting state. In order to check the possible effects of nonlinear interaction of the nonradiatively produced and heater-produced phonons, the laser intensity was reduced by half. No change in the spectral distribution was observed. Moreover, the AS signal vanished in the absence of any heat pulses. Anti-Stokes signals could be observed from 0.5 to 4.0 THz. The lower limit is due to the intrinsic width of the zero phonon line, and the upper limit results only from the weakness of the signal at these power levels and averaging times.

3 RESULTS AND DISCUSSION

Vibronic sidebands arise from the interaction between the time-varying crystal field associated with phonons and an electronic transition localized at the probe ions. The Stokes sideband intensity at frequency ω is

$$S(\omega) = e^2\hbar/(4\omega)[n(\omega)+1]\Sigma_\Gamma F(\Gamma)\rho(\omega,\Gamma)F'(\Gamma) \quad (3)$$

and the anti-Stokes sideband intensity is given by

$$AS(\omega) = e^2\hbar/(4\omega)n(\omega)\Sigma_\Gamma F(\Gamma)\rho(\omega,\Gamma)F'(\Gamma) \ , \quad (4)$$

where $n(\omega)$ is the phonon distribution function, $\rho(\omega,\Gamma)$ is the phonon density of states projected onto the irreducible representations Γ, consistent with the electronic transitions, and $F(\Gamma)$ is the electron-phonon interaction operator. $S(\omega)$ was experimentally determined at 1 K, where $n(\omega)$ is small, and denoted by $D(\omega)$. From this, the primary experimental information, $n(\omega)$ can be determined directly and is given by [1]

$$n(\omega) = AS(\omega)/D(\omega). \quad (5)$$

A typical $n(\omega)$ is shown in Fig.1, for a delay time of 80 ns between the applied heat pulse and the observation gate, and a heater power of 4.3 KW. A best-fit Bose-Einstein distribution is also shown for comparison and the fitted temperature is displayed. $n(\omega)$ leads, via the Bose-Einstein distribution, to the so-called mode temperature $T(\omega)$. Figure 2 shows $T(\omega)$ plotted from the data of Fig.1. The interesting observations are that under most conditions, the mode temperatures vary only a few degrees with frequency, which is indicative of a quasiequilibrium among the non-equilibrium phonons. The quasiequilibrium is observed at power levels of 1 to 4 KW, at distances from the film of 0 to 0.3 mm, and at delay times of 80 to 5000 ns. The fitted temperatures are seen to decrease with distance and delay time, and increase with power. These results imply that phonon-phonon interactions are important processes, in addition to elastic scattering, during the observation times. Otherwise, no quasiequilibrium could exist among the higher frequency phonons [2,3].

Two [4,5] theoretical calculations have been attempted in order to, at least qualitatively, explain the presence of quasiequilibrium phonon distributions. The first of these [4] allows for a hot, thermalized film to inject phonons into an anharmonic, diffusive dielectric, initially at very low temperature. It self-consistently determines the time dependence

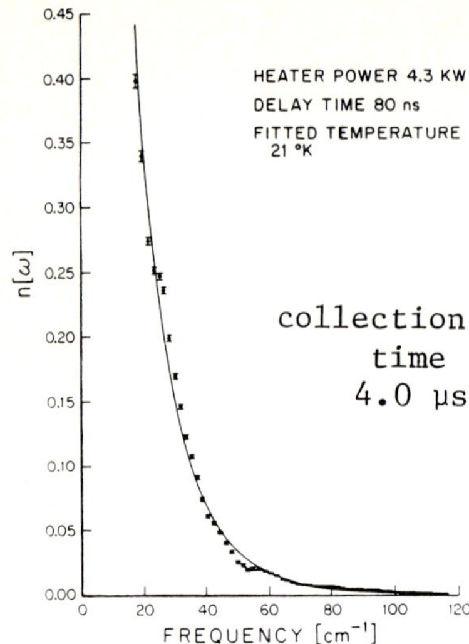

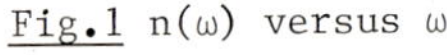
Fig.1 n(ω) versus ω

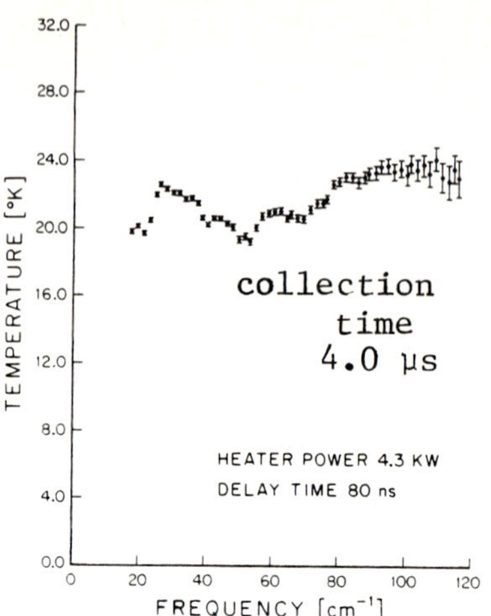

Fig.2 Mode temperature versus ω

of the film temperature from the power applied to the film and the energy flux into the crystal. It is shown that the anharmonic processes thermalize the phonon frequencies above 1 THz and that the lower frequency phonons are spatially segregated away from the film. The low-frequency mode temperatures are enhanced relative to the temperature of the high-frequency quasiequilibrium. The second approach [5] uses the theory of LEVINSON [6] to calculate the space-and time-dependent bulk temperature of an anharmonic, diffusive dielectric, subjected to a phonon source from a thermalized heater film whose temperature is raised only an infinitesimal amount above the ambient crystal temperature T_0. Again, the results give a quasiequilibrium phonon distribution in the crystal for high frequencies ($\omega > T_0$), but the mode temperatures of the low-frequency phonons ($\omega \ll T_0$) are predicted to be higher than that of the quasiequilibrium. Although both models qualitatively explain the observation that the crystal exhibits a quasiequilibrium, they fail to explain the observation that the mode temperatures remain essentially constant with frequency; i.e., even for the low-frequency components. The model calculations can be modified to reduce the low-frequency mode temperatures by assuming less than perfect transmission through the interface [1,7,8], as well as less than complete thermalization in the heater film [9,10].

†Research supported by the U.S. Army Research Office, Contract No. DAAG29-80-C-0085

1. W.E. Bron and W. Grill: Phys Rev. B16, 5303, 5315 (1977).
2. W.L. Schaich: J. Phys. C11, 4341 (1978).
3. D.V. Kazakovtsev and Y.B. Levinson: J. Low Temp. Phys. 45, 49 (1981).
4. W.L. Schaich: to be published.
5. T.E. Wilson: to be published.
6. Y.B. Levinson: Zh. Eksp. Teor. Fig. 79, 1394 (1980); Sov. Phys. J.E.T.P. 52, 704 (1980).
7. W.A. Little: Can. J. Phys. 37, 334 (1959).
8. O. Weis: Z. Angew. Phys. 26, 325 (1969).
9. N. Perrin and H. Budd: Phys Rev. Lett. 28, 1701 (1972).
10. N. Perrin and H. Budd: J. Physique 33, C4-33 (1972).

Infrared Excitation of High-Frequency Phonons by Multiphonon Absorption

U. Happek, R. Baumgartner, and K.F. Renk

Institut für Angewandte Physik, Universität Regensburg
D-8400 Regensburg, Fed. Rep. of Germany

A new method for generation of high-frequency phonons is reported. Phonons were generated in CaF_2 crystals by multiphonon absorption of infrared radiation of a CO_2 laser. The new method was applied to study transient broadband spectra of nonthermal phonons.

In earlier studies nonequilibrium phonons at frequencies above 1 THz were generated using heat pulse techniques or by nonradiative electronic transitions [1]. We report on a new method: it is based on multiphonon absorption and has the important advantage that phonons can be generated in pure crystals without use of defects or electronic excitations. Compared to heat pulse techniques surface effects are avoided. We have applied the method to study transient spectra of nonequilibrium phonons in a CaF_2 crystal surrounded by liquid helium.

The principle of our experiment is shown in Fig. 1a. CO_2 laser radiation at a frequency $\nu_L \simeq 30$ THz is absorbed by three-phonon absorption [2]. According

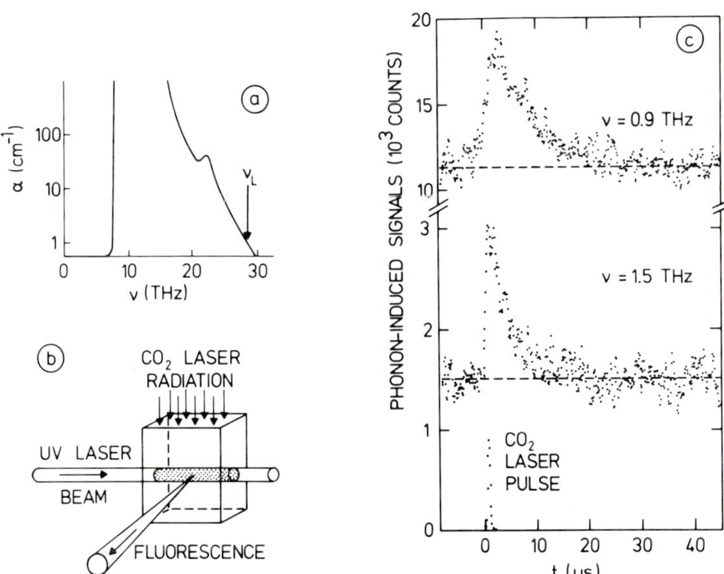

Fig. 1 Phonon generation by multiphonon absorption; (a) absorption of CO_2 laser radiation (at ν_L) in CaF_2; (b) experimental configuration; (c) signal curves

to the dispersion curves of CaF_2 [3] the generated phonons are mainly optical phonons which decay very fast into acoustic phonons. We have detected acoustic phonons by vibronic sideband spectroscopy observing anti-Stokes emission from the lowest excited state of Eu^{2+} ions in CaF_2. For the purpose we used a crystal weakly doped with Eu^{2+} ions (0.01 mol%). The experimental configuration is shown in Fig. 1b. A CaF_2 crystal (size $8\cdot 8\cdot 10$ mm^3), immersed in liquid helium at a temperature of 1.6 K, is illuminated with radiation from a Q-switch CO_2 laser (pulse width 0.5 μs, pulse energy 10^{-3} J, repetition rate 100 cps). The beam diameter of the laser was chosen as large as the crystal surface area and, since the absorption coefficient for infrared radiation at 10.6 μm is small (1 cm^{-1}), phonons were generated in the whole crystal.

Experimental signal curves for two different detector frequencies ν are shown in Fig. 1c. At $\nu = 1.5$ THz, the signal increases fast and decreases with a time constant much longer than the duration of the CO_2 laser pulse (lowest curve). The signal decay is due to anharmonic phonon decay as already observed for phonons generated by nonradiative electronic transitions [4]. At $\nu = 0.9$ THz (upper curve) we observe a rise time which is longer than the duration of the laser pulse. This is attributed to phonons generated by anharmonic splitting processes of phonons at higher frequencies. The decay of phonons at 0.9 THz is slower than for phonons at 1.5 THz.

From the signal curves we have determined phonon occupation numbers $p(\nu) = I_{AS}(\nu)/I_S(\nu)$ where $I_{AS}(\nu)$ is the intensity of phonon-induced anti-Stokes radiation at a frequency distance ν from the zero phonon line and $I_S(\nu)$ is the fluorescence intensity in the Stokes sideband. In our experiment the phonon occupation numbers in the signal maxima (Fig. 1c) varied, at constant CO_2 laser pulse energy, between 10^{-2} (at 0.5 THz) and $2\cdot 10^{-4}$ (at 1.8 THz). From $p(\nu)$ we calculated the spectral energy density $\rho(\nu) = p(\nu)D(\nu)h\nu$, where $D(\nu) = 12\pi\nu^2/c^3$ is the density of states of acoustic phonons, c a mean velocity of sound and $h\nu$ the phonon energy.

We have studied the spectral distribution of phonons at different times t after pulsed laser excitation. A spectrum is shown in Fig. 2a for a time, $t = 1$ μs, comparable to the time of ballistic flight of the phonons through the crystal. Because of diffusion, phonons did not yet escape at this time. The spectrum is strongly non-Planckian; in case of thermalisation of the absorbed energy, we would expect an equilibrium distribution corresponding to a

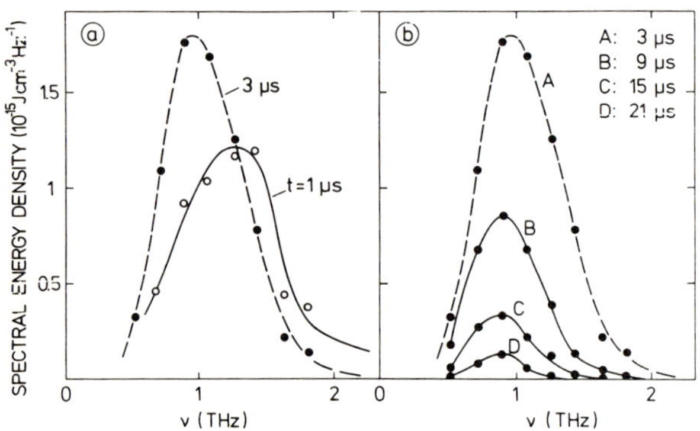

Fig. 2 Transient spectra of nonequilibrium phonons for different times t after pulsed laser excitation

temperature of about 8 K with a maximum at 0.5 THz. At $t = 3\ \mu s$, the energy density has decreased in the high-frequency range, while an increase is observed at lower frequencies. This result shows directly that high-frequency phonons decay by splitting processes into low-frequency phonons. Anharmonic decay is also responsible for the decrease of $\rho(\nu)$ in the high-frequency range found for later times (Fig. 2b).

Most interesting is the development of the phonon spectrum at late times after the pulsed phonon generation (Fig. 2b). We find that with increasing time, the phonon spectrum narrows and changes its shape. It is interesting that the maximum of the distribution at a frequency ν_{max} remains unchanged. According to our result the phonon population at ν_{max} has a maximum lifetime τ_{max}, while phonons at higher and lower frequencies have shorter lifetimes. For a description of our result we introduce an average phonon decay rate $\tau^{-1}(\nu) = \tau_a^{-1}(\nu) + \tau_d^{-1}(\nu)$, where $\tau_a(\nu)$ is the anharmonic decay time and $\tau_d(\nu)$ the average diffusion time of phonons at frequency ν for spatial escape from the crystal. With $\tau_a^{-1}(\nu) = A\nu^5$ where $A = 8\cdot 10^{-56} s^4$ [4] and assuming $\tau_d^{-1}(\nu) = B\nu^{-4}$ we calculate a maximum lifetime at a frequency $\nu_{max} \simeq (B/A)^{1/9}$. With a value ν_{max} = 0.95 THz (Fig. 2b) we obtain the constant $B = 5\cdot 10^{52} s^{-5}$. The value of B is in agreement with the experimental decay time τ_{max} observed directly at the frequency ν_{max} (Fig. 1c). The diffusion is consistent with Rayleigh scattering at Eu^{2+} impurities [5]. We have also performed a quantitative calculation of the temporal development of the phonon spectrum. We found that the diffusive escape cannot be described by a ν^2 dependence of $\tau_d(\nu)$ as assumed for phonon Rayleigh scattering at Eu^{2+} ions in SrF_2 [6].

In conclusion, we have demonstrated a new very effective method for generation of high-frequency phonons and we have shown that an originally broadband phonon spectrum in a CaF_2 crystal narrows with increasing time after pulsed phonon generation.

The work was supported by the Deutsche Forschungsgemeinschaft.

1. K.F. Renk: "Far Infrared Volume-Generation and -Detection of Phonons"; this volume
2. W. Kaiser, W.G. Spitzer, R.H. Kaiser, L.E. Howarth: Phys. Rev. 127, 1950 (1962)
3. M.M. Elcombe, A.W. Pryor: J. Phys. C, Solid State Phys. 3, 492 (1970)
4. R. Baumgartner, M. Engelhardt, K.F. Renk: Phys. Rev. Lett. 46, 1210 (1981)
5. K.F. Renk: "A Survey on the Optical Detection of Terahertz Phonons", in IEEE 1979 Ultrasonics Symposium Proc., p. 427
6. W.E. Bron, W. Grill: Phys. Rev. B 16, 5303, 5313 (1977)

Zero-Field Hyperfine Splitting of $Al_2O_3:V^{3+}$ by Josephson Phonon Spectroscopy

A. Schick, P. Berberich. W. Dietsche, and H. Kinder

Physik Department der Technischen Universität München
D-8046 Garching, Fed. Rep. of Germany

1. Introduction

Recently, phonon emission spectra of Josephson junctions at a variety of fixed operating points were studied by using a phonon spectrometer [1]. A monochromatic peak at the Josephson frequency was found which was as sharp as the spectrometer resolution of several percent. To use this peak for phonon spectroscopy, the Josephson frequency must be continuously tunable over the frequency range of interest, and, simultaneously, it must be as stable as possible. For an application, we have chosen $Al_2O_3:V^{3+}$, a system well studied by spin resonance [2,3] and by far infrared [4] and phonon [5,6] spectroscopy, whose hyperfine structure in zero magnetic field was not yet known.

2. Tunablility and Stability

The Josephson frequency is given by $f = 2eV/h$, and hence depends solely on the dc voltage V applied to the junction. With low impedance junctions, required for maximum phonon output, the voltage cannot be controlled directly, but is rather given by the impressed current, and by the junction characteristic at the operating point. For resonant excitation of electromagnetic waves, the junction must always operate near the top of the "Eck bump" whose position depends on the external magnetic field (0 to .3mT) [7]. It is therefore preferable to keep the current fixed and to change the voltage by the position of the Eck bump, i.e. by the external magnetic field.

Unfortunately, the bump itself is not always continuous, but it has a fine structure, the Fiske steps [7], due to the electromagnetic cavity modes of the junction. This structure can lead to small hysteretic voltage jumps under the impressed current condition. Only when the density of modes per frequency is greater than the lifetime of the individual modes, the characteristic is still monotonous and hence can be used for spectroscopy.

The stability of the voltage at a given magnetic field depends on the stability of both dc current and field:

$$\delta V/V = (\partial \ln V/\partial \ln I)\, \delta I/I + (\partial \ln V/\partial \ln B)\, \delta B/B\,. \qquad (1)$$

From the characterisics of the Sn junction used for the experiments described below, we got the average values $\langle \partial \ln V/\partial \ln I\rangle = 0.06$. and $\langle \partial \ln V/\partial \ln B\rangle = 1.4$. Therefore, more effort was to be taken to reduce $\delta B/B$. Because previous measurements have shown that any contribution of electromagnetic waves was negligible [1], and positions of absorption lines were of prime interest here, pulse measurements were not required. We have therefore used square waves and lock-in techniques. For optimum stability, the magnetic field coil was supplied with a constant main current source and an auxiliary current sweep. With this arrangement, we have reached $\delta I/I = 10^{-3}$ and $\delta B/B = 10^{-4}$ so that $\delta V/V = 2.10^{-4}$ should be the achievable frequency resolution.

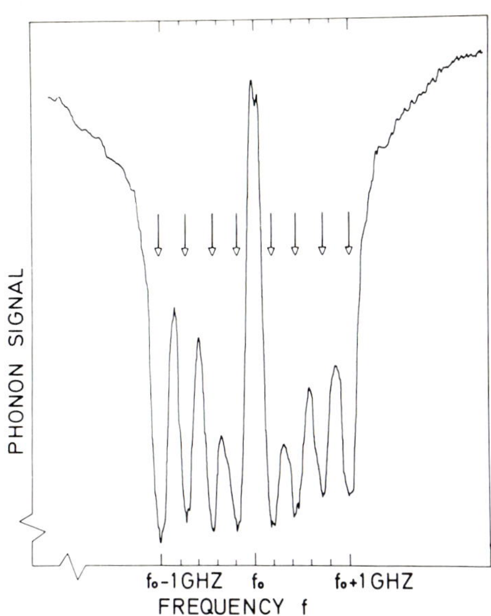

Fig. 1. Phonon absorption line multiplet of V^{3+} in Al_2O_3

3. V^{3+} in Sapphire

The electronic ground state of V^{3+} in Al_2O_3 can be described by an effective spin Hamiltonian of the form [8]

$$H = DS_z^2 + g_{\parallel}\beta H_z S_z + g_{\perp}\beta(H_x S_x + H_y S_y) + E(S_x^2 + S_y^2) + AI_z S_z + B(I_x S_x + I_y S_y). \quad (2)$$

Here, D is the second-order splitting due to the trigonal field plus the spin-orbit interaction, E is a rhombic field coefficient, and the constants A and B represent the hyperfine interaction. The magnetic field required to control the Josephson effect is of the order of 3×10^{-4}T so that $g_{\parallel}\beta H/A \simeq 0.03$ in our experiments. Hence, the magnetic field can be neglected. Neglecting also terms of the order B/D results in a level scheme [3]

$$\varepsilon_m = D + 2Am + E^2/Am\,. \quad (3)$$

Here, m runs from -7/2 to +7/2 according to the nuclear spin I = 7/2 for ^{51}V. The third term is suppressed in magnetic fields $g\beta H/A > 1$, and was therefore not yet measured before, although E is, on the other hand, responsible for the magnitude of spin resonance signals [2].

4. Results

For samples of Al_2O_3 with low V^{3+} concentration of 74 ppm [9], and orientation in a-direction, we have observed absorption line multiplets as shown in Fig. 1. The center frequency was f_0 = 248±1 GHz, which may be measured with much greater precision if required. The eight hyperfine lines are obviously not equidistant in zero magnetic field, and the rhombic field coefficient E manifests itself primarily by the central gap. The best fit of (2) to the data, as indicated by arrows, was obtained with A/h = (288±2) MHz and E/h = 110 MHz. The physical significance of this latter coefficient is not clear yet. If it were due to random internal strains or electric fields, however, the gap would not be so well defined.

In fact, the gap is the sharpest feature of the multiplet and was therefore used to obtain an upper limit for the frequency resolution. By fitting error functions to the gap edges, we obtained a frequency resolution (stan-

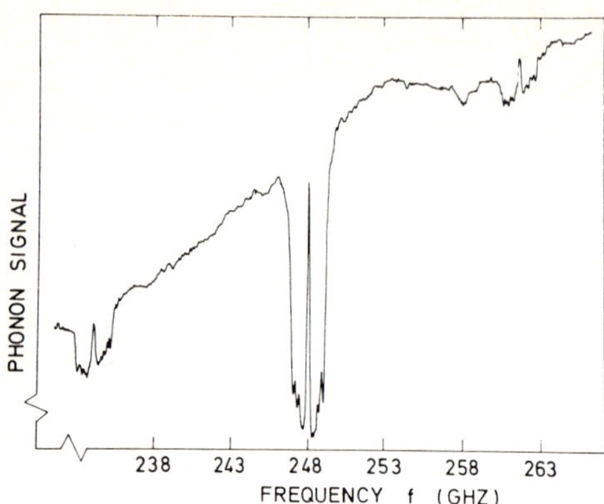

Fig. 2. Absorption lines of the same sample in a larger frequency range

dard deviation) of 50 MHz or 2×10^{-4} of the frequency. This is just the resolution expected from section 2, so that the stability still seems to be the limiting factor.

In this same sample of low concentration, we have also observed satellite lines, 14 GHz away from the center line, see Fig. 2. The satellite lines have the same hyperfine structure as the main line and hence are certainly also due to V^{3+}. Satellite lines were also observed by Challis et al.[10] in a sample cut from the same boule as ours, in a magnetic field of 3 T. Surprisingly, these were only 0.29 cm^{-1} = 8.7 GHz away from the main line. Their hyperfine structure was not resolved. Challis et al.[10] attributed these lines to V^{3+} pairs, and assumed exchange interaction as being the cause of the splitting. These data and ours can be reconciled, if a magnetic field dependence of the splitting is assumed, but we have no justification for this, nor for the fact that the hyperfine structure was essentially unaffected, on the basis of the existing model.

References

1. P.Berberich, R. Buëmann, H.Kinder: Phys. Rev. Lett. 49, 500 (1982)
2. G. M. Zverev, A. M. Prokhorov: Zh. Eksp. Teor. Fiz. 34, 1023 (1958) (Engl. trans.: Sov. Phys. JETP 7, 707 (1958))
3. J. Lambe, C. Kikuchi: Phys. Rev. 118, 71, (1960)
4. E. A. Vinogradov, N. A. Irisova, T. S. Mandel'stam: ZhETF Pis'ma 4, 373 (1966) (Engl. trans.: Sov. Phys JETP Lett. 4, 252 (1966)); R. R. Joyce, P. L. Richards: Phys. Rev. 179, 375 (1969)
5. H. Kinder: Z. Phys. 262, 295 (1973)
6. M. N. Wybourne, L. J. Challis, A. A. Ghazi: J. Phys. C 13, 6495 (1980)
7. For a review, see D. N. Langenberg, D. J. Scalapino, B. N. Taylor: Proc. IEEE 54, 560 (1966)
8. A. Abragam, M. H. L. Pryce: Proc. Roy. Soc. 205, 135 (1951)
9. N. Devismes, A. M. de Goer: J. Phys. C 11, 3805 (1978)
10. L. J. Challis, A. A. Ghazi, D. J. Jefferies, D. L. Williams, M. N. Wybourne: Phys. Rev. Lett. 40, 519 (1978)

Subnanosecond Bolometry Using Niobium Films

J.P. Maneval, J. Desailly, and B. Pannetier

Groupe de Physique des Solides de l'Ecole Normale Supérieure
24 rue Lhomond, F-75231 Paris Cêdex 05, France

Niobium and compounds NbN, Nb_3Sn, Nb_3Al..., are being intensively studied as high T_c superconductors [1]. However, these materials were little used so far in high-frequency phonon work, although their inherent strong electron-phonon coupling promises fast bolometric operation [2].

In this communication, we report on the transient response of Nb films excited by recurrent light pulses (0.15 nsec duration) from a mode-locked argon-ion laser. The films are deposited on sapphire substrates by RF sputtering at room temperature [3] and mounted inside a semi-rigid coaxial cable which, looked at from one or the other side, serves either as a 50-ohm source or as a 50-ohm monitoring device. Un-amplified signals are fed into a sampling scope.

According to thickness d, the films fall into two categories .

1 Superconducting Films (d > 5 nanometers)

The transition temperature T_c can be tailored from 9.2 K (bulk value) downwards by controlling the thickness, assumed to be proportional to the sputtering time. These low-impedance samples are current-biassed and the voltage across them is recorded as a function of the time following the laser pulses (Fig.1). The response time τ is evaluated through the convolution between the signal from a fast photodiode and a causal exponential $\exp(-t/\tau)$, where τ is adjustable (dashed curves).

The narrowness of the transition ($\simeq$ 0.1 K) makes Nb granular films suitable for heat pulse detection. Upon omission of phonon diffusion in the substrate, it can be anticipated that the combination with a Nb emitter allows a time resolution of about 1 nsec at 4.2 K. On the other hand, tunability can be achieved by application of a magnetic field H (Table 1).

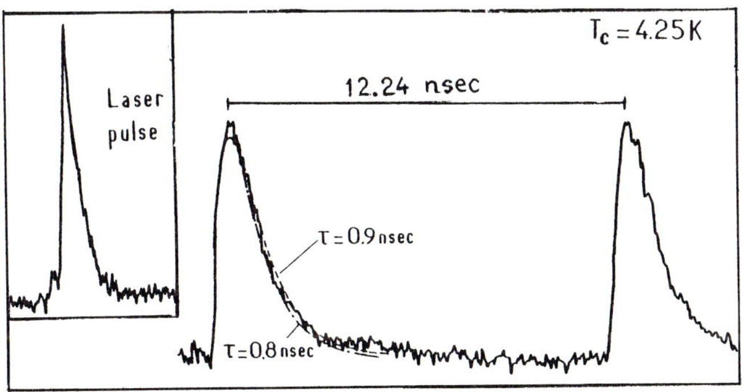

Fig.1 Transient responses of photo-diode (upper left) and of film S-29. Dashed curves are from computation

Table I : Data for superconducting (S) and exponential (L^*) films. $R_\square$ is a sheet resistivity per square at room temperature

Sample	d [nm]	$R_\square$ [Ω/□]	T_c [K]	H [T]	T_{el} [K]	R_{Nb} [K]	τ [nsec]
S-26	7.5	183	1.95	0			25
S-30	10	85	3.80	0			1.35
			4.25	0			0.85
S-29	12	51	2.19	1			14
			1.58	1.5			163
					2.1	15.1	17
L^*-34	2.5	7600			3.1	6.3	3.5
					3.6	5.0	2.1

2 Semiconducting films (d ≃ 2.5 nm)

Superconductivity is suppressed at reduced thickness, and weak localization takes place [3]. Exponential conductivity $\sigma \sim \exp(-T_A/T)$ is obtained with only small adjustability around 2.5 nm. The activation temperature for sample L*-34 is about 6.9 K.

Due to their high impedance, these semiconducting samples are mounted in series inside the coaxial line and voltage-biassed. For reasons of sensitivity, they had to be investigated in non-ohmic conditions, in fields of a few tens of V/cm. The corresponding electron temperature T_{el} is derived from the ohmic conductivity versus T. A comparison of τ and of the sample resistance R_{Nb} (Table I) shows that τ is not the charging time of some external capacitor.

3 Electron-phonon collision time τ_{e-p}

To summarize the results : (i) subnanosecond response times at T > 4.2 K, (ii) little or no material dependence for our films and in contrast (iii) an extremely fast temperature dependence. It is now well established [4] that interelectronic relaxation takes place on the picosecond time scale, so that our response times are governed by the inelastic τ_{e-p} or eventually by the phonon escape time to the substrate τ_{esc}. As a matter for comparison, the response at T_c of 100 nm Pb films on quartz was found to be limited by phonon trapping [5].

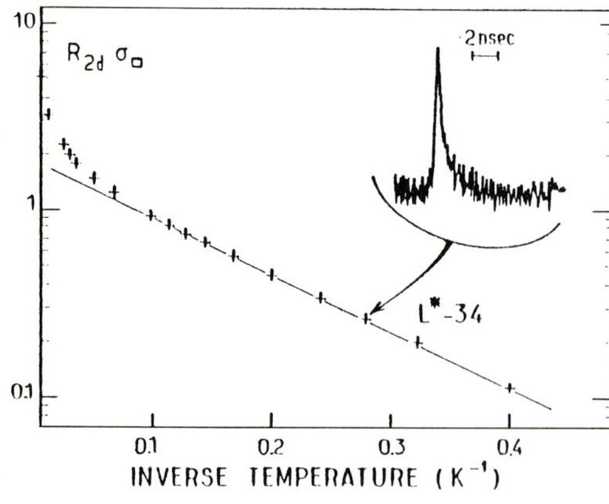

Fig.2 Sheet conductivity per square of sample L^*-34 versus T^{-1}. Normalization factor R_{2d} = 40.7 kΩ. Insert shows the response signal at T_{el} = 3.6 K

From the present data, τ_{esc} < 0.5 nsec for Nb on sapphire (d < 12 nm). Actually, DC experiments of another type [6] allow to set an even lower limit. So, τ here measured is the intrinsic time τ_{e-p}.

For ideal bulk metals, one expects a T^{-3} dependence of τ_{e-p}. Due to disorder, a further factor $q\ell$, where q is the relevant phonon wave number and ℓ the elastic electron mean free path, leads to a T^{-4} law [7], as was substantiated by experiment [6]. Our results at 4.2 K agree reasonably with the extrapolation of [6]. However, the temperature dependence at lower temperatures is steeper than T^{-4}, and is reminiscent of earlier results on Ni-Cr films [8]. Although the microscopic structure of our films is still uncertain, this behaviour may be related to a weaker electron-phonon coupling at long wavelengths, reported in amorphous transition metals [1].

References

1. D.B. Kimhi and T.H. Geballe, Phys. Rev. Letters, 45, 1039 (1980).
2. K. Weiser, U. Strom, S.A. Wolf and D.U. Gubser, J. Appl. Phys. 52, 4888 (1981).
3. J. Desailly, B. Pannetier and J.P. Maneval, to appear in Physics Letters (1983).
4. E. Abrahams, P.W. Anderson, P.A. Lee and T.V. Ramakrishnan, Phys. Rev. B 24, 6783 (1981).
5. C.C. Chi, M.M.T. Loy and D.C. Cronemeyer, Phys. Rev. B 23, 124 (1981).
6. M.E. Gershenson and V.N. Gubankov, Physica 108B, 971 (1981).
7. A. Schmid, Z. Für Physik, 271, 251 (1974).
8. S.J. Rogers, C.J. Shaw and H.D. Wiederick, J. de Physique (Paris) C 6, 317 (1981).

Generation of High-Frequency Phonons by Metallic Point Contacts

R.J.G. Goossens, J.I. Dijkhuis, and H.W. de Wijn

Fysisch Laboratorium, Rijksuniversiteit, P.O. Box 80.000
NL-3508 TA Utrecht, The Netherlands

A.G.M. Jansen and P. Wyder

Research Institute for Materials, University of Nijmegen, Toernooiveld
NL-6525 ED Nijmegen, The Netherlands

The nonlinear relation between current and voltage for a metallic point contact has recently been used for spectroscopic studies of the electron-phonon interaction in metals [1,2]. The conduction electrons accelerated over the contact by an applied voltage V spontaneously emit phonons up to frequencies eV/h. Here, we present a novel technique for injection of high-frequency phonons in insulating crystals using metallic point contacts. The contacts were made by pressing a Au whisker to a thin Au film (< 500 Å) evaporated on the surface of sapphire or ruby crystals. The phonons were detected with either a broadband bolometer or a frequency-selective optical detector.

With a superconducting Sn bolometer we observed the phonons generated in a point contact after ballistic transport along the c axis of a sapphire crystal. In Fig.1 we have plotted the phonon signal versus time for various voltages of the pulse (width 170 ns) applied to the contact. The difference in arrival time (380 ns) between the two peaks agrees with the difference in sound velocities of the degenerate transverse and longitudinal modes. The signal is linear with applied power up to 3 mW.

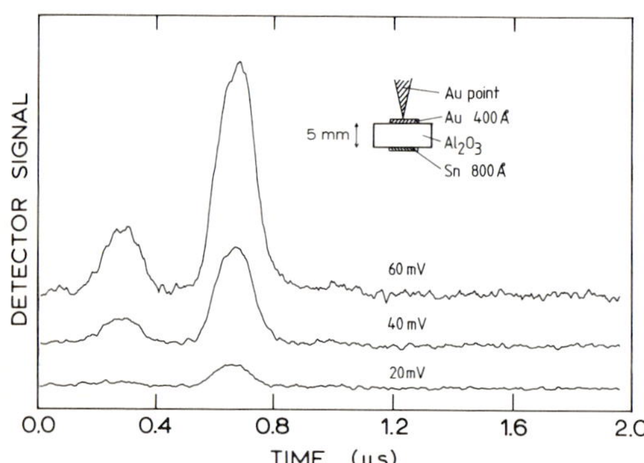

Fig.1 Time of flight of phonons generated by a Au-Au point contact in a sapphire crystal (Al_2O_3) at 1.2 K with broadband detection. The superconducting Sn bolometer is operated at the critical magnetic field. The resistance of the contact is 5 Ω

In ruby, the Cr^{3+} ions allow frequency-selective determination of the phonon occupation by measuring the R_2 fluorescent intensity from the $2\bar{A}(^2E)$ crystal-field level following population of the $\bar{E}(^2E)$ level by optical pumping. This scheme is sensitive to 3.61-meV (29-cm^{-1}) phonons by the direct transition $\bar{E} \rightarrow 2\bar{A}$ [3], but by Raman processes the observed signal is mainly due to phonons above this frequency when the latter are sufficiently abundant [4-6]. Figure 2 shows the R_2 intensity induced by a pulsed point contact. The signal-to-noise ratio was improved by switching off the optical pumping for 10 μs in every 50 μs, and detection after decay of the $2\bar{A}$ population

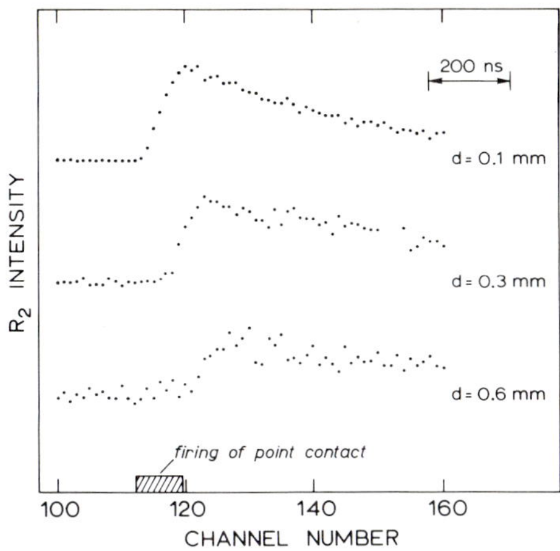

Fig.2 Optical detection of phonons generated by a Au-Au point contact via the R_2 fluorescence in ruby (Al_2O_3:Cr^{3+}, 700 ppm) at various distances d from the contact. Horizontal axis represents time. The volume of detection (cylinder of 50 μm diameter and 100 μm length) is prepared by excitation with an argon laser, and is further selected by the optics. The resistance of the contact is 0.5 Ω, the applied voltage is 190 mV

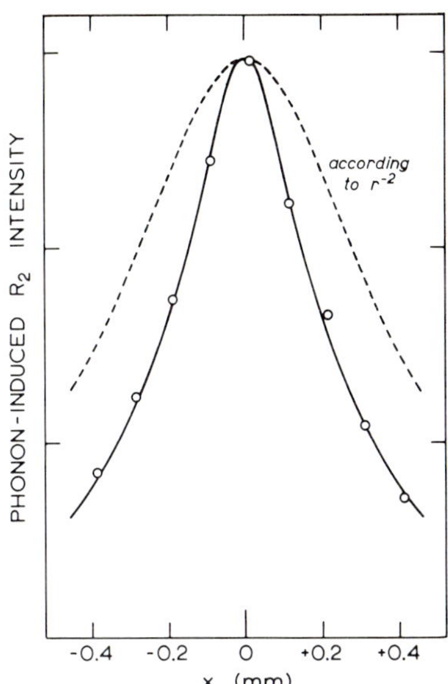

Fig.3 Dependence of the phonon-induced R_2 peak intensity (cf. Fig.2) on the position x along the laser beam. The distance d is 0.4 mm. Dashed curve is the fall with x according r^{-2}, with $r^2 = d^2 + x^2$

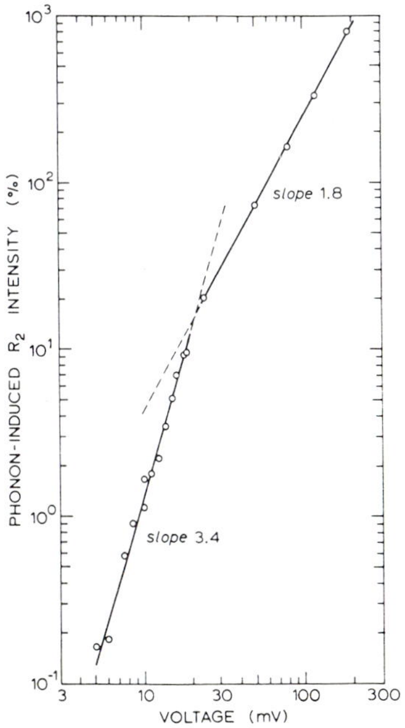

Fig.4 Dependence of the phonon-induced R_2 peak intensity (cf. Fig.2) on the voltage applied to the point contact. Allowance for the slight variation of the resistance with voltage yields a strictly linear dependence of the R_2 intensity on the dissipated power above 30 mV. The slope 3.4 at lower voltages is suggestive for the point contact acting as a Planck's radiator, and the ruby detector being sensitive to high-frequency phonons via Raman processes

(~ 1 μs). In Fig.3 the optical system selects the position of detection along the laser beam, parallel to the Au film. The experimental result is compared to the r^{-2} law expected for a point source of phonons and ballistic propagation. Apparently, due to their small dimensions (~ 100 Å), point contacts make it possible to examine phonon focusing and decay over small distances. In this context it is noted that the times of flight in Fig.2 are longer than according to the sound velocity.

In Fig.4 we have plotted the optical signal versus the applied voltage. Although theories exist for the spectral distribution of phonons in the point contact area [2], we approximate our generator by a phonon system in thermal equilibrium at a temperature T (~ 100 K at 30 mV). Above, say, the Debye temperature of Au the temperature of the point contact and accordingly the phonon occupation scale with the absorbed power, as observed above 30 mV. At lower temperatures, the point contact acts as a Planck's radiator, i.e., T^4 is proportional to the power, or $T \propto V^{1/2}$. For a Raman process the probability of a transition $\bar{E} \rightarrow 2\bar{A}$ goes with T^7. Upon eliminating T, the phonon signal thus depends on $V^{7/2}$, as observed. Note that direct absorption of 3.61-meV phonons ($\propto$ T) would yield $V^{1/2}$.

1 I.K. Yanson: Zh. Eksp. Teor. Fiz. 66, 1035 (1974)
2 A.G.M. Jansen, A.P. van Gelder, P. Wyder: J. Phys. C 13, 6073 (1980)
3 K.F. Renk, J. Deisenhofer: Phys. Rev. Lett. 26, 764 (1971)
4 A.P. Abramov, I.N. Abramova, I.Ya. Gerlovin, I.K. Razumova: Zh. Eksp. Teor. Fiz. 79, 1303 (1980)
5 S.A. Basoon, A.A. Kaplyanskii, V.L. Shekhtman: Sov. Phys. Solid State 24, 1093 (1982)
6 R.J.G. Goossens, J.I. Dijkhuis, H.W. de Wijn, to be published

Nonequilibrium Phonon Distribution in a Quantizing Magnetic Field: A Tunable GHz to THz Phonon Generator?

G.W. Slater and A.-M.S. Tremblay

Département de Physique et Groupe de Recherche sur les Semiconducteurs et les Diélectriques, Université de Sherbrooke
Sherbrooke, Québec, Canada J1k 2R1

Although the main body of the quantum transport literature is concerned with electronic properties, the nonequilibrium phonon spectrum under hot electron and quantizing magnetic field conditions turns out to be most interesting. The scattering rate due to electron-phonon interactions being very much dependent on $\vec{q}$ and $\vec{H}$, one finds intriguing angular and frequency dependences of $N(\vec{q},\vec{H})$, the nonequilibrium phonon population.

The physics of our problem is as follows. If one overpopulates (heats) the upper electronic Landau levels which appear when $H \neq 0$, it is plausible to expect that many phonons will be emitted with $\omega = \omega_c$ since the density of states at the bottom of the levels is singular.

To have an appreciable cyclotron output while staying in the quantum regime, we want the electronic temperature T_E to be of the order of ω_c. This forces us to sum the contribution to the phonon scattering of a large number of Landau levels, not only one or two. To eliminate the background, we want the lattice temperature T_0 to be much lower than ω_c.

We use a hot Maxwell-Boltzmann electronic distribution function both for mathematical convenience, and because we want to show that any featureless electronic distribution leads to a usefully structured phonon spectrum. For such a distribution, the phonon transport equation may be solved exactly to yield [1,2]

$$N(\vec{q},\vec{H}) = \frac{N^0(q)\,\tau_{ep}(\vec{q},\vec{H}) + N^E(q)\,\tau_{ne}(q,T_0)}{\tau_{ep}(\vec{q},\vec{H}) + \tau_{ne}(q,T_0)} \tag{1}$$

where N^0 and N^E are the Bose-Einstein distributions for $T = T_E$ and $T = T_0$, while τ_{ne} and τ_{ep} are the total non-electronic and electron-phonon relaxation times, the latter being given, for $\vec{H} = H\hat{z}$, by

$$\tau_{ep}^{-1}(\vec{q},\vec{H}) = \frac{n_0\,\hbar^{-1} q E_1^2}{N^E(q)\rho s|q_z|}\left(\frac{\pi\, m^*}{2k_B T_E}\right)^{\frac{1}{2}} \exp -\left\{\frac{1}{2}\lambda^2 q_\perp^2 \coth\left(\tfrac{1}{2}\hbar\omega_c/k_B T_E\right) + \frac{\hbar\omega_q + \hbar^2 q_z^2/4m^*}{2k_B T_E}\right\} \times \sum_{-j_{min}}^{+j_{max}} I_j\left(\tfrac{1}{2}\lambda^2 q_\perp^2 / \sinh\left(\tfrac{1}{2}\hbar\omega_c / k_B T_E\right)\right) \times \exp -\left\{\frac{m^*(\omega_q - j\omega_c)^2}{2k_B T_E q_z^2}\right\} \tag{2}$$

where n_0 is the electronic density, s is the sound velocity, ρ and E_1 are the density and the deformation potential constant of the material, and $\lambda = (\hbar c/eH)^{\frac{1}{2}}$ is the magnetic length. Each modified Bessel function I_j sums all contributions from downward electronic jumps of j levels.

The rate τ_{ep}^{-1}, although singular at $q_z = 0$ for $\omega_q = j\omega_c$, is a rather weak function of $\theta \equiv \arcsin\ (q_z/q)$. Only very near $\theta = 0^0$ do we have either a singularity giving $N = N^E$ (for $\omega_q = j\omega_c$) or a dip giving $N = N^0$ (for $\omega_q \neq j\omega_c$), the latter occurring because of conservation laws. The typical $\theta = 90^0$ phonon spectrum of Fig. 1 is independent of $\vec{H}$, and represents very well the not too small θ directions as well. The cut-off frequency ν^* is not fixed by $s\lambda^{-1}$ here, but one has instead

$$\frac{(h\nu^*)^2}{8\ m^*s^2} = k_BT_E\left[\frac{h\nu^*}{k_BT_o} + \frac{1}{2}\ \ell n\left(\frac{\pi\ m^*\ n_o^2\ \tau_{ne}^2\ E_1^4}{2\hbar^2\ k_BT_E\ \gamma^2\rho^2 s^2}\right)\right] \qquad (3)$$

where the Casimir relation is used for τ_{ne} because of the low ($\leq$ 10 K) temperatures we have, and γ is the definition of the cut-off (γ=N-N^0/N).

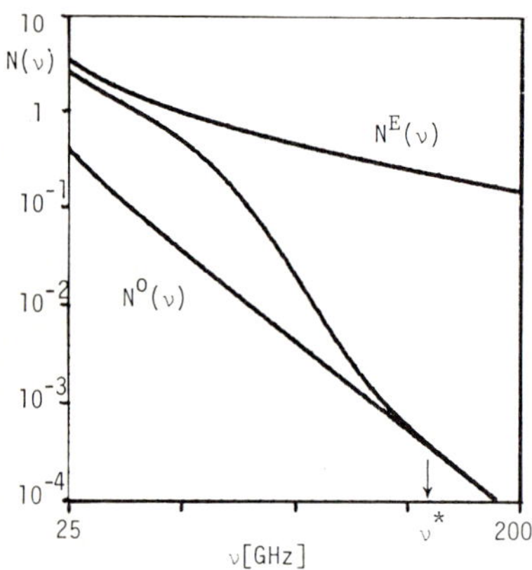

Fig. 1 Parallel phonon distribution $N(\nu)$ for GaAs with the following experimental conditions: $T_E = 4.8$ K, $T_0 = 1$ K, $n_0 = 7\times10^{14}$ cm^{-3}, $\tau_{ne} = 0.4$ µsec. Also shown: the $T = T_E$ (upper curve) and $T = T_0$ (lower curve) Bose-Einstein distributions. For $\nu > \nu^*$, the phonons are at the lattice temperature. According to (3), ν^*=163 GHz for $\gamma = 0.1$

If one chooses the cyclotron frequency in the cold region $\nu_c > \nu^*$, one obtains the function $N(\nu,\theta)$ pictured in Fig. 2. There is a weak but very monochromatic cyclotron peak in the $\theta = 0^0$ direction. One may imagine ways to use it as a highly monochromatic phonon source, but since the large - θ background is clearly dominating, efficiency should be rather low.

More promising is the situation where ν_c is near the (ν_m) maximum of the $\theta = 90^0$ output. Figure 3 shows that there is no frequency shift as the ν_c peak becomes wider. Although monochromaticity has been reduced, a large fraction of the total power is produced nearby the cyclotron mode. Using a $\simeq 2^0$ - aperture detector, this set up may give a strong 80 ± 7 GHz phonon flux at the crystal boundaries. The interplay between the different parameters of (3) may allow higher T_E and higher power.

For comparable temperatures and a different material, $h\nu^*$ scales roughly like m^*s^2. For example, ν^* is of the order of 10 GHz for InSb, and one is reduced to the case of Fig. 2 for almost all the GHz - THz range. To obtain the $\nu_c \simeq \nu_m < \nu^*$ case, the material, or more precisely

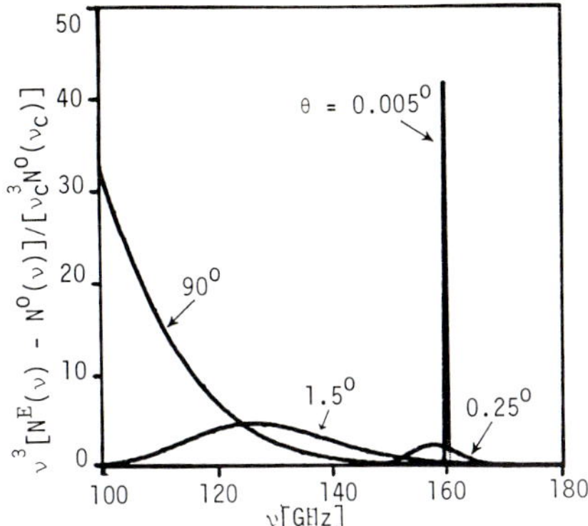

Fig. 2 Relative power density of the nonequilibrium phonons for various angles θ. This is for GaAs with parameters of Fig. 1, and $\nu^* \lesssim \nu_c = 160$ GHz. For $\gtrsim$ 155 GHz, only a narrow beam of cyclotron phonons is produced

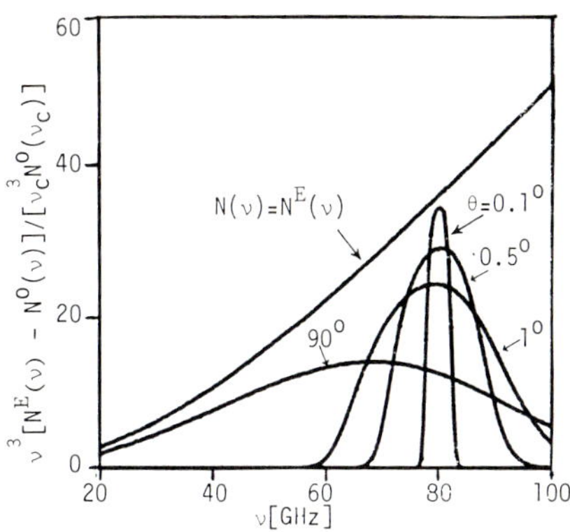

Fig. 3 Same as in Fig. 2, but with $\nu^* > \nu_m \sim \nu_c = 80$ GHz. The upper curve shows for reference the $N = N^E$ curve

its m^*s^2, should be chosen for the frequency range we are interested in. GaAs is well suited to the 100 GHz region and seems to be a good candidate to try to build the first phonon generator using the ideas developed here.

1. L.E. Gurevich and T.M. Gasymov: Sov. Phys. Sol. St. 9, 78 (1967).
2. Id. 10, 2577 (1969).

Relaxation Processes and Response Time of Superconductors to a Periodic Excitation

N. Perrin

Groupe de Physique des Solides de L'Ens, 24 Rue Lhomond
F-75231 Paris Cedex 05, France

C. Vanneste

Laboratoire de Physique de la Matière Condensée, Université de Nice
F-06034 Nice Cédex, France

The knowledge of the response time of superconductors to a periodic excitation as a result of the different relaxation processes is important because it leads to a better understanding of nonequilibrium states ; it also has a practical interest in the development of superconducting electronic devices where fast switches are required [1]. Some experiments have been done in Pb to determine the response time : the result is not as short as expected [2]. We consider here a superconducting film of thickness $d = 10^3$ Å excited by optical irradiation. With the assumption of a BCS quasiparticle (q-p) distribution at a temperature T^*, the coupled phonon Boltzmann equation and q-p energy balance are solved numerically, the time-dependent gap parameter Δ being determined by the instantaneous nonequilibrium energy distribution function $f(E)$

$$\Delta/\Delta_0 = \exp - \left\{ \int_{-\infty}^{+\infty} d\varepsilon \; f(E)/E \right\} . \tag{1}$$

With the T^* assumption, the Boltzmann equation is simply written with phonon pair breaking and scattering relaxation times τ_{PB} and τ_{PS}. The substrate at T_b = 1.5 K is introduced through a phonon escape time τ_{es}. With the drive frequency ω_i for the input power P_i, the q-p energy balance equation gives :

$$C_e dT^*/dt = P_i(t) - P_N(t) = P_S(1+\sin\omega_i t) - \sum_{\vec{q}\,\alpha} \hbar\omega\left(\overline{N}(T^*) - N_\alpha\right)/\tau_{P\alpha} \tag{2}$$

$\tau_{P\alpha}$ being the α mode-phonon relaxation rate ($\tau_{P\alpha}^{-1} = (\tau_{PB}^{-1} + \tau_{PS}^{-1})_\alpha$). A typical T^* time behaviour is shown in Fig.1 for a Sn film. Though the frequency f = 1 GHz is smaller that most of the characteristic frequencies in the system, the q-p temperature modulation depth is about 5 %. This is due to the ratio of the q-p and phonon specific heat $C_e/C_p \sim 1.5$ [3]. Moreover this ratio partly explains the rather small amplitude response of $P_N(t)$, which leads to small deviations from a phase quadrature between T^* and P_i.

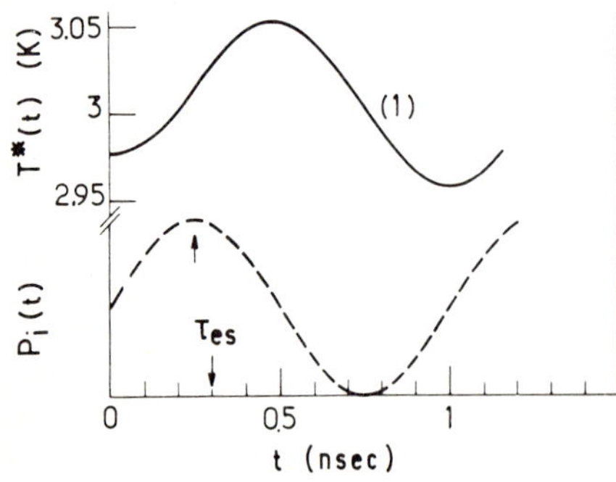

Fig.1 : Quasiparticle temperature $T^*(t)$ and input power P_i for a drive period f^{-1} = 1 ns and a mean power $P_S = 14.1\ 10^4\ \mathrm{W\ cm^{-3}}$

To obtain the characteristic times of the q-p pair-phonon system, let us now study the frequency dependence of its time behaviour. The phase shift ϕ_N of $P_N(t)$ is shown in Fig.2. It exhibits two kinds of behaviour : a monotonically decreasing one (curve b2 ; $\tau_{es} \sim \tau_p$) and a nonmonotonic behaviour, either with $\tau_{es} > \tau_p$ (curve a3 ; $\tau_p \sim 0.12$ ns) or with $\tau_{es} \lesssim \tau_p$ (curve a4 ; $\tau_p \sim 0.12$ ns). However, according to the different relaxation times, the positions of the maximum and minimum are different. The curves a and b are related to systems which differ essentially through the ratio C_e/C_p with $C_e/C_p \sim 16$ for curves a and $C_e/C_p \sim 1.6$ for curve b. The non-monotonic behaviour observed makes us guess that there are characteristic times in the system that are different from the basic times τ_{es} and τ_p. The $P_N(t)$ modulation depth however decreases monotonically with f in the frequency range considered : it is larger for curve b than for curves a .

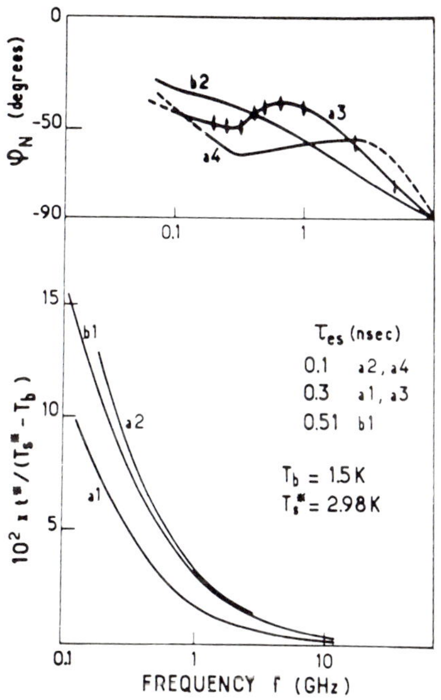

Fig.2 : Phase shift ϕ_N of $P_N(t)$ and q-p modulation depth $t^*/(T_s^*-T_b)$ with $t^* = T_{s\,max}^* - T_s^*$, T_s^* being the mean q-p temperature

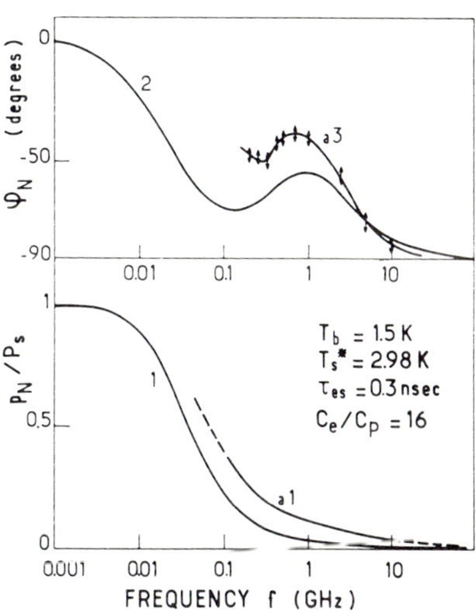

Fig.3 : $P_N(t)$ modulation depth and phase shift ϕ_N for the analytic model (curves 1 and 2) and for our numerical calculations (curves a1 and a3)

The q-p phase shift relative to P_i shows a deviation from $-\pi/2$ which decreases as f increases between 0.1 and 1 GHz. For $1 < f < 10$ GHz, T^* is in phase quadrature with P_i ; this points out the rather small value of $P_N(t)$ in the latter conditions. The q-p temperature modulation depth is seen in Fig.2 to decrease with the excitation frequency but to remain rather small, even for the smallest frequencies considered. Furthermore, it is larger for $\tau_{es} = 0.1$ ns (curve a2) as expected [4] . The curve b ($\tau_{es} \sim \tau_p$) is below the curve a2 ($\tau_{es} < \tau_p$) : this results from the larger phonon specific heat. However, the q-p response is more sensitive to the ratio τ_{es}/τ_p whereas the phonon response is essentially determined by the ratio C_e/C_p. Moreover, a large ratio C_e/C_p leads to a small response of the system due to the small amount of power $P_N(t)$ transferred to the phonons.

We now use a simple analytical model with a single phonon temperature and energy-independent relaxation times τ_e and τ_p respectively for q-p and phonons, and we show that there are two characteristic times in the system, τ_1 and τ_2 with

$$\tau_{1,2}^{-1} = \frac{1}{2\tau} \times \{1 \pm \sqrt{1 - 4 \frac{\tau^2}{\tau_{es}\tau_p} \frac{C_p}{C_e}}\} \qquad (3)$$

where $\tau^{-1} = \tau_{es}^{-1} + \tau_p^{-1} + \tau_e^{-1}$. Both, or one of them only, may be relevant in the superconductor time behaviour. A comparison of the P_N results obtained from this model with the previous results shows a rather good agreement (Fig.3). Therefore this model can be used to give a good insight of the behaviour of any superconducting film. In the case of a Sn film, a maximum response is obtained up to frequencies $\omega_M = \tau_2^{-1} < \tau_{es}^{-1}$; then, the Sn film response is not limited by the phonon escape time [2].

Other types of excitation and different superconductors leading to either a q-p injection term or a phonon injection term, or both can be considered. In table I, we give some values of the relevant characteristic times for some materials near T_c, i.e., either τ_2 or τ_1 and τ_2.

Table I : Relevant characteristic times

	q-p injection		phonon injection	
Pb	$\tau_1 \simeq 20$ ps	$\tau_2 \simeq \tau_{es}$	$\tau_2 \simeq \tau_{es}$	
Sn	$\tau_1 \simeq 900$ ps	$\tau_2 \simeq 2\tau_{es}$	$\tau_2 \simeq 2\tau_{es}$	
Nb	$\tau_1 \simeq 50$ ps	$\tau_2 \simeq 3\tau_{es}$	$\tau_2 \simeq 3\tau_{es}$	$\tau_{es} = 1$ ns
Al		$\tau_2 \simeq 175$ ns	$\tau_2 \simeq 175$ ns	
Hg	$\tau_1 \simeq 10$ ps	$\tau_2 \simeq \tau_{es}$	$\tau_2 \simeq \tau_{es}$	

These values can be improved by the use of different geometries like microbridges [2] in which the direct coupling of the q-p and the substrate can be described by a relaxation time τ_d which may be as small as 10 ps.

1. S.M. Faris, S.I. Raider, W.I. Gallagher and R.E. Drake, App. Superconductivity Conf., Knoxville (1982).
2. C.C. Chi, M.M.T. Loy, D.C. Cronemeyer and M.L. Thewalt, IEEE Trans. Mag. 17, 88 (1981).
3. N. Perrin and C. Vanneste, to be published in Phys. Rev.
4. U. Eckern and G. Schön, J. Low Temp. Phys. 32, 821 (1978).

Amplitude Modulated Heat Pulses

L. Hirschbiegel, M. Siemon, and W. Grill

Physikalisches Institut der Universität Frankfurt
Robert Mayer Straße 2-4, D-6000 Frankfurt am Main, Fed. Rep. of Germany

The generation of acoustic phonons by a metal heater evaporated on a crystal has found a wide range of applications since the original work of von Gutfeld and Nethercot [1]. In dielectric crystals with a sufficiently low number of dislocations and impurities a ballistic propagation of acoustic phonons of all modes can be observed. Thereby the heater is excited by a current pulse no longer than the time of flight in the crystal. The phonons can be detected with a superconducting bolometer evaporated on the surface of the crystal opposite to the heater. The system is usually described by the acoustic mismatch model [2,3], leading to time constants of the heater and detector typically only little more than 1 ns [4]. The observed risetimes are furthermore disturbed by geometric effects [5] and in the case of crystals with a diffusive propagation of high frequency phonons, it can be difficult to resolve the ballistic signal caused by phonons of lower frequency from the background.

We have developed a method which selectively detects any signal due to ballistic phonon transport. To achieve this, the phonon generator is driven by a rectangular shaped current pulse to which an oscillatory pulse is added (Fig.1). This pulse is generated with the help of a diode switch from a continuously running hf-sinus generator. Such one can observe an amplitude modulated phonon pulse in foreward direction, whereas the modulation will disappear due to geometric effects for a direction with group velocities other than normal or for any signal due to

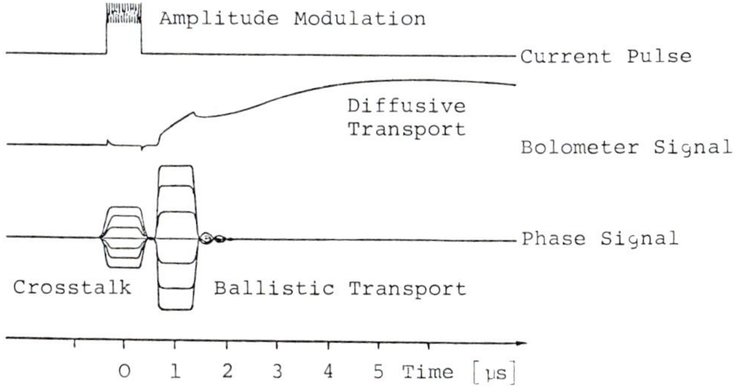

Fig. 1 Principles of amplitude modulated heat pulse generation and detection as observed in ruby (propagation along Z axis)

phonons scattered in the volume of the crystal. The bolometer with a surface parallel to the heater will detect a modulated heat pulse, that can be compared in phase with the hf generator used for the modulation of the phonons. The phase shift will thereby depend on the frequency and the time of flight in the crystal. A phase signal will only appear if ballistic phonon transport is present (Fig.1). By this method we can reach the resolution typical in ultrasonic experiments for phonons with frequencies above 1 THz which do not allow for a direct phase sensitive detection, but which can be modulated in amplitude with frequencies ranging from 10 MHz to 1 GHz.

A first application of this method is the determination of the group velocity of thermally excited phonons with a frequency distribution peaked around 2 THz for the results presented here. For this experiment we have chosen a laser quality ruby crystal with 0.03 mol % chromium ions to demonstrate the selectivity of this detection scheme with respect to the phonon signal due to scattering at impurity sites. With modulation frequencies of 50 MHz we measure a group velocity along the Z axis of the crystal of (11.22±0.08) km/s. Corrections for the electronic delay are taken into account by monitoring the frequency dependence of the phase of the electrical crosstalk from the heater to the detector. Further corrections for the time constant of the heater and bolometer can be applied but are usually of the same size as the statistically determined error. This experimental result clearly deviates from values ranging from 10.7 to 12.0 km/s calculated from the elastic constants and used for the computation of phonon focussing in sapphire [6,7], but there is good agreement with a more recently calculated value by Bialas, Weis and Wendel [8].

Further applications of this method result from the observation of the amplitude of the modulation of the phonon signal (video signal). For the purpose of noise reduction we have developed a technique using a feedback circuitry to stabilize the hf generator on an average frequency leading to a zero phase signal which is picked up by a boxcar. We further modulate the frequency of the hf generator with a frequency high with respect to the inverse time constant of the stabilization. This modulation is picked up by a lock-in at the output of the boxcar, producing a sensitive offset free video signal. We have now monitored the modulation amplitude of the phonon signal while varying the heater power. At modulation frequencies of 7 MHz we observe a reduction of the modulation beginning at heater powers of 1 W/mm^2 (Fig.2). This results from the onset of phonon-phonon interaction taking place inside the crystal right behind the heater, even though the system is, up to 100 W/mm^2, still well described by an acoustic mismatch model [9]. This indicates that the scattering processes do not lead to a significant backscattering of the phonons into the heater, but do change the effective time constant of the system. The observed effects could then be explained by small-angle scattering as to be expected from 4-phonon processes, which in fact would not significantly change the height of the ballistic pulse, but would influence the rise time and as thus the amplitude modulation by geometric effects.

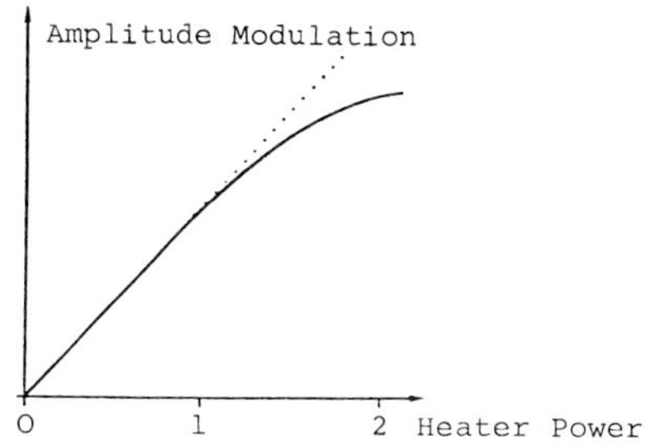

Fig. 2 Amplitude modulation of longitudinal phonons as a function of heater power (dotted line: expected signal in the absence of scattering)

The amplitude modulation of phonon pulses which can also be applied to optically generated and detected phonons allows the application of detecting schemes used in ultrasonic experiments. With this method group velocities can be determined with an error below 1% for phonons with frequencies in the THz regime. The selective detection of ballistically propagating phonons can also be used to study phonon-phonon scattering. Experiments where 2 phonon generators are used simultaneously to observe such effects resolved with respect to phonon modes and scattering angle are in progress.

This work has been supported by the Deutsche Forschungsgemeinschaft; SFB 65 Darmstadt-Frankfurt.

References

1 R.J.von Gutfeld, A.H.Nethercot: Phys.Rev.Lett. 12,641(1964)
2 O.Weis: Z.Angew.Phys. 26,325(1969)
3 N.Perrin, H.Budd: Phys.Rev.Lett. 28,1701(1972)
4 P.Herth, O.Weis: Acoustica 21,162(1969)
5 F.Rösch, O.Weis: Z.Physik B 29,71(1978)
6 F.Rösch, O.Weis: Z.Physik B 25,101(1976)
7 G.W.Farnell: Can.J.Phys. 39,65(1961)
8 H.Bialas, O.Weis, H.Wendel: Phys.Lett. 43A,97(1973)
9 P.Herth, O.Weis: Z.Angew.Phys. 29,101(1970)

Thermal Conductivity Measurement Under Applied Uniaxial Stress

B. Salce

Service des Basses Températures, Laboratoire de Cryophysique
Centre d'Etudes Nucléaires de Grenoble, 85 X
F-38041 Grenoble Cédex, France

1. Introduction

The application of a uniaxial force $\vec{f}$ to a material provides a very convenient way to investigate its electronic properties, as the stress $\vec{\sigma} = \vec{f}/s$ (where s is the cross-section of the sample) produces changes in lattice parameters and perturbes local electric fields. For many years, uniaxial stress has been introduced as an external parameter in techniques like optical spectroscopy, EPR, APR, etc., which need isothermal samples. However, difficulties arise if parasitic heat leaks must be avoided, this being the case for thermal conductivity K(T) measurements which, on the other hand, are a valuable spectroscopic technique in view of the large energy range available (5 → 2000 GHz).

2. Background to the problem

K(T) is measured using the steady state heat flow method. In the classical arrangement($\vec{\sigma} = 0$), the power Q generated in an electrical heater flows to the thermal bath through the sample. The temperature difference ΔT is measured between 2 thermometers attached at a distance l apart. Provided $\Delta T \ll T$ (the average temperature) :

$$K(T) = (1/s) \times (Q/\Delta T).$$

Applying an external stress introduces a heat leak Q" (Q = Q' + Q") through the end of the sample opposite to the end usually linked to the cryogenic bath so that only part of the power Q' flows through the sample. The ratio Q"/Q' depends on the ratio of the sample conductance and the stress apparatus conductance.

From a mechanical point of view, performing tensile measurements is the best way to get a good alignment of the force relative to the sample but bounding the specimen to the stress mechanism is a major problem which prevents the use of this mounting in many cases (insulating samples).

Until now, these problems have been partly solved by using an experimental arrangment needing long samples ($\sim$ 3 cm) and several calibrated thermometers [1], or by neglecting the heat leak Q" (< Q/100) [2], [3], [4], but in this case, only materials with K higher than $2 \times 10^{-2} W.cm^{-1}.K^{-1}$ at T = 1.5 K could be measured with an acceptable precision.

3. Experimental set-up

We propose a compressional set-up able to measure K(T) as low as $7 \times 10^{-4} W.cm^{-1}.K^{-1}$ at T = 1.5 K, for samples as short as 5 mm, using only 1 calibrated thermometer, under forces up to 500 kgf if $T_{bath} > 4.2$ K, with "in situ" measurement of the exact force experienced by the sample.

The experimental cell and the equivalent thermal circuit are shown in Fig.1. It consists of a cage made of two parallel plates connected by 3 copper rods (Ø = 8 mm) ensuring the rigidity of the whole assembly. The force is applied by a piston (cross section $\sim$ 11.5 cm^2) enclosed in stainless steel bellows operated at low temperature with pressurized He^4 gas and surrounded by vacuum. The heat leak Q" goes through a stainless steel cylinder and is balanced by an additional heater driven by a thermocouple used as a zero detector : when $\Delta T'' = 0$, the compensation heater supplies a power Q" and the heat flow through the sample is equal to Q, the measured power supplied by the main heater.

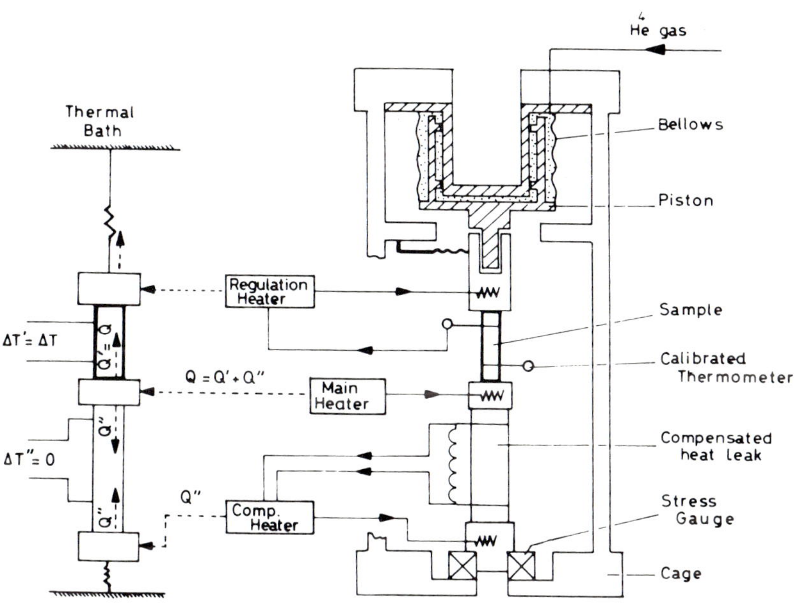

Fig.1 Experimental cell and equivalent thermal circuit

The exact force experienced by the sample is measured using a piezoelectric ring-shaped stress gauge located between the cage and the stainless steel cylinder. This gauge (supplied by KISTLER) has been proved to work at very low temperature. The generated electric charge is measured by a bridge giving the force on the sample. Using this system provides a major advantage : we are able to cool down the sample without any applied force, thus preventing accidental breaking due to uncontrolled stress from thermal contraction of the cage, and to detect when the piston comes into contact with the sample by introducing He^4 into the bellows when the lowest temperature is reached. If T bath < T_λ = 2.17 K, a maximum He^4 pressure of 25 bars can be used (from the He^4 phase diagram). If T bath > T_λ, the allowed He^4 pressure rapidly increases with T and the force is restricted by the mechanical design of the cell (a 500 kgf force was applied on a Cr-doped GaAs sample, which resulted in a $\sim$ 4.5 kbar stress).

4. Results

We report in Fig. 2 measurements performed on a stainless steel sample. The broken curve is the lowest K(T) limit available with the previous stress apparatus [4]. At $\sigma = 0$ and down to 1.7 K, it can be seen that results from the stress arrangement (crosses) and from a classical arrangement (open circles) agree very well. Setting to 0 the compensation heater introduces an error on K(T) of a factor of 2 (triangles). As an example, we show measurements under a 1 kbar applied stress. No detectable effect is observed.

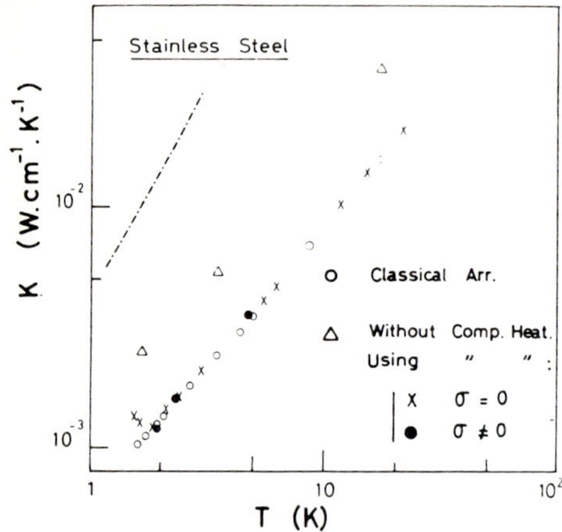

Fig.2 Test measurements on a stainless-steel sample

At this time, the smallest thermal conductivity which has been measured (within an error less than $\pm$ 3 x 10^{-2}) was 7 x 10^{-4} W.cm^{-1}.K^{-1}. By improving the thermal linkages and the electronic regulation systems, we plan to reach $\sim 10^{-4}$ W.cm^{-1}.K^{-1} to study amorphous solids under uniaxial stress.

1 A. ADOLF : Rev. Sci. Instrum. 51(9), 1217 (1980)
2 R.W. KEYES and R.J. SLADEK : Phys. Rev. 125, 478 (1962)
3 B. SALCE and A.M. de GOER : J.Phys.C : Solid State Phys. 12, 2081 (1979)
4 A. RAMDANE, B. SALCE and L.J. CHALLIS : Phys. Rev.B 27 n° 4, 2554 (1983)

Thin Piezoelectric PVDF-Layers as Ultrasonic Transducers

A. Ambrosy, K. Holdik, W. Scheitler, and H. Schulze

Universität Stuttgart, Physikalisches Institut, Teil 1
Pfaffenwaldring 57, D-7000 Stuttgart 80, Fed. Rep. of Germany

High-frequency coherent phonons may be generated by piezo-materials. Bonding of thin crystal platelets or epitaxial deposition by evaporating or sputtering are convenient methods. The disadvantages of these methods are the bond problem for high frequency and low temperature, orientation, complicated technology and the time consumption. A new, simple, fast method with a small thermal load of the substrate is the use of the piezoelectric polymer **P**oly (**v**inyli**d**ene **f**luoride) [Fig.1]. The bulk material constants in longitudinal mode of PVDF ($k=0.12$, $c=2200$ m/s, $Z=2.5\cdot10^6$ kg/s m^2) have similar values as CdS ($k=0.17$, $c=4500$ m/s, $Z=21.5\cdot10^6$ kg/s m^2) [1,2].
PVDF is well soluble in N,N-Dimethylformamide C_3H_7NO, N,N-Dimethylsulfoxide $(CH_3)_2SO$ or similar solvents. The substrate is covered with an evaporated 1µm Al counter electrode. Substrate and solution are warmed up. At about 60^0C a fixed amount of solution is dropped onto this surface. The film quality and thickness can be controlled by the concentration of the solution, the temperature and the angle velocity of the centrifuge, which is used to flatten the drying film [Fig.2]. Thereafter the piezoelectric formation of the layer is achieved by a corona discharge [Fig.3]. We used a high voltage of about ±10kV, a distance of ~1cm and 10 min application time.

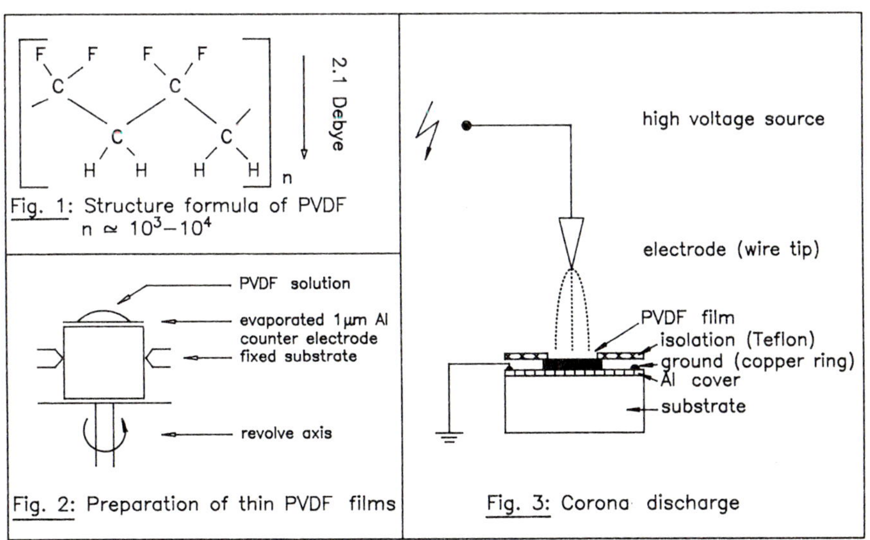

Fig. 1: Structure formula of PVDF
$n \simeq 10^3 - 10^4$

Fig. 2: Preparation of thin PVDF films

Fig. 3: Corona discharge

Another possibility is the preparation of a bonded PVDF foil. A very good means of bonding is one of the above-mentioned solvents. The substrate is warmed up as before. Some drops of the solvent between the substrate and the PVDF foil dissolve the foil a little. By application of warm air the solvent evaporates and the PVDF layer sticks as well as the film above.

The **principle of measurement** is based upon the pulse-echo method [Fig. 4]. Microwave power is generated by a travelling wave tube (TWT,P_p < 1kW). The microwave cavity is a coaxial resonator [3,4], which is tunable up to 12 GHz [Fig.5]. The resonator is mounted into a cryostat and can be cooled down to 1K. Measurements to compare between CdS transducers and the novel PVDF piezo-layers were taken at 4.2 K [Fig.6].

The insertion loss is determined by extrapolating to the onset of the echo train. The rf pulse is switched directly from transmission range via a calibrated step attenuator to the detector. Thereby the insertion loss can be read at the attenuator, when the original signal is damped to the beginning of the extrapolated echo train.

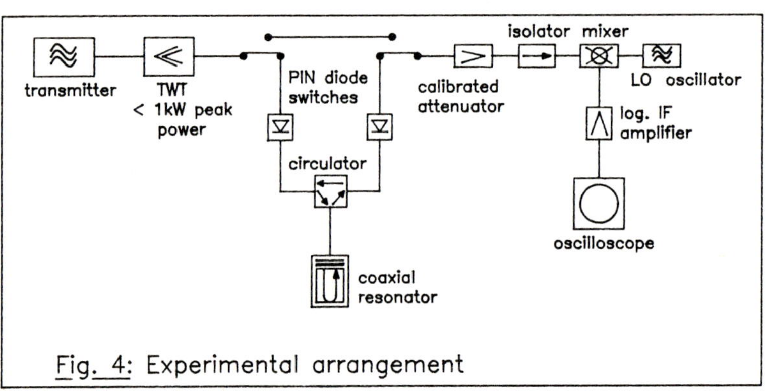

Fig. 4: Experimental arrangement

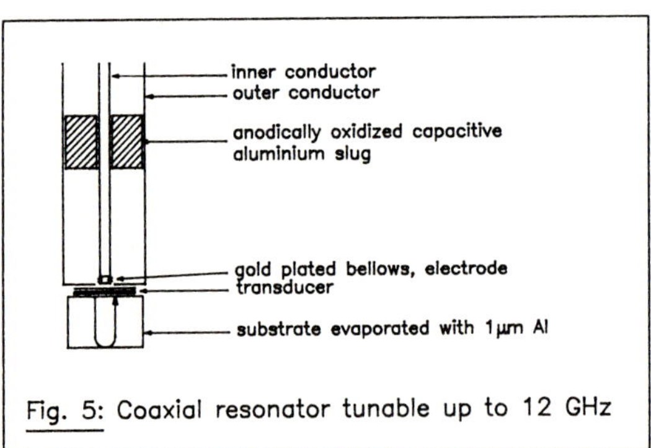

Fig. 5: Coaxial resonator tunable up to 12 GHz

Comparison between PVDF and CdS transducers

- Usually only one transducer is necessary for each frequency range (1...8)GHz and (8...12)GHz, i.e. they are broadband [Fig.6].
- Both materials have comparable efficiency even at 24 GHz.
- Both tranducer types were found to be not affected by temperature in the range below 50 K or magnetic field up to 4.5 Tesla.

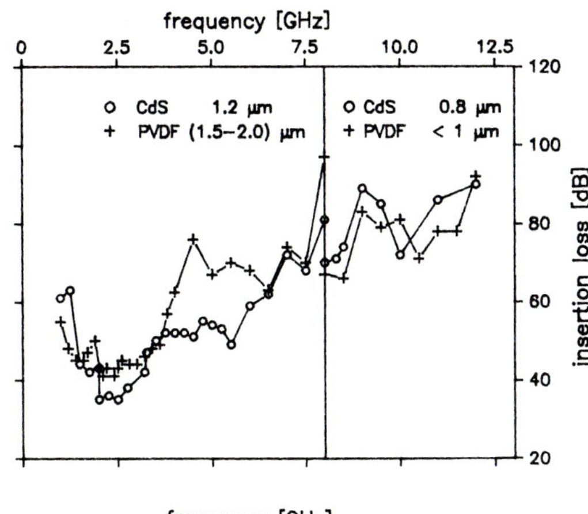

Fig. 6: Insertion loss for CdS and PVDF transducers at 4.2 K

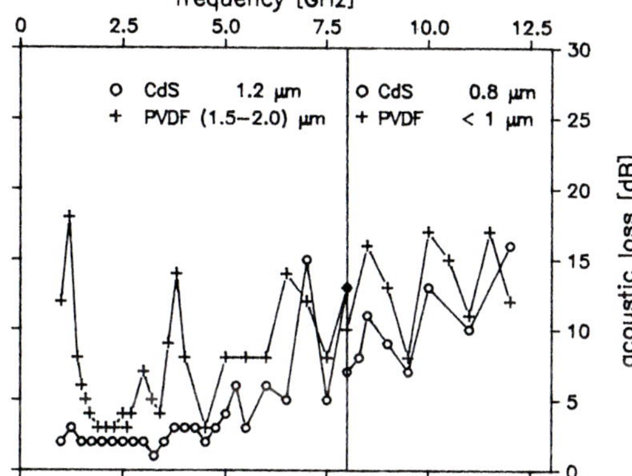

Fig. 7: Acoustic loss for CdS and PVDF transducers at 4.2 K

- The acoustic loss measured between the first and the second echo is larger in PVDF than in CdS due to stronger damping in the polymer material [Fig. 7].
- PVDF transducers can withstand a higher rf power than CdS.
- The PVDF films show extraordinarily good adhesion to the substrate even after many cooling cycles.
- The most outstanding advantage of PVDF transducers is the extremely time (1:10 to CdS) and cost (no complicated apparatus) saving way of preparation.

1 M. Haardt, Thesis, Univ. Stuttgart
2 A. J. Bahr, I. N. Court, J. Appl. Phys., **39**,2863,(1968)
3 H.Kinder, Thesis, Univ. Göttingen
4 H. Zeile et al., Phys. Stat. Sol. (b) **111**,213,(1982)

Discussions

CROSSING EFFECTS IN PHONON SCATTERING page 2
L.J. Challis

W. Dietsche: Do you think that annealing at 1775 ^oC could change the state of the impurities in your sample? Did you find changes of your resonant scattering?

L.J. Challis: V^{3+} is the majority ion in Al_2O_3 and measurements by de Goër et al. have shown that annealing has a negligible effect on its concentration. Fe^{2+} is a minority ion and preliminary experiments were carried out on an Al_2O_3 (Fe) sample using different atmospheres. These showed that annealing in pure argon did not appear to change the Fe^{2+} concentration significantly, although it may of course have changed those of any trace impurities present.
The frequency crossing signals were all somewhat larger after annealing, including those due to V^{3+} - V^{3+} where the concentration is almost certainly very similar. This might indicate a decrease in the overall background scattering rate, but these results are very recent and are still being analysed.

W. Eisenmenger: How can you distinguish experimentally structures eventually caused by the phonon analogue to the Hanle effect from other frequency crossing signals.

L.J. Challis: Theoretically there is a small difference from the usual frequency-crossing line shape caused by the Hanle effect. In practice, to detect this would require very detailed information about the resonant scattering line shapes of the two processes so that one could calculate the frequency-crossing line shape without the Hanle effect. So I think it is very unlikely one could demonstrate these coherence effects in this way. To do this one will need to look at the angular distribution in a situation where one avoids multiple scattering,which will not be easy!

W.E. Bron: In surface scattering one should consider the presence of small randomly orientated crystallites caused by typical polishing techniques. Such crystallites will scatter phonons strongly when the crystallite size approaches that of the phonon wavelengths.

L.J. Challis: That is an interesting suggestion. Presumably it is more likely to lead to elastic scattering than to inelastic scattering. For inelastic scattering I suppose one is thinking of dynamic defects such as two level systems or fluttering dislocations.

FAR INFRARED VOLUME GENERATION AND DETECTION OF PHONONS page 10
K.F. Renk

W. Bron: In order to fit the general requirements of the quasidiffusion process, your spectral diffusion process would need to have a strong frequency dependence. Can you, in fact, justify this for your proposed mechanism?

K.F. Renk: In our model for Al_2O_3 : V^{4+} we assumed, for simplification, that the inelastic scattering time is independent of frequency. This model seems to predict an approximately linear dependence of the effective diffusion time on the dimension of the resonance volume. We do not know, however, whether the inelastic scattering rate depends as ν^4 or in a more complicated way on frequency.

J.I. Dijkhuis: You have said that in the 29 cm^{-1} phonon bottleneck experiments broadband phonons are involved. Can you explain our experiments on the magnetic field dependence of the bottleneck?

K.F. Renk: I guess that your results on the magnetic field dependence give some evidence for trapping of broadband phonon radiation. While it is expected for monochromatic phonons, that, at large N^*, the R_2 signal decreases to a value of 0.25 compared to the signal at zero field, you have found a smaller decrease and the amount of decrease was dependent on N^*. Such a behaviour is expected if in addition to monochromatically generated phonons also broadband phonons are trapped.

W. Eisenmenger: Are there experimental ways to determine the increased width of the phonon frequency distribution, perhaps by hole burning experiments?

K.F. Renk: We have not yet a reasonable method for determination of phonon frequency distributions. We have also thought about the possibility of hole burning experiments: The technique should be well suited to study frequency distributions in systems that have inhomogeneously broadened resonance lines. The data I have presented were obtained for crystals with homogeneously broadened lines and hole burning techniques are not applicable. However, in Al_2O_3 crystals doped with very large concentrations of impurity ions inhomogeneous lines are observed and hole burning techniques should be very useful.

R.S. Meltzer: I think your new ruby experiments provide a valuable new piece of information concerning the decay of 29 cm^{-1} phonons. However, to claim your explanation makes clear the ruby problem will require a detailed quantitative analysis of all the data. For instance, some of our experiments were performed with $2\bar{A}$ resonant excitation where broadband phonons are not produced. In addition, the effective lifetime decreases as the ground state is depopulated. This must be explained in any complete model.

K.F. Renk: The inelastic impurity scattering which I discussed as a phonon-loss process for phonons trapped in ruby is not necessarily the most important loss process. There is no doubt that other processes discussed in the past can also occur and can be the dominant loss processes, depending on sample geometry, on sample quality, on the concentrations of Cr_2O_3 and of N^*. It is therefore, as you suggest, necessary to perform a detailed analysis of all the data. With respect to your experiments on $2\bar{A}$ resonant interaction you may have, in addition to generation of monochromatic phonons, also generation of broadband phonons due to excited-state absorption. The effect could lead to an apparent decrease of the lifetime with increasing N^*.

J.E. Rives: Both your and our model involve spectral diffusion, by energy transfer, which can fit the data at large N^*. Can your model fit our data at large N^*?

K.F. Renk: Our data for $Al_2O_3 : Cr^{3+}$ were obtained for a sample containing 0.05 mol % Cr_2O_3 and with excited state concentrations up to $N^{\times} \approx 3.10^{18} cm^{-3}$. Under these conditions we found no evidence for inelastic phonon scattering by energy transfer processes in which excited Cr^{3+} ions are involved. It is well possible that at larger N and eventually larger Cr_2O_3 concentrations nonresonant energy transfer is a dominant process.

THE JOSEPHSON JUNCTION – A NEW TUNABLE PHONON SOURCE WITH HIGH-FREQUENCY RESOLUTION page 18

P. Berberich, H. Kinder

W. Eisenmenger: Did you check the possibility of direct deformation potential coupling of phonons to Josephson waves?

H. Kinder: Yes we did. Theoretically the effect was predicted to be extremely small. Experimentally we know that the Josephson phonon emission is maximum if we bias the junction to the top of the Eck bump where the electromagnetic wave generation is maximum. From this we know for sure that we are generating electromagnetic waves first, and only these are then converted into phonons.

W. Eisenmenger: If the oxide is amorphous and the reason for the almost one to one photon-phonon conversions, is there also the possibility of generating high-energy phonons with dirty glass films by IR laser radiation?

H. Kinder: Yes, in principle. But the IR absorption of the oxide in our junctions is only of the order of 1 cm^{-1}, so a glass film will absorb only a very small fraction of the incident IR energy. The advantage of the Josephson junction is that you have the electric field just there where you want to generate the phonons. In addition, the Josephson junctions have the practical advantage of being tunable continuously.

B. Stock: Is phonon generation very sensitive to the junction quality?

H. Kinder: Not really. We quote the quality of our junctions by the ratio of the current above 2Δ to that below 2Δ. This ratio is typically 50 for the junctions used, which means just average quality. It is important, however, to make junctions of very low impedance to get Eck bump heights of typically 100 mA for sufficient phonon output.

TWO APPLICATIONS OF MICROWAVE ACOUSTICS IN LIQUID HELIUM: HIGH RESOLUTION MICROSCOPY AND DIRECT MEASUREMENT OF PHONON DISPERSION page 26

D. Rugar

W. Eisenmenger: I wish to congratulate you for your beautiful results. With respect to the resolution or intensity limitations by the anharmonicity in ultrasonic propagation in liquid ^4He I wish to ask whether the application of static pressure appears feasible to reduce this influence.

D. Rugar: The application of static pressure will improve the signal-to-noise ratio of the microscope by over 10 dB, according to my calculations. We have not tried this, however, since it is not compatible with the sample changing feature of our dilution refrigerator.

K. Dransfeld: How do you achieve the surface polish of your lenses with the 500 Å accuracy?

D. Rugar: We use a mechanical polishing technique with, I believe, 0.1 μm diamond grit.

M.A. Brown: To what extend would you be able to observe subsurface damage brought about by mechanical polishing of surface, as for example in the heat pulse propagation experiment which you mentioned?

D. Rugar: If the phonon scattering were large enough to appreciably affect phonon transmission into the helium, we would expect to observe subsurface damage.

R.P. Huebener: Have you used your technique in terms of a highly localized phonon source for studying the two-dimensional variation of phonon transmission through a sample (phonon focusing; phonon scattering by inclusions)?

D. Rugar: Yes, we have considered using the microscope as a source and receiving transmitted signals using incoherent detection techniques. J. Fostu and B. Hadimioglu are working on such a system.

W. Dietsche: How does the contrast achieved with the acoustic microscope using liquid He compare with the contrast achieved by using water?

D. Rugar: The helium reflection microscope is primarily sensitive to topographical features and not very sensitive to the elastic properties of the object. Because of the much greater acoustic impedance of water compared to helium, the water microscope is much more sensitive to the elastic properties of the object.

HEATER FILM DYNAMICS? PHONON DIFFUSION AND PHONON DECAY page 34
T. Wilson, W.E. Bron, F.M. Lurie, W.L. Schaich

G. Northrop: Have you compared the temperatures measured at early times and short distances to the expected film temperature?

T. Wilson: Treating the constantan film as a Debye solid and using the sound velocities for constantan found in the literature, the expected film temperature should reach more than 100 K for applied powers of 1 KW or more. We do not find any temperatures in the bulk to be this large. We believe this to be due to the very slow cooldown of the heater film so that only a small portion of the energy in the film gets transferred to the substrate during fractions of the cooldown time.

J.K. Wigmore: You assume that thermalisation is total within the heater film. This is an interesting problem that has never been completely resolved, since it depends on the electron mean-free path in the metal. Do your experiments allow you to draw any conclusions about the phonon distribution within the heater?

T. Wilson: The two aftermentioned models of Schaich and Levinson perhaps indirectly give us information about the film phonon distribution. There, if one assumes a thermalized film, the predicted mode temperatures in the bulk are enhanced at low frequencies. This implies indirectly that one actually may have a film phonon distribution correspondingly depleted at low ($<$ 1 THz) frequencies.

INFRARED EXCITATION OF HIGH-FREQUENCY PHONONS BY MULTIPHONON ABSORPTION page 37

U. Happek, R. Baumgartner, K.F. Renk

K. Laßmann: What were the surface conditions of the crystal?

R. Baumgartner: The crystal was surrounded by superfluid helium. We have the impression that phonons at the crystal surface were mainly transmitted into liquid helium. We could not directly observe an influence from the surfaces. We found that the phonon spectrum near the surface has the same spectral shape as in the center of the crystal but was, at fixed time after phonon excitation, lower than in the center. This gives some evidence that the spectrum around 1 THz is not strongly influenced by phonons reflected at the surfaces.

J.J. Kim: Doesn't the lifetime of the zero-phonon line affect the decay time of your signal of vibronic side bands?

R. Baumgartner: There is no influence since our crystal is continuously optically pumped and contains a stationary population of excited Eu^{2+} ions for phonon detection.

W. Dietsche: Is your detector also sensitive to multiphonon processes?

R. Baumgarnter: Yes, it is. At a detector frequency ν phonons also of $\nu/2$ are detected, however, with a quadratic dependence on the occupation number. By performing the experiment at small occupation numbers, i.e., at low CO_2 laser power, the influence of two-phonon and multiphonon processes was most likely negligible.

M. Lax: Doesn't the diffusion rate go as ν^4 not as ν^{-4}? Perhaps the escape rate is 1/diffusion rate.

R. Baumgartner: Yes, you are right; I apologize for using the expression in an unusual way.

ZERO-FIELD HYPERFINE SPLITTING OF $Al_2O_3:V^{3+}$ BY JOSEPHSON PHONON SPECTROSCOPY page 40

A. Schick, P. Berberich, W. Dietsche, H. Kinder

L.J. Challis: May I first congratulate you on your very nice experiments. The fact that your J values and ours are different is very interesting and presumably indicates that the pair is more complicated than we'd thought. I hope we can both do some more on this. May I correct your remark that we have not seen hyperfine splitting of pair lines. We published our work on pairs before we had improved our resolution but after we had, we looked again at a few pair lines and we did see some hyperfine structure. I do not remember now which lines we looked at but it is in the thesis by M.N. Wybourne (Nottingham). Could I ask if you have seen any of the other pair lines we saw corresponding to higher J value and/or any of those seen by Hasan and King, Rampton and Murphy?

SUBNANOSECOND BOLOMETRY USING NIOBIUM FILMS page 43

J.P. Maneval, J. Desailly, B. Pannetier

M. Meißner: Did you find experimental evidence that the $\tau \propto T^{-4}$ dependence holds well below 1K? We can show from our experiments that granular carbon films at ~ 0.2 K have response times of less than 50 ns.

J.P. Maneval: We have no experimental data below 1.5 K. Our results suggest that the temperature dependence can be even faster than T^{-4} below 2 K. A depletion of the long wavelength density of states could account for it. A difficulty with granular films is to keep them in a near-equilibrium state. (Hot electrons are produced in rather low fields). If this is not the case of your carbon films, your method of preparation should be published.

R.P. Huebener: Your results on superconducting films seem to indicate that the order-parameter relaxation time in these films must be smaller than the response times of the mechanisms you are considering. Can you comment? In a forthcoming paper we report on such response of superconducting films to light pulses (J. Low Temp. Phys. 53, 613 (1983) A.F. Görlach and R.P. Huebener.

J.P. Maneval: I was not aware of the paper you mention. However, in our view, the properties of the electrons in the middle of the resistive transition are essentially those of the normal state. They also turn out to be similar in the other conduction regimes, as far as the electron-phonon interaction is concerned.

K. Weiss: With regard to the question raised by Dr. Huebener, I think that it is not very likely that the order parameter relaxation is too fast to be of any influence (these times characteristically are rather slow) but that Maneval's results indicate that superconductivity is not involved in the mechanisms which he observes.

J.P. Maneval: I agree with this statement (see answer to Prof. Huebener).

U. Strom: Anodized NbN bolometers have been used at NRL in the past as ballistic phonon detectors. These bolometers have an estimated response time of $\lesssim 0.1$ ns and an effective operating range of 2-13 K. The effective thickness of these bolometers is ~50 Å, consisting of interconnected islands of superconducting NbN interspersed in NbO. Is there a physical similarity between your bolometer and ours?

J.P. Maneval: The authors are aware of this work (see references). We suggest that response time measurements be actually performed on NbN films to investigate the electron-phonon relaxation, especially at low temperatures.

W.E.Bron: Is it possible that the active elements in your detector are actually superconducting or semiconducting junctions?

J.P. Maneval: It is quite possible that absorption of energy from light or from electric current takes place in such junctions. Not so much can be said for relaxation to phonons.

S.P. Harrison: Can you compare the magnitude of the T^{-4}-dependent time constant with that expected from Pippard theory?

J.P. Maneval: By deriving the electron mean free path from conductivity in the normal state (in superconducting samples) and computing τ_{ep} by using the Pippard formula, one obtains order of magnitude agreement with experiment. However, a large uncertainty remains in the assignment of coupling constants.

NONEQUILIBRIUM PHONON DISTRIBUTION IN A QUANTIZING MAGNETIC FIELD: A TUNABLE GHz TO THz PHONON GENERATOR? page 49

G. Slater, A.-M.S. Tremblay

V. Narayanamurti: Have you taken focusing into account in your calculations?

G. Slater: No. Would it be important here? If one chooses correctly his sample orientation, the effect of focusing on a 2°-aperture beam should be rather small.

W. Dietsche: Did you also consider (two-dimensional) inversion layers as phonon generators?

G. Slater: Yes and no. From simple arguments, one may expect to obtain similar qualitative results but with higher monochromaticity. Since the 3D case would certainly lead to more powerful generators, we have studied this problem first. Moreover, 3D experiments should be easier.

RELAXATION PROCESSES AND RESPONSE TIME OF SUPERCONDUCTORS TO A PERIODIC EXCITATION page 52

N. Perrin, C. Vanneste

W. Dietsche: Would the use of the μ^* model lead to different results?

N. Perrin: The μ^* model would probably lead to different results. But it is not appropriate when the q.p. scattering relaxation time is larger than the recombination time, which occurs in the nonequilibrium situation considered.

Part II

Phonon Focusing

Chairman: **O. Weis**

Phonon Focusing in Germanium Imaged by Electron-Beam Scanning *

W. Metzger, R. Eichele, H. Seifert, and R.P. Huebener

Physikalisches Institut II, Universität Tübingen, Morgenstelle
D-7400 Tübingen, Fed. Rep. of Germany

Recently we have shown that anisotropic phonon propagation and phonon focusing in single crystals can be investigated using electron-beam scanning for imaging the anisotropic energy flux of ballistic phonons [1]. These experiments were performed in a scanning electron microscope equipped with a low-temperature stage. The sample arrangement was such that the top side of the crystal was exposed to the vacuum in the sample chamber of the microscope and could directly be irradiated with the electron beam generating the phonons at the upper sample surface. The bottom side of the crystal was in direct contact with the liquid He bath. A small-area bolometer at the center of the bottom surface served for phonon detection. From the time-resolved images of the phonon-energy flux the contributions of the different acoustic phonon branches could be studied separately. Further details are given in Ref. 1.

In this paper we report on similar experiments performed with single-crystalline germanium. The sample is a disk of 21 mm diameter and 3 mm thickness, its axis pointing in [110] direction. For eliminating electric charging effects due to the beam, the top side of the Ge crystal was covered with a granular Al film of 0.5 μm thickness. The phonons were detected with a thin-film superconducting bolometer (granular Al) of 2 μm x 2 μm sensitive area, deposited on the bottom side of the crystal. During the experiments the He-bath temperature was 1.85 K.

In Fig. 1 we show the two-dimensional image of the time-integrated bolometer signal containing the sum of the contributions of the different acoustic phonon modes. The bright regions indicate high values of the phonon-energy flux. The scanned sample area corresponds to an angle of about 110° around the crystal [110] axis. Figure 2 shows the time-resolved bolometer signal recorded while the electron-beam focus was kept fixed at the points marked A (T1 mode) and B (T2 mode) on the intensity pattern given in the inset. These measurements were carried out using a 5 ns time window. The temporal location of the electron-beam pulse is also indicated. The relatively sharp peaks in Fig. 2 confirm the ballistic nature of the phonon-energy flux detected by the bolometer. From the peak positions of the curves A and B in Fig. 2, we find the same value of the sound velocity v = 3500 m/s. This value agrees

* Supported by a grant of the Deutsche Forschungsgemeinschaft

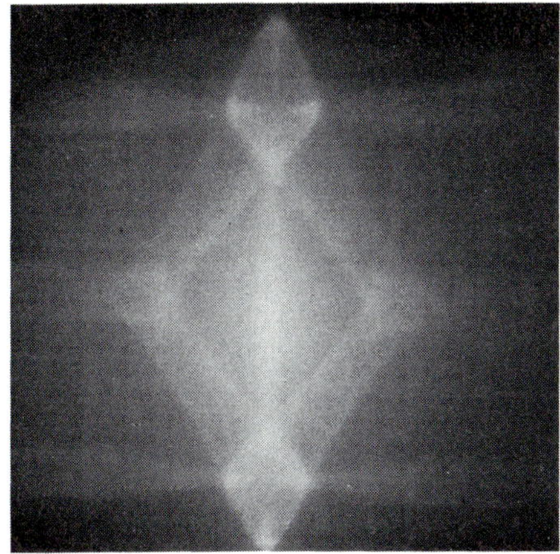

Fig.1. Two-dimensional image of the time-integrated bolometer signal. Electron-beam parameters: 26 kV, 0.1 μA

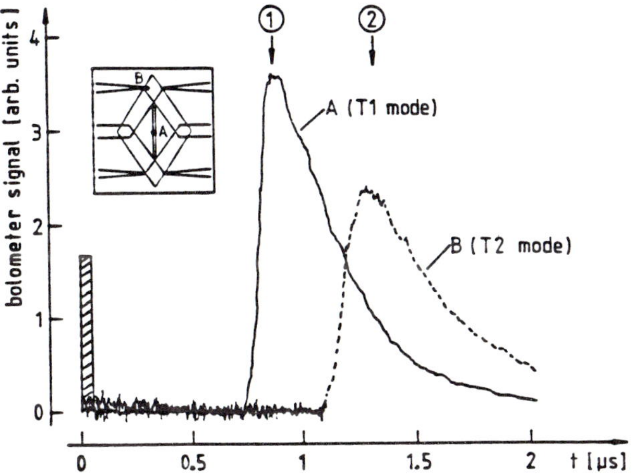

Fig.2. Time-resolved bolometer signal for the two points of the electron-beam focus indicated on the inset. The shaded area shows the electron-beam pulse. The numbers (1) and (2) indicate the position of the 20 ns time windows at which the images of Fig. 3a and 3b, respectively, were obtained. Electron-beam parameters: 26 kV; 1 μA

reasonably with the velocity of the T1 and the T2 mode for the [110] and the [100] direction, respectively [2].

Two-dimensional images of the time-resolved bolometer signal are shown in Fig. 3. Here we have used the 20 ns time windows marked (1) and (2) in Fig. 2, corresponding to the temporal peak position for the T1 and the T2 mode, respectively. Hence, Fig. 3 displays predominantly the image of the T1 mode (part a) and of the T2 mode (part b). In Fig. 3b the faster T1 mode mixes with the image of the T2 mode due to the increased path length between phonon source and detector (horizontal bright region at

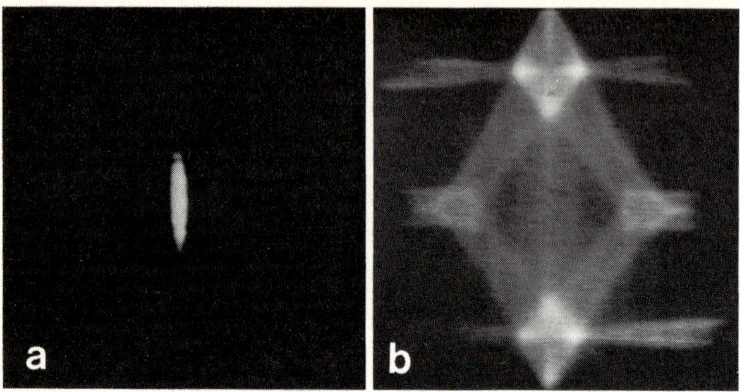

Fig.3. Two-dimensional image of the time-resolved bolometer signal for time window (1) (part a) and (2) (part b) marked in Fig. 2. Scanned area same as in Fig. 1. Electron-beam parameters: 26 kV, 1 μA

the top and bottom starting from the outer corners of the rhombus). Of course, this portion of the T1 mode is missing in Fig. 3a.

The two-dimensional images shown in Fig. 1 and 3 agree well with earlier measurements by Northrop and Wolfe [3]. These authors used a technique similar to ours except that two-dimensional laser-beam scanning has been employed for phonon generation. Our results also agree with the calculations of Rösch and Weis [4] based on the elasticity theory for continuous anisotropic media. The images presented in Fig. 1 and 3 display a slight nonsymmetry, which can be noticed most clearly from the difference in the size of the rhombus at the top and at the bottom. It appears that this nonsymmetry results from a small misalignment (by a few degrees) of the axis of the Ge disk from the [110] direction.

The spatial resolution of the method for two-dimensional imaging of the anisotropic phonon propagation by means of electron-beam scanning will be limited by the areal extension of the phonon source and of the phonon detector. The diameter of the phonon source is approximately given by twice the thermal healing length for the particular geometry. It appears that by means of high-frequency electron-beam modulation (thermal skin effect), the diameter of the phonon source can be reduced to about 1 μm [5]. Of course, in order to benefit from the thermal skin effect for increasing the resolution, the modulated signal must be detected with the bolometer. On the other hand, an effective bolometer area of 1 μm x 1 μm appears feasible utilizing standard lithographic methods.

References:

1. R. Eichele, R.P. Huebener, and H. Seifert: Z. Phys. B - Condensed Matter 48, 89 (1982)
2. A.G. Every: Phys. Rev. B24, 3456 (1981)
3. G.A. Northrop and J.P. Wolfe: Phys. Rev. B22,6196 (1980)
4. F. Rösch and O. Weis: Z. Physik B25, 115 (1976)
5. H. Pavlicek, L. Freytag, R.P. Huebener, and H. Seifert: Proceed. of the 1982 Applied Supercond. Conf., Knoxville, Tennessee, USA

Nonlinear Phonon Focusing

D. Armbruster, G. Dangelmayr, and W. Güttinger

Institute for Information Sciences, University of Tübingen
Köstlinstraße 6, D-7400 Tübingen, Fed. Rep. of Germany

Acoustic phonon propagation in a cold anisotropic crystal is dominated by focusing which typically occurs on structurally stable caustics [1]. By applying singularity theory, the forms of these caustics and the associated high-intensity diffraction patterns can be classified into a few topological types [2], [3]. Suppose a monochromatic point source of frequency ω generates phonons with wave vectors $\underline{k}$ that propagate ballistically in a crystal whose anisotropy is described by a dispersion relation $\omega=\Omega(\underline{k})$. Then only those $\underline{k}$ contribute to the phonon field $u(\underline{r},\omega)$ at a point $\underline{r}$ in space which make up the constant-frequency surface $S:\omega=\Omega(\underline{k})=\text{const}$, i.e.,

$$u(\underline{r},\omega) \propto \int d\underline{k}\,\delta(\omega-\Omega(\underline{k}))e^{i\underline{k}\underline{r}} = \int dS\frac{e^{ir\phi}}{|\nabla\Omega(\underline{k})|} \tag{1}$$

for a given polarization mode. Here, $\underline{r}=r\hat{\underline{r}}$ with unit vector $\hat{\underline{r}}$, $\phi=\hat{\underline{r}}\cdot\underline{k}$ and the second integral is taken over S. Suppose first that the phonon's group velocity $\underline{v}=\nabla\Omega(\underline{k})$, with $\hat{\underline{v}}=\underline{v}/|\underline{v}|=\underline{n}$ the unit normal to S, has no zeros. Then the phonon flux is in the directions $\hat{\underline{r}}=\hat{\underline{v}}(\underline{k})$. The corresponding wave vectors $\underline{k}$ on S are (for large r) those for which ϕ is stationary, $\underline{t}\cdot\nabla_k\phi=0$ for vectors $\underline{t}$ tangent to S. Phonon focusing directions, i.e., angular caustics, come from the inflection points of S along a principal curvature line where the Gaussian curvature vanishes. These are the stationary points where the Hessian determinant of ϕ vanishes. Since the caustics are structurally stable, i.e., insensitive to small perturbations, ϕ is equivalent to a Thom catastrophe polynomial $\phi=\phi_T$. This implies that the caustics can be classified by the topological singularities of the Gaussian map G of the frequency surface S on the multiply covered sphere S^2 of the group velocity directions $\hat{\underline{v}}(\underline{k})$ on which the caustic images are observed.

The fact that the structurally stable caustics are the images on S^2 of all points on S with zero Gaussian curvature has the following consequences. By Whitney's theorem, the only generic singularities on S^2 are folds and cusps [4]. They are produced, respectively, by the line L on S with zero Gaussian curvature and its points of contact with the principal curvature line P that has zero curvature on L. When the fold lines, which separate dark from bright regions on S^2, are crossed, the number of phonon $\underline{k}$ vectors changes by two while three $\underline{k}$ vectors coalesce at a cusp point. If the dispersion relation is linear, $\omega=c(\hat{\underline{k}})k$ (nearest-neighbor interaction), the surfaces S are similar for different ω, so that folds and cusps are the only possible caustics. If, however, the dispersion relation $\omega=\Omega(\underline{k})$ is nonlinear (long-range forces), a variation of ω may change the topological type of S and that of the caustics when ω crosses a critical value ω_c. These sudden transitions are called critical events. Parametrizing G by $\omega-\omega_c$ the following events can occur.

(i) Swallowtail: Here, L has a second-order contact with P (Fig. 1b) which splits into three intersections (Fig. 1a) and one intersection (Fig. 1c) of L with P for $\omega<\omega_c$ and $\omega>\omega_c$, giving rise to the caustics of Fig. 2.

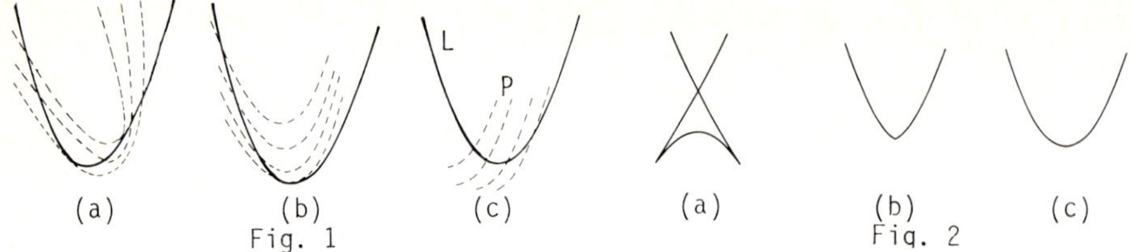

Fig. 1 Fig. 2

(ii) Hyperbolic and elliptic umbilics: S possesses an umbilic point which is a point $\underline{k}$ with equal principal curvatures where the surface is locally spherical. In the case of a hyperbolic umbilic the curvature vanishes for $\omega=\omega_C$ and the locally flat S produces the middle caustic in Fig. 3b. For $\omega \gtrless \omega_C$, a tangential contact of L and P produces the cusps in the two other caustics in Fig.3b. An elliptic umbilic point appears when a circular L shrinks to a point for $\omega \to \omega_C$ from below and expands into another circle for $\omega > \omega_C$. The resulting caustic is shown in Fig. 3a. The three cusps in Fig. 3a shrink to a point for $\omega \to \omega_C$.

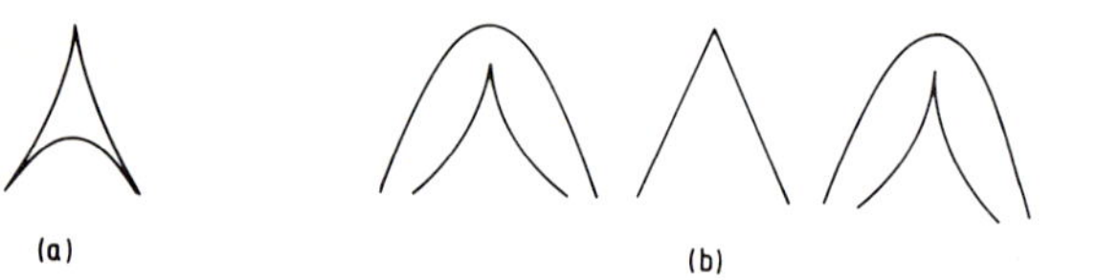

Fig. 3

(iii) There are but two possible degeneracies in a line L for $\omega=\omega_C$, viz., one that produces a caustic beak-to-beak event (Fig. 4a) and one giving rise to a caustic lip event (Fig. 4b). These singularities exhaust all stable possibilities when no symmetries are involved. On a crystal's reflection planes and rotation axes new types of caustics are generated. They have been described in [2]. A C_6-symmetric singular event is shown in Fig. 5. The Airy, Pearcey, etc., diffraction patterns around the caustics follow by substituting ϕ_T into (1) and evaluating the integral asymptotically [3]. By scaling one finds that on the caustic $|u| \propto r^{-\beta}$ for large r with the singularity indices β=(5/6, 3/4, 7/10, 2/3) for the fold, cusp, swallowtail and umbilic events and β=3/4 for lips and beak-to-beak events.

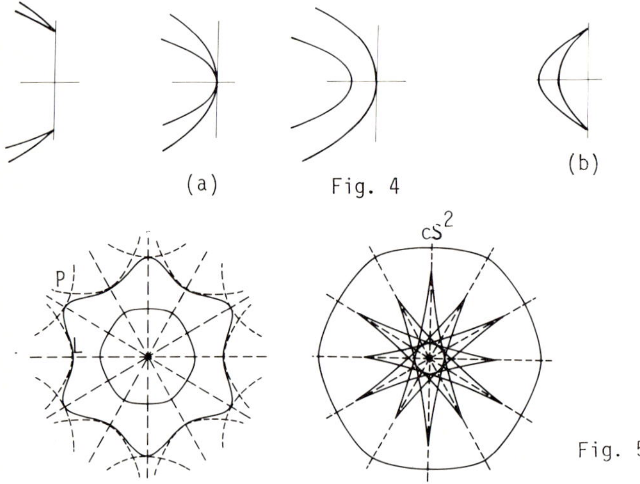

Fig. 4

Fig. 5

If the dispersion relation $\omega=\Omega(\underline{k})$ has stationary (Morse) points, where the group velocity $\underline{v}$ vanishes, a new phenomenon springs up. Suppose that in Fig. 6 $(\omega_1,\underline{k}_1)$ is a minimum and $(\omega_2,\underline{k}_2)$ a saddle. Then the constant-frequency surface develops a second disconnected sheet when ω goes through ω_1, which gives rise to an overall flux enhancement. If ω goes through ω_2, the two surfaces coalesce in one of the two ways shown in Fig. 7. The caustic originating from the vertex of the cones is a great circle on S^2, but since $\underline{v}$ is zero at the point $\underline{k}_2$, the intensity is zero and, therefore, this circle must appear as a dark ring ("anticaustic"). On either side of this dark line one expects two diffuse bright small circles whose positions correspond to the semiangles of the distorted cones near the vertex.

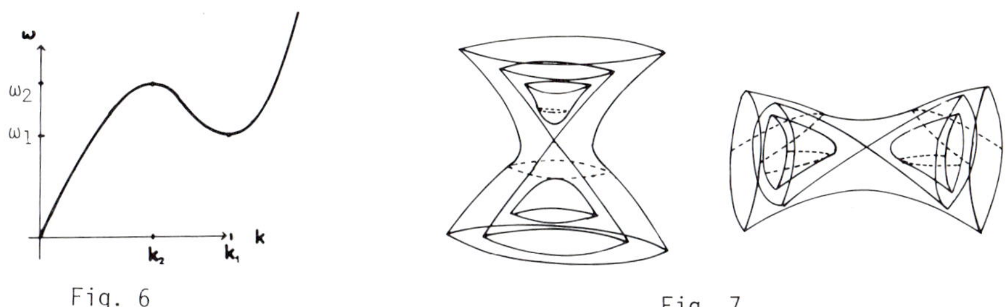

Fig. 6

Fig. 7

We expect that with the recent development of tuneable phonon sources the caustic singularities occurring at the specific critical frequencies ω_c will become accessible to detailed experimental investigation.

Acknowledgments:
We thank Dr. F.J. Wright for valuable suggestions. The support by the Stiftung Volkswagenwerk is gratefully acknowledged.

References:

1 Wolfe, J.P.: Phys. Today 33, 44 (1980)
2 Armbruster, D., Dangelmayr, G.: Z. Phys. B 52 (1983)
3 Dangelmayr, G., Güttinger, W.: Geophys. J.R.Astr.Soc. 71, 79 (1982)
4 Taborek, P., Goodstein, D.: Solid State Commun. 33, 1191 (1980)

Phonon Focusing in Highly Dispersive and Isotopically Impure Crystals

S. Tamura

Department of Engineering Science, Hokkaido University
Sapporo 060, Japan

Abstract

It is demonstrated theoretically that the dispersive effects and isotope scattering are substantial to understand the ballistic phonon focusing of near 1-THz TA phonons in highly dispersive and isotopically impure crystals.

Transverse-acoustic (TA) phonons in some of group-IV elements and III-V compounds are highly dispersive at 1-THz frequency range characterized by the marked flattening of dispersion curves especially in the [100] and [111] directions. The focusing due to crystal anisotropy of large-wave-vector TA phonons in these substances is thereby expected to change considerably from that of the low-frequency limit. As an illustration, we show in Fig.1 the constant-frequency surfaces (ω surfaces) of the upper-TA (T2) branch of Ge at 0.3 and 1.0 THz. (Above 0.3 THz the acoustic dispersion is discernible.) Since the propagation direction of a phonon in the real space is given by by the vector normal to the ω surface, the deformations of the ω surface as viewed in Fig.1 will lead to noticeable changes of the focusing properties of the T2 phonons of THz frequencies.

In Fig.2 we plot phonon distributions in the real space at 0.3 and 1.0 THz of Ge. (Arrows indicate the locations at which caustics intersect with the {110} plane.) As anticipated, the extension of the T2-phonon focusing region near the (100) plane ($\phi=0^\circ$) is striking in the map of 1.0 THz. Also the focusing structure of the lower-TA (T1) branch which extends from the

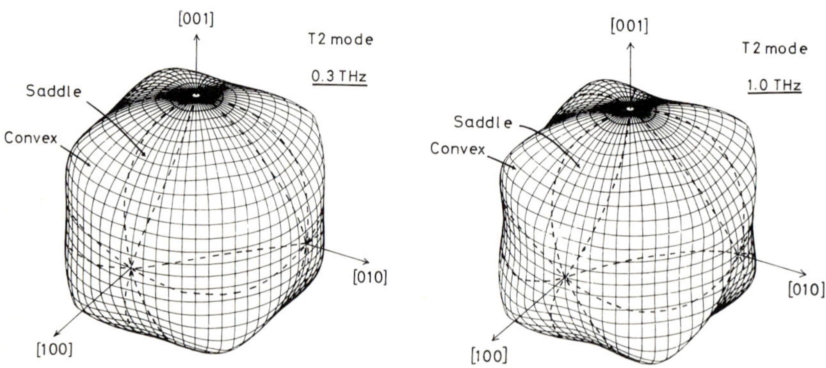

Fig. 1 ω surfaces of T2 phonons

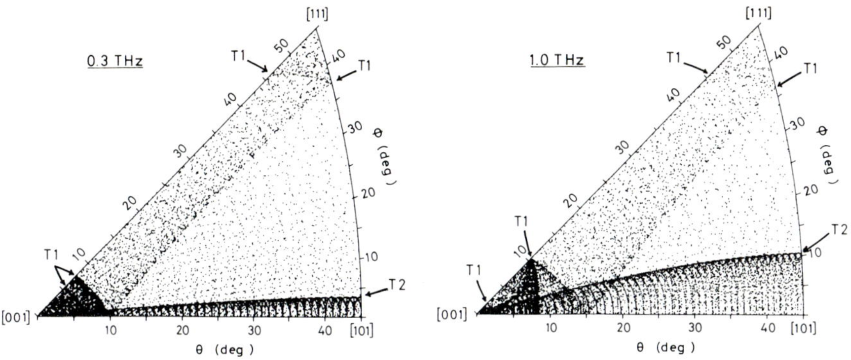

Fig. 2 Polar plots of phonon distributions in the real space

[001] toward [111] directions changes as well. In particular, the focusing in the vicinity of the [111] direction which makes a three-fold cusp structure becomes gentle at higher frequencies. All these results are consistent with the findings of DIETSCHE, NORTHROP and WOLFE [1]. That is, comparing with the image of low-frequency phonons obtained by the Al bolometer, remarkable alterations of ballistic flux patterns have been detected by the Pb-oxide-Pb tunneling junction which responds selectively to phonons higher than 0.7 THz. At present their observations are qualitatively interpreted in terms of phonon focusing effects taking account of the dispersion [2,3].

Here we note, however, that natural Ge consists of a mixture of five isotopes. The isotope scattering of phonons is highly frequency-dependent and accordingly should severely affect the ballistic transport, or the focusing of high-frequency, near-zone-edge phonons. (Indeed, DIETSCHE et al. [1] had to prepare thin Ge samples of 0.5-mm thickness to observe sharp images of ballistically transmitted phonons higher than 0.7 THz.) In Fig.3 we show the frequency distributions of the energy density of unscattered phonons which hit the detector separated a distance d from the phonon source. In this example, the Planck distribution with local source temperature T=12.5 K has been assumed. (The shaded region indicates the frequency range sensitive to the tunnel-junction detector.) An important effect of the isotopes is to reduce an extremely broad frequency distribution of the phonons at the excitation region to a very narrow spectrum when the ballistic components are detected after the travel over the distance d. Because the dispersive effects are rather small up to 0.4 THz, the focusing pattern to be observed by an Al bolometer in the experimental configuration with d=2 mm should be almost identical to that of the low-frequency limit. We also expect that the ballistic phonons which respond to the Pb-oxide-Pb junction detector in the case of d=0.5 mm are essentially those confined in a narrow frequency band 0.7 to 0.8 THz.

Now, plotted in Fig.4 are the angular dependences of the ballistic TA-phonon intensity integrated over $\pm 0.25^\circ$ and $\pm 2.5^\circ$ parallel and perpendicular to the $(1\bar{1}0)$ plane. In obtaining these results the isotope effects are properly taken into account. The trace A shows the structures very similar to those of the low-frequency limit published in the literature [4,5]. Comparing with the trace A, the trace B exhibits the shifts of locations of focusing singularities as well as the reduction of overall magnitudes of

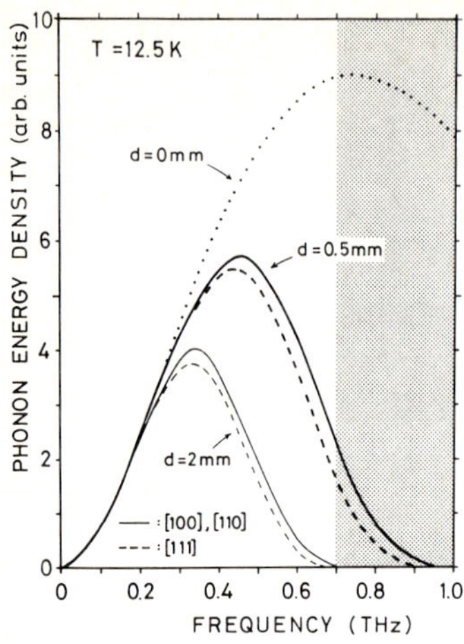

Fig. 3 Energy spectrum at T=12.5K

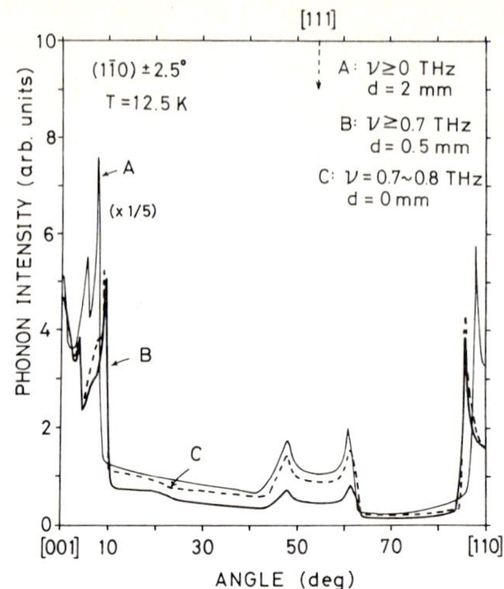

Fig. 4 Angular dependences of ballistic phonon intensity

the intensity. The sharp features of the directional property of this trace is clearly due to the effects of the isotopes as described above, i.e., the selection of quasi-monochromatic phonons. In order to comprehend the isotope effects on the angular dependence of the phonon intensity near 1-THz frequencies, we have also displayed in Fig.4 the intensity integrated over the frequency range 0.7 to 0.8 THz under the condition d=0 mm (trace C). (The magnitude is adjusted to the trace B in the [110] direction.) In fact we find that the isotope scattering acts to diminish the ballistic component of phonon intensities preferentially in the [111] direction as envisaged from Fig.3.

To summarize, we have carried out a comprehensive study of the TA-phonon focusing near 1-THz frequency range. The acoustic dispersion and the isotope scattering are found to be crucial in understanding the transport of ballistic phonons in such a highly dispersive and isotopicaly impure crystal as Ge. In particular, at frequencies near 1 THz the TA-phonon intensity in the three-fold [111] direction is reduced by both the dispersive and isotope effects.

References

1. W. Dietsche, G. A. Northrop, and J. P. Wolfe: Phys. Rev. Lett. 47, 660 (1981)
2. S. Tamura: Phys. Rev. B25, 1415 (1982) and ibid. B28, 897 (1983)
3. G. A. Northrop: Phys. Rev. B26, 903 (1982)
4. J. C. Hensel and R. C. Dynes: Phys. Rev. Lett. 43, 1033 (1979)
5. S. Tamura and Y. Nakane: Phys. Rev. B24, 4317 (1981)

Spatial Variation of Phonon Distribution in Thermal Conduction

F.W. Sheard, G.A. Toombs, and S.R. Williams

Department of Physics, University of Nottingham, University Park
Nottingham, NG7 2RD, England

Thermal conduction is usually studied for a homogeneous solid in which there is a uniform temperature gradient dT/dx, which is sufficiently small that the linear relation $Q = -KdT/dx$ is valid, where Q is the heat current and K is the thermal conductivity. However there is increasing interest in inhomogeneous situations in which heat is injected into a solid from a heater or flows across an interface between two solids as in Kapitza conduction [1]. We shall only consider insulating solids in which heat conduction is by lattice phonons. Moreover, at liquid helium temperatures, where such experiments are often performed, phonons are principally elastically scattered by defects or boundaries. The existence of a local phonon temperature cannot then be assumed since inelastic processes are necessary to equilibrate phonons of different frequencies.

We shall illustrate some of the features mentioned above by means of a simple model of lattice conduction, in which the solid is terminated by heat reservoirs at temperatures $T + \Delta T$ and T ($\Delta T \ll T$) whose surfaces, at $x = 0$ and $x = d$ respectively, behave as black bodies for phonon emission and absorption. In the solid, since the phonon frequency is unchanged by the scattering process, the phonon distribution function N_q cannot relax to the Bose-Einstein function $N_q^o(T) = 1/\{\exp(\hbar\omega_q/kT) - 1\}$. Instead, the distribution will relax to an average value $\overline{N}_q$, where the average is over all directions of propagation for a given frequency ω_q. We write $N_q = N(\omega,\mu,x)$, $\mu = \cos\theta$, where θ is the angle between the phonon propagation vector q and the x axis which is the direction of heat flow. The Boltzmann equation can then be written

$$\mu v \frac{\partial N}{\partial x} = -\frac{N(\mu,x) - \overline{N}(x)}{\tau}, \quad \overline{N}(x) = \frac{1}{2}\int_{-1}^{1} N(\mu,x)d\mu,$$

where v is the phonon velocity and τ the relaxation time. The ω variable may be suppressed since it is the same in all terms. The boundary conditions at $x = 0$ and $x = d$ are

$$N(\mu,0) = N^o(T + \Delta T) \ (\mu > 0), \ N(\mu,d) = N^o(T) \ (\mu < 0).$$

A formal integration of the Boltzmann equation then gives

$$N(\mu,x) = \begin{cases} N^o(T + \Delta T)e^{-x/\mu\ell} + \dfrac{1}{\mu\ell}\displaystyle\int_0^x e^{-(x - x')/\mu\ell}\overline{N}(x')dx' & (\mu > 0) \\ \\ N^o(T)e^{(d - x)/\mu\ell} - \dfrac{1}{\mu\ell}\displaystyle\int_x^d e^{(x' - x)/\mu\ell}\,\overline{N}(x')dx' & (\mu < 0) \end{cases}$$

where $\ell = v\tau$ is the phonon mean free path.

Parrott [2] recently solved the above integral equation for $N(\mu,x)$ by iteration using as zeroth-order approximation the first term on the right-hand side, which is a black-body distribution decaying exponentially with distance from each surface (at $x = 0$ or d) as the emitted phonons are scattered out of the direction μ. But this ignores diffusion of phonons back into the direction μ and is only a good approximation when the scattering is very weak, $\ell \gg d$. However Parrott used this method in the opposite limit $\ell \ll d$ and obtained the ratio of the thermal conductivity $K = Q/(\Delta T/d)$, to the value $K_o = (1/3)\int C(\omega)\ell v d\omega$ given by the usual kinetic formula. Here $C(\omega)$ is the specific heat per unit frequency range. Carrying the iteration to third order gave $K/K_o \simeq 0.675$ and Parrott suggested, as a reasonable extrapolation, that for this model K is only about 70% of the standard value K_o.

Because the iterative method relies on such a poor initial approximation we have investigated this problem more fully using numerical methods starting from the integral equation for $\overline{N}(x)$ which is easily obtained from the above expressions:

$$\overline{N}(x) = \frac{1}{2} E_2\left(\frac{x}{\ell}\right) N^o(T + \Delta T) + \frac{1}{2} E_2\left(\frac{d - x}{\ell}\right) N^o(T) + \frac{1}{2\ell}\int_0^d E_1\left(\frac{|x - x'|}{\ell}\right) \overline{N}(x')dx',$$

where the $E_n(x) = \int_0^1 \exp(-x/\mu)\mu^{n-2}d\mu$ are exponential integrals. Using the method of Loyalka [3] developed for neutron transport, $\overline{N}(x)$ was computed for a range of values of ℓ/d and is shown in Figure 1 for $\ell/d = 0.1$. It is seen that $\overline{N}(x)$ varies linearly with x over most of the range but some curvature is found within a mean free path of the thermal reservoirs. At each boundary the difference between $\overline{N}$ and the thermal-equilibrium distribution is analogous to

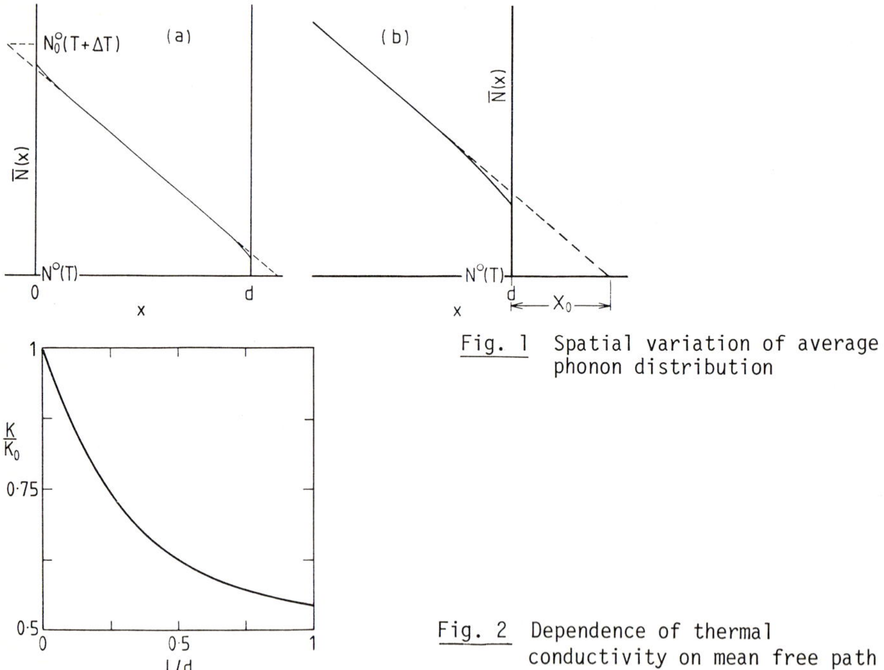

Fig. 1 Spatial variation of average phonon distribution

Fig. 2 Dependence of thermal conductivity on mean free path

the temperature difference at the interface between two media in the Kapitza problem. The thermal conductivity ratio K/K_0 is shown in Figure 2. As $\ell \to 0$, $K \to K_0$ so that the standard value is indeed recovered in this limit. The linear decrease of K/K_0 for small ℓ/d may be interpreted in terms of the extrapolation length X_0 shown in Figure 1(b). The temperature difference ΔT effectively occurs over a distance $d + 2X_0$ rather than d so that the thermal conductivity is given by $K/K_0 = d/(d + 2X_0)$. For $\ell/d \ll 1$ it is found that $X_0 \simeq 0.71\ell$ so that $K/K_0 \simeq 1 - 1.4(\ell/d)$. This value of the extrapolation length is well known in the theory of neutron diffusion [4].

Solutions have also been obtained for a bicrystal system containing an interface separating regions with different elastic scattering strengths. No nonlinear variation of $\overline{N}(x)$ was found near the bicrystal interface. However the interpretation of experiments on such a system requires consideration of inelastic processes [5].

1. L.J. Challis: J. Phys. C 7, 481 (1974).
2. J.E. Parrott: J. Phys. C 15, 6919 (1982).
3. S.K. Loyalka: Nucl. Sci. Eng. 56, 317 (1975).
4. J.L. Duderstadt and W.R. Martin: Transport Theory (Wiley, New York 1979) p.110.
5. L.J. Challis, A.A. Ghazi and M.N. Wybourne: Phys. Rev. Lett. 48, 759 (1982).

Discussions

PHONON FOCUSING IN GERMANIUM IMAGED BY ELECTRON-BEAM SCANNING page 72
W. Metzger, R. Eichele, H. Seifert, R.P. Huebener

K. Dransfeld: Do the electrons penetrate the aluminium film and create an electron hole plasma inside the germanium? If they do, how long does it take until the phonons are actually emitted from the surface?

W. Metzger: These are interesting questions which we have not yet studied in detail. We hope to do so in the near future.

M. Maris. If you can make detectors and generators as small as 1μ you should be able to detect the effect of finite phonon wavelengths on phonon intensity near to cusps and folds. I have recently calculated how large these effects are for germanium and silicon.

W. Metzger: This would be interesting, indeed, and we should look for such an effect.

J.P. Wolfe: 1. What is the rather large broadening (~.5 s) in your heat pulses due to? 2. Have you measured the spatial resolution of your method from the sharpness of the focusing pattern?

W. Metzger: 1. Possibly the thermal response time of our evaporated Al film on the top of the crystal is involved. 2. Detailed measurements of this kind have not been done so far.

J.C. Hensel: Isn't the multiple scattering of electrons a factor in setting the limits of resolution for this technique? Such is the case for electron beam lithographic procedures in VLSI technology.

W. Metzger: Ultimately, this mechanism will limit the spatial resolution. One expects this mechanism to take over if the thermal healing length for the particular geometry becomes sufficiently small.

NONLINEAR PHONON FOCUSING page 75
D. Armbruster, G. Dangelmayr, W. Güttinger

A.S. Every: There are hazards in applying catastrophe theory to phonon focusing. For instance, in the [100] directions of cubic crystals the transverse phonon energy surfaces are degenerate, and hence the "generating function" is non-analytic. This can give rise to four intersecting fold lines. How can this catastrophe be classified? Secondly, for some cubic crystals the FT fold lines connecting different [100] directions disappear over their entire length as the elastic constants are varied - a "global catastrophe".

D. Armbruster: 1. Non-analytic behaviour occurs typically for critcal frequencies at isolated points on S. Near such points the generating function is not analytic and hence not tractable by a catastrophe analysis of caustics. However, the behaviour near such points can be analyzed by topological methods as shown in the last Fig. of the paper. Concerning interesting fold lines Fig. 5 shows that C.T. properly generalized to include symmetry can give more than two intersecting fold lines (cf. Ref. 2). 2. From a topological point of view this is a nongeneric phenomenon. We expect that with higher angular and frequency resolution one will see a gradual merging of these two lines.

PHONON FOCUSING IN HIGHLY DISPERSIVE AND ISOTOPICALLY IMPURE CRYSTALS page 78
S. Tamura

W. Dietsche: Disregarding isotope scattering, which experimental frequency resolution would be required to observe a high-frequency dispersive focusing pattern,let's say at 1.5 THz?

S. Tamura: I cannot answer quantitatively. In general, as the frequency increases the high-frequency resolution is needed because the dispersive effects become very conspicuous near the zone boundary frequency.

SPATIAL VARIATION OF PHONON DISTRIBUTION IN THERMAL CONDUCTION page 81
F.W. Sheard, G.A. Toombs, S.R. Williams

M. Wagner: I have not understood how you can simulate a heat current at the boundaries with your kind of boundary conditions.

G.A. Toombs: The boundary condition only specifies phonons entering the sample. Phonons leaving the sample are not in equilibrium with the boundaries. Therefore, there is a net heat current.

T. Nakayama: You considered elastic scattering. Do you need a big modification in your results if you assume inelastic scattering like TLS in glasses?

G.A. Toombs: There are some formal similarities between elastic and inelastic scattering with $\overline{N}$ replaced by a local thermal equilibrium distribution and the local temperature is determined by an integral equation. We plan to include inelastic scattering in future calculations. But for purely elastic scattering a local temperature is not defined.

Part III

Large Wave-Vector Phonons

Chairmen: **W. E. Bron A. G. Every J. E. Rives R. G. Ulbrich**

Observation of a Quasi-Diffusive Phonon Propagation Mode

W.E. Bron and J.M. O'Connor

Department of Physics, Indiana University, Bloomington, IN 47401, USA

Y.B. Levinson

L.D. Landau Institute for Theoretical Physics, The Academy of Sciences of the USSR, SU-117334 Moscow, USSR

We report the first direct experimental observation of quasidiffusional propagation of nonequilibrium phonons in solids.

Phonon transport is said to be "diffusive" if the time of arrival t_A for a phonon packet to traverse L is proportional to L^2/D, where D is the diffusion constant. Such propagation is normally observed in the presence of elastic scattering sites [1]. If phonon decay through anharmonic interaction is simultaneously present, a "quasidiffusive" propagation mode [2,3] should be discernible [4] whose qualitative features can be understood as follows [2]. We assume, at $t = 0$, an injection at one point in a solid of a monochromatic packet of acoustic phonons of frequency ω_0. If the lifetime against elastic scattering, $\tau^*(\omega)$, is short compared to the lifetime against anharmonic decay, $\tau(\omega)$, then during $\tau(\omega)$ propagation proceeds diffusively. For simplicity, we limit anharmonic processes to the spontaneous decay of ω_0 phonons into two lower energy phonons with $\omega_1 \simeq \omega_0/2$ [5]. This process repeats itself, producing new "generations" of phonons with $\omega_2 \simeq \omega_0/4$, ..., $\omega_i \simeq \omega_0/2^i$ ..., until the transformed packet reaches the detector. For our purposes it suffices to express the frequency dependence [6] as $\tau(\omega) = \tau_0(\omega/\omega_0)^{-5}$ and as $\tau^*(\omega) = \tau_0^*(\omega/\omega_0)^{-4}$.

The strong dependence of $\tau^*(\omega)$ and $\tau(\omega)$ on ω has two major consequences; namely, (i) every new generation has a much longer lifetime and (ii) occupies a much larger spatial volume (as defined by, say, the root mean displacement) compared to that of its predecessor. A measurement of the temporal evolution of the phonon population at a distance L_A from the site of injection reflects, then, almost exclusively the transport of the last generation since, compared to this generation, all previous ones occupy essentially delta functions in space and time. Thus the distance traveled to the detector $L_A \simeq C(D_A t_A)^{1/2}$ and $t_A \simeq \tau(\omega_A)$, where D_A is the effective diffusion constant of the generation which arrives at the detector, and C is a factor of order unity. But from kinetic theory, $D_A \simeq \langle \upsilon^2 \rangle \tau^*(\omega_A)$, such that $\langle L_A^2 \rangle \simeq \langle \upsilon^2 \rangle \tau^*(\omega_A) t_A$, where υ is the sound velocity. Accordingly, $L_A \simeq \upsilon\, [\tau^*(\omega_A)\tau(\omega_A)]^{1/2}$. Moreover, $\omega_A \simeq \omega_0 (t_A/\tau_0)^{-1/5}$. By direct substitution, one finds that

$$t = C' t_b (t_b \tau_0^4/\tau_0^{*5})^{1/9}, \qquad (1)$$

where $t_b = L_A/\upsilon$ (the ballistic arrival time) and C' is another factor, of order unity, which contains all the geometric and other proportionality factors which have so far been neglected. It is assumed that $\tau_0^* < \tau_0 < t_b$. It follows from Eq.(1) that $t_A \propto L^{10/9}$ and $t_A > t_b$.

To summarize, the distinguishing features which characterize quasidiffusional propagation are that t_A is nearly linearly related to L_A and not to L^2 as in pure diffusion. On the other hand, $t_A > t_b$. Moreover, the time-of-flight spectrum has a long tail similar to pure diffusion [7], but the empirically observed diffusion constant increases as L increases.

The experimental conditions and the sample of Al_2O_3:V have been described elsewhere [8]. In the present case 28.8 cm^{-1} light, from a CO_2-laser-pumped CH_3NH_3 laser, was resonantly absorbed to excite V^{4+} ions from their $E_{3/2}$ ground state to the ${}_1E_{1/2}$ first excited state. Since at this frequency the density of one-phonon states greatly exceeds that in the photon field, the most likely deexcitation of the electronic system is the emission of phonons of ω_Q = 28.1 cm^{-1}. A superconducting tin bolometer, 1 x 1 mm^2 in cross section, was evaporated on the doped end of the crystal (see inset, Fig.1) and the center of the focused laser beam was positioned at distances L from the bolometer. In this way the beam passes through the width of the sample which is 7 mm. The full width at half maximum of the laser beam was approximately 1 mm with peak power of about 100 W and a pulse width of 250 ns. The time-of-flight spectrum (arbitrarily normalized to the value at t = 80 μs) as obtained at various fixed values of L is shown in Fig.1. If a diffusion profile [9] is forced to fit the data, then the empirically obtained diffusion constant is found to increase with L as shown in Fig.2, which also contains a plot of the arrival time t_A as determined from the leading edge of each spectrum. The ratio L/t_A is found to be 2.3 x 10^5 cm/s which is some 5 times smaller than the longitudinal and some 3 times smaller than the transverse sound velocity.

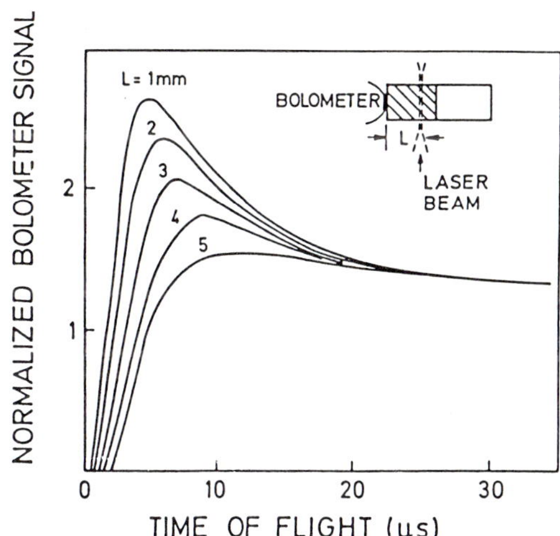

Fig.1: Time-of-flight spectrum for various propagation distances. The hatched area in the inset represents the vanadium-doped part of the crystal

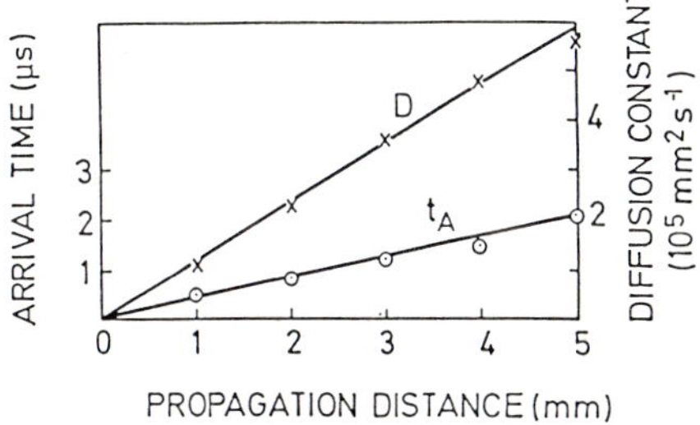

Fig.2: Arrival time and diffusion constant vs. propagation distance

It is seen that the experimental results fit the characteristics of quasidiffusional phonon propagation [10]. An alternate explanation has been suggested by others [11]. These authors suggest the possibility of spectral diffusion, rather than anharmonic decay, as the source of the migration of the ω_0 phonons out of the excitation volume. Such a spectral diffusion process could be the basis for an alternative explanation of the results presented here.

This work was supported in part by the US Army Research Office, Contract No. DAAG29-80-C-0085.

REFERENCES

1. R.J. von Gutfeld, in Physical Acoustics: Principles and Methods, edited by W.P. Mason (Academic, New York, 1969), Vol.5, p. 233.
2. D.V. Kazakovtsev and Y.B. Levinson: Pis'ma Zh. Eksp. Teor. Fiz.27, 194 (1978) [JETP Lett.27, 181 (1978)].
3. D.V. Kazakovtsev and Y.B. Levinson: Phys. Status Solidi (b) 96, 117 (1979).
4. A "quasiballistic" mode has also been predicted. See, e.g., Y.B. Levinson, Mol.Cryst.Liq.Cryst.57, 23 (1980), and references cited therein.
5. See, e.g., R. Orbach and L.A. Vredevoe: Physics (N.Y.)1, 19 (1964).
6. J.W. Tucker and V.W. Rampton: Microwave Ultrasonics in Solid State Phgysics (North-Holland, Amsterdam, 1973).
7. For a more detailed treatment see Ref.3.
8. W.E. Bron and W. Grill: Phys.Rev.Lett.40, 1459 (1978).
9. J.M. O'Connor: Ph.D.thesis, 1977 (unpublished).
10. A more detailed analysis of these results appears in W.E. Bron, Y.B. Levinson, and J.M. O'Connor: Phys.Rev.Lett.49, 209 (1982).
11. K.F. Renk and J. Pekenzell: J. Phys. (Paris), Colloq.33, C4-103 (1972). See also K.F. Renk: in Proceedings of the 1979 Ultrasonic Symposium of the IEEE, edited by J. de Klerk and E.R. McAvoy (IEEE, New York, 1979). Also see A.A. Kaplyanskii, S.A. Basoon, and V.L. Shekhman: J. Phys. (Paris), Colloq.42, C6-439 (1982).

Propagation of Near Zone Boundary Acoustic Phonons in Solid (hcp) ^{4}He

T. Haavasoja, V. Narayanamurti, and M.A. Chin

Bell Laboratories, Solid State Electronics, Research Laboratory
Murray Hill, NJ 07974, USA

The transport properties of large wave vector near zone boundary transverse acoustic phonons in solids has attracted a good deal of experimental attention over the last few years.[1] In order to explain the long energy relaxation times for phonons of energy $\hbar\omega > kT$, LAX, HU and NARAYANAMURTI[2] showed that irrespective of crystal anharmonicity, anisotropy and symmetry a phonon of a given phase velocity cannot decay into phonons of higher phase velocity. The theorem applies to both N and U processes.

Recently[3] we reported on experiments on the propagation of near zone boundary transverse acoustic phonons in solid (hcp) ^{4}He. We showed that even in this highly anharmonic solid, phonons of wavelength 10-20Å could propagate macroscopic distances ~1mm at T ~ 0.1K. In this paper, we summarize these earlier results and present the temperature dependence of the phonon decay which shows quite dramatically the effect of thermal excitations on the propagation of high-frequency phonons.

The experimental conditions have been described earlier.[3] For covering the energy range 0.36 to 1.2 meV, superconducting tunnel junctions of Aℓ, granular Aℓ and Sn were used as phonon generators and detectors. The propagation distance was ~1.1mm. Measurements were made as a function of temperature using a dilution refrigerator.

Figure 1 shows the range of energies achievable with the tunnel junctions used in these measurements. Also shown are the dispersion curves for solid He measured by neutron scattering[4] by the open circles and the solid curves are those calculated theoretically.[5] It is clear that zone boundary phonon transport can be readily probed using superconducting tunnel junctions.

Figure 2 shows the derivative of detector signal ($\frac{dS}{dI_G}$) as a function of generator current (I_G). Here both generator and detector are bulk Aℓ with a gap $2\Delta_{A\ell} \approx 0.38$ meV. Besides the direct recombination peak near 2Δ we see a step-like increase at 4Δ corresponding to propagation of relaxation phonons of energy 0.38 meV. Use of Sn generators and granular Aℓ detectors confirms that propagation up to at least 0.7 meV and probably to 1.1 meV occurs as discussed earlier[3] at T ~ 0.1K.

Figure 3 shows time-of-flight spectra using a granular Aℓ detector whose gap was tuned with magnetic field. Curve (a) corresponds to $2\Delta = 0.46$ meV whilst (b) corresponds to 0.52 meV. The leading edge group velocity is shown as a function of detector gap. The solid line in the figure corresponds to a dispersion curve $\hbar\omega(q) = 1.2 \sin(a_1 q/2)$ (meV) with $a_1 = 2.6$Å, values close to that expected for dispersive wave propagation in our crystal of solid He.[3]

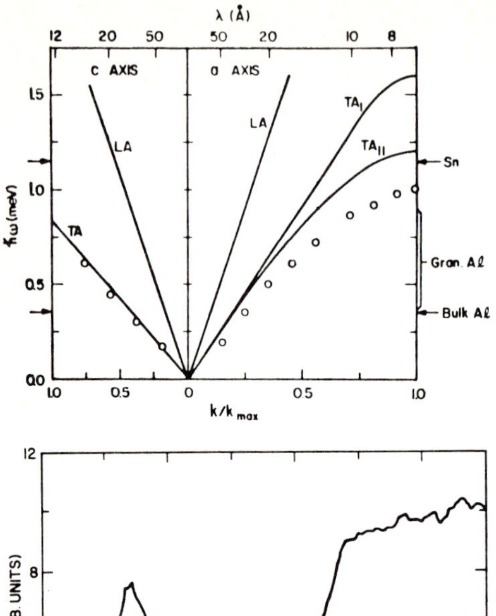

Fig. 1. Dispersion curve for solid ^{4}He. See text

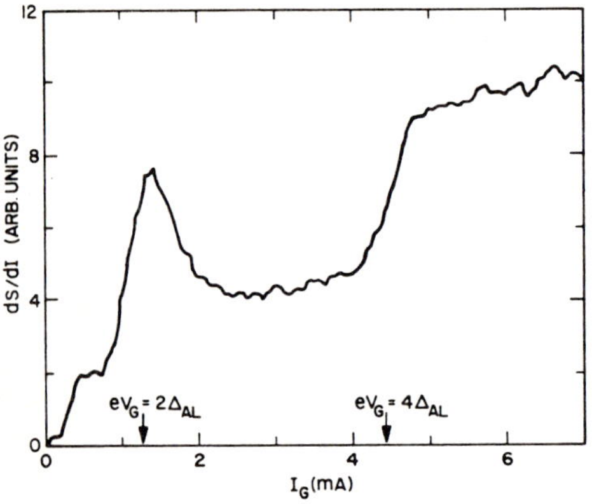

Fig. 2. Derivative of detector signal vs. generator current

The data presented so far were all for temperatures $T \lesssim 0.1K$. It is clear that at these low temperatures nearly lossless transmission of phonons of wavelength 10 to 20Å can occur over macroscopic distances ~1mm. The step-like increase of the signal at the detector gap combined with the measurements of the leading edge velocities provide convincing evidence of large wave vector dispersive TA phonon propagation at T = 0.1K in solid ^{4}He. We now show that this propagation changes markedly as the temperature is raised and attenuation processes set in. Figure 4 shows the heights of the 4Δ step as a function of temperature. Here $2\Delta_{A\ell} \sim 0.38$ meV. Above about 200 mK the step-like signal disappears as strong attenuation and thermalization processes begin to dominate for propagation lengths of 1mm. In fact, the attenuation observed in the solid is considerably stronger than that observed in liquid helium just below the melting pressure by us and previously by DYNES and NARAYANAMURTI.[6] In liquid He complete thermalization occurs only at T ~ 0.45K.

The strong temperature dependence of the attenuation suggests that phonon difference processes involving scattering off thermal phonons can become rapidly important. From the data shown in Figure 4 we estimate the lifetime of 0.38 meV phonons at T = 0.2K to be about 3 μs. From previous second sound measurements[7] involving thermal phonons, the lifetime of 0.2K phonons in solid He is estimated to be about 60 μs. The superthermal phonons studied

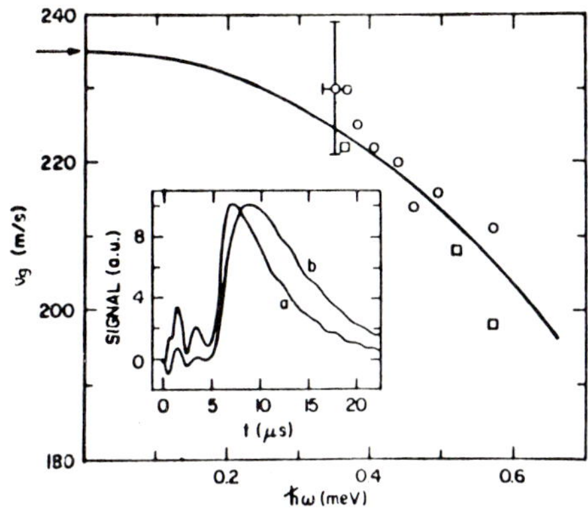

Fig. 3. Group velocity vs. phonon energy. Inset shows time of flight data for two detector gaps

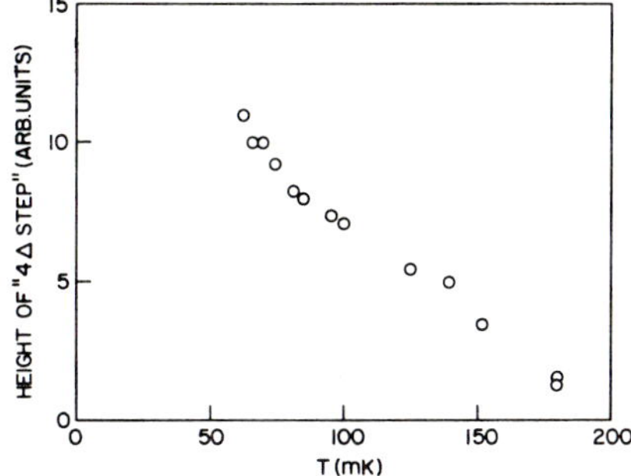

Fig. 4. Temperature dependence of the height of the "4Δ step "

here, however, correspond to a temperature of about 4 [K]. These data then strongly suggest that the tails of the thermal distribution are already sufficient at 0.2K to cause phonon difference processes to affect the propagation of superthermal phonons at only moderately elevated temperatures. Theoretical computation of the phase space for phonon difference processes involving the scattering of zone boundary phonons by lower frequency thermally generated phonons would be of great interest.

In summary, we have shown that near zone boundary phonons in solid He have long lifetimes ($\gtrsim$ 10 μs) provided T $\lesssim$ 0.1K. Scattering off thermal phonons becomes important once the temperature is raised and becomes dominant above about 0.2K. The measurements suggest that the propagation of such high-frequency phonons is fundamentally inaccessible to heat pulse or thermal and conductivity type measurements but may be explored by any narrow-band high-frequency excitation technique at low temperatures.

References

1. R. G. Ulbrich, V. Narayanamurti, and M. A. Chin, Phys. Rev. Lett. 45, 1432 (1980); see also J. Phys. (Paris) Colloq. 42, C6 (1981).
2. M. Lax, P. Hu and V. Narayanamurti, Phys. Rev. B23, 3095 (1981).
3. T. Haavasoja, V. Narayanamurti and M. A. Chin, Phys. Rev. B27, 2767 (1983).
4. V. J. Minkiewicz, T. A. Kichens, F. P. Lipschultz, R. Nathans, and G. Shirane, Phys. Rev. 174, 267 (1968).
5. N. S. Gillis, T. R. Koehler and N. R. Werthamer, Phys. Rev. 175, 1110 (1968).
6. R. C. Dynes and V. Narayanamurti, Phys. Rev. B12, 1720 (1975).
7. V. Narayanamurti and R. C. Dynes, Phys. Rev. B12, 1731 (1975).

Scattering of Debye Phonons by Substitutional Defects in Solids

V.P. Srivastava

Physics Department, University of Zambia, Lusaka, Zambia

We consider a three-dimensional Bravais crystal with volume V containing N atoms such that p lattice sites are occupied by substitutional defects each of mass M', while the remaining (N-p) lattice sites are occupied by identical host atoms of mass M. The introduction of impurities leads to simultaneous changes in mass and force constants between the host atoms and defect atoms. It is assumed that the defects are distributed randomly and their concentration (p/N) is small. The changes in the force constant between the impurity and the host atoms may be assumed to be significant only to nearest neighbors. The Hamiltonian of such a system(defect crystal) in the harmonic approximation can be written as

$$H = \sum_{n\alpha} \frac{p_\alpha^2(n)}{2M} + \frac{1}{2} \sum_{n\alpha} \sum_{n'\beta} \Phi_{\alpha\beta}(n,n')\, u_\alpha(n)\, u_\beta(n')$$

$$+ \sum_{i\alpha} \left(\frac{1}{2M'} - \frac{1}{2M} \right) p_\alpha^2(i) + \frac{1}{2} \sum_{n\alpha} \sum_{n'\beta} \Delta\Phi_{\alpha\beta}(n,n') u_\alpha(n)\, u_\beta(n') \qquad (1)$$

where n denotes the position of an atomic site and i the position of the impurity; $u_\alpha(n)$ and $p_\alpha(n)$ are the α-Cartesian components of the displacement and momentum vectors of the nth atom, M and M' are the masses of the normal and impurity atoms, $\Phi_{\alpha\beta}(n,n')$ and $\Phi'_{\alpha\beta}(n,n')$ are the harmonic force constants for the pure and impure crystal respectively and $\Delta\Phi_{\alpha\beta}(n,n') = \Phi'_{\alpha\beta}(n,n') - \Phi_{\alpha\beta}(n,n')$.

If we define the weighted harmonic mean M_o of the masses of all the atoms in the system by

$$\frac{1}{M_o} = \frac{f}{M'} + \frac{1-f}{M}, \quad f = p/N, \quad \mu = M\,M'/(M'-M) \qquad (2)$$

and express the components $u_\alpha(n)$ and $p_\alpha(n)$ in terms of phonon operators as usual, the Hamiltonian of the system becomes

$$H = H_O + H_D \qquad (3)$$

$$H_O = \sum_\kappa \hbar\omega_\kappa \left(a_\kappa^+ a_\kappa + \frac{1}{2} \right) \qquad (3a)$$

$$H_D = - \sum_{\kappa_1,\kappa_2} C(\kappa_1,\kappa_2)\, B_{\kappa_1} B_{\kappa_2} + \sum_{\kappa_1,\kappa_2} D(\kappa_1,\kappa_2)\, A_{\kappa_1} A_{\kappa_2} \qquad (3b)$$

$C(\kappa_1, \kappa_2)$ in (3b) is given by

$$C(\kappa_1, \kappa_2) = (\hbar\, M_o/4N\mu)\, (\omega_{\kappa_1}\, \omega_{\kappa_2})^{\frac{1}{2}}\, \vec{e}(\kappa_1).\vec{e}(\kappa_2)$$

$$\times \left[\sum_n^N f e^{i(\vec{k}_1+\vec{k}_2).\vec{R}(n)} - \sum_i^p e^{i(\vec{k}_1+\vec{k}_2).\vec{R}(i)} \right] \qquad (3c)$$

$$D(\kappa_1,\kappa_2) = (\hbar/4NM_o)(\omega_{\kappa_1}\omega_{\kappa_2})^{-\frac{1}{2}} \sum_{n\alpha}\sum_{n'\beta} \Delta\Phi_{\alpha\beta}(n,n')\, e_\alpha(\kappa_1)\, e_\beta(\kappa_2)$$

$$\times\, e^{i[\vec{k}_1.\vec{R}(n) + \vec{k}_2.\vec{R}(n')]} \qquad (3d)$$

$$A_\kappa = a_\kappa + a^+_{-\kappa} = A^+_{-\kappa} \; ; \; B_\kappa = a_\kappa - a^+_{-\kappa} = -B^+_{-\kappa} \, . \qquad (3e)$$

H_o is the Hamiltonian of the perfect crystal and H_D is the perturbation part which gives rise to the scattering of phonons. $C(\kappa_1,\kappa_2)$ vanishes when p is either zero or N. For the sake of convenience, here and in what follows, we use one index κ as short for pair of indices $\vec{k}j$.

Now, we define an imaginary time thermodynamic Green's function by

$$G(\kappa ,\kappa',u) = \langle T_d\, a_\kappa(u)\, a^+_{\kappa'}(o)\rangle \qquad (4)$$

and use the perturbation expansion

$$G(\kappa, \kappa';u) = \langle T_d\, \tilde{a}_\kappa(u)\, \tilde{a}^+_{\kappa'}(o) \sum_{m=o}^{\infty} \frac{(-)^m}{m!} \int_o^\beta d\beta_1 \ldots \int_o^\beta d\beta_m$$

$$\times\, \tilde{H}_D(\beta_1)\, \tilde{H}_D(\beta_2) \ldots\ldots \tilde{H}_D(\beta_m)\rangle_{oc} \qquad (5)$$

to evaluate it.

We are interested in obtaining an expression for the inverse relaxation time for the phonons scattered by (3b). This relaxation time is related to the imaginary part of the Self-energy associated with the Green's function (4). As is evident from (5), contribution to the self-energy comes from different order terms in the expansion (5). Zero- and first-order terms do not contain any imaginary parts and so do not contribute to the life time of the phonons. From the second-order term, the inverse relaxation time is found to be

$$\tau^{-1}(\omega_\kappa) = (12V/\pi c^3)[\omega^4 C_I^2(\kappa, \kappa) + D_I^2(\kappa, \kappa)]_A \qquad (6)$$

where the suffix A denotes the directional average of the quantities in the bracket.

Equation (6) gives the inverse relaxation time for phonons scattered by substitutional defects in a cubic Bravais crystal. We note that though the Hamiltonian takes into account the mass changes and force constant changes explicitly and in the second-order self-energy interference terms do appear but their contribution is found to be zero. This implies that mass changes and force constant changes do not show any interference effects in the expression for $\tau^{-1}(\omega)$. KRUMHANSL and MATTHEW [1] have shown that for a linear chain of atoms, atomic mass changes and force constant changes cause reinforcing or cancelling contributions to the relaxation rate according to whether they are of equal or of opposite signs, respectively. We find that their result is not true in three dimensions. Further, if we scale the change in the force constant, then following ELLIOT and TAYLOR [2], we find that $D_I(\kappa, \kappa)$ is proportional to ω^2 and thus τ^{-1} becomes proportional to ω^4.

JOSHI and VERMA [3] have wrongly concluded that KRUMHANSL and MATTHEW's result is true in three dimensions. They expressed τ^{-1} as a sum of two terms τ_1^{-1} and τ_2^{-1} given by (14), (18), and (19) of their paper. It seems that they derived their conclusion from τ_1^{-1} while it should be from $\tau^{-1} = \tau_1^{-1} + \tau_2^{-1}$. In fact, if we estimate τ^{-1} using their expressions, we see that no interference terms appear.

SHARMA and BAHADUR [4] have found that τ^{-1} consists of three types of terms, one proportional to ω^4; the other proportional to ω^2 and the third frequency independent. For this result they treated $D_I(\kappa, \kappa)$ as frequency independent. Their expression gives well-known KLEMENS [5] result for the mass defects case ($\Delta\Phi$ =o) but for the force constant change case ($\Delta\Phi \neq o$), it predicts a frequency-independent scattering contrary to KLEMENS'[5] result. In fact, $D_I(\kappa, \kappa)$ is not frequency independent but varies as ω^2 if the change in the force constant is scaled[2]. Then their expression too reduces to KLEMENS'[5] result.

References

1. J.A.Krumhansl, J.A.D.Matthew:Phys.Rev.A140,1812 (1965)
2. R.J.Elliot, D.W.Taylor: Proc.Roy.Soc.(Lond.),A296, 161(1967)
3. Y.P.Joshi,G.S.Verma:Phys.Rev. B3,501(1971)
4. P.K.Sharma,Rita Bahadur:Phys.Rev.B12,1522 (1975)
5. P.G.Klemens:Proc.Phys.Soc. A68, 1113(1955)

Direct Observation of Ballistic Large-Wavevector Phonon Propagation in Gallium Arsenide

B. Stock, R.G. Ulbrich, and M. Fieseler

Universität Dortmund, Institut für Physik, Postfach 500 500
D-4600 Dortmund 50, Fed. Rep. of Germany

We present phonon time-of-flight measurements which give direct evidence of ballistic dispersive transport of near-zone-edge TA phonons at T=1.4 K in GaAs. Our results show that large wavevector phonons generated in the process of non-radiative e-h pair recombination are able to propagate ballistically over a distance of 1.2 mm along the <1,1,1> direction.

I. INTRODUCTION

Experimental phonon time-of-flight studies in high purity zinc blende semiconductors at low temperatures showed the possibility of ballistic phonon transport in the dispersive regime of the lowest TA branch. Experiments in GaAs {1},InP {1}, and GaP {2} revealed surprisingly long energy <u>and</u> momentum relaxation times of μsec corresponding to mean free paths of several mm for TA phonons with frequencies of 1 ... 2 THz and wavevectors 0.3 ... 0.7 q_{max} in these materials {1,2}. Within these experiments the characteristics of ballistic dispersive transport, i.e. linear scaling in time and space,were obtained by plotting the detector signals as a function of t/r {3}. We present experiments which directly demonstrate ballistic propagation properties of large wavevector phonons at T=1.4 K over a distance of 1.2 mm along the <1,1,1> direction in GaAs. Above band gap photoexcitation of e-h pairs and their non-radiative recombination generates large wavevector TA phonons via decay of optical excitation. For 1.2 mm free propagation distance we found our time-of-flight signals to consist mainly of ballistically propagating phonons with group velocities down to 0.3 v_{TA} which corresponds to phonon frequencies of approximately 1.8 THz. Here v_{TA} is the low-frequency group velocity. Our experimental setup also allowed the seperate observation of the diffusive component in the detector signals. The results are compared to low-frequency phonon propagation generated by Joule heating of a metal film in the same experimental setup.

II. EXPERIMENT

The experiments were performed in a thin GaAs crystal (7x7x1.2 mm^3) held in HeII at 1.4 K. The crystal was cut in the <1,1,1> direction with both sides carefully polished by a combined chemical and mechanical etch. Additionally a small slit was cut into the crystal (1.5 mm deep,0.25 mm wide) parallel to the polished sides. A chemical etch was applied to deepen the slit furthermore (100μm) in order to remove imperfections and damaged crystal areas induced by the mechanical cutting. These defects act

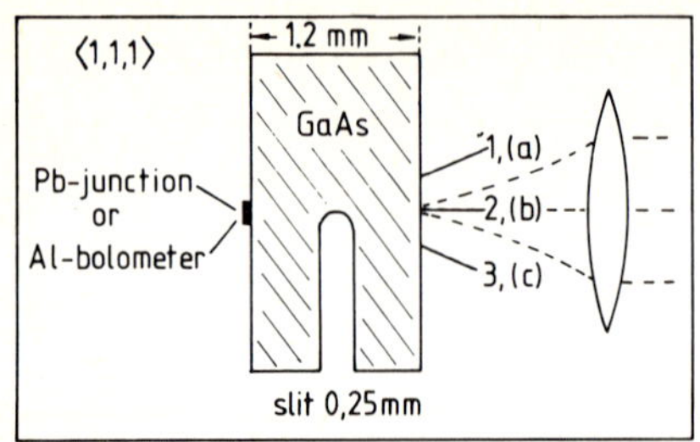

Fig.1: Experimental setup and crystal geometry. Sample thickness is 1.2 mm and phonon propagation occurs along the <1,1,1> direction. Excitation positions on the crystal surface are numbered by 1,2,3; those on the metal film are denoted by (a), (b),and (c)

as scatterers and severely affect the generated phonon distribution by rapid down-conversion of high-frequency phonons. The latter effect was observed in the experiments when chemical etching was not applied. For phonon detection a Pb tunnel junction (0.2x0.2 mm^2) or alternatively a superconducting aluminum bolometer (40x150 μm^2) was evaporated on one side of the crystal. Phonon excitation was done with an argon laser on the opposite side of the sample. Light pulses of 50 nsec duration and a few kilohertz repetition rate were produced by an electro-optical modulator. For a good resolution in time and space the beam was focussed to a spot size of 40 μm and could be moved deliberately on the <1,1,1> surface. In addition one part of this sample side was covered with an evaporated constantan film to generate also low-frequency phonons via direct optical excitation and subsequent down-conversion in the metal film.

III. RESULTS AND DISCUSSION

In our experiments phonon generation is done by pulsed photoexcitation of e-h pairs above the direct gap resulting in a relatively small penetration depth (∿10 μm). They subsequently return to thermal equilibrium through energy exchange with the lattice. Most of the energy goes into multiphonon emission (predominantly LO phonons) involved in the non-radiative capture process into deep impurity and/or defect states. These LO phonons decay quickly into LA phonons {4} which in turn decay into TA phonons via anharmonic processes {4,5}.
Figure 2 shows time-dependent phonon signals in GaAs along the <1,1,1> direction for three different excitation positions (see fig.1) on the crystal surface (left hand) and on the metal film (right hand). For convenience excitation positions on the crystal surface are numbered by 1,2,3 while those on the metal film are denoted by (a),(b),and (c) throughout the text. This sample is 1.2 mm thick and a Pb junction is used for phonon detection. Curve 1,fig.2 shows the detector response for a phonon pulse propagating through the crystal <u>without</u> being influenced by the slit. More detailed phonon time-of-flight studies in this crystal preceeding the actual experiment revealed very similar results to those published by Ulbrich et al.{1,3} and indicated that energy transport is dominated by large wavevector TA phonons. Peak velocity of the signal at 1 is about $1x10^5$cm/sec which is considerable smaller than the low-frequency group velocity of $2.8x10^5$cm/sec and corresponds to a phonon wavevector of 0.6 q_{max}. For comparison curve (a) represents the detector response after photoexcitation of the metal film. The <u>sharp</u> distinct phonon pulse with much narrower width is characteristic for ballistic transport in the region of no dispersion. Peak velocity is now

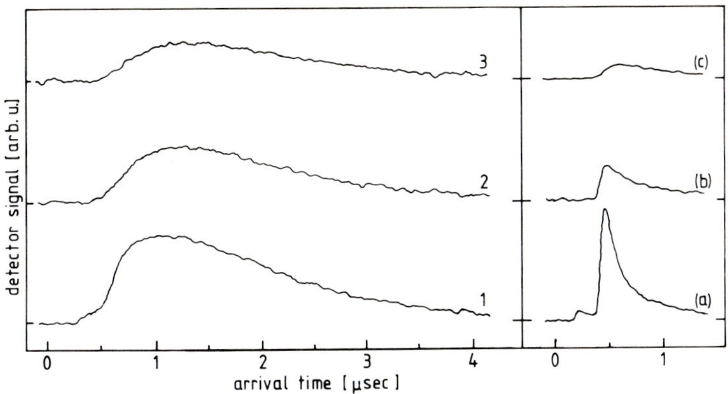

Fig.2: Measured phonon signals after pulsed photoexcitation. Numbers 1,2,3 and letters (a),(b),(c) denote position of the laser spot according to fig.1

2.7×10^5cm/sec and agrees reasonably well with the velocity of sound in this crystal direction.
In the next step the laser spot is moved to position 2,(b) (see fig.1) where the direct path between detector and excitation region is partially interrupted by the slit. The peak of the low-frequency phonon pulse at (b) is now used as a measure of amplitude reduction in the case of purely ballistic propagation. For excitation at position 2 phonon pulses generated in the process of non-radiative e-h pair recombination are obviously less affected by the slit compared to the low-frequency heat pulse at (b). But a direct comparison of signal 1 which contains both ballistically and diffusively propagating phonons with signal 3 where only the diffusive component is present reveals that the major contribution to signal 1 comes from ballistically propagating phonons because the integrated detector signal decreases more than 50%. Energy transport through liquid helium inside the slit can be neglected because measurements with the crystal cooled by helium gas gave the same results.

In summary, we have observed large wavevector phonons with frequencies up to 1.8 THz propagating ballistically a distance of 1.2 mm. The threshold energy of the Pb junction ensured that only phonons with frequencies above 700 GHz contributed to the detector response. Our results have further shown that the phonon time-of-flight signals possess a diffusive component which has to be taken into account if experimental data are analysed for detailed and quantitative information about energy transport by near-zone-edge phonons.

REFERENCES

{1} R.G. Ulbrich, V.Narayanamurti, and M.A. Chin, Phys. Rev. Lett. 45,1432 (1980)
{2} B.Stock and R.G. Ulbrich, Physica 117B-118B, 540 (1983)
{3} R.G. Ulbrich, V. Narayanamurti, and M.A. Chin, J. Phys. (Paris) 42, Colloque C6,226 (1981)
{4} R. Orbach, Phys. Rev. Lett. 16, 15 (1966)
{5} R. Orbach and L.A. Vredevoe, Physics 1, 92 (1964)

Search for Large k-Vector Phonons in GaAs

J.P. Wolfe and G.A. Northrop

Physics Department and Materials Research Laboratory, University of Illinois at Urbana-Champaign, 1110 W. Green Street
Urbana, IL 61801, USA

Photoexcitation of semiconductors has recently been explored as a source of high-frequency phonons. Ulbrich, Narayanamurti, and Chin [1] have shown that the heat-pulse propagation resulting from direct photoexcitation of GaAs is much different from that obtained with a metal film heater. The photo produced heat pulse is much broader than the laser pulse and arrives after the low-frequency time of flight. Because the width and delay of this pulse scaled linearly with path length they attributed it to ballistic propagation of a distribution of large wave-vector slow transverse (ST) phonons at dispersive velocities. We have also observed broad heat pulse signals in photoexcited GaAs, but by making use of additional information provided by phonon imaging [2], we conclude that the increased propagation times are due primarily to bulk scattering, not dispersion.

The samples used in our experiment are high-purity (undoped with $SN_A+N_b < 10^{16}$ from NRL) wafers .2 to 2 mm thick with a small Al bolometer evaporated on one face. They were immersed in liquid He at 2K and a pulsed argon ion laser focused to 30μm produced the heat pulse on the surface opposite the detector. The pulses are repeated at 100 kHz and sampled by a gated integrator. In this geometry scanning the source is equivalent to a fixed source and a mobile detector.

A typical phonon image is shown in Fig. 1a. Spatial scans of phonon flux for several delay times are shown in Fig. 1b corresponding to the white line in the image. At 560ns, the low-frequency transverse time of flight, the intensity shows the large angular anisotropy due to phonon focusing [2]. As the gate delay is increased these sharp features diminish without shifting or broadening and, at twice the ballistic time of flight, they are replaced by a smooth broad background. A comparison of heat pulse shapes for a 1.9mm thick sample and a 250μm sample in Fig. 2a shows that the pulse shape scales linearly with path length. Thus the length of the tail is not a lifetime, but changes with the source-to-detector distance.

The behavior of the phonon focusing pattern indicates that the long tail of the heat pulse is due to scattering in the bulk. If propagation were ballistic but dispersive, the sharp singularities should remain but shift as lower velocities and higher frequencies are sampled [3]. On the other hand, if the phonon transport is diffusive one expects the propagation time to change with the square of the distance and the pulse shape to change dramatically with distance, neither of which occurs.

A model which may resolve the above paradox (scattering but no diffusive scaling) involves diffusive propagation of a spectrum of decaying phonons by a diffusion constant which increases as the phonons decay. Such a model has been proposed by Guseinow and Levinson [4]. Assuming three-

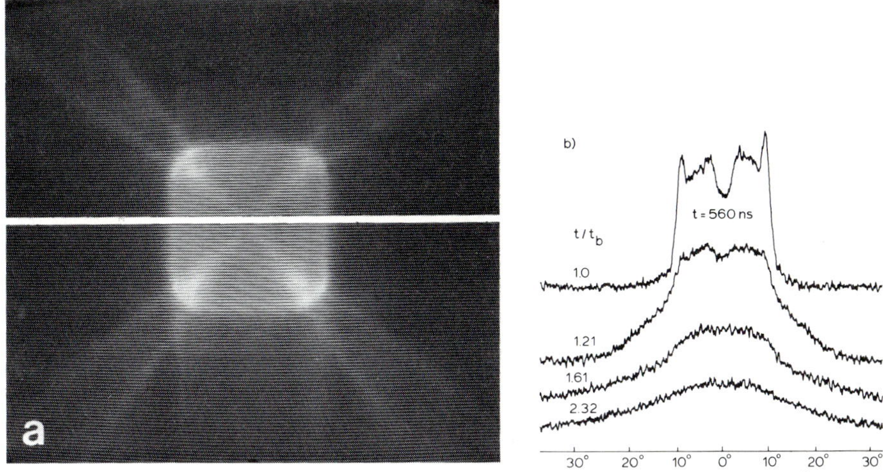

Fig. 1 a) Phonon image for photo excitation of a [100] oriented 1.9mm thick GaAs sample. b) Heat pulse intensity vs. propagation direction for different delay times. The low-frequency ballistic time of flight is t_b

phonon decay with a lifetime $\tau \propto \omega^{-5}$ and elastic scattering with a mean time $\tau^* \propto \omega^{-4}$, and that $\tau^* < \tau$, they showed that the propagation time should scale nearly linearly with distance. A closer correlation of this model to our data requires a calculation of actual heat pulse shapes. We have performed Monte Carlo simulations of phonon decay and propagation for an infinite, isotropic, dispersionless medium, with two acoustic branches (F and S) having a velocity ratio $v_F/v_S = 1.1$. The initial state was a set of phonons of frequency ω_o in the upper branch at the spatial origin. These phonons were allowed to propagate, decaying at a rate $\tau\omega^{-1} = \tau_o^{-1}(\omega/\omega_o)^5$ and scattering elastically and isotropically at a rate $\tau^{*-1} = \beta\tau_o^{-1}(\omega/\omega_o)^4$. The phonons were weighted by frequency and tabulated in two

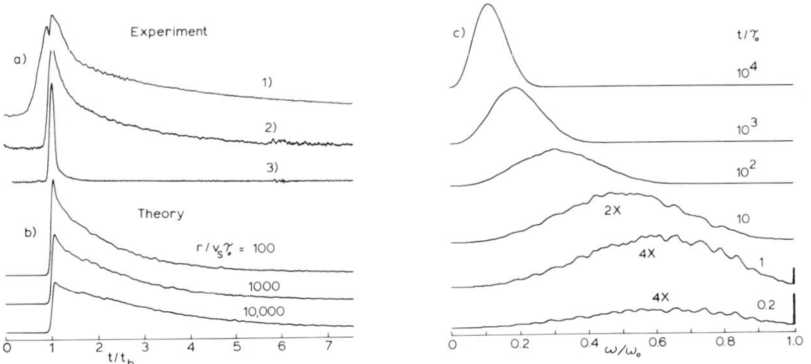

Fig. 2 a) Heat pulses with time scale normalized to t_b: 1) 250 μm sample with photo-excitation, 2) 1.9mm sample with photo excitation, 3) 1.9mm sample with metal film excitation. b) Monte Carlo simulated heat pulses for $\beta = 3$ and various propagation lengths. c) Phonon spectra from the Monte Carlo simulation in b) at various times t/τ_o

forms: as frequency spectra at various times, and as time spectra at various propagation distances with the time scale normalized to distance.

Figure 2c shows the time-varying frequency spectrum for $\beta = 3$ from $t/\tau_o = 10^{-1}$ to 10^4. The result is a broad, roughly Lorentzian distribution with a peak frequency that varies as $t^{-1/5}$. It is clear that with elastic scattering proportional to ω^4 the mean free path may vary by several orders of magnitude between the high- and low-frequency ends of these spectra. Thus a heat pulse composed of a decaying phonon spectrum may exhibit both a ballistic onset and diffuse tail. Time spectra were tabulated from $r = 10^{-1} v_s \tau_o$ to $r = 10^4 v_s \tau_o$ and this was repeated for several values of β. For $r > v_s \tau_o$ the pulse shapes varied only slightly with r, verifying the linear nature of the propagation. For $\beta \leqslant 100$ these pulses exhibit a ballistic-like onset at $t/r = 1/v_s$, and a tail which grows in length and intensity as β increases. For $\beta \geqslant 100$ there is almost no ballistic-like arrival, and the tail becomes a broad pulse lasting much longer than the ballistic time. A set of pulse shapes for $\beta = 3$ is shown in Fig. 2b. These calculated pulse shapes are quite similar to the experimentally observed pulses in Fig. 2a.

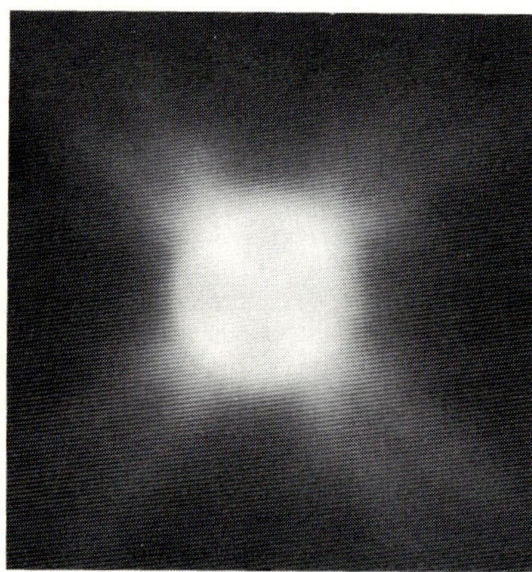

Fig. 3 Phonon image for the 250 μm thick sample showing ballistic propagation of dispersive phonons. The four pairs of V-shaped phonon focusing singularities, known as the FT ridges, have an opening angle 50% wider than in Fig. 1a. The square structure in Fig. 1a, due to the ST mode, has become nearly round at this shorter path length. When compared to calculated phonon focusing images based on a dispersion model, frequencies of 1.3 -1.5 THz for the ST and 0.9 -1.1 THz for the FT are indicated

Thus the above model qualitatively agrees with our observations of the temporal and spatial behavior of a photo produced heat pulse in GaAs over the millimeter distances. The question remains whether at small enough propagation distances, it is possible to observe true ballistic motions of dispersive phonons in this material. By phonon imaging of the 250 μm thick sample, we have obtained strong evidence that the phonon spectrum arriving at the detector extends significantly above 1 THz. Figure 3 shows that the phonon focusing pattern is substantially different from the longer path image of Fig. 1a. Comparison with images generated by a dispersion model imples that the detected phonons have frequencies in the range 1.3 -1.5 THz (ST), and 0.9 to 1.1 THz (FT). Thus at sufficiently short path lengths, a significant fraction of a photo produced heat pulse consists of ballistically propagating dispersive phonons.

References

1. R.G. Ulbrich, V. Narayanamurti, and M.A. Chin, in Proc. Int. Conf. Phonon Physics, edited by W.E. Bron, (J. Phys. (Paris) 42, C6, 226 (1981)).
2. G. A. Northrop and J. P. Wolfe, Phys. Rev. B22, 6196 (1980).
3. W. Dietsche, G. Northrop and J. Wolfe, Phys. Rev. Lett. 47, 660 (1981).
4. N.M. Guseinow and Y.B. Levinson, Sol. St. Comm. 45, 371 (1983); W.E. Bron, Y. Levinson, and J.M. O'Connor, Phys. Rev. Lett. 49, 209 (1982).

On the Propagation of Long-Lived Short Wavelength TA Phonons in GaAs

M. Lax[a)], V. Narayanamurti, and R. Ulbrich

Bell Laboratories, Solid State Electronics, Research Laboratory
Murray Hill, NJ 07974, USA

N. Holzwarth[b)]

Physics Department, City College of Cuny
New York, NY 10031, USA

There has been a great deal of recent interest in the propagation of large wave-vector phonons. Much of the work has been stimulated by the observation by Ulbrich, Narayanamurti and Chin [1] of energy transport in high purity GaAs, via dispersive large wave-vector TA phonons which possess a long lifetime against anharmonic decay [2]. The long energy relaxation time of zone boundary TA phonons has since been confirmed experimentally through Raman scattering [3] and X-ray scattering [4] experiments.

Even though the long *energy* relaxation time is now well established, there remains an open question regarding the nature of the propagation of such phonons because the time-of-flight experiments also imply a long *momentum* relaxation time (isotope scattering) as discussed earlier [1]. In the presence of strong isotope scattering – which is expected for Ga^{69}/Ga^{71} nuclei in GaAs – the propagation should be diffusive over distances of the order of 1 mm (Ref. 1). In addition, the effective energy relaxation time would then be governed through interbranch scattering to shorter lived longitudinal acoustic (LA) phonons which can decay anharmonically [1,2]. Recently Guseinov and Levinson [5] have proposed that under certain conditions the propagation of non-decaying TA phonons could show a linear scaling of mean traveled distance with time provided many frequency down conversions have occurred.

This explanation appears to be inadequate for our experiments in GaAs because it rests on the assumption of many repeated down-conversions. In the actual experiment, however, a detector sensitive only to phonons *above* 0.7 Thz was used and the initial phonon frequencies were around 1.5 Thz on the [111] TA branch. Thus, after more than one down-conversion the phonons would not have been observed in the experiment. Since only zero or at most one (or two) down-conversions can occur, (with the phonons remaining above the frequency threshold of the detector) the diffusion equation approach appears inappropriate. We have therefore embarked on Monte Carlo calculations adequate to handle the initial regime before any appreciable down-conversion occurs. These calculations are not subject to the Guseinov-Levinson assumptions: They are applicable to general times and distances, and valid when an arbitrary (and in particular a small) number of down-conversions have occurred.

To conform with the Guseinov-Levinson model, we have made the non-essential approximation of replacing GaAs by an isotropic solid obtained by averaging $\omega(q)$ over known results for three symmetry directions [q00], [qqq] and [qq0]. The rate of isotope TA→TA, TA→LA and LA→TA scattering was taken from Lax, Narayanamurti, Hu and Weber [6]. For this purpose, the correct density of states, weighted by the fraction of kinetic energy associated with the Ga, was obtained from the (non-isotropic) bond charge model calculations of ref. [6]. There is an additional factor associated with the Ga fractional kinetic energy associated with the initial phonon. For simplicity, this factor has been set to 1/2 (although it can be of order 1/3 for some TA phonons, and actually goes to zero for [100] LA phonons). A plot of this isotope scattering rate $1/\tau$ for the three choices of final state is shown in Fig. 1. The plotted values constitute an overestimate of the isotope scattering rate. The rate of down-conversion was taken from GL. A plot, on the same scale, of the rates of down-conversion LA → LA + TA and LA → TA + TA is also shown in Fig. 1.

At each initial frequency ω_0, phonons are assumed created at a surface point in the crystal sample in a $\delta(t)$ pulse. A large number (e.g. 200,000) of starting TA phonons are tracked for each ω_0. The phonons are

---

a) Also at City College of New York. Work at CCNY supported in part by ARO and DOE.

b) Now at Wake Forest University, Winston-Salem, NC 27109

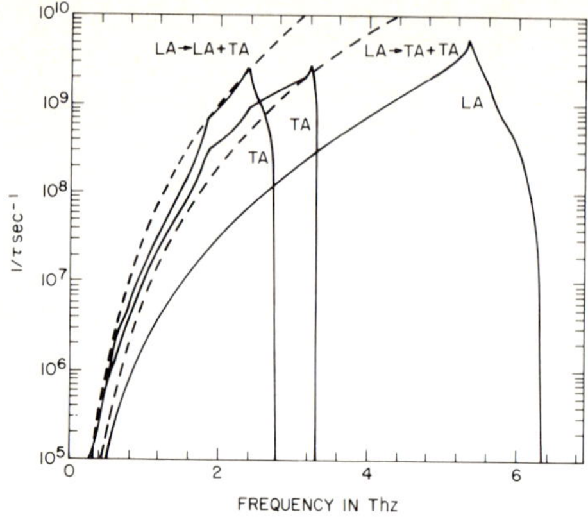

FIG 1. Scattering rate due to isotope scattering (solid curves) separated into contributions by type of final state phonon branch. Rate of down-conversion (dashed curves) based on Guseinov-Levinson isotropic formula

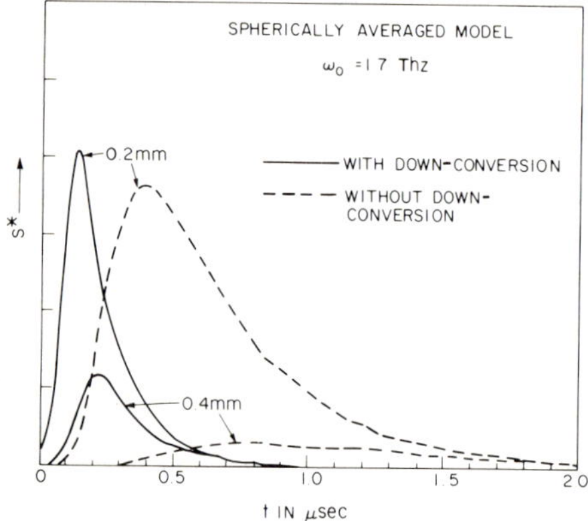

FIG 2. Strength of signal at detector (after renormalizing to eliminate $1/r^2$ effects. Results are shown for an initial frequency of 1.7 Thz at two detector distances 0.2mm and 0.4mm. Results are shown with and without the inclusion of down conversion. The latter case is essentially diffusive

assumed emitted into the crystal with an isotropic distribution. The total probability per unit time for isotope scattering is then used to generate the time for the first collision. The group velocity is used to get the new particle position. The appropriate branching ratio is then applied to determine the probability that the scattered phonon is longitudinal or transverse. The scattering is assumed isotropic. If the scattered phonon is transverse the process is repeated. If the scattered phonon is longitudinal, it is given a direction and tracked to the next event. The latter can be either a down-conversion, or an isotope scattering. If down-conversion occurs, all resulting phonons are tracked.

If a phonon pierces the starting surface or the narrow edge surfaces it is discarded. If it pierces the opposite surface, it is regarded as detected, and we record, among other things, the time of arrival and the "apparent" distance of travel. We have separately recorded all the phonons, and just the phonons above the 0.7 Thz threshold.

A typical plot of detected flux versus time at a given distance is shown in Fig. 2. The results are shown for detectors at several distances, with and without inclusion of down-conversion. The latter case is essentially diffusive. In Fig. 3, we show comparable results (with down-conversion) if one uses an $\omega(q)$ based on the [qqq] spectrum, rather than one overaged over three orientations. We also show results for two input frequencies.

We have also made a plot in Fig. 4 of the time t of the peak versus distance r. An examination of Fig. 4 shows ballistic propagation at 1.2 Thz to 1mm. At 1.5 and 1.7 Thz the phonons start out ballistically, then propagate with a much higher velocity, roughly that appropriate to the group velocity of phonons in the vicinity of the detector threshold. Thus one gets neither the diffusive $\tau \propto r^2$ behavior, nor the apparent ballistic behavior $t \propto r$.

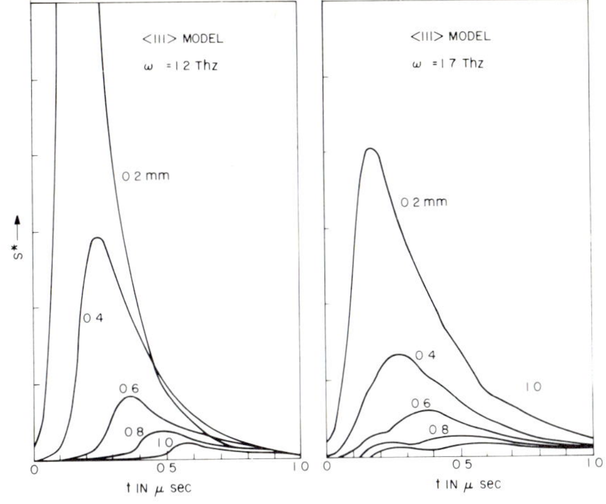

FIG 3.
Renormalized signal strength at two frequencies using group velocities based on the values in the [qqq] direction. Results for different detector distances show how the pulse moves and spreads

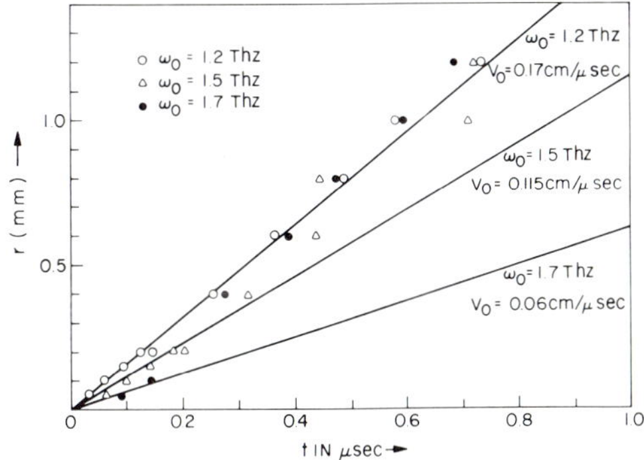

FIG 4. Signal peaks taken from Fig. 3 are plotted versus detector distance and time. The straight lines have a slope equal to the initial group velocity of the starting phonon. At 1.5 Thz this initial velocity is obeyed to about 0.3mm. The later increase in group velocity is due to down-conversion. Average velocities at large distances approximate the group velocity of the down-converted phonon that is detected

We believe these results demonstrate the difficulty of explaining the ballistic velocity of 0.9×10^5 found in the UNC experiments by the Guseinov-Levinson down-conversion mechanism, or indeed by any mechanism if appreciable isotope scattering occurs.

[1]R. G. Ulbrich, V. Narayanamurti and M. A. Chin, Phys. Rev. Lett. 45, 1432 (1980).
[2]M. Lax, P. Hu, and V. Narayanamurti, Phys. Rev. B 23, 109 (1981).
[3]K. T. Tsen, D. A. Abramsohn and R. Bray, Phys. Rev. B 26, 4770 (1982).
[4]D. B. McWhan, P. Hu, M. A. Chin and V. Narayanamurti, Phys. Rev. B 26, 4772 (1982).
[5]N. M. Guseinov and Y. B. Levinson, Sol. St. Comm. 45, 371 (1983); see also D. V. Kazakovtsev and Y. B. Levinson, Phys. Status Solidi (b) 96, 117 (1979).
[6]M. Lax, V. Narayanamurti, P. Hu and W. Weber, Colloque de Physique C6, Tome 42, 161 (1981).

Temperature Dependence of Optical Phonon Lifetimes

J. Kuhl, B.K. Rhee*, and W.E. Bron*

Max-Planck-Institut für Festkörperforschung, Heisenbergstraße 1
D-7000 Stuttgart 80, Fed. Rep. of Germany

We report an application of a picosecond laser system to the measurement of the temperature dependence of the relaxation time of LO phonons in GaP and ZnSe [1]. The results are compared to the traditional (indirect) determination of the relaxation time of both LO and TO phonons from spontaneous Raman linewidths, [2,3] and to the theoretically predicted temperature dependence of the relaxation time.

Two synchronised picosecond lasers are used to produce packets of coherent LO phonons and the dephasing of these phonons is observed by direct measurement of the intensity of a time delayable CARS signal [4]. Figure 1 is an illustration of the typical temporal variation of the CARS signal intensity from LO phonons at an ambient temperature of 5 K for a GaP crystal with a carrier concentration of less than 10^{16} cm^{-3}. The component of the signal from ~ -40 to ~ +20 ps is primarily the result of nonlinear mixing of the laser beams, whose full contribution is shown by the dashed curve. When the system is tuned to the phonon frequency (solid line) the signal beyond 20 ps becomes a measure of the temporal evolution of the coherent population intensity of the excited phonon mode. The intensity is seen from Fig.1 to decay exponentially over several orders of magnitude with a time constant $\tau = T_2/2$. Here T_2 is the traditional dephasing time. The

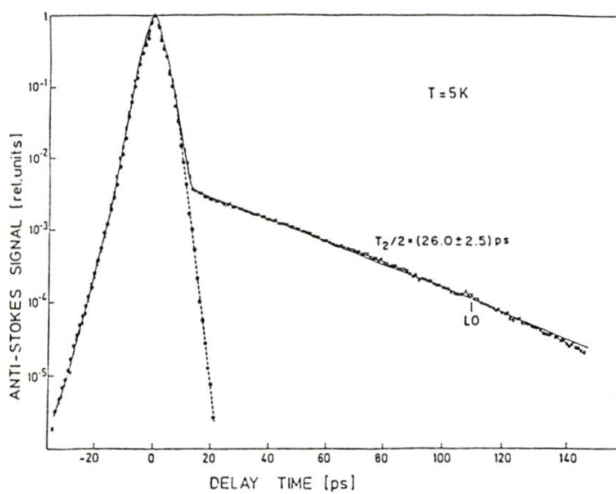

Fig.1: CARS signal for LO phonons in GaP at 5 K as a function of delay time

* Permanent address: Department of Physics, Indiana University, Bloomington, Indiana 47405, U S A.

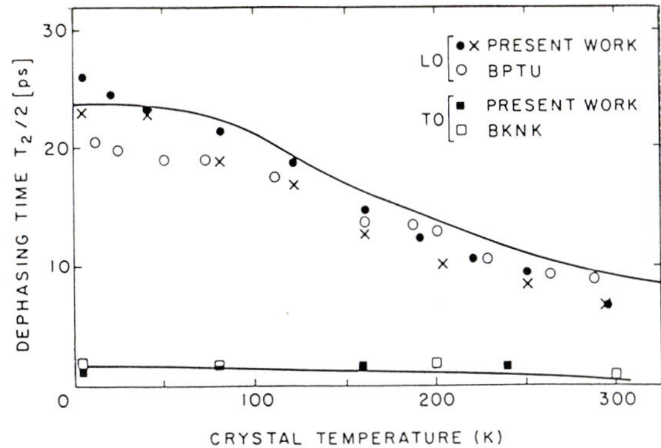

Fig.2: Temperature dependence of LO (upper values) and TO (lower values) phonon dephasing times as obtained, respectively, from CARS and Raman linewidth measurements

temperature dependence of τ is shown in the upper part of Fig.2 (full circles). Note that for GaP the observed dephasing time varies from (26.0 ± 2.5)ps at 5 K to (6.7 ± 0.3)ps at 300 K. For ZnSe the dephasing time is (4.7 ± 0.3)ps at 5 K and (3.6 ± 0,.3)ps at 100 K.

The linedwidth $\delta\nu$ (in cm^{-1}) of the same Raman mode, ω_R, as observed from spontaneous Raman scattering is related to T_2 by $\delta\nu = (\pi c T_2)^{-1}$ provided that the Raman line is homogeneously broadened [5]. Linewidth measurement of the LO phonon (at $\vec{k} \sim 0$) have previously been reported by Bairamov et al. [2,3]. The corresponding decay times are designated as BPTU in Fig.2 (open circles). We have repeated these measurements using our sample and a standard Raman double monochromator spectrometer with a resolution of 0.15 cm^{-1}. Phonon dephasing times were obtained by deconvoluting the instrumental profile from the measured Raman spectrum assuming a Gaussian profile for the spectrometer function and a Lorentzian for the natural Raman line. The results are displayed as crosses in Fig.2.

In contrast to the lifetimes of LO phonons in GaP and ZnSe, the corresponding lifetimes of TO phonons near $\vec{k} \sim 0$ at 300 K as well as at 5 K are found to be somewhat less than the smallest lifetime measureable with the current CARS apparatus. However, the linewidth of the corresponding spontaneous Raman scattering can be readily resolved by a normal double monochromator. Such measurements for GaP have been carried out by us and by Bairamov et al. (BKNK) [3].

In the analysis of these results we concentrate on the case of GaP for which the available data are more extended. The dephasing time T_2 includes contributions from all dephasing phenomena such as elastic scattering from surfaces, from impurities, from imperfections, and from electronic carriers, inhomogeneous line broadening, and also includes phonon-phonon scattering which leads to phonon population decay. The population decay time is conventionally referred to as T_1. It is not possible at this time to calculate the magnitude of T_1, although it is possible to predict theoretically its temperature dependence. We concentrate here on the latter contributions to T_2 arising from depopulation processes. At very low ambient temperatures, such that $\hbar\omega_R \gg kT$, the three-phonon decay of $\vec{k} \sim 0$, optical phonons into two acoustic phonons should dominate all other phonon-phonon interactions [6].

We assume that only spontaneous and thermally stimulated three-phonon decay processes contribute to the temperature dependence. Accordingly, the temperature-dependent part of the expression for the inverse lifetime of the LO phonons is [7]

$$[N(N'+1)(N''+1)] = N[1 + 1/(\exp(x)-1)]^2 \qquad (1)$$

in which the N's are the phonon occupation numbers and $x = \hbar\omega_{LA}/kT$. Expression (1) with its magnitude fitted to the low-temperature data appears as the solid line in the upper part of Fig.2. Comparison with the data shows indeed, for $\hbar\omega_R \gg kT$ ($T \sim 250$ K), that the temperature dependence of T_2 is dominated by three-phonon decay. Beyond this temperature additional processes, including higher order phonon-phonon interactions and scattering from thermally activated carriers, must also be considered. Moreover, the converted linewidth data (open circles and crosses) yield a temperature-dependent dephasing time in fair agreement with that obtained by the CARS technique.

The observed linewidths for TO phonons appear at first glance to be in contradiction to the results for LO phonons, since due to dispersion the joint density of states for $\omega_{LA} = \omega_{TO}/2$ phonons should be less than that for $\omega_{LA} = \omega_{LO}/2$ phonons. Thus a narrower Raman linewidth would be predicted for TO as compared to LO phonons. It has, however, been known for some time [8] that for GaP the TO spontaneous Raman line shape is an asymmetric non-Lorentzian one as a result of the presence of a second decay channel which involves decay to LA and TA zone boundary phonons at the X point. It is, therefore, risky at best to assign a "linewidth" to the asymmetric Raman line of the TO(Γ) phonon and to derive from it a dephasing time. If one nevertheless persists and converts the FWHM linewidths of the spontaneous Raman line to apparent decay times, the values displayed in the lower part of Fig.2 are obtained. Moreover, evaluation of a suitably modified expression (1), for either decay channel, leads to a temperature dependence as shown by the lower solid line in Fig.2. Again, the observed temperature dependence and the predicted one are found to be in satisfactory agreement.

Two of us (W.E.B and B.K.R) acknowledge support through ARO DAAG 29-80-C-0085 and to ARO DAAG 29-83-K-0091.

REFERENCES

1. Preliminary results for GaP were presented at the 1982 Semiconductor Conference, Montpellier, France. See J. Kuhl and W.E. Bron: Physica 117/118B+C, 532 (1983).
2. B.Kh. Bairamov, D.A. Parshin, V.V. Toporov, and Sh.B. Ubaidullav: Sov.Tech.Phys.Lett. 5, 466 (1979).
3. B.Kh. Bairamov, Yu.E. Kitaev, V.K. Negoduiko, and Z.M. Khashkhozhev: Sov. Phys. Solid State 16, 1323 (1975).
4. J. Kuhl and D. von der Linde: Picosecond Phenomena III, Ed. K.B. Eisenthal, R.M. Hochstrasser, W. Kaiser and A. Laubereau (Springer: Berlin) 1982, p. 201-4.
5. A. Laubereau and W. Kaiser: Rev. Mod. Phys. 50, 607 (1978).
6. R. Orbach and L.A. Vredevoe: Physics 1, 91 (1964).
7. See e.g., P.G. Klemens: Phys. Rev. 148, 845 (1966).
8. A.S. Barker, Jr.: Phys. Rev. 165, 917 (1968).

Anharmonic Decay of High-Energy LA Phonons

S. Tamura and K. Okubo

Department of Engineering Science, Hokkaido University
Sapporo 060, Japan

Abstract

We discuss the effects of lattice dispersion upon the anharmonic decay of high-energy LA phonons in Ge. At frequencies higher than 2 THz the density of two-phonon final states increases more rapidly than ν^2, suggesting that the spontaneous decay rate of the LA phonons depends on the frequency more strongly than ν^5. The spatial anisotropy of the decay time is also remarked.

ORBACH and VREDEVOE [1] and KLEMENS [2] showed theoretically that the lifetime of high-energy LA phonons (i.e., $h\nu \gg k_B T$) against anharmonic three-phonon processes decreases with increasing the frequency in proportion to ν^{-5}. Recently, their predictions have been verified experimentally by BAUMGARTNER, ENGELHARDT and RENK [3] for CaF_2 crystals over the frequency range 1.5 to 3 THz. The theories [1,2] are based on the isotropic, continuum approximation and the experiments [3] have been performed with weakly dispersive and quasi-isotropic samples. The recent experiments on the propagation of non-equilibrium phonons attract our attention to dispersive effects on the transport of near-zone-boundary phonons [4]. In this paper we discuss the anharmonic decay of LA phonons in Ge at THz frequencies. It should be noted that TA phonons which may be produced by the LA phonon decay are highly dispersive in Ge characterized by the notable flattening of dispersion curves especially in the [100] and [111] directions. Accordingly, the decay of the near-zone-boundary LA phonons should be affected by the lattice dispersion mainly through the density of final states which comprise the TA phonons.

In a simplified Grüneisen approximation, the spontaneous decay rate of a j-mode phonon with wave vector $\vec{q}$ via three-phonon processes can be written as [5]

$$\tau^{-1}(\vec{q},j) = A\nu^3(\vec{q},j)D_2(\vec{q},j)/N, \qquad (1)$$

where A is a constant determined by the Grüneisen constant (γ), the atomic mass and the phonon velocity, and N is the total number of unit cells in a crystal. The density of two-phonon final states D_2 is given by

$$D_2(\vec{q},j) = \sum_{\vec{q}'\vec{q}''}\sum_{j'j''} \Delta(\vec{q}-\vec{q}'-\vec{q}'')\delta[\nu(\vec{q},j)-\nu(\vec{q}',j')-\nu(\vec{q}'',j'')], \qquad (2)$$

where $\Delta(\vec{Q})=1$ if $\vec{Q}$ is zero or a reciprocal-lattice vector and $\Delta(\vec{Q})=0$ otherwise.

The calculation of D_2 is based on the dynamical matrix constructed very elaborately for Ge. The comparison of theoretical phonon frequencies with the experimental data [6] is shown in Fig.1. In order to calculate D_2 we

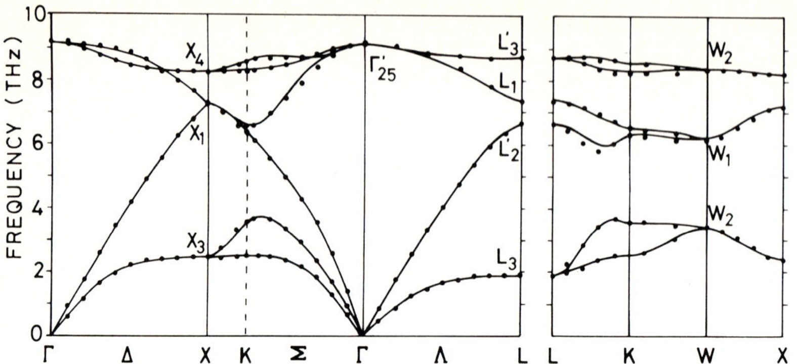

Fig. 1 Dispersion curves of Ge: dotts indicate experimental values [6] and solid lines represent our calculations

have further employed the linear-analytic method devised by GILAT and RAUBENHEIMER [7]. The results for j=LA in the [100] direction are given in Fig.2 together with contributions to D_2 of various combinations of phonons in the final states. (T1 and T2 denote the lower and upper TA branches, respectively.) Also plotted by dotted curve is the two-phonon density of states extrapolated from the low-frequency limit exhibiting the ν^2 behavior. From Fig.2 we see that the introduction of the lattice dispersion enhances D_2 at frequencies higher than 2 THz. We also recognize that the predominant contributions to the final two-phonon states come from those consisting of only TA phonons except at frequencies very close to the zone boundary.

These observations suggest that at frequencies higher than 2 THz the lifetime of the LA phonons becomes much shorter than that predicted by the continuum elasticity theory. In Fig.3 we plot the decay rate versus frequency (heavy line) in the THz frequency region deduced from (1). (We find $A=4.5\times10^{-16}$ sec with $\gamma=2$.) The deviation of the decay rate from the ν^5 behavior is clearly seen above 2 THz. Here it may be of interest to compare

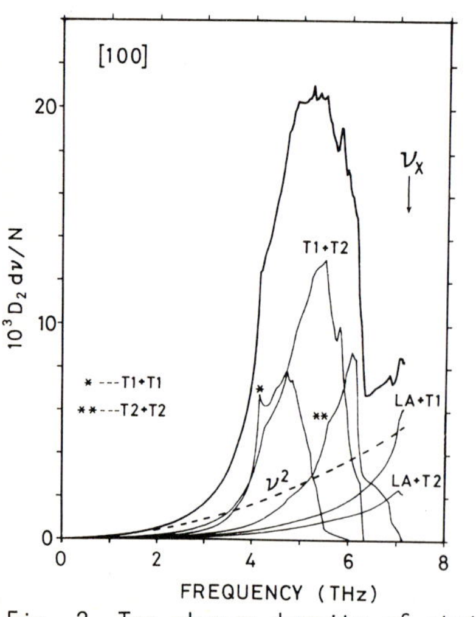

Fig. 2 Two-phonon density of states

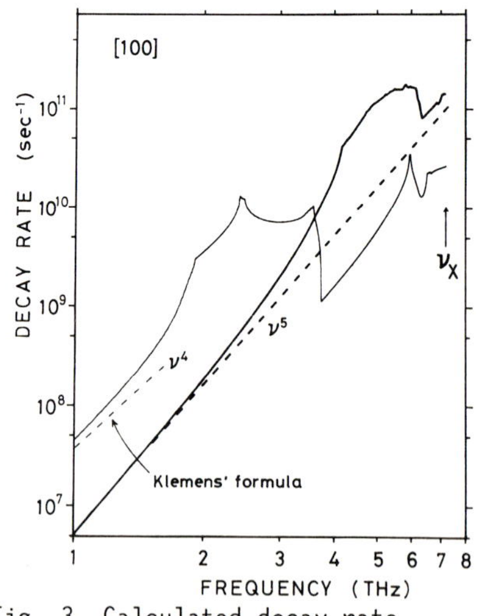

Fig. 3 Calculated decay rate

the magnitude of the spontaneous decay rate with the scattering rate by the isotopic disorder in Ge. The theoretical scattering rate of the phonons by naturally occurring isotopes is also shown in Fig.3 by thin solid line. (Note that the isotope scattering in Ge is isotropic and independent of the polarization of initial phonon. Dispersive effects are discernible at frequencies higher than 0.5 THz [8].) According to these results the anharmonic decay dominates the elastic scattering by the isotopes at frequencies higher than 3.5 THz. This implies that the lifetime of the LA phonons in Ge is governed by the isotope scattering up to the frequencies near the midway of the first Brillouin zone. Incidentally, we note that in the presence of the isotopes the high-energy TA phonons which are stable against the anharmonic decay are mode-converted into the LA phonons. (The highest frequency of the TA modes is 3.66 THz.) Then, in the frequency range where the isotope scattering dominates, the LA phonons are highly populated and the effective time for the down-conversion (which should be observed experimentally) becomes much longer than that shown in Fig.2.

Finally, we briefly remark on the anisotropy of the anharmonic decay. The lifetimes calculated at 1 THz are 2.9×10^{-7}sec, 2.8×10^{-7}sec and 1.7×10^{-7} sec for the [100], [111] and [110] propagations, respectively. At this frequency the dispersion has very small effects on the interaction. Hence, in the calculation we have employed the long-wavelength approximation instead of (1) with recently published values of third-order elastic constants [9]. Thus, in Ge the decay rate depends rather weakly on the propagation direction. These results should be compared with the case of CaF_2, for which TUA and MAHAN [10] have reported the anisotropy of lifetimes of a factor of about five between propagations in the [100] and [111] directions.

To summarize, we have demonstrated the effects of lattice dispersion on the anharmonic LA-phonon decay through the study of two-phonon density of states. The dispersive effects are found to be crucial at frequencies higher than 2 THz where they act to reduce the lifetime of the LA phonons compared with that predicted for dispersionless solids.

References

1. R. Orbach and L. A. Vredevoe: Physics 1, 92 (1964)
2. P. G. Klemens: J. Appl. Phys. 38, 4573 (1967)
3. R. Baumgartner, M. Engelhardt, and K. F. Renk: Phys. Rev. Lett. 47, 1403 (1981)
4. See, e.g. V. Narayanamurti: J. Phys. (Paris) Colloq. 42, C6-221 (1981)
5. K. Okubo and S. Tamura: to be published in Phys. Rev. B (1983)
6. G. Nilsson and G. Nelin: Phys. Rev. B3, 364 (1971)
7. G. Gilat and L. J. Raubenheimer: Phys. Rev. 144, 390 (1966)
8. S. Tamura: Phys. Rev. B27, 858 (1983)
9. J. Philip and M. A. Breazeale: J. Appl. Phys. 54, 752 (1983)
10. P. F. Tua and G. D. Mahan: Phys. Rev. B26, 2208 (1982)

Lifetime and Linewidth of Resonant 40-cm^{-1} Phonons in Alexandrite

R.J.G. Goossens, J.I. Dijkhuis, and H.W. de Wijn

Fysisch Laboratorium, Rijksuniversiteit, P.O. Box 80.000
NL-3508 TA Utrecht, The Netherlands

The lifetime and linewidth of 40-cm^{-1} phonons resonant between the optically excited $\bar{E}(^2E)$ and $2\bar{A}(^2E)$ states of Cr^{3+} in alexandrite ($BeAl_2O_4:Cr^{3+}$; about 600ppm) have been determined by observing the bottlenecking of the phonons via the R_1 and R_2 fluorescent lines under the conditions of both intense cw and pulsed optical excitation into the broad bands and at temperatures where thermalization is absent (1.5 K). $BeAl_2O_4$ is of the orthorombic structure [1], with the Al ions occupying sites with either mirror or inversion symmetry. The active Cr^{3+} ions in the present study substitutionally replace Al at mirror-symmetry sites. The relevant part of the energy-level diagram closely resembles that of ruby. The broad bands occur around 590 nm above the ground state for the $^4T_2$, and 420 nm for the $^4T_1$; at 1.5 K the R_1 fluorescence is at 679 nm and the $2\bar{A}$-$\bar{E}$ splitting is 40 cm^{-1}.

Resonant 40-cm^{-1} phonons are generated by first optically exciting the Cr^{3+} into the $^4T_1$ and $^4T_2$ bands with a focused argon-laser beam followed by fast nonradiative decay to $2\bar{A}$ and the subsequent direct decay $2\bar{A} \rightarrow \bar{E}$ emitting the phonon. When the pumping is sufficiently intense, the 40-cm^{-1} phonons are multiply absorbed and emitted by the metastable Cr^{3+} prior to their escape out of the zone or anharmonic decay. Under these conditions the 40-cm^{-1} phonons are strongly bottlenecked, i.e., they form a hot spike above the thermal background of the phonon spectrum. For their occupation

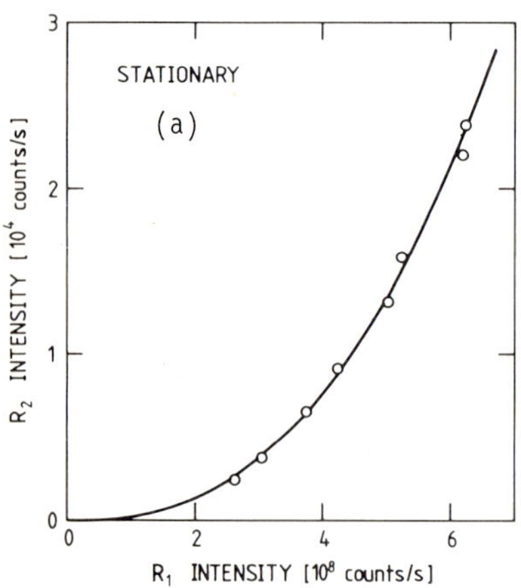

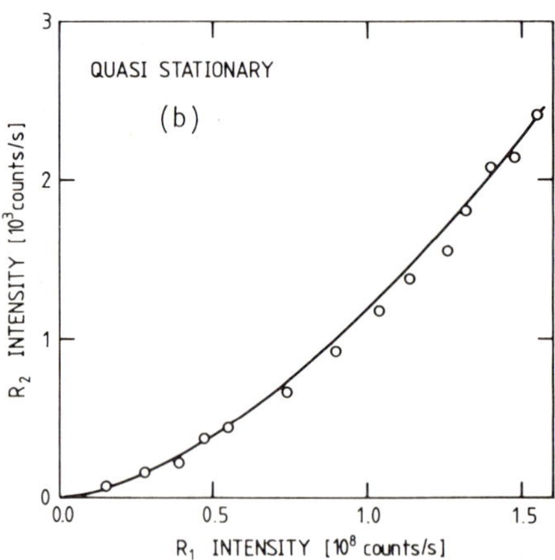

Fig.1 (a) R_2 fluorescent intensity vs the R_1 intensity with stationary optical pumping at 1.5 K. The focused laser beam (diameter 0.1 mm) is along the a axis, the detection along the b axis. (b) Same, but with quasi-stationary pumping

number p we have $p = N_{\overline{2A}}/N_{\bar{E}}$ ($N_{\overline{2A}} << N_{\bar{E}}$), which is directly accessible to experiment by the ratio of the fluorescent intensities R_2/R_1 [2, 3].

In case of stationary pumping, R_2 indeed depends nonlinearly on R_1 (Fig. 1a). We find $R_2 \propto R_1^{2.5}$, which implies $p \propto N^{*1.5}$, or the effective phonon lifetime τ to scale with $N^{*0.5}$ (N^* is the metastable population). Essentially the same result is obtained for lower bottlenecking from quasi-stationary experiments (Fig. 1b) [4]. Here, advantage is taken of the slowness of the typical time associated with the build-up of the metastable population (2 ms) relative to the time constants with which the phonons and excited Cr^{3+} equilibrate (< 0.5 μs). Full optical pumping is applied at t=0 by switching the Ar laser with an acousto-optical modulator, leading to a gradually growing N^*. As distinct from the stationary case, the phonon generation is maximum at all N^* instead of increasing linearly with N^*. We observe $R_2 \propto R_1^{1.5}$, again leading to $\tau \propto N^{*0.5}$.

More direct access to τ is provided by observing the decay of $\overline{2A}$-$\bar{E}$ through the temporal development of the R_2 fluorescence after switching off the optical feeding. From rate equations for $N_{\bar{E}}$, $N_{\overline{2A}}$, and p we have

$$T^{eff}_{\overline{2A}-\bar{E}} = \tau\left(1 + \frac{N^*}{\rho\Delta\nu}\right) , \tag{1}$$

in which $\rho\Delta\nu$ is the number of resonant phonon modes. The experiment (Fig.2) yields $T^{eff}_{\overline{2A}-\bar{E}}$ proportional to N^*.

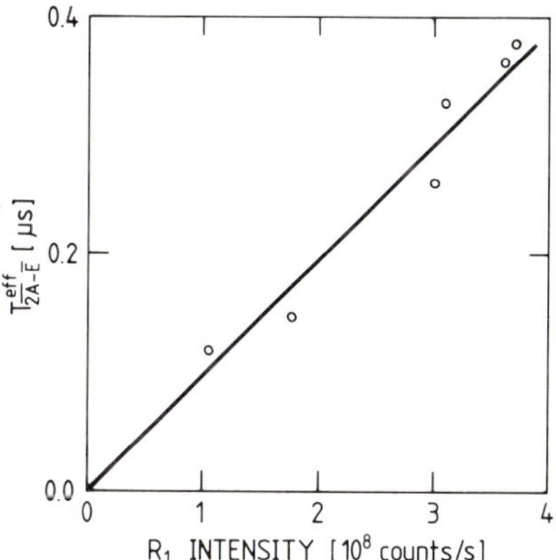

Fig.2 The effective relaxation time of $\overline{2A}$, as measured via the decay of the R_2 fluorescence after switching off the optical pumping vs the R_1 intensity. The experimental conditions are the same as in Fig.1

To arrive at absolute values for τ we need an estimate for the bandwidth of the resonant phonons, $\Delta\nu$, which was obtained by observing the fall of the bottleneck in an external magnetic field parallel to the c axis (Fig. 3) [2]. The field gradually splits the $\overline{2A}$-$\bar{E}$ transition into four components, thereby quadrupling the number of phonon modes involved in the bottleneck, and accordingly reducing the bottleneck. The halfwidth of the fall clearly is a direct gauge for $\Delta\nu$, yielding $\Delta\nu \approx 0.07$ cm^{-1}, substantially larger than in ruby of the same concentration. With a phonon density of states according to Debye, we thus have $\rho\Delta\nu \approx 3 \times 10^{17}$ cm^{-3}. N^* may be tied to the observed R_1 intensity by the onset of R_1 laser action. At the highest laser powers used in Fig. 2, we estimate $N^* \approx 10^{18}$ cm^{-3}, or $N^*/\rho\Delta\nu \approx 3$ and $\tau \approx 0.1$ μs. Adopting $\tau \propto N^{*\frac{1}{2}}$ from the above stationary measurements, the right-hand side of (1) is within errors proportional to N^*, as observed. Both the magnitude of τ and its dependence on N^* point to limitation of τ by spatial diffusion out of the excited zone rather than by pure anharmonic decay or ballistic flight [5].

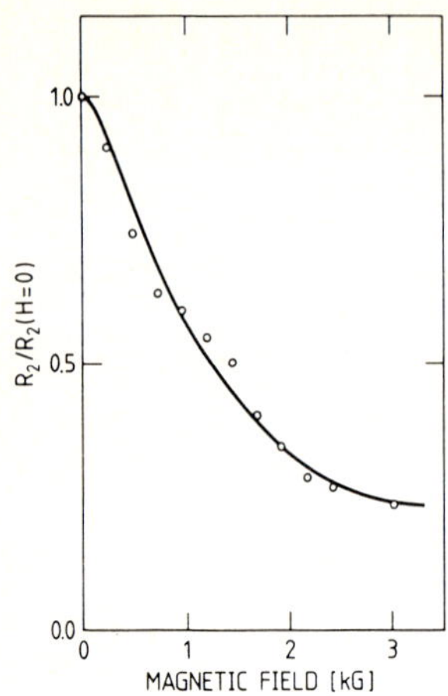

Fig.3 R_2 fluorescent intensity normalized to zero field vs a magnetic field parallel to the c axis at 1.5 K. The excited population N* is about 10^{18} cm^{-3}

1 E.F. Farrell, J.H. Fang, R.E. Newnham: Amer. Mineralogist 48, 804 (1963)
2 J.I. Dijkhuis, A. van der Pol, H.W. de Wijn: Phys. Rev. Lett. 37, 1554 (1976)
3 J.I. Dijkhuis, H.W. de Wijn: Phys. Rev. B 20, 1844 (1979)
4 J.I. Dijkhuis, H.W. de Wijn: Solid State Commun. 31, 39 (1979)
5 J.A. Giordmaine, F.R. Nash: Phys. Rev. 138, A 1510 (1965)

High-Frequency Phonon Dynamics in LaF_3 Using Monoenergetic Optical Detection Methods

R.S. Meltzer, J.E. Rives, and D.J. Sox

Department of Physics and Astronomy, University of Georgia
Athens, GA 30602, USA

G.S. Dixon

Oklahoma State University, Stillwater, OK 74078, USA

Much of the recent work on dynamics of phonons has involved heat pulses generated by resistive heating of metallic thin films or by optical absorption on highly absorbing surfaces. It is well known, but little understood, that weak absorption of light also takes place on the surfaces of nominally transparent solids. We have measured this at 1.06μm and 0.53μm in LaF_3 and have found that it is possible to detect the resulting phonons in the bulk with measurements of excited state population dynamics of Zeeman-split sublevels. The phonon dynamics are interpreted in terms of anharmonic decay, Raman scattering, and elastic scattering from defects.

The surface absorption is measured with low-temperature calorimetry of thin samples of $LaF_3{:}Er^{3+}$ 0.1%. The samples are mechanically polished and mounted on a cold finger in thermal contact with 1.1K liquid helium through a stainless steel tube. The temperature rise on the stainless steel tube near the sample is first measured with the optical beam and then with power applied to an electrical heater near the sample until an equivalent temperature rise is produced. This electrical power is identical to the optical heating. We obtained a fractional absorption of 5×10^{-4} at 1.06μm and 1×10^{-3} at 0.53μm. This is comparable to surface absorption seen at ∿5μm[1] and ∿3μm[2] in IR laser window materials.

The surface-generated heat pulse is observed from excited-state population dynamics of Er^{3+} in a 28kG magnetic field. Ions in a small cylindrical volume immediately behind the laser-heated surface are excited to the lower Zeeman level of $^4S_{3/2}(I)$ with a pulsed tunable dye laser coincident with the heater laser (5ns duration). The geometry is shown in the insert of Fig. 1. The temporal behavior of the fractional population of the upper Zeeman level n_2 is shown in Fig. 1 for several 1.06μm pulse energies. At low energies, n_2 rapidly rises, followed by a decay at the Zeeman direct process decay rate T_D^{-1}=0.16/μs. As the optical heater energy E_H is increased, the peak in the temporal response of n_2 saturates at a value of 0.3. For E_H=5mJ, n_2 remains at this value for up to 15μs before decaying at T_D^{-1}. The temporal behavior of n_2 is independent of detector energy (usually 0.5mJ).

The initial rise of n_2 is much too rapid to be explained by T_D^{-1} and must involve higher frequency phonons through two-phonon Raman scattering involving $^4S_{3/2}(II)$ at an energy Δ=30cm^{-1} above $^4S_{3/2}(I)$ as an intermediate state. This process is indicated in Fig. 2. It shows strong resonant enhancement (Orbach process).

As the detector is moved away from the heater the initial rise in n_2 occurs more slowly as shown in Fig. 3. The time to attain 90% of the peak value is nearly linear in the distance d between heated surface and detector and has the value of 2μs/mm. We attribute this to diffusive propagation of these high-frequency phonons. This diffusion rate implies a mean free path of ≃0.15mm or a total scattering rate T_S^{-1} = 20/μs.

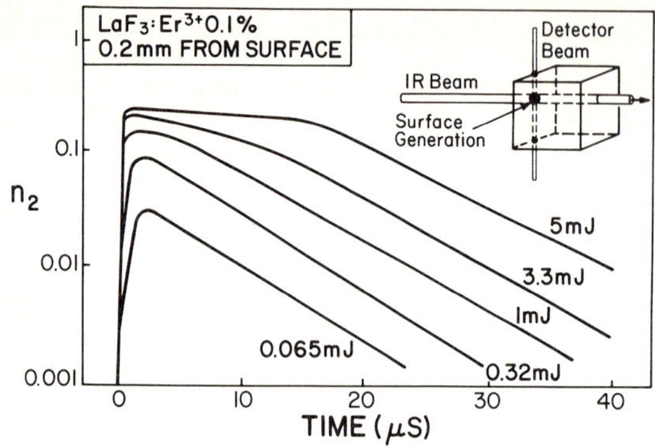

Fig. 1. Temporal dependence of n_2 as a function of optical heater energy E_H for detector 0.2mm from heater

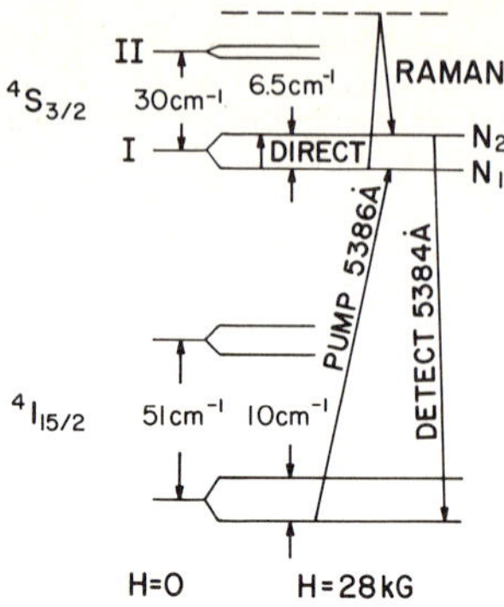

Fig. 2. Low-lying crystal field states and Zeeman splittings at 28kG for Er^{3+}

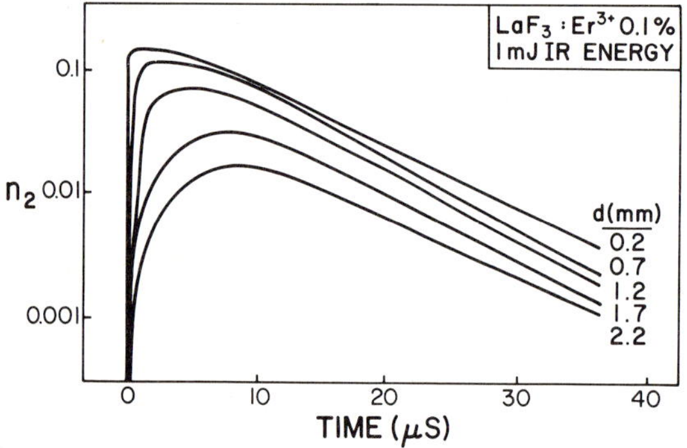

Fig. 3. Temporal dependence of n_2 for a 1mJ heater energy as a function of distance d between heater and detector

The electron-phonon dynamics is governed by three processes: (1) anharmonic decay, (2) two-phonon Raman and (3) elastic defect scattering. WILL's [3] experimental results for the frequency dependence of the anharmonic decay in LaF_3 are summarized in the first column of Table 1. The two-phonon Raman scattering rates can be estimated from $T_1^{-1}(^4I_{15/2})$, the direct process relaxation rate connecting the lowest two crystal field levels separated by $51cm^{-1}$ (see Fig. 2). Homogenous linewidth measurements of fluorescence to the $51cm^{-1}$ state give a rate of $7x10^4/\mu s$ [4]. Raman rates T_R^{-1} are estimated in Table 1 using ORBACH's [5] expression but avoiding the usual assumption that $\hbar\omega \ll \Delta$ since we are interested in the near-resonant phonons.

Table 1. Frequency dependence of estimated phonon scattering rates in $(\mu s)^{-1}$ and phonon mean free paths

$\hbar\omega(cm^{-1})$	T_{AH}^{-1}	$T_R^{-1}(^4I_{15/2})$	T_{DEFECT}^{-1}	λ(mm)
5	0.0004	—	0.02	200
10	0.013	0.0002	0.18	16
20	0.4	0.05	3	0.9
30	3	2	15	0.15
40	13	70	45	0.02
50	40	—	150	—

The anharmonic decay removes the high frequency phonons so that after 100ns only phonons with $\hbar\omega<40\text{cm}^{-1}$ remain. Under these conditions the resonant Raman (Orbach) process with $\hbar\omega=30\text{cm}^{-1}$ dominates the relaxation in the excited state and produces the initial growth of n_2. We therefore associate T_S^{-1} with the 30cm^{-1} phonons. Since this is greater than T_R^{-1}, we attribute the remainder to defect scattering at 30cm^{-1}. We scale the defect scattering rate by ω^4 to complete column 4 based on the total scattering rate. The Raman process dominates the scattering at 40cm^{-1} and is non-negligible at 30cm^{-1}. ENGELHARDT [6] has recently observed Raman scattering of phonons in Al_2O_3:Cr.

The saturation of $n_2=0.3$ can be understood by the dominance of the Orbach process in the rate equation for n_2. A quasi equilibrium is established when $dn_2/dt=0$. This occurs when the terms depopulating the lower and upper Zeeman levels are equal: i.e., $n_1\bar{p}_+ = n_2\bar{p}_-$. $\bar{p}_+$ and $\bar{p}_-$ are occupation numbers for resonantly absorbed phonons from the lower and upper Zeeman levels of $^4S_{3/2}$, respectively. It follows that $n_2 = \bar{p}_+/(\bar{p}_+ + \bar{p}_-)$. At the estimated initial temperature of the detector, shortly after surface generation ($T\simeq15K$), we calculate $n_2=0.35$, close to the observed value of 0.3.

As the electron-phonon system relaxes, n_2 continues to equal $\bar{p}_+/(\bar{p}_+ +\bar{p}_-)$ until the Orbach rates up and down fall below T_D^{-1}. This occurs when $\bar{p}_+T_1^{-1}(^4S_{3/2})=T_D^{-1}$ or $\bar{p}_+ \simeq 10^{-4}$ to 10^{-5} since $T_1^{-1}(^4S_{3/2})=3\times10^4/\mu s$ [7]. For $\bar{p}_+<10^{-5}$ n_2 decays at T_D^{-1}. This is observed in Fig.1 where at E_H=5mJ (2.5μJ absorbed), n_2 remains nearly constant for 15μs and then abruptly decays at T_D^{-1}. As E_H is reduced, the break in slope occurs at earlier times. Since $T_{AH}^{-1}(30\text{cm}^{-1})=3/\mu s$, these long saturation times imply that $\bar{p}_+$ and $\bar{p}_-$ are being fed by inelastic scattering from lower frequency phonons. The Orbach resonant phonon modes must in fact be in good communication with the non-resonant phonons since the major fraction of resonant energy (electron + phonon) is stored in the electronic system, which in the absence of communication, would decay at T_D^{-1}.

Optical surface heating of a transparent material has been observed directly in the phonon system. We have shown that inelastic Raman scattering is a major scattering mechanism for the high-frequency phonons. The dynamics of ions excited to Zeeman-split states is a sensitive monoenergetic probe for resonant Raman (Orbach) phonons. We are presently attempting to model the electron-phonon dynamics in a realistic manner using rate equations for the several phonon systems at different ω. Preliminary results indicate that the essential features described above are corroborated.

This work is supported by the U.S. Army Research Office

1. S.D. Allen: Appl. Opt. 16,2914 (1977)
2. A. Hordvik and L. Skolnik: Appl. Opt. 16,2919 (1977)
3. J.M. Will, W. Eisfeld and K.F. Renk: to be published
4. W.M. Yen, W.C. Scott and P.L. Scott: Phys. Rev. 137,1109 (1965)
5. R. Orbach: Proc. Roy. Soc. (London) A264,458 (1961)
6. M. Engelhardt, U. Happek and K.F. Renk: Phys. Rev. Lett. 50,116 (1983)
7. H.P. Sandner, H. Wolfrum, W. Eisfeld and K.F. Renk: Phys. Rev. 27,79 (1983)

Raman Spin-Lattice Relaxation Induced by Optically Generated Zone-Boundary Phonons in Ruby

J.G.M. van Miltenburg, J.I. Dijkhuis, and H.W. de Wijn

Fysisch Laboratorium, Rijksuniversiteit, P.O. Box 80.000
NL-3508 TA Utrecht, The Netherlands

This investigation is to show that at low temperatures zone-boundary phonons generated in the non-radiative decay following optical pumping of a paramagnetic center induce spin-lattice relaxation when the pumping is sufficiently intense to maintain an appreciable nonthermal phonon occupation. The system chosen is the optically excited Kramers doublet $\bar{E}(^2E)$ of 130 ppm ruby at 1.5 K and a magnetic field of about 1 T applied along the trigonal crystal axis. Under these conditions the relaxation processes connecting the substates of $\bar{E}$, viz., the direct relaxation, the Orbach relaxation via $\overline{2A}$, and Raman processes, are all slow compared to the radiative decay (τ_R = 3.8 ms) [1,2]. The Cr^{3+} ions are excited with stationary pumping into the $^4T_1$ and $^4T_2$ bands using an unfocused argon laser operating at all lines. The populations N_+ and N_- of the upper and lower $\bar{E}$ level, respectively, are monitored through appropriate components of the Zeeman-split R_1 luminescence. The spin-lattice relaxation time may then be extracted from the recovery of the populations following removal of microwave saturation (Fig.1). For an optically excited system another method to determine the relaxation is to measure the departure of the population ratio N_-/N_+ from the value according to the spin memory in the optical feeding [3,4].

The first question to be answered is which phonons cause the optically induced $\bar{E}$ spin-lattice relaxation. In the $\overline{2A}$-$\bar{E}$ decay resonant 29-cm^{-1} phonons are produced, which induce Orbach processes with $\overline{2A}$ as intermediate level.

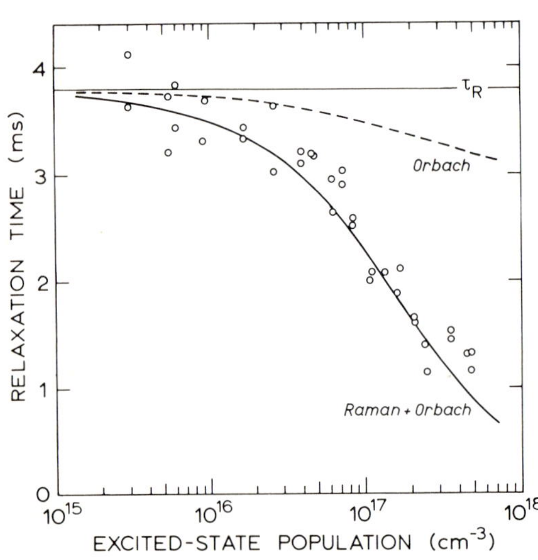

Fig.1 Effective relaxation of $\bar{E}_+$, i.e., the spin-lattice relaxation within $\bar{E}$ in parallel to the radiative decay vs the excited-state population as measured from the recovery following microwave saturation. An acceleration by a factor of over 3 relative to τ_R is observed. The microwave frequency is 32.74 GHz. The maximum laser power is 1.5 W

Calculations based on rate equations unambiguously show, however, that the rate of these processes does not exceed $\phi_{\overline{2A}}\tau_R^{-1}$, with $\phi_{\overline{2A}}$ the fraction of the non-radiative decay to ^2E which passes $\overline{2A}$. With $\phi_{\overline{2A}} = 0.28$ [5], we then find for the Orbach process in parallel to τ_R the dashed curve in Fig.1, which levels off at 0.78 τ_R. The optically induced Orbach process thus is too slow by almost an order of magnitude. A similar calculation for N_-/N_+ yields the same result (Fig.2). Phonons, but nonresonant ones, are also produced in the fast non-radiative decay $^4T_{1,2} \rightarrow {}^2E$. The energy released amounts to about 5500 cm^{-1}, which ultimately has to be taken up by the acoustic phonons. These phonons when sufficiently abundant may induce Raman spin-lattice relaxation within $\bar{E}$. The effect of Raman transitions from $\bar{E}$ to $\overline{2A}$ followed by direct decay to $\bar{E}$ is under the present conditions negligible. Both the transition probability of Raman relaxation and the density of the phonon states, which of course is some measure of the optical feeding, heavily emphasize the phonons near the zone boundary. It turns out that only the part of the phonon spectrum $0.90\ \omega_m < \omega < \omega_m$, with $\hbar\omega_m \approx 200\ cm^{-1}$ the energy of the transverse zone-boundary phonons, is pertinent to Raman relaxation. Only the transverse phonons at the zone boundary are sufficiently long-lived by lack of anharmonic decay mechanisms. Experiments are suggestive for lifetimes of the order of 1 µs [6-8].

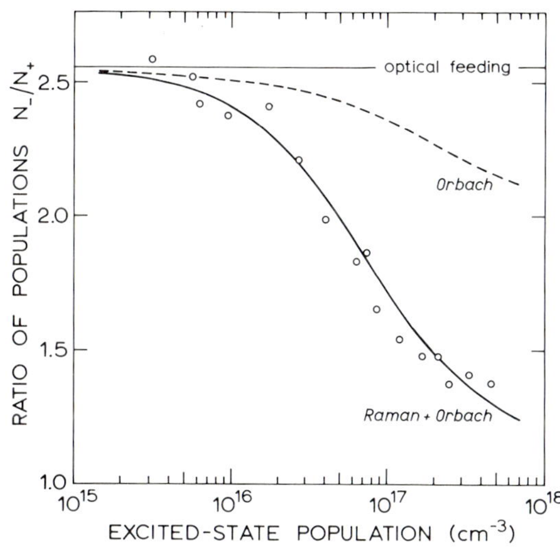

Fig.2 Ratio of the populations N_-/N_+ vs the excited-state population

The phonon occupation number of the near-zone-boundary phonons equals $p = \phi_{ph}\tau$, where ϕ_{ph} is the optical feeding rate per mode and τ the lifetime. A reasonable estimate for ϕ_{ph} for a branch is given by

$$\phi_{ph} = \left(\frac{1}{3N_0} \right) \left(\frac{5500}{200} \right) \left(\frac{N^*}{\tau_R} \right), \tag{1}$$

where 5500/200 represents the number of near-zone-boundary phonons generated in the decay $^4T_{1,2} \rightarrow {}^2E$, $N^* = N_+ + N_-$ is the excited-state population, and $3N_0$ is the number of acoustic phonon modes. With $\tau = 1$ µs and $N^* = 5 \times 10^{17}$ cm^{-3} we find $p = 5 \times 10^{-8}$. This corresponds to a 'temperature' of 17 K, a regime of temperatures where Raman processes have become effective.

For a more quantitative calculation of the optically induced Raman relaxation, we rely on the standard theory for a Kramers doublet, but with the essential modification of the insertion of the above p in place of the thermal one. The level providing the necessary admixing into the Kramers doublet $\bar{E}$ is taken to be $2\bar{A}$. For the dispersion relation we adopt $\omega = \omega_m \sin\frac{1}{2}ka$, which is a fair representation of the true dispersion at high frequencies [9]. We then find for the Raman relaxation rate, accounting for the two transverse polarizations,

$$\frac{1}{T_{Raman}} = \frac{8\omega_m^5}{9\pi\omega_o^6 T_1 T_2} \, p \int \frac{x \arcsin^4 x \arcsin^4(x-x_1)}{(x-x_1)(x^2-x_o^2)(1-x^2)^{\frac{1}{2}}[1-(x-x_1)^2]^{\frac{1}{2}}} dx \,, \tag{2}$$

where x, x_o, and x_1 are the phonon frequency, the $2\bar{A}$-$\bar{E}$ splitting, and the $\bar{E}$ Zeeman splitting in units of ω_m, respectively. T_1 and T_2, the spin-flip and spin-non-flip $2\bar{A} \to \bar{E}$ spontaneous decay times, are measures for the spin-phonon interaction. Limiting the numerical integration to the region near the zone boundary ($0.9 \leq x \leq 1$) we find

$$T^{-1}_{Raman} = 4 \times 10^{-10} \, N^*\tau. \tag{3}$$

To compare this result with experiment, we solve the rate equations for N_+ and N_- including the Raman relaxation, optical feeding and radiative decay for stationary optical pumping. This yields for the relaxation time of an $\bar{E}$ Zeeman state

$$1/T_{eff} = 1/\tau_R + 2/T_{Raman} \,, \tag{4}$$

and for the population ratio

$$N_-/N_+ = (\phi_- + T^{-1}_{Raman}) \,/\, (\phi_+ + T^{-1}_{Raman}) \,, \tag{5}$$

where ϕ_- and ϕ_+ normalized to $\phi_- + \phi_+ = \tau_R^{-1}$ represent the optical feeding into $\bar{E}_-$ and $\bar{E}_+$, respectively. The full drawn curves in Figs. 1 and 2 are fits of (4) and (5) to the data with τ as fitting parameter. Here, a small correction for the optically induced Orbach process has been accounted for. We find good agreement up to the highest N* for both types of experiments for a single value of τ, viz., τ = 3 µs. In consideration of the approximative theory and the uncertainties in the calibration of N*, the output value for τ is of course uncertain by up to an order of magnitude. It is finally noted that the few other experiments available on the lifetime of zone-boundary phonons [6-8] also point to a surprising longevity.

1 S. Geschwind, G.E. Devlin, R.L. Cohen, S.R. Chinn: Phys. Rev. 137, A 1087 (1965)
2 J.I. Dijkhuis, K. Huibregtse, H.W. de Wijn: Phys. Rev. B 20, 1835 (1979)
3 H.W. de Wijn, R. Adde: Solid State Commun. 27, 1285 (1978)
4 J.I. Dijkhuis, H.W. de Wijn: Phys. Rev. B 20, 3615 (1979)
5 J.E. Rives, R.S. Meltzer: Phys. Rev. B 16, 1808 (1977)
6 H. Lengfellner, K.F. Renk: Phys. Rev. Lett. 46, 1210 (1981)
7 P. Hu, V. Narayanamurti, M.A. Chin: Phys. Rev. Lett. 46, 192 (1981)
8 W. Grill, O. Weis: Phys. Rev. Lett. 35, 588 (1975)
9 H. Bialas, O. Weis, H. Wendel: Phys. Lett. 43A, 97 (1973)

Raman Probe of the Brillouin Zone for Nonequilibrium Phonons in GaAs

R. Bray, K.T. Tsen, and K. Wan

Physics Department, Purdue University*, West Lafayette, IN 47907, USA

1. Introduction

Very large wavevector phonons - their properties and methods of generation and detection-- constitute a subject of long standing interest [1]. Of special interest have been the slow TA phonons which may exhibit a bottleneck in their decay at sufficiently low temperatures [2,3]. Of the several techniques developed recently for observing such phonons [2-5], second-order Raman scattering [5] is the most comprehensive probe, uniquely capable of providing simultaneous and correlated information on phonons throughout the Brillouin zone and also on the electrons. We have previously reported on nonequilibrium, slow TA phonons at the zone boundary in GaAs, based on a study of a limited portion of the second-order Raman spectrum [5]. Here, we review and expand on the properties of these phonons, and also present an extended study of the full Raman spectrum which identifies all other nonequilibrium phonons.

Our measurements were carried out with GaAs samples, ~ 3 mm cubes, in a closed cycle cryogenic refrigerator which facilitated very long Raman scans. Weakly absorbed Nd:YAℓG laser radiation was used in a 90°, bulk scattering configuration. The equilibrium phonon spectrum was determined, as a base, at low cw intensity. High intensity, Q-switched operation was used both for the generation of the excess phonons in the path of the focussed beam (150 μm diameter), and with gated detection, for in situ Raman measurements. The nonequilibrium phonons with group velocities too small to escape from the radiation channel during the 100 nsec wide pulses (1000 Hz repetition rate) are preferentially observed. At pulse intensities ~ 3 MW/cm^2, saturable extrinsic excitation of electrons (~ $5 \times 10^{15}/cm^3$) from deep traps was observed [5]; the excess phonon generation is attributed to multiphonon emission in the concurrent recombination processes, and/or in intracenter absorption and relaxation processes [6].

2. Probe and Analysis of the Extended Raman Spectrum

The full Raman spectrum on the Stokes side is shown in Fig. 1, for (a) cw and (b) Q-switched operation, with the intensities normalized to their respective integrated TO lines. The dashed curve in (b) is a repeat of the cw curve. The differences between the two curves establishes where, and by how much, the phonon population is changed. The scattering configuration $[\Gamma_1 + \frac{4}{3}\Gamma_{15}]$ was chosen to permit both overtones and sum and difference frequency combinations to contribute to the spectra, with the overtones strongly dominant. The anti-Stokes side is not shown because it was completely

*Work supported by NSF Grants 79-14618 and 82-17442.

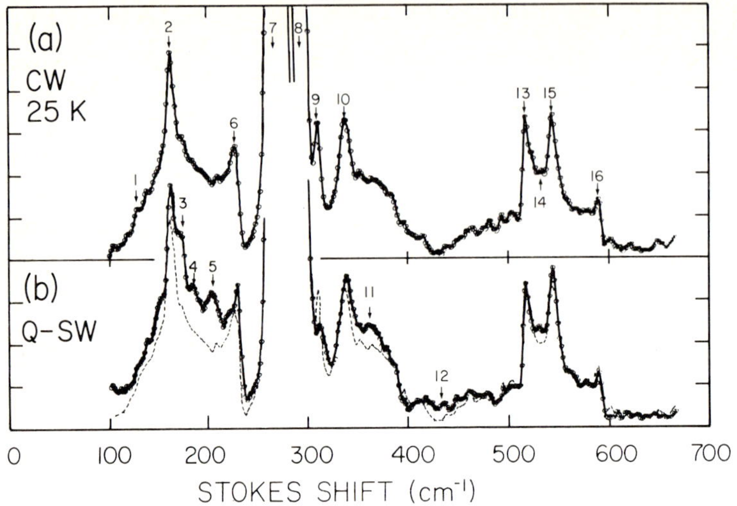

Fig. 1. Comparison of normalized cw and Q-switched Raman spectra for GaAs at ~ 25 [K], demonstrating enhancement of phonon population

frozen out for both laser modes at the low temperature (~ 25 [K] in the irradiated portion of the sample). The data were corrected for fluorescent background and instrumental response, and binomially smoothed.

In the range 0-100 cm^{-1} (not shown), there is a large increase (25×) in electron population. The enhancement in peaks (1) and (2) for the 2TA(L) and 2TA(X,K) overtones, demonstrate an increase in the occupation number, $\Delta n(TA) \sim 0.1$. The sharp peaks (3), (4) and (5) appear uniquely in the Q-switch spectrum; (3) and (5) represent the difference frequency combinations TO-TA, (i.e., TO emission and slow TA phonon absorption), respectively at (X,K) and at (L), while (4) represents the combination of the same TA(K) phonon and another of the optical modes at (K). The phonon occupation factor in these combinations is given by $[1+n(TO)]n(TA)$ and the peaks are attributed solely to the increase in n(TA). These combinations are unobservable in the cw spectrum, at the very small (0.01) equilibrium value of n(TA). The sharp structure of the combination peaks (3-5) is most interesting. It is in marked contrast to the relatively smooth spectrum observed [5] for both the equilibrium phonons (cw spectrum at 300 [K]), and the calculated joint density of states for these difference frequency combinations [7]. This contrast suggests that the sharp combination peaks reflect preferential enhancement of the slow TA population at the high-symmetry points X,K and L. This preferential aspect can be due to specificity of the multiphonon emission process associated with the particular defects in the sample, combined with density of final states considerations [6,8]. Finally, it follows that if indeed the slow TA phonons can be observed to be concentrated at the high-symmetry points, they can not be undergoing substantial momentum randomization, e.g., by isotope scattering, during the 100 ns pulse duration. This gives an isotope scattering time substantially larger than the most recent theoretical estimate (∿ 10 ns) for zone boundary 2 THz phonons.

Further in the spectrum, we find confirmatory evidence of slow TA phonon generation, as well as generation of other phonons. The overtone peak (6) demonstrates an enhancement of the fast TA(K) phonons, comparable to that for the slow TA. Peaks (7) and (8) represent the zone center TO and LO phonons. Their relative strengths are unchanged by Q-switch operation. The absence of these signals in the anti-Stokes spectrum puts an upper limit of $\sim 10^{-4}$ for Δn(LO,TO) at these laser intensities. At (9), we find a peak that can not be ascribed to phonons, and is actually substantially diminished in the Q-switch mode! It would appear therefore to be an electronic Raman contribution, with its decrease related to the depopulation of electrons in traps. Peak (10) corresponds to sum frequency combinations TO+TA, with unresolvable contributions from the X, K and L points. The enhancement here, determined by the factor [1+n(TO)][1+n(TA)], is attributable primarily to Δn(TA). At (11), we find a broad interval which contains several sum frequency combinations of different transverse optical phonons with both slow and fast TA phonons and possibly a 2LA contribution, all near the K point. Again, the enhancement is probably mainly due to the TA phonons.

The first definitive evidence of enhancement of other than TA modes is obtained in the broad regime designated (12); here, from the joint density of states calculations in Ref. [7], we can expect contributions from sum frequency combinations of optical and longitudinal acoustic modes, such as TO+LA(K) and LO+LA(L). However, it is strange that any enhancement should be observed here, since the corresponding cw equilibirum signal is barely observable. At (13), the 2TO(X,K) peak shows no change; this indicates that in previous combinations involving TO(X,K), the latter did not contribute to the enhancement. Final examples of enhancement occur at (14) and (15), where there may be overlapping contributions from optical phonons combinations and 2TO(L) overtones. Finally, for the 2LO peak at (16), which is attributed to iterative zone-center LO phonon scattering, no enhancement is observed.

In summary, we have found confirmatory evidence at many points in the Raman spectrum, for the generation of excess populations of zone-edge, slow and fast TA phonons, and also optical phonons and possibly LA phonons. The increase in population for the zone-edge phonons is much greater than for the zone-center optical phonons. We can conclude that the generation rate and/or the lifetime is much greater for the former than for the latter, whose lifetime is $\sim$ 10 ps. The observation of preferential generation of zone-edge phonons is consistent with the results of electron capture cross-section studies for defects in GaAs [8]. A dominant role is expected for zone-edge TA phonons for vacancy-type defects, particularly for the dominant deep donor EL2 [6]. Raman scattering studies are well suited for probing such effects, and it can be expected that the results will be sample-dependent.

1. W.E. Bron, Rep. Prog. Phys. 43, 303 (1980).
2. R.G. Ulbrich, V. Narayanamurti, and M.A. Chin, Phys. Rev. Lett. 45, 1432 (1980).
3. H. Lengfellner and K.F. Renk, Phys. Rev. Lett. 46, 1210 (1981).
4. D.B. McWhan, P. Hu, M.A. Chin and V. Narayanamurti, Phys. Rev. B26, 4774 (1982).
5. K.T. Tsen, D.A. Abramsohn and Ralph Bray, Phys. Rev. B26, 4770 (1982).
6. M. Kaminska, M. Skowronski, J. Lagowski, J.M. Parsey and H.C. Gatos, Appl. Phys. Lett. 43, 302 (1983).
7. M. Lax, V. Narayanamurti, R. Bray, K.T. Tsen and K. Wan, this issue.
8. D.V. Lang, J. Phys. Soc. Japan 49, Suppl. A, 215 (1980).

Phonon Decay in X-Ray Irradiated Ruby Crystals

M. Engelhardt and K.F. Renk

Institut für Angewandte Physik, Universität Regensburg
D-8400 Regensburg, Fed. Rep. of Germany

Experiments with monochromatically generated phonons in Al_2O_3 are reported. We found that the lifetime of phonons, trapped in a ruby crystal because of resonance scattering at excited Cr^{3+} ions, is strongly reduced after X-ray irradiation of the crystal indicating inelastic scattering at impurity ions generated by the X irradiation. Furthermore, direct observation of inelastic scattering of phonons, trapped in Al_2O_3 because of resonance scattering at V^{4+} ions, is reported.

In this paper we present experimental results which show that inelastic scattering at impurity ions can play an important role for the decay of high-frequency phonons resonantly trapped in optically excited ruby and we report evidence that inelastic scattering is also important for the escape of phonons resonantly trapped in Al_2O_3 containing V^{4+} ions. In our experiments phonons were generated monochromatically by far infrared excitation and detected, also monochromatically, with an optical method. Preliminary results were reported in [1], for a survey on phonon trapping we refer to [2].

For generation of phonons in ruby the separation between the $\bar{E}_-$ and $2\bar{A}_+$ levels of excited Cr^{3+} ions was magnetically tuned to the emission frequency ν_{FIR} (891 GHz) of a far infrared cw HCN laser (Fig. 1). By far infrared absorption and relaxation phonons at two frequencies, ν_{++} and $\nu_{+-} = \nu_{FIR}$, are generated [2]. The concentration N^* ($\simeq 10^{16}$ cm^{-3}) of excited Cr^{3+} ions obtained by continuous optical pumping with radiation of an argon-ion laser was suffi-

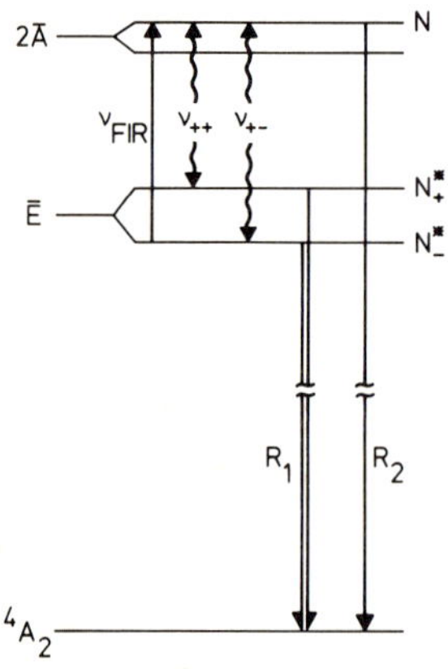

Fig. 1 Monochromatic generation and detection of phonons in Al_2O_3:Cr^{3+}

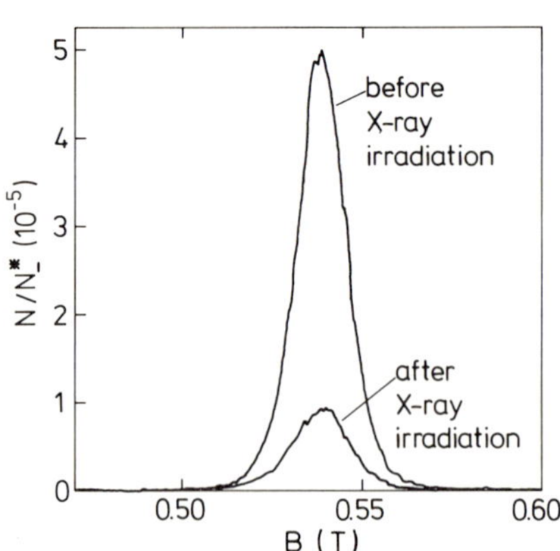

Fig. 2 Phonon-induced $2\bar{A}$ population for a ruby crystal before and after X-ray irradiation

cient for total spatial trapping of both phonon packets in our crystal (0.05% Cr_2O_3; thickness 5 mm), leading to a relative $2\bar{A}_+$ population $N/N^*_\pm$ which was determined from the intensity ratio of R_2 and R_1 fluorescence.

We found that the phonon-induced population is strongly reduced (by a factor of five) after X-ray irradiation of a crystal (Fig. 2). We conclude that inelastic scattering occurred due to the impurity ions generated by the irradiation; after an inelastic scattering process occurring within an average inelastic scattering time τ_{in} a phonon is no longer reabsorbed resonantly by excited Cr^{3+} ions. We guess that, before irradiation, the crystal contained also impurities acting as inelastic scattering centers and therefore we attribute the decay of trapped phonons in the irradiated as well as in the unirradiated crystal to inelastic phonon scattering. Using the relation $N/N^*_- = q\,\tau_{in}/2D(\nu)\Delta\nu_{det}$ where q is the $2\bar{A}$ population rate due to the far infrared absorption, $D(\nu)$ the density of states of phonons at the frequencies $\nu \simeq \nu_{FIR}$ and $\Delta\nu_{det}$ the halfwidth of the $\bar{E} \rightarrow 2\bar{A}$ resonance line we obtain $\tau_{in} \simeq 10^{-6}$s for phonons in the unirradiated crystal and $2 \cdot 10^{-7}$s in the irradiated crystal.

Inelastic phonon scattering was directly observed for phonons in an X-ray irradiated Al_2O_3 crystal that contained originally Cr_2O_3 (0.05%) and V_2O_3 (0.3%). By the X-irradiation V^{4+} ions were obtained. By use of HCN laser radiation phonons were generated monochromatically by $E_{3/2} \rightarrow E_{1/2}$ excitation and subsequent $E_{1/2} \rightarrow E_{3/2}$ relaxation of V^{4+} ions, whereas phonon detection was again performed via excited Cr^{3+} ions (inset of Fig. 3). The detector responds to phonons in a narrow frequency interval at the detector frequency ν_{det} (874 GHz). Since ν_{FIR} and ν_{det} differ (by 17 GHz), our technique allowed for observation of a frequency shift of the monochromatically generated phonons. We obtained a strong detector signal (Fig. 3) and found that it depended strongly on magnetic field strength B for an orientation ($B \perp c$) for which a magnetic field has almost no influence on the generator levels of V^{4+} ions and on the detector levels of Cr^{3+} ions because of very small g values. The observation of a detector signal indicates that inelastic scattering occurred. The magnetic field dependence suggests that impurities, possibly Cr^{2+} ions which are generated by the X-irradiation or V^{3+} ions, that have low-lying energy levels were responsible for inelastic scattering.

We introduce a mean Raman shift ν_R for a single inelastic scattering process and describe the experiment by the rate equations

$$\dot{Z}(\nu) = -Z(\nu)\,[\tau_d^{-1}(\nu) + 2\,Z\,\tau_{in}^{-1}] + [Z(\nu+\nu_R) + Z(\nu-\nu_R)]\tau_{in}^{-1} = 0 \qquad (1)$$

where $Z(\nu)$ is the stationary density of phonons in an interval of width ν_R at frequency ν, $\tau_d(\nu) \simeq N\tau_F^2 D^{-1}(\nu)[1+4(\nu-\nu_o|^2/\Gamma^2]^{-1}$ is the diffusion time for spatial phonon escape from the crystal volume due to resonance scattering at V^{4+} ions, N ($\simeq 10^{19}$ cm^{-3}) the concentration of V^{4+} ions, τ_F ($\simeq 2 \cdot 10^{-7}$s) an average ballistic time-of-flight through the crystal, Γ (= 21 GHz) the halfwidth of the resonance line and ν_o (= 843 GHz) the center frequency of the V^{4+} line. For $\nu = \nu_{FIR}$ a source term has to be added to eq.(1). Assuming that the inelastic scattering time τ_{in} is the same for phonons in the X-irradiated Al_2O_3:(Cr^{3+},V^{4+}) crystal as in the X-irradiated Al_2O_3:Cr^{3+} crystal we have solved the coupled eqs.(1). Taking into account that $N/N^* = Z(\nu_{det})/D(\nu_{det})$ and that the total phonon escape rate is equal to the (estimated) generation rate of phonons at ν_{FIR} we obtained ν_R = 4 GHz for $B = 0$.

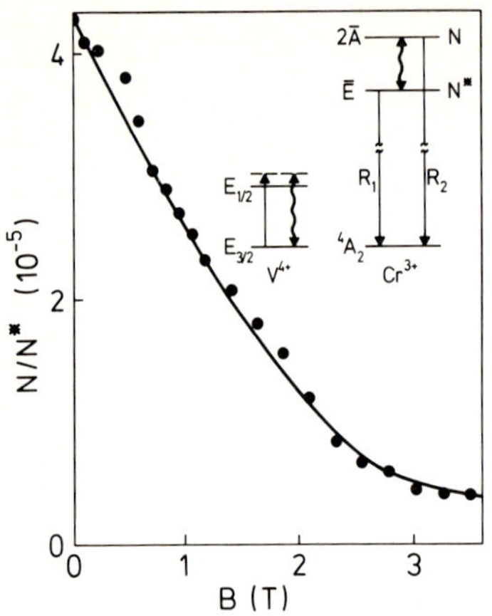

Fig. 3 Phonon-induced $2\bar{A}$ population for an Al_2O_3 (V^{4+}, Cr^{3+}) crystal and principle of phonon generation and detection (inset)

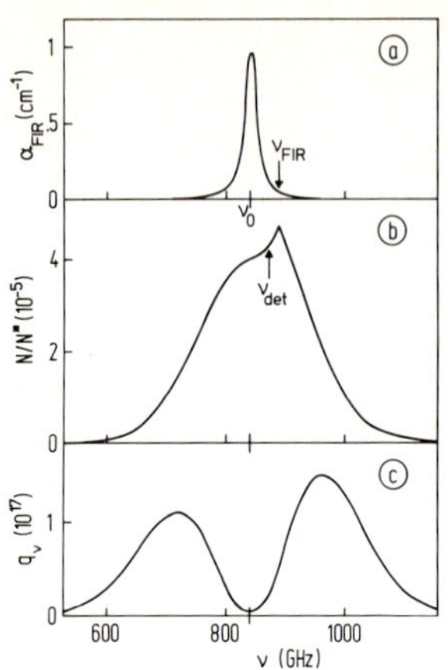

Fig. 4 Phonon trapping in Al_2O_3:V^{4+} (a) resonance line (b) calculated phonon distribution (c) calculated spectral escape rate

Figure 4 shows the $E_{3/2} \rightarrow E_{1/2}$ resonance line, determined by a far infrared experiment, and the spectrum of nonequilibrium phonons, calculated using eq. (1), and also the calculated spectral phonon escape rate $q_\nu(\nu) = Z(\nu)/\tau_d(\nu)$. The analysis shows that phonons are distributed over a range (200 GHz) much larger than the halfwidth of the resonance line and that phonon escape occurs preferentially via the far wings of the resonance line and, furthermore, that the average lifetime of a single phonon against diffusive escape is $\langle\tau_d(\nu)\rangle \simeq 10^{-4}$s. We believe that the dominant process of phonon escape is the spectral-spatial diffusion caused by multiple elastic and multiple inelastic scattering. The decrease of N/N* in a magnetic field (Fig. 3) is most likely due to an increase of the Raman shift ν_R causing a faster spectral-spatial diffusion. We guess that the phonons (near 900 GHz) have an anharmonic lifetime larger than $\langle\tau_d(\nu)\rangle$ in agreement with a recent estimate [1]. We note that "quasi-diffusion" [3] is consistent with spectral-spatial diffusion.

The work was supported by the Deutsche Forschungsgemeinschaft.

1. M. Engelhardt, U. Happek, K.F. Renk: Phys. Rev. Lett. 50, 116 (1983)
2. K.F. Renk: "Far Infrared Volume-Generation and -Detection of Phonons"; this volume
3. W.E. Bron, Y.B. Levinson, J.M. O'Connor: Phys. Rev. Lett. 49, 209 (1982)

Decay of a Highly Excited Phonon Mode

P. Ullersma

Fysisch Laboratorium, Rijksuniversiteit Utrecht, P.O. Box 80.000
NL-3508 TA Utrecht, The Netherlands

A phonon mode, highly excited above thermal equilibrium, will decay to equilibrium by anharmonic interaction with the thermal phonons. This decay may be described by a master equation, which is not exactly solvable because the transition probabilities are nonlinear functions of the occupation number N. Following a method developed by VAN KAMPEN [1] one arrives in a systematic way at an approximate solution. A nonlinear rate equation for the average value $\langle N \rangle$ of N is obtained, and for the stochastic variable $n = N - \langle N \rangle$ a Fokker-Planck equation with time-dependent coefficients results.

Hamiltonian, initial conditions and transition probabilities. We consider a phonon system for which the Hamiltonian $H = H_0 + V$ contains an unperturbed part, $H_0 = \Sigma \hbar\omega_k a_k^+ a_k$, and the lowest order anharmonic approximation of the potential energy, e.g., LEIBFRIED [2],

$$V = \frac{1}{3!v^{1/2}} \sum_{kk'k''} \Phi_{kk'k''} (a_k - a_{-k}^+)(a_{k'} - a_{-k'}^+)(a_{k''} - a_{-k''}^+).$$

v is the volume of the crystal, a_k^+ respectively a_k are creation and annihilation operators for phonons with wavevector k and energy $\hbar\omega_k$, and $\Phi_{kk'k''}$ represent the interaction constants. For convenience polarisation vectors and indices for the phonon branches are omitted. One mode, excited far beyond the equilibrium state, is selected and characterized by its frequency Ω, and its wavevector K. The eigenstates of H_0 will be denoted by $|N;n_1,\ldots\rangle$, where N is the occupation number of the selected Ω mode and n_j that of the mode with frequency ω_j. At an initial time the Ω mode is excited very strongly [$N(0) = N_0$, $N_0 \gg 1$], whereas the rest of the system is in thermal equilibrium at temperature T. The 'temperature' of the Ω mode may be of the order of 100,000 K, e.g., the stimulated emission and decay of phonons resonant between the Zeeman states of $\bar{E}(^2E)$ in ruby as discussed by VAN MILTENBURG et al. [3]; see Fig.1. As we are interested in the decay of the Ω mode the transition probabilities $W_{MN} = W_{N \to M}$ for changes in N have to be calculated. For a given temperature T W_{MN} reads [$\beta = \hbar(k_B T)^{-1}$]

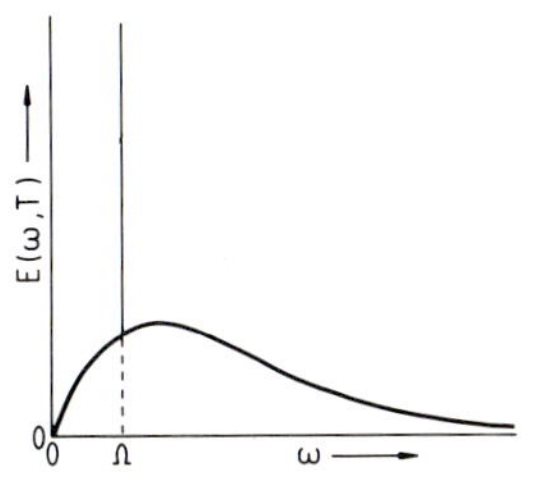

Fig.1 The Planck distribution $E(\omega,T)$ for the phonon modes at temperature T; the mode at Ω is very strongly excited

$$W_{MN} = \frac{\sum\limits_{\{n_j\}}\sum\limits_{\{m_j\}} |\langle m_1,\ldots;M|V|N;n_1,\ldots\rangle|^2 e^{-\beta\Sigma n_j\omega_j}\delta(N\Omega+n_1\omega_1+\ldots-M\Omega-m_1\omega_1-\ldots)}{\hbar \sum\limits_{\{n_j\}} e^{-\beta n_j\omega_j}} \cdot (1)$$

Due to the δ function in (1) only the terms a^+a^+a and $a^+a\, a$ in V will contribute to W_{MN}. Hence, only the processes for which $N \rightleftarrows N \pm 1$ and $N \rightleftarrows N \pm 2$ need to be taken into account. One obtains

$$W_{N \to N-1} = (\alpha_1 + \frac{\alpha_2}{v})N,\quad W_{N \to N-2} = \frac{\gamma}{v} N(N-1),$$

and according to the principle of detailed balance $W_{n-1 \to N} = e^{-\beta\Omega}W_{N \to N-1}$, $W_{N-2 \to N} = e^{-2\beta\Omega}W_{N \to N-2}$. Here α_1, α_2 and γ depend on $\Phi_{kk'k''}$ and T but not on the volume v. They correspond to the various scattering processes, formally described as follows: α_1 : $K + k_1 \rightleftarrows k_2$, $K \rightleftarrows k_1 + k_2$; α_2 : $K/2 + K/2 \rightleftarrows K$; γ : $K + K \rightleftarrows 2K$. In the limit of a continuous spectrum α_1 reads (α_2 and γ will be dealt with elsewhere)

$$\alpha_1 = \frac{1}{3!(2\pi\hbar)^2}\{\int d_3k\Phi^2_{Kk\,-K-k}(e^{\beta\omega_k}-1)^{-1}(e^{-\beta\omega_{K+k}}-1)^{-1}\delta(\Omega+\omega_k-\omega_{K+k})$$

$$+ \tfrac{1}{2}\int d_3k\Phi^2_{Kk\,-K-k}(e^{-\beta\omega_k}-1)^{-1}(e^{-\beta\omega_{K+k}}-1)^{-1}\delta(\Omega-\omega_k-\omega_{K+k})\}.$$

Master equation and its solution. Assuming that the time evolution of the probability distribution $P_N(t)$ for the occupation numbers N may be described by a master equation

$$\frac{dP_N(t)}{dt} = \sum_M \{W_{NM}P_M(t) - W_{MN}P_N(t)\}\,,$$

one obtains

$$\frac{dP_N}{dt} = (\alpha_1 + \frac{\alpha_2}{v})\{(N+1)P_{N+1} + Ne^{-\beta\Omega}P_{N-1} - (N+1)e^{-\beta\Omega}P_N - N\,P_N\}$$

$$+ \frac{\gamma}{v}\{(N+2)(N+1)P_{N+2} + N(N-1)e^{-2\beta\Omega}P_{N-2} - (N+2)(N+1)e^{-2\beta\Omega}P_N - N(N-1)P_N\}. \qquad (2)$$

Because of the nonlinearity of the coefficients of the P_N in the term proportional to γ, (2) is not exactly solvable, neither are the equations for the moments of N, $\langle N(t)\rangle = \sum_N NP_N(t)$, $\langle N^2(t)\rangle = \sum_N N^2P_N(t)$, etc. These equations are not allowed to be solved by linearization around the equilibrium values ($t\to\infty$), $P_N(\infty) = (1 - e^{-\beta\Omega})e^{-N\beta\Omega}$ and $N(\infty) = (e^{\beta\Omega}-1)^{-1}$, because the initial state of the system is too far from the equilibrium state. If one assumes that $P_N(t)$ is a slowly varying function of N the occupation number may be considered as a continuous variable. This enables us to expand the r h s of (2) formally in a power series

$$\frac{\partial P(N,t)}{\partial t} = \sum_{k=1}^{\infty} \frac{(-)^k}{k!}\frac{\partial^k}{\partial N^k}[c_k(N)P(N,t)] \qquad \text{with} \qquad (3)$$

$$c_k(N) = (\alpha_1 + \frac{\alpha_2}{v})\{(-)^kN+(N+1)e^{-\beta\Omega}\} + \frac{2^k\gamma}{v}\{(-)^kN(N-1)+(N+2)(N+1)e^{-2\beta\Omega}\}.$$

This is the Kramers-Moyal expansion. Expansion up to second order yields the Fokker-Planck equation. However, this equation will not be a reliable approximation of (2) because (3) is not a systematic expansion in powers of a small variable. In the Ω-expansion method developed by VAN KAMPEN [1] one first looks for such proper variable. For the underlying system this will be $v^{-1/2}$. Secondly, one anticipates the physical situation that the average of N will be of the order v, and its fluctuations of the order $v^{1/2}$. This leads to the Ansatz $N = v\phi(t) + v^{1/2}x$. Here N is split up into a semi-macroscopic part $v\phi(t)$ and a stochastic part $v^{1/2}x$. x is a stochastic variable, and $\phi(t)$ a time-dependent function which still has to be determined. The distribution function P(N,t) transforms into a function $\Pi(x,t)$ according to $P(N,t) = P[v\phi(t) + v^{1/2}x,t] = v^{-1/2}\Pi(x,t)$. Substitution into the Kramers-Moyal expansion yields

$$\frac{\partial\Pi}{\partial t} - v^{1/2}\phi'(t)\frac{\partial\Pi}{\partial x} = \sum_k \frac{(-)^k}{k!} v^{-k/2} \frac{\partial^k}{\partial x^k} [c_k\{v\phi(t) + v^{1/2}x\}\Pi]. \tag{4}$$

In (4) terms of the order $v^{1/2}$ appear, which would make the expansion an improper one. However, if one chooses ϕ such that

$$\frac{d\phi(t)}{dt} = - \alpha_1(1 - e^{-\beta\Omega})\phi(t)\ \{1 + \frac{2\gamma\phi(t)}{\alpha_1}(1 + e^{-\beta\Omega})\}, \tag{5}$$

these terms, which both contain Π as a factor $\frac{\partial\Pi}{\partial x}$, will cancel one another. Consequently, a proper expansion in powers of $v^{-1/2}$ is left. Equation 5 may be interpreted as a semi-macroscopic rate equation which describes the decay of the excited Ω mode from its initial value N_0 to equilibrium. This equation is nonlinear, and has one stable solution: $\phi(\infty) = 0$. Substitution of the solution of (5) with the initial value $\phi(0) = N_0$ into (4) results in a partial differential equation with time-dependent coefficients for $\Pi(x,t)$, which describes the fluctuations of N around its average $v\phi(t)$. From this resulting equation one derives rate equations for the moments $\langle x\rangle$, $\langle x^2\rangle$, etc., which may be solved systematically by writing these moments as series in powers of $v^{-1/2}$.

1 N.G. Van Kampen: Can. J. Phys. 39, 551 (1961); also in Stochastic Processes in Physics and Chemistry (North-Holland, Amsterdam, 1981)
2 G. Leibfried in Encyclopedia of Physics, VII/I, 104, ed. by S. Flügge (Springer, Berlin, 1955)
3 J.G.M. van Miltenburg, G.J. Jongerden, J.I. Dijkhuis, H.W. de Wijn, these proceedings

Stimulated Emission and Decay of Phonons Resonant Between the Zeeman States of $\bar{E}$ (^2E) in Ruby

J.G.M. van Miltenburg, G.J. Jongerden, J.I. Dijkhuis, and H.W. de Wijn

Fysisch Laboratorium, Rijksuniversiteit,
P.O. Box 80.000, 3508 TA Utrecht, The Netherlands

Stimulated emission of acoustic phonons has been observed following selective excitation of the upper Zeeman-split $\bar{E}(^2E)$ crystal-field Kramers doublet in 700 ppm ruby at 1.5 K. The phonon amplification manifests itself through a substantial acceleration of the direct decay [1] within the inverted $\bar{E}$ system (Fig.1). The spontaneous decay time of the transition is fairly long ($\approx$ 250 μs). Because of the narrow bandwidth (0.002 cm^{-1}), however, such high monochromatic phonon densities could be attained (equivalent to 'temperatures' of $\sim 10^5$ K) that decay processes of the amplified phonons among themselves become the limiting factor in the phonon occupations reached. Bottlenecking of the direct decay under noninverting conditions has earlier been observed [2].

The effect of the stimulated emission was studied by measuring the time evaluation of the populations of the levels following the excitation, as observed through appropriate components of the Zeeman-split R_1 luminescence. The excitation is from the lowest-lying $^4A_2$ state directly to the upper Zeeman level of $\bar{E}(^2E)$ using an excimer-pumped dye laser with 25 ns pulse length and 30 kW peak power (upper part Fig.1). The exciting beam, focused

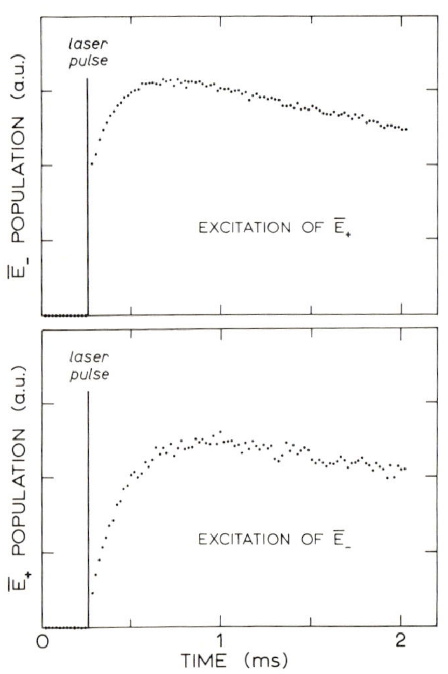

Fig.1 Example of the measured development with time of the populations of the lower ($\bar{E}_-$) and upper ($\bar{E}_+$) Zeeman components of $\bar{E}(^2E)$ after pulsed optical excitation (excimer-pumped dye laser with 25 ns pulse length and 30 kW maximum power) of 700 ppm ruby from the $^4A_2$ ground state to $\bar{E}_+$ and $\bar{E}_-$, respectively, at 1.5 K. The upper graph shows a strong acceleration of the decay $\bar{E}_+ \rightarrow \bar{E}_-$ immediately following the pulse due to stimulated emission of phonons. In the lower graph, for which experimental conditions are the same except for the absence of population inversion, no speed-up is observed, ruling out heating.

to 200 μm, was perpendicular to the c axis, and the detection was at right angles to the beam with a time resolution of up to 1.2 μs. The external magnetic field (5.85 T) was at 77° to the c axis, which compromises between a minimum direct relaxation time (55°) and a maximum transition probability for excitation (90°). To rule out the effects being due to heating $\bar{E}_-$ was also excited instead of $\bar{E}_+$ under otherwise similar conditions (lower part of Fig.1).

Graphs as the top one of Fig.1 for various laser powers have been summarized in Figs. 2 and 3, which emphasize the effects at times short to the normal decay and times of the order of the decay, respectively. In both figures stimulated emission is seen to occur above a threshold at, say, a total excited-state population $N^* = N_+ + N_- = 1.5 \times 10^{16}$ cm^{-3}. To analyze these results we resort to rate equations for the populations $N_\pm$ of $\bar{E}_\pm$, and the occupations p of the resonant phonons. That is,

$$dN_-/dt = [(\varepsilon p+1)N_+ - \varepsilon pN_-]/T_1 - N_-/\tau_R \ , \qquad (1)$$

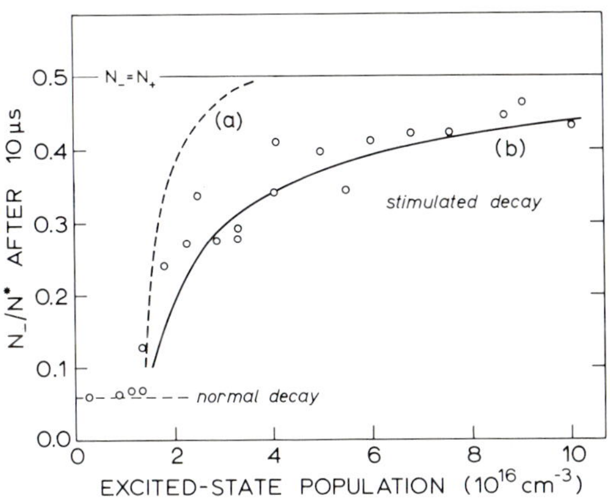

Fig.2 The ratio N_-/N^*, i.e., the fraction of the initial population of $\bar{E}_+$ decayed, vs the excited-state population $N^* = N_+ + N_-$ at 10 μs following excitation of $\bar{E}_+$.

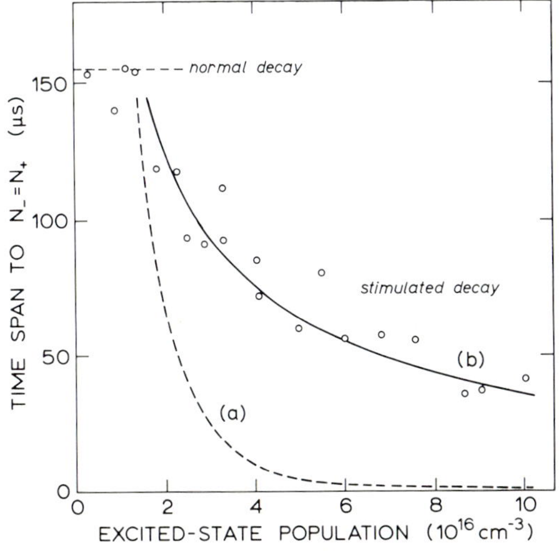

Fig.3 Time span to $N_- = N_+$, i.e., the point where the inversion ceases to exist, vs the excited-state population N^* with excitation of $\bar{E}_+$.

and similarly for N_+, and

$$dp/dt = [(p+1)N_+ - pN_-]/\rho\Delta\nu T_1 - (p-p_o)/\tau_b - p^2/\tau_a \ , \qquad (2)$$

in which T_1 is the spontaneous decay time of $\bar{E}_+ \to \bar{E}_-$, τ_R the radiative lifetime, $\rho\Delta\nu$ the number of resonant phonon modes, and τ_b the phonon lifetime against transport out of the excited volume (cylinder of length 4 mm and diameter 0.2 mm). Anticipating to the discussion below we further inserted a term $-p^2/\tau_a$, describing decay of the hot phonons among themselves. The quantity ε accounts in a heuristic way for the directionality of phonon amplification (for normal decay $\varepsilon = 1$), and is set to $1/4\pi$ times the solid angle spanned by the excited cylinder ($\sim 10^{-3}$). The broken curves in Figs.2 and 3 represent the solution of (1) and (2) but without the term $-p^2/\tau_a$ [model (a)] for reasonable values of the parameters involved [2,3]. In both cases it appears that model (a) overestimates the stimulated emission. The departures, which become more severe with N*, cannot be removed by adjustment of the parameters. Apparently, at high N* an additional loss mechanism becomes operative. Such a mechanism may be represented by $-p^2/\tau_a$. Equations (1) and (2) including this term [model (b)] indeed yield an excellent fit to the data of Figs. 2 and 3 for τ_a = 6 ms. The p's reached in the hot spike are of the order of 10^5.

We finally point out that the proposed mechanism has a physical background, viz. the nearly collinear three-phonon process. This process is indeed favored by the geometry and the high directionality inherent to phonon amplification. The corresponding decay rate of a mode with wave vector $\vec{k}$ may be calculated from [4]

$$\frac{1}{\tau_a} = \frac{2\pi}{\hbar^2} \frac{V}{(2\pi)^3} \int |V_{\vec{k}\,\vec{k}'\,\vec{k}''}|^2 \, \delta_{\vec{k}+\vec{k}',\vec{k}''} \, \frac{\tau}{\pi} \, \frac{1}{1+\Delta\omega^2\tau^2} \, d\vec{k}'d\vec{k}'' \ , \qquad (3)$$

where $|V_{\vec{k}\,\vec{k}'\,\vec{k}''}|^2 = A\,k\,k'\,k''$ in the long-wavelength limit, $\Delta\omega$ is the energy difference, and τ describes the broadening by the spin-phonon interaction. With appropriate values for A, τ, and the dispersion, numerical evaluation of (3) on a computer yielded τ_a = 20 ms, which is of the right order of magnitude.

1 W.J. Brya, P.E. Wagner: Phys. Rev. Lett. 14, 431 (1965)
2 J.I. Dijkhuis, K. Huibregtse, H.W. de Wijn: Phys. Rev. B 20, 1835 (1979)
3 S. Geschwind, G.E. Devlin, R.L. Cohen, S.R. Chinn: Phys. Rev. 137, A 1087 (1965)
4 A.J. Leggett, D. ter Haar: Phys. Rev. 139, A 779 (1965)

Raman Scattering and the Two-Phonon Density of States in GaAs

M. Lax

Physics Dept., City College* of CUNY, New York, NY 10031 and
Bell Laboratories, Murray Hill, New Jersey 07974

V. Narayanamurti and R. C. Fulton

Bell Laboratories, Murray Hill, New Jersey 07974

R. Bray, K. T. Tsen and K. Wan

Purdue University*, W. Lafayette, Indiana 47907

1. Introduction : A knowledge of the vibration spectrum in GaAs is of importance for explaining a variety of phonon transport and scattering experiments. [1] Recently two-phonon Raman Scattering [2] has been used to probe the non-equilibrium phonon population of optically excited GaAs. This work has led to the most precise determination of the phonon spectrum at the critical points of GaAs. It is therefore appropriate to fit the parameters of a lattice vibration model by comparing the computed two-phonon density of states against the experimental Raman data. The comparison is presented here for both the overtones and the sum and difference frequency combinations. Although calculations have been previously presented [3] for overtones in GaAs, we are not aware of any calculations for the combinations. The latter, particularly the difference frequencies, provide the most sensitive and rigorous test of the analysis. Finally, the model can then be used to compute the one-phonon spectrum for comparison with neutron scattering results and for calculating transport properties.

2. Force Constant Models: Standard Born-Von Karman models have been found to be inadequate in the past since they required fifth neighbor forces to explain the neutron data [4]. More recent work by Tamura has used eight neighbor forces with 31 parameters to fit the neutron spectrum of Ge [5]. Long range forces were shown to arise from (electrostatic) quadrupolar interactions by Lax [6], and from ion-shell forces by Cochran [7]. The most physically appealing model, which gives an excellent fit to the neutron determined phonon spectrum is the adiabatic bond charge model applied by Weber [8,9] to Ge and GaAs. In this model, the bond charge is regarded as a point charge of essentially zero mass that resides not halfway between neighboring ions but at an instantaneous position determined by adiabatic response to the neighboring ion positions. The appropriate electrostatic forces then arise automatically.

In the adiabatic bond charge model for GaAs, there are six nearest-neighbor ion-bond charge and ion-ion mechanical forces, plus a bond charge and a charge asymmetry. Although a fit to one percent would require the introduction of second neighbor forces, we have chosen to restrict ourselves to this nearest-neighbor model. The relatively good fit to experiment described below, to Tamura's one phonon density of states in Ge, and to Trommer and Cardona's [3] overtone spectrum, can then be attributed to the essential, physical correctness of the model rather than the multitude of parameters. Since the room-temperature difference frequency data is extremely sensitive to zone boundary phonon frequencies, we have used these to determine our eight parameters. [Elastic constants are not so well reproduced, but these are sensitive to second neighbor forces which have not been introduced.]

---

*Work at City College supported in part by ARO and DOE. Work at Purdue University supported in part by NSF Grant 82-17442.

3. Comparison of Experiment and Theory: The Raman measurements were made at 25 K and 300 K. Only the latter are presented here, since the difference frequency combinations are frozen out at low temperature. A 90 degree bulk scattering configuration was used with 1.06 μm Nd:YAIG laser light, which is weakly absorbed in GaAs. The data were taken with a Spex double grating monochromator, at 2Å intervals with 4Å resolution, corrected for instrumental response, and binomially smoothed. Very long scans with photon counting were necessary to obtain good spectra for the very weak sum and difference frequency combinations.

The comparison of calculated joint density of states and experimental Raman spectra are shown in Figs. 1 and 2 for overtones and combinations, respectively. The calculated curve in Fig. 1(a) is limited to overtones; the experimental data in 1(b) is for the $\Gamma_1 + 4\Gamma_{12}$ spectrum, where the Γ_{12} contribution is negligible and the very strong Γ_1 contribution is dominated by overtones. However, there is a small leakage of the forbidden one-phonon TO and LO peaks, at 268 and 292 cm^{-1}, and also of sum

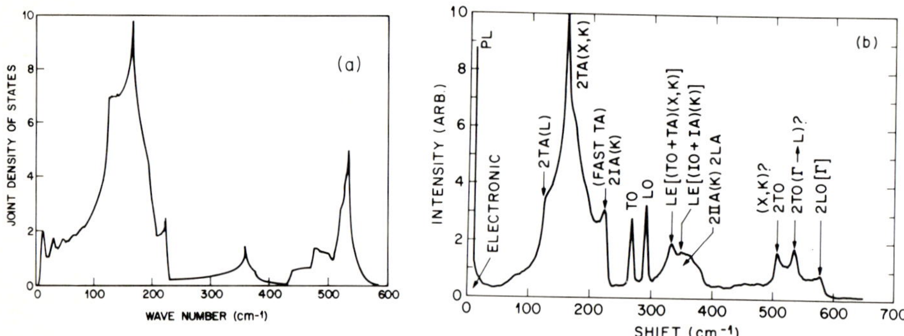

FIGURE 1. (a) Calculated joint density of states for overtones multiplied by thermal factor for Stokes scattering (b) Stokes Raman spectrum for GaAs at room temperature overtones are predominant for the $\Gamma_1 + \Gamma_{12}$ spectrum. The one phonon lines marked LO and TO, and the sum frequency combinations marked LE (leaked into this spectrum) are to be ignored.

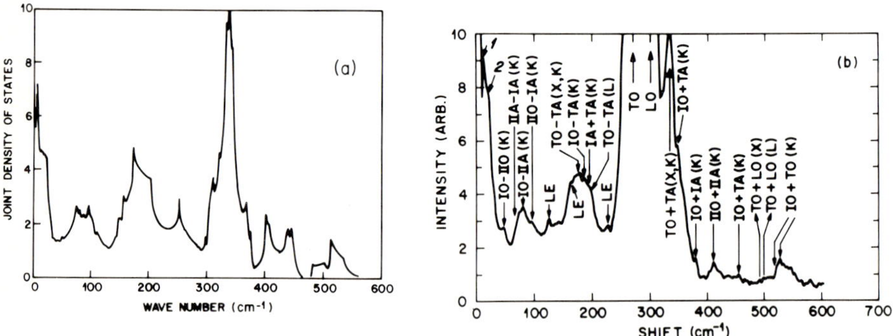

FIGURE 2. (a) Calculated joint density of states for sum and difference frequency combinations (without overtones) multiplied by thermal factors for 300 K (b) Stokes Raman spectrum for the same GaAs sample. Combinations are dominant for this Γ_{15} spectrum. The strong allowed TO and LO lines should be ignored. Peak 1: TO-LO(X); IO-TO(K). Peak 2: TO-LO(L); IIO-IIA(K).

frequency combinations, primarily between 330 and 430 cm^{-1}. The latter obscures the calculated 2LA overtone peak at 360 cm^{-1}. With the leakage contributions discounted, there is a reasonably good fit obtained, both in frequencies of the peaks and general shapes of the structures. Weaknesses in the calculation are demonstrated by the insufficient strength of the 2TA(X,K) peak, and the absence of a well defined 2TO(X,K) peak at 505 cm^{-1}. The oscillations in the calculated curve at very low frequencies is an artifact of the calculation. The absence of the 2LO peak at 588 cm^{-1} in the calculated curve is not a fault of the calculation; it has been shown [3] that the peak does not arise from two-phonon scattering, but rather from iterated one-phonon scattering of the zone center LO phonon.

Details of the shape depend sensitively on how flat the frequencies are over regions of the Brillouin zone. Since this flatness is not easily controlled with only 8 parameters, the parallelism of the gross features indicates a qualitative correctness of the bond charge model. A less severe test is the comparison between the experimental neutron-derived spectrum and the theoretical results obtained using parameters based on the Raman data.

Except for greater sharpness and better resolution, we have had no new features to report for the overtone spectrum compared to previous back scattering measurements with visible light in the opaque regime [3]. However, we find significant new details for the combination spectrum, particularly for the difference frequency structure. The calculated curve for the combinations is shown in Fig. 2(a). It is to be compared with the Γ_{15} experimental spectrum in 2(b), which contains primarily combinations, plus the allowed one-phonon TO and LO peaks which tower over the two-phonon spectrum, and obscure the calculated combination peak near 250 cm^{-1}. There is a noticeable, but remarkably small, leakage of the 2TA overtone spectrum between 120 and 230 cm^{-1}. The combination spectrum is particularly rich in structure. The detailed spectrum, particularly below 200 cm^{-1}, has not previously been presented. The interesting new features include the sharp peak and shoulder at 12 and 22 cm^{-1} in the wing of the Rayleigh line. (The peak is particularly distinct because the parasitic scattering is especially weak in this sample, which was selected for its minimal scattering of phonons in heat pulse measurements). The computed curve reproduces this structure particularly well, as it also does the difference frequencies in the broad structure centered around 80 cm^{-1} and in the band between 170 and 200 cm^{-1}. The sum frequency band near 340 cm^{-1} is also well reproduced, however the structure above 400 cm^{-1} appears much stronger than is warranted by the experimental data.

Since there are, in fact, no highly localized charges, why does the bond charge model work as well as it does? (1) Because the long-range electrostatic forces are similar whether the charge is strongly or only partially localized, and (2) because the forces between ions and bonds automatically give rise to exponentially decaying forces of moderate range - as in the shell model. [10]

[1]R. G. Ulbrich, V. Narayanamurti and M. A. Chin, Phys. Rev. Lett. **45**, 1432 (1980).
[2]K. T. Tsen, D. A. Abramsohn and R. Bray, Phys. Rev. **B26**, 4770 (1982).
[3]R. Trommer and M. Cardona, Phys. Rev. **B17**, 1865 (1978).
[4]F. Herman, J. Phys. Chem. Solids **8**, 405 (1959).
[5]S. Tamura, Phys. Rev. **B28**, 897 (1983).
[6]M. Lax, Phys. Rev. Lett. **1**, 133 (1958).
[7]W. Cochran, Phys. Rev. Lett. **2**, 495 (1959).
[8]W. Weber, Phys. Rev. B **15**, 4789 (1977).
[9]K. Rustagi and W. Weber, Solid State Comm. **18**, 673 (1976).
[10]M. Lax, "Comments on the Shell Model for Lattice Vibrations," in *Lattice Dynamics* (Pergamon Press, Oxford, 1965) and J. Phys. Chem. Solids, Suppl. I, 179-187 (1965).

Propagation of Phonons Generated by Nonradiative Transitions in KCl-NO_2

I. Sildos, G. Zavt, and I. Dolindo

Institute of Physics, Estonian S.S.R., Academy of Sciences, Riia 142
SU-202400 Tartu, USSR

1. The absorption and emission of the light by an impurity centre are accompanied by the creation of phonons, which, in turn, can drastically change the optical response of the centre. In recent years, various phenomena of such kinds have been studied by using the impurity luminescence spectra for the detection of nonequilibrium phonons [1-3]. In the present work another kind of detection based on the changes after a laser excitation of impurity *absorption* spectra, as well as the peculiarities of phonon propagation in KCl-NO_2^- are reported.

The following properties of KCl-NO_2^- [4] are essential here: (i) the system is an effective phonon source since at least 98% of the energy absorbed in impurity bands (350-400 nm) is converted into phonons; (ii) in both ground $^1A_1$ and excited singlet $^1B_1$ electronic states the NO_2^- molecule possesses low-lying rotational levels which are revealed in a well-resolved structure of zero-phonon absorption lines (see Fig.1). It is the changes of this structure that are used for the detection of nonequilibrium phonons.

2. The crystals containing $N_d \sim 3\cdot10^{19}$ cm^{-3} NO_2^- impurities were immersed in a helium bath at T_B=1.8 K and excited in the vibronic band v=2 with dye-laser pulses (duration 10 ns, energy 0.1 mJ, repetition rate 10 Hz) focussed to a spot size 0.2 mm in diameter. The probe beam from a xenon lamp was parellel to the exciting one and could be focussed either within or outside the excitation track (ET). The rovibronic absorption spectra with a time gate (5 μs minimal) were measured with a time discrimination synchronized with the laser pulse. Figure 1 shows the time evolution of the spectra taken within the ET. The enhancement of R(1) and P(1)+P(2) lines together with the drop of the total intensity are clearly detectable. The latter decreases by 35% whereas only ∿1% of the centres having been excited by the laser. The spectrum maintains changes for a long period of time after the centres have returned to the ground state. In Fig.2 the dashed lines show the evolution of the absorption coefficient κ for separate lines measured within the ET, whereas the solid lines, the behaviour of κ[R(0)] for different distances r between the exciting and probing beams. One can see rather slow (∿1 ms) propagation of the main part of a phonon packet, which is related to the resonance scattering of phonons by the impurity rotational levels and which correlates with the drastic decrease of the stationary thermal conductivity in 1-10 K interval [5]. However, much more fast (<5 μs) mode of phonon propagation accompanied by a long tail is also detectable in Fig.2.

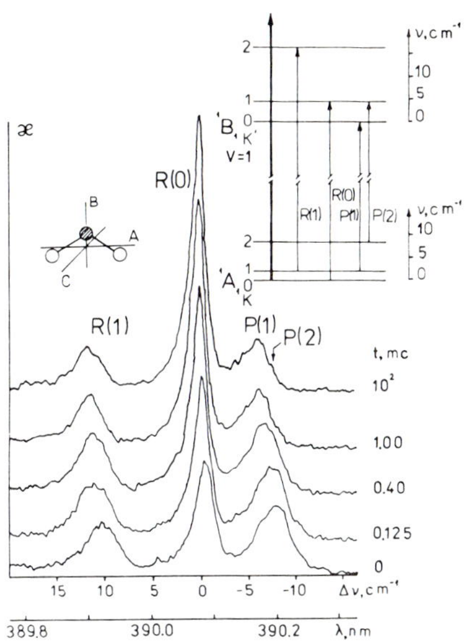

Fig.1. Time evolution of the KCl-NO_2 absorption spectrum after laser excitation. The spectrum is measured within the excitation track with the time gate 50 μs. The excitation (heavy arrow) and detection scheme are shown in the upper part. The rotational levels K=0,1,2 correspond to a weakly hindered rotations around A axis

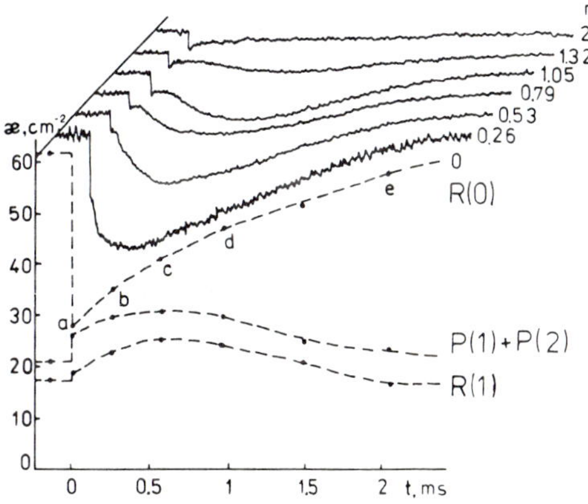

Fig.2. Time dependence of the intensities of separate lines in absorption spectra measured within (dashed lines) or outside (solid lines) the excitation region

3. Immediately after nonradiative transitions a broad spectral distribution of phonons (SDP) arises. Its evolution is controlled mainly by the anharmonic decay and impurity resonance scattering processes, the relaxation times being $\tau_A^{-1} \sim \omega^5$ and $\tau_R^{-1} \sim N_d \omega^4 (\omega^2 - \omega_o^2)^{-2}$ ($\omega_o \sim 20$ cm^{-1} for KCl-NO_2 [5]), respectively. For high-frequency phonons $\tau_A < \tau_R$, therefore, the SDP maximum shifts fast to the region $\omega \sim \omega_o$ where the decay slows down abruptly. Thus, shortly after the excitation a locally equilibrium phonon distribution with the temperature $T^* \gg T_B$ arises within the ET. The SDP does not change noticeably until phonons begin to leave the ET (compare with [3]). Later the diffusion develops and the SDP maximum shifts slowly to lower frequencies. Owing to the strong frequency dependence of $\tau_R(\omega)$ the diffusion 'constant' D should be time and space dependent. Indeed, from the time behaviour of $\kappa[R(0)]$ at $r=0$ the value $D \sim 0.1$ cm^2/s can be deduced,

whereas the spatial shift of κ minimum (which scales roughly as t^2) leads to $D \sim 40$ cm^2/s. Such type of propagation was termed in [6] as a quasidiffusional one.

The evolution of T^* within the ET can be estimated from the comparison of the spectra shown in Fig.1 with the stationary ones. For the intrinsic 'thermometer' both the Debye-Waller factor, which 'feels' SDP as a whole, and the ratio $\kappa[R(1)]/\kappa[R(0)]$,which tests SDP in the region of K=1 rotational level can be used. The T^* values obtained by these two methods (in brackets, the second one) in the points a, b, c, d, e of Fig.2 are: 22(6), 17(7), 12(5), 10(4), 2(2) K. Thus, the low-frequency part of SDP is considerably impoverished as compared to the Plank distribution. The reson is the fast escape of low-frequency phonons from the ET, which at the present stage may by treated either as a quasiballistic [6] or spectral diffusion type of propagation.

The authors are indebted to K.K.Rebane, L.A.Rebane and O.I.Sild for useful discussions.

References

1. V.L.Broude et al.: JETP 74, 314 (1978).
2. S.A.Basoon, A.A.Kaplyanskii and V.L.Shekhtman: FTT 24, 1913 (1982).
3. R.Baumgartner, M.Engelhardt and K.F.Renk: Phys. Lett. 94A, 55 (1983).
4. R.Avarmaa amd L.Rebane: phys. stat. sol. 35, 107 (1969).
5. V.Narayanamurti, W.D.Seward and R.O.Pohl: Phys. Rev. 148, 181 (1968).
6. D.V.Kazakovtsev and I.B.Levinson: JETP Letters 27, 194 (1978).

Discussions

OBSERVATION OF A QUASI-DIFFUSIVE PHONON PROPAGATION MODE page 88
W.E. Bron, J. O'Connor, Y.B. Levinson

R. Stock: Have you repeated your experiment at lower T with an Al bolometer? We have found a strong dependence of scattering on temperature in the course of our experiments.

W.E. Bron: No, we have not.

K. Renk: I think that your value for the anharmonic lifetime is much too short compared to values concluded from other experiments. It follows, e.g.. from time-of-flight experiments, performed by Kaplyanskii and co-workers, that the anharmonic lifetime is larger than 1 μsec, for longitudinal and transverse phonons.

W.E. Bron: I agree. This point was raised by us in the Phys.Rev.Letters paper which we have published on this work. The time source of the decay process remains uncertain to us.

M. Lax: Was your experiment done with superconducting detectors as well as bolometers? In our work on GaAs, to be presented later today, the presence of a threshold for phonon detection invalidates the Levinson mechanism.

W.E. Bron: The experiments were performed solely with Sn superconducting bolometers.

PROPAGATION OF NEAR ZONE BOUNDARY ACOUSTIC PHONONS IN SOLID (hcp) ^{4}He page 91
T. Haavasoja, V. Narayanamurti, M.A. Chin

W. Eisenmenger: How can you determine the orientation of your crystals?

V. Narayanamurti: We determined this indirectly through the velocity and expected phonon focussing intensities from low-frequency heat pulse studies. These indicate that the vital orientation is $\sim 45^0$ relative to the c axis for most of the crystals.

P. Leiderer: Has the influence of isotopic scattering from ^{3}He impurities been taken into account at low temperatures?

V. Narayanamurti: Their concentration (~/ppm) is so small that even if they got incorporated into the crystal they would have no influence.

J.P. Wolfe: I agree that an observation of phonon focusing would help to corroborate your conclusion that the high-frequency phonons are propagating ballistically. How do you propose to do this type of experiment under such extreme conditions.

V. Narayanamurti: I want to emphasize that phonon focusing would be interesting but the double junction experiment combined with observation of well-defined steps corresponding to relaxation phonon emission provides very direct and unequivocal proof of large wave-vector phonons in solid He. It is possible to introduce light into a dilution fridge and do the type of experiment you have done.

W. Dietsche: Have you seen any effect of reaching the maximum phonon frequency in your Sn generator -Al detector experiment?

V. Narayanamurti: 1.1$\pm$.1 meV is right near the zone boundary. It would be nice to go somewhat higher with a Pb junction and a sensitive Sn detector.

SCATTERING OF DEBYE PHONONS BY SUBSTITUTIONAL DEFECTS IN SOLIDS page 94

V.P. Srivastava

M. Wagner: I am somewhat astonished about your result. If you would diagonalise the problem exactly, having mass and spring constant changes, you always would get a result involving a complicated combination of the two kinds of changes. So I cannot understand that you get a simple separation in the transport properties.

S.V. Srivastava: Our result is obtained from second order-self-energy. Though the combination of two kinds of changes do appear in the self-energy their contributions to the phonon relaxation rate is found to be zero. Further, our result agrees with results obtained by earlier workers using perturbation theory and so there is nothing to be astonished about.

DIRECT OBSERVATION OF BALLISTIC LARGE WAVEVECTOR PHONON PROPAGATION IN GALLIUM ARSENIDE page 97

B. Stock, R.G. Ulbrich, M. Fieseler

K. Dransfeld: Do imperfections in the crystal diminish the effect of down-conversion due to the destruction of coherence between the three phonons participating in the process of down-conversion?

B. Stock: We observed a very rapid down-conversion of large K-vector phonons when the slit in the crystal is not etched furthermore after mechanical cutting.

SEARCH FOR LARGE k-VECTOR PHONONS IN GaAs page 100

J.P. Wolfe, G.A. Northrop

J.C. Hensel: Why do you choose to work in the [100] direction? Wouldn't the [111] direction be more favorable for the propagation of high-frequency dispersive phonons?

J.P. Wolfe: The spatial scans extend over a range of propagation directions away from [100], typically extending as far as [110]. The focusing effect is much stronger along these directions than along [111]. giving us much better signal-to-noise. Future experiments are planned for a [111] scanning surface as well, which does have a somewhat different dispersion.

V. Narayanamurti: 1. In your short sample how far in length do you have to go to get enough of an angle? 2. You must average the differences in frequency when you do your experiment at a fixed velocity. This is very important when you have dispersion.

J.P. Wolfe: 1. The maximum scan angle is typically 30^0 from the center of the pattern, corresponding to a 15-20% larger path length than the crystal thickness. 2. The phonon focusing pattern in the dispersive regime will indeed be different for constant velocity compared to constant frequency. Sharp caustics are expected in both cases.

H. Kinder: Was your sample in vacuum or in He? This makes a difference by a factor 10 in the detector signal. 9/10 of the intensity is lost to the helium according to our experiments in silicon.

J.P. Wolfe: Our sample was immersed in superfluid helium. We have not observed (or looked for) the interesting effect you mention.

TEMPERATURE DEPENDENCE OF OPTICAL PHONON LIFETIMES page 106

J. Kuhl, B.K. Rhee, W.E. Bron

K. Renk: I did not understand why you have omitted process (c) in the description of the phonon decay; in my opinion it should be included.

W.E. Bron: The reason for neglecting recombination processes is simply that the occupation probability of the acoustic (decay) phonon is very small; much smaller (by some 10^8) than the occupation probability of the coherently pumped optical phonons.

H. Maris: It seems that with your geometry the phonons will be scattered significantly at the walls. One should consider the possibility that inelastic processes occur there.

W.E. Bron: Yes, I agree. Whatever the mechanism for population decay will be, however, it needs to have a strong frequency dependence.

ANHARMONIC DECAY OF HIGH-ENERGY LA PHONONS page 109

S. Tamura, K. Okubo

B. Perrin: You assume that the Grüneisen parameter γ is dispersionless but it has been shown by neutron scattering measurements that γ may be strongly dispersed in Ge. What would be the consequences of this fact?

S. Tamura: The Grüneisen parameter may depend on the wave vector near the zone boundaries. If so, we may not place a great reliance on the absolute values of the decay rate. But the point I want to stress in my talk is to make clear the effects of lattice dispersion on the anharmonic decay through the two-phonon density of states, because Bron suggested a few years ago that the lifetime of zone-boundary LA phonons in quartz may be very long if the dispersion effects to the density of final states are taken into account. I showed, however, that this is not valid for the decays in Ge and possibly in GaAs.

K. Renk: The lifetimes are very sensitively dependent on the value of the third-order elastic constants. At which temperature were the values you mentioned measured?

S. Tamura: We used the values at 0 K which are obtained by extrapolating the values measured at low temperatures.

LIFETIME AND LINEWIDTH OF RESONANT 40-cm^{-1} PHONONS IN ALEXANDRITE page 112
R.J.G. Goossens, J.I. Dijkhuis, H.W. de Wijn

K. Renk: Do you have an explanation for the $N^{*1/2}$ dependence of the lifetime τ? I guess that this dependence gives evidence that your phonon source is non-monochromatic, and that you see trapping of broadband phonons. These phonons may be generated by the nonradiative transitions.

R.J.G. Goossens: The data itself gives no direct information about the nature of τ other than the $\sqrt{N^*}$ dependence itself (onset of diffusion?). Together with the broadband phonons one should mention a process to couple these phonons to the spins. It will be hard to explain the magnetic field dependence of the R_2 signal if one ascribes this R_2 signal to broadband phonons.

RAMAN SPIN-LATTICE RELAXATION INDUCED BY OPTICALLY GENERATED ZONE-BOUNDARY PHONONS IN RUBY page 118
J.G.M. van Miltenburg, J.I. Dijkhuis, H.W. de Wijn

K. Renk: I wonder about your long lifetime for zone boundary phonons. Because of weaker dispersion than in GaAs, I would expect a much shorter time.

J.G.M. van Miltenburg: I don't think the dispersion is the dominant factor in determining the lifetime of the TA phonon. The value of 3 s is to be considered as an order of magnitude estimate but is in agreement with measurements in other crystals. A simple model for the mode conversion from TA to LA phonons (Klemens) leads to about the same lifetimes for $Al_2O_3:Cr^{3+}$ (130 ppm) as for GaAs and TlCl. I think that the reduction of the scattering rate for ZB phonons because of the band mode appearing as a consequence of the introduction of a heavier mass into the crystal leads to a universal low scattering rate of ZB phonons. For Ca F_2 : Eu the short lifetimes of TA phonons suggested in the experiment could be caused by the fact that the frequency involved lies closer to the maximum amplitude of the lattice excitations on the mass defects caused by the band mode character.

RAMAN PROBE OF THE BRILLOUIN ZONE FOR NONEQUILIBRIUM PHONONS IN GaAs page 121
R. Bray, K.T. Tsen, K. Wan

W.E. Bron: Linewidth measurements suggest that LO and TO $k \sim o$ phonons have lifetimes of the order of 10^{-12} sec. How do you correlate this fact to apparently much larger lifetimes of the phonon near the zone boundary?

R. Bray: Our "steady-state" populations of zone-edge optical phonons greatly exceeds that due to zone-center LO phonons. Either our generation rate is much greater for the former (a preferential aspect of the defect generation process) and/or the lifetime of the zone-edge phonons is much greater than

10 psec. This could be due to missing or weaker decay channels (lower density of final states) for the zone-edge phonons. Analyses of the final density of state for the various optical phonons at Γ, X, K, L are in progress by M.Lax.

R.G. Ulbrich: Does Two-Photon-Absorption occur at the light intensity used in your experiments?

R. Bray: Our 1-photon extrinsic excitation mechanism (from deep levels) predominates at our laser intensities. We expect that as we got to higher laser intensities the two-photon absorption mechanism would eventually dominate. This could introduce new features in the Raman spectrum, including stronger zone-center LO phonon excitation, hot carrier effects and changes in the 2-phonon spectrum.

PHONON DECAY IN X-RAY IRRADIATED RUBY CRYSTALS page 124

M. Engelhardt, K.F. Renk

M.A. Brown: Very interesting paper. Many years ago Lawrie Challis and I did some work on the properties of γ-irradiated ruby. We found (a) that there was always some residual conc. of Cr^{2+} which increased with total Cr concentration and (b) that the conc. of Cr^{2+} could be systematically removed by u-v radiation with $\lambda \leq 365$ nm. Questions are: (1) have you tried removing the Cr^{2+} systematically and progressively by u-v bleaching? and (2) what is the total chromium concentration of your samples?

M. Engelhardt: (1) We did not try to remove the Cr^{2+} ions after X-ray irradiation. (2) The Cr^{3+} concentration in the samples was 0.05 %.

R.S. Meltzer: (1) Did the irradiation change the resonance shapes of either the optical or FIR absorption? (2) Have you considered the possibility of inelastic scattering from the ground state splitting of the V^{3+} ions present in your sample?

M. Engelhardt: (1) We did not measure the lineshape of the optical absorption, however, with respect to the far infrared absorption measurements before and after X irradiation gave the same absorption lines. (2) It is well possible that we have also inelastic scattering at V^{3+} ions. We think that the analysis with only one Raman shift is too simple. For a detailed analysis it should be included that it is necessary to know more about the different inelastic scattering centers and the corresponding transitions.

K. Weiss: Did you try to analyse the fascinating interplay you observe between elastic and inelastic scattering in terms of random walk theory (in frequency space)?

M. Engelhardt: Our analysis with an average Raman shift is much too simple to account for the real dynamics. It would be very interesting to perform theoretical and experimental studies on the problem of the random walk in our system.

H.W. de Wijn: At this point it is probably useful that we assess the ruby problem, now ruby is one of the best studied systems. Over the years a variety of mechanisms have been proposed to be important to the dynamics of the phonons. These include anharmonic decay, surface decay, decay at centers, spatial diffusion out of the excited zone, spectral diffusion by exchange, dipolar interaction etc., and spin-lattice relaxation by direct and Raman processes. All of these mechanisms are presumably important under favorable conditions. One example we have seen today: Centers created on purpose speed up the decay. Another example, on which I know Renk and I agree, is narrowing down the excited zone. We have seen under these conditions, in the language used today, the phonon decay time to scale with the excited-state population, i.e., we have a pure case of decay determined by spatial diffusion. I think we have to continue to vary the experimental conditions.

W. Eisenmenger: Do you think it possible to influence the phonon decay rate by applying high intensity microwave or microwave phonon radiation to the sample during the measurement, thereby reducing the ground level population of the the Cr^{2+} ions?

M. Engelhardt: This is a very interesting suggestion. We have also thought about this possibility. We found a temperature dependence of the signal in the range of 2 K to 4 K giving evidence that population changes occur for low-lying energy levels. Therefore, we think that with microwaves and microwave phonons it should be possible to influence the population of low-lying energy levels and eventually of the inelastic scattering.

L.J. Challis: (1) Could I ask if the 10 GHz phonons can be associated with a particular transition of Cr^{2+}? (2) Some years ago, Mike Brown and I studied the removal of Cr^{2+} (produced by γ irradiation) by uv light. This produced the reverse process $Cr^{2-} \rightarrow Cr^{3+}$. Did you observe any similar effects, i.e. a decrease in the inelastic scattering during the length of the experiment? Finally could I suggest that as the penetration depth of X rays may be less than the thickness of your sample you would get a more homogeneous distribution of Cr^{2+} if you use V rays.

M. Engelhardt: (1) It is an interesting question which electronic transitions are involved in the inelastic phonon scattering. Our experiment gives us only the overall magnitude for a mean Raman shift. (2) To confirm that Cr^{2+} ions are really responsible for the inelastic scattering, removal of the ions would be most useful. Thank you for the suggestion. During measurements over a longer period (1 year to 10 years) we did not find noticeable changes of the signals of irradiated crystals indicating that the scattering centers were stable. We tried to reduce the spatial inhomogeneity of the Cr^{2+} distribution by exposing the crystal used for our experiment to V rays for a long time, and we irradiated from different surfaces. Your suggestion to use γ rays is most important for a more quantitative analysis.

DECAY OF A HIGLY EXCITED PHONON MODE page 127

P. Ullersma

K. Weiss: Did you try to start from a Boltzmann equation instead of a master equation and to use (for instance) a relaxation time approximation (with frequency-dependent relaxation time)? It would be easier to discuss the physics of your system!

P. Ullersma: No, I did not. The equation for $\varphi(t)$ may be solved for a large range of values for t_1, t_2 and γ. Comparing the outcome with the experimental values for $\varphi(t)$ one would gain insight in the interaction. The disadvantage is that one has to do the calculations by computer. Applying your idea, this may be probably avoided.

M. Lax: Did you end up with a Bose distribution as $t \rightarrow \infty$?

P. Ullersma: Yes. From the solution of the rate equation for $\langle \varkappa \rangle$ one obtains exactly $\langle \varkappa(\infty) \rangle = V^{*1/2} (e^{\beta\Omega} - 1)^{-1}$,

and consequently $\langle N(\infty)\rangle = (\exp(\beta\Omega) - 1)^{-1}$.
So the Bose distribution originates from the stochastic part of N.

M. Wagner: In using a master equation approach you have assumed that your initial state is a completely dephased one (without coherent part). Do you think that this is an experimentally easily feasible situation?

P.Ullersma: In the underlying system we are dealing with a small subsystem in interaction with the phonon bath. I think it is a reasonable assumption that these interactions will afford theoretically as well as experimentally during the whole process repeated dephasing states.

Part IV

Surfaces, Interfaces, Kapitza Resistance

Chairmen:
J.P. Harrison A.J. Ikushima K. Laßmann P. Leiderer
A. Levelut K. Weiss A.F.G. Wyatt

Phonon-Induced Desorption of Helium

P. Taborek

Bell Laboratories, 600 Mountain Avenue, Murray Hill, NJ 07974, USA

Phonons from a solid substrate can interact with the atoms in an adsorbed film and cause them to be desorbed; understanding the microscopic mechanism of this process is one of the basic goals of surface science. Desorption of helium at low temperatures is a particularly simple and interesting system for testing our understanding of the phonon-adsorbate interaction. The binding energy of helium to a substrate is smaller than any other adatom, and the magnitude and shape of the binding potential is well known for many surfaces. Because helium is so weakly bound to a surface, it also has the unique property that it is energetically possible for a single substrate phonon to desorb an atom in a photo-electric effect type of process. The helium system also offers experimental advantages over more traditional desorption investigations using heavier adsorbates. The low-temperature techniques of ballistic phonon scattering which provide a means of controlling the frequency, intensity, duration, polarization and wave vector of the desorbing phonons provide a much more detailed probe of the desorption process than high-temperature experiments can possibly achieve. Using these techniques, one can generate both equilibrium and non-equilibrium substrate phonon distributions and detect the phonon-induced desorption of sub-monolayer helium films. The experimental apparatus and a typical desorption signal are illustrated in Figs. 1 and 2. Previous experimental work is described in references [1-4].

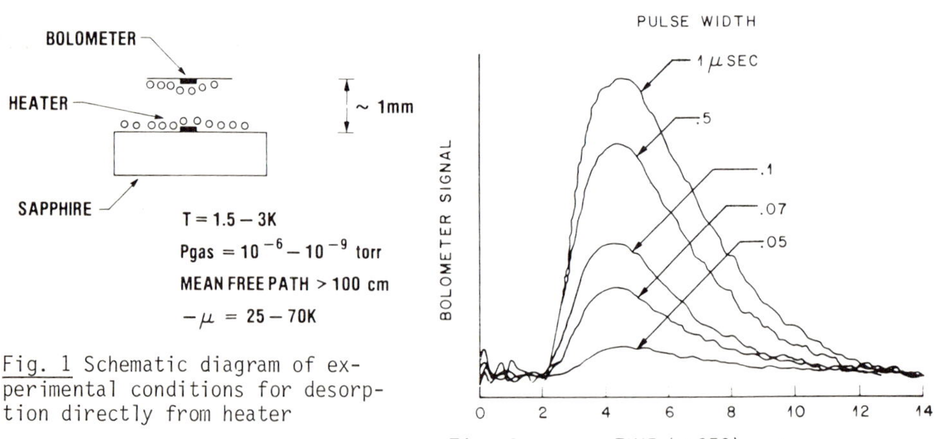

Fig. 1 Schematic diagram of experimental conditions for desorption directly from heater

Fig. 2

Fig. 2 Desorption of He^3 for several values of the heater pulse width. T_h = 4K, μ = -23K ℓ = 0.9 mm

The most important characteristics of the desorption process are the rate at which it occurs and the velocity distribution of the desorbed atoms; both of these properties can be measured directly. These results can then be compared to theoretical predictions of the rate and velocity distribution. Two models of desorption based on opposite points of view have been particularly helpful in guiding the experimental work [3-8]. The first model is based on an equilibrium thermodynamic description of the adsorbate/gas system, while the second model is based on a perturbation theory calculation using a quantum mechanical Hamiltonian to describe the phonon-adatom interaction. Experimentally distinguishing between these ostensibly quite different theories has been surprisingly difficult. I will first outline the basic features of the two models, including the expected form of the rate and velocity distribution, and then compare each with the experimental results.

1. Thermodynamic Model

In a container in which the gas pressure is P and, therefore, the chemical potential $\mu = T \log\left(\frac{P}{kT}\left(\frac{2\pi h^2}{mkT}\right)^{3/2}\right)$, the number of atoms with velocity v and angle θ from the normal which strike the wall per unit time and area is given by:

$$J(v) = \frac{m^3}{(2\pi\hbar)^3} e^{\mu/T} v^3 \cos\theta \exp\left(\frac{mv^2}{2T}\right) dv d\Omega. \tag{1}$$

Following reference [8], we assume that some fraction α of the incident atoms sticks to the surface, while the remainder are specularly reflected. In order to maintain detailed balance in equilibrium, the velocity distribution of atoms leaving the surface must be of the same form as (1) with an additional factor of the sticking probability α. The nonequilibrium situation of interest in a flash desorption experiment can be analyzed using eq. 1 if we assume that the adatoms always behave as if they were in equilibrium with a hypothetical gas; the only difficulty is in estimating the temperature and chemical potential of this gas (and, therefore, the desorbing film). Since the thermal time constant of a monolayer film of helium $\tau = RC \sim 10$ nsec (with Kapitza resistance $R \sim 10/T^3$ cm^2/watt K and C $\sim 4\times10^{-10}T^3$ J/cm^2 layer) one expects, for desorption times longer than τ, that the desorbed gas has a maxwellian velocity distribution with characteristic temperature equal to the substrate temperature, T_S, which can be estimated reliably using acoustic mismatch theory [8]. The angular distribution is expected to be $\sim \cos\theta$ unless α has a pronounced angular dependence. The initial chemical potential of the adatoms is the same as the chemical potential of the background gas. As desorption proceeds, the chemical potential of the adatoms can increase or decrease, depending on interactions in the film, but for many models it never deviates significantly from its initial value. If we adopt this simple assumption, the desorption rate R can be computed by integrating eq (1) over velocity and angle to yield:

$$R = \alpha T \frac{2\pi m(kT)^2}{(2\pi\hbar)^3} e^{\mu/T} . \tag{2}$$

Assuming $\alpha = 1$, $T = 10$, this gives a characteristic time for desorbing a monolayer of $2 \times 10^{-11} e^{-\mu/T}$ sec. A more detailed description of this theory which utilizes a specific model for the film valid at low temperatures is presented in reference [9].

1-Phonon Theory

In the quantum theory of phonon-induced desorption, which dates from the 1930s [10], the phonon interacts with an adatom by displacing the surface from its equilibrium position. The energy associated with this displacement is $H = \frac{\partial V}{\partial z} \cdot u$, where $V(z)$ is the surface potential, $u = (h/2nM_s\omega)^{1/2}$ is the phonon displacement (typical value $u \sim 3\times10^{-12}$ cm) and M_s is the mass of a substrate atom. The desorption rate is given by a Golden Rule expression:

$$R = \frac{2\pi}{\hbar} \sum_{p,k} |\langle|H|\rangle|^2 \delta(E_i - E_f) \tag{3}$$

$$|\langle|H|\rangle|^2 \sim e^{-\omega/T} \left(\frac{Eb}{a}\right)^2 \frac{1}{\omega} \tag{4}$$

where E_b and a are the depth and width respectively of the binding potential $V(z)$. (Typical values are $Eb = 50k$, $a = 2Å$.) Theories of this sort have been worked out in detail using various approximations [11-14]; the form of the result is dictated, however, by the density of states factors for the atoms and phonons and the conservation conditions:

$$\hbar\omega = \frac{p^2}{2m} + E_b \tag{5}$$

$$p_{||} = \hbar k_{||}. \tag{6}$$

The first condition implies that phonons with energy less than the binding energy cannot interact with the adatoms; the second condition is a consequence of the fact that the surface potential is translation invariant. These two conditions completely specify the momentum of an outgoing atom in terms of the k vector of the phonon that desorbed it, as shown in the graphical construction of Fig (3). The conservation conditions thus inevitably lead to a highly anisotropic distribution of desorbed atoms [11, 13], which emerge from the surface within a narrow cone. It is also easy

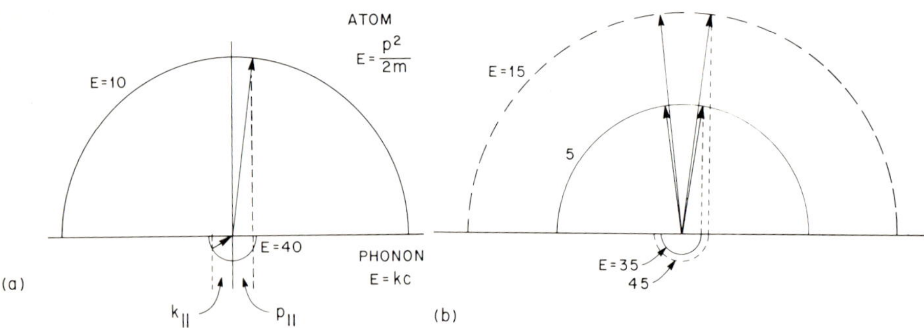

Fig 3. Curves of constant energy in momentum space for atoms above the interface, and phonons below, drawn to scale with $C = 3\times10^{-5}$ cm/sec and $E_b = 30$ K. a) Geometric construction which shows that parallel momentum conservation leads to atoms emitted in a highly focused cone. b) The angular width of the cone depends on the energy, with high-energy atoms emitted into a narrower cone

to see that a desorption process which involves the absorption of many phonons will be constrained to emit atoms within the same 1-phonon critical cone, since the sum of the parallel momentum of N phonons is always less than or equal to the sum in the case that all N k vectors are collinear and parallel to the surface. In addition to this strong forward focusing effect, the conservation conditions also imply a correlation between velocity and angle in which atoms which leave the surface at large angles have a lower average velocity; the kinematic reasons for this effect are illustrated in Fig (3). The velocity distribution which results from a more complete calculation using the matrix element of eq. 4 and the conservation conditions eqs. 5 and 6 has the form:

$$J(v) \sim \frac{v^3 e^{-E_b/T} \exp\left(\frac{-mv^2}{2T}\right) \cos\theta}{\frac{mv^2}{2} + E_b} \, dv d\Omega \qquad (7)$$

if $mv^2/2 + E_b > mcv\sin\theta$, and zero otherwise. The total rate, given by the integral of equation (7), leads to a characteristic desorption time $\tau = \tau_o e^{E_b/T}$, but unfortunately the values of τ_o in the literature vary between 10^{-5} to 10^{-12} seconds, and cannot be used as a definitive test of the theory.

Experiments using Equilibrium Phonon Distributions

Desorption in the normal direction directly from a substrate (see Fig. 1) with a known temperature T_s is the simplest case to analyze using the above models. This can be realized using a thin film heater on a phonon-transparent substrate such as sapphire. Acoustic mismatch theory relates the electrical power dissipated in the heater to the temperature. Then

$$S(t) \sim J\left(\frac{\ell}{t}\right) * \frac{\ell}{t^2} * \left(\frac{m\ell^2}{2t^2} + E_b\right) \qquad (8)$$

where ℓ is the distance between the heater and bolometer. The particle beam deposits both kinetic energy and the energy of adsorption on the bolometer surface; the term ℓ/t^2 comes from $dv = -(\ell/t^2)dt$. The bolometer signals expected from the desorption models discussed above are:

$$S(t) \sim t^{-5}\left(\frac{m\ell^2}{t^2} + E_b\right) \exp \frac{-m\ell^2}{2t^2 T} \qquad \text{(thermodynamic)} \qquad (9)$$

$$S(t) \sim t^{-5} \exp \frac{-m\ell^2}{2t^2 T} \qquad \text{(1-phonon)}. \qquad (10)$$

The absolute magnitude of the bolometer signal can be estimated using a simple thermal model which relates the heat flux Q(J/sec cm^2) carried by the gas to the temperature rise δT of the bolometer. The energy balance in the bolometer with temperature T is:

$$C \frac{dT}{dt} = QA - \frac{A}{R}(T - T_s) \qquad (11)$$

where C is the heat capacity of the bolometer, A is its area, and R is the

energy. Assuming that 10^{15} atoms/cm^2 are desorbed, and that they arrive at the bolometer during a time interval of 10 μsec gives $Q \sim 4\times10^{-2}$, so $\delta T \sim 10^{-1}$ K. The sensitivity of our bolometers is typically .1-1.0 V/K so the bolometer signal for desorption of a monolayer is expected to be of order 10^{-2}V. Actual signals are within a factor of 5 of this estimate, so detection of the desorption of 1% of a monolayer is possible using these devices.

The chemical potential $\mu \cong E_b$ appears as a factor in the bolometer signal, in expressions for the desorption rate, and is an important parameter which specifies the thermodynamic state of the film (and the background gas) before it is desorbed. The bolometer signal depends on the repetition rate of pulses to the heater since for high repetition rates, the film does not have enough time to reform on the heater surface. The critical repetition rate crr (Hz) at which the signal is one-half of its maximum value is related to the pressure in the gas by P(torr) = crr* 5.7 × $10^{-8}\sqrt{T}$; pressures as low as 10^{-9} torr can be measured to within a factor of 2 in this way. The chemical potential of He^4 is related to P by $\mu(K) = T \log(8.4\, P/T^{2.5})$; since μ depends only on ln(P), μ can be determined accurately. See reference [8] and [9] for further details.

The most basic characteristic of the experimental bolometer signal is the time of the maximum tmax. For models of the signal of the form $S(t) \sim t^{-n}\exp(-E/T)$, tmax is related to the temperature by tmax = $(m\ell^2/nkT)^{1/2}$; for models such as eq (9) which explicitly contain the binding energy the result is only slightly different. Experimental studies [8] of the dependence of tmax on heater power using both ^{3}He and ^{4}He show that the temperature of the gas deduced from tmax agrees with the substrate temperature calculated from acoustic mismatch theory to within 15% of either model of the signal. Thus the desorbed atoms are a good thermometer which measures the substrate phonon temperature, but one cannot distinguish between the 1-phonon and thermodynamic models simply using the velocity spectrum.

Since thin film heaters on sapphire substrates have thermal relaxation times ~10 nsec, the desorption kinetics can be studied by varying the heater input pulse width but keeping the power and therefore Ts constant, as shown in Fig (1). The time required to desorb ≈1 monolayer varies between 30 nsec and 2 μsec. By repeating the measurements shown in Fig. 1 for various heater powers, the dependence of the desorption time constant τ can be determined. The results can be parameterized by $\tau = \tau o \exp(Q/T)$ with $\tau o \sim 10^{-10}$-10^{-9} sec and $Q \sim -2\mu/3$. Although the values of τo are reasonably consistent with the thermodynamic model [9] and versions of the 1-phonon model [12,14], the relationship between Q and μ is surprising and so far unexplained.

The angular distribution of the desorbed flux is the only property of desorption induced by a thermal phonon distribution for which the two models discussed above give clearly different predictions. An experiment designed to measure the angular distribution of desorption from a nichrome heater film is described in reference [6]. The results show that monolayer films of helium desorb preferentially in the normal direction within a cone of half-angle ~15°. Also, the atoms leaving the surface at large angles were found to have lower average velocities than the atoms desorbing in the normal direction. These results are in qualitative agreement with the 1-phonon theory and the conservation conditions (5) and (6), but other explanations based on a nonequilibrium classical mechanical model [15] and collisions within the desorbed gas [16] have been suggested. Although the mean free path in the background gas is very large, the instantaneous local density of the desorbing gas can be sufficiently high that collisions can become important. A simple model which accounts for the decreasing density of the puff of desorbed atoms as it propagates away from the substrate shows that the number of collisions per atom is approximately $\ell n(T/\tau)/C$, where C is the initial coverage in layers, τ is the desorption time and T is the transit time; for most of the experiments reported here, this number is ~ 3.

Experiments using Nonequilibrium Phonons

As discussed above, many of the predictions of the equilibrium thermodynamic theory and the 1-phonon theory are very similar if the desorbing phonons are themselves in equilibrium with temperature Ts. The predictions of the two models differ significantly, however, if a nonequilibrium distribution of phonons causes desorption. If the adatoms behave as a thermodynamic system in internal thermal equilibrium, the desorption process depends only on the power density of the phonons which heat the film, not on their frequency distribution. If a 1-phonon process subject to the conservation conditions (5) and (6) causes desorption, then the velocity spectrum of the desorbed atoms directly reflects the energy distribution of the phonons; in particular, if the phonons have $h\omega < E_b$, there should be no desorption at all.

A simple nonequilibrium phonon distribution can be produced by injecting phonons with a Bose distribution at temperature Tp into a phonon-transparent crystal. As the phonons reach the opposite surface of the crystal, they still have an equilibrium frequency distribution characterized by Tp, but because of geometrical spreading the intensity is much lower than if the crystal temperature were Tp. Reference [7] describes the results of desorption experiments using this type of phonon distribution. The principle result is that two groups of atoms with distinctly different temperatures are desorbed from the surface; approximately 5% have a temperature equal to the heater temperature while the rest emerge more slowly with a temperature slightly above the ambient. See Fig. 4. One possible explanation of this result is that both direct 1-phonon desorption and thermalization in the film with subsequent desorption occur in parallel, but the direct process is approximately 20 times slower.

The "phonon-fluorescence" device described in reference [17] provides another technique for generating nonequilibrium phonon distributions with

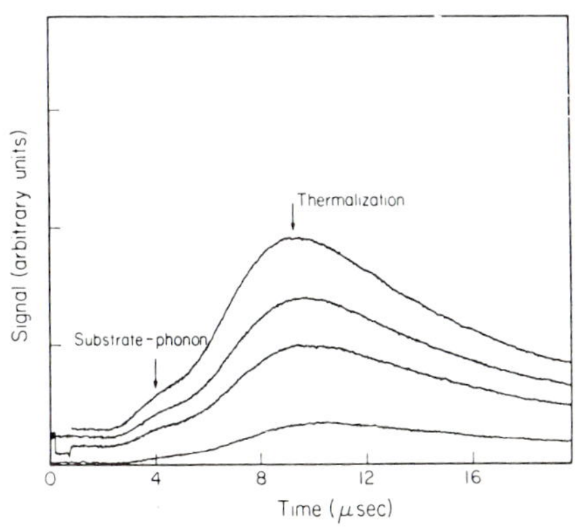

Fig. 4 Desorption induced by low intensity, high-frequency phonons

sufficient power to cause experimentally observable desorption signals. In this geometry, a superconducting film acts as a low pass filter ($h\omega<2\Delta$) for thermal phonons which desorb atoms from the surface of the superconductor. It is possible to arrange that $-\mu>2\Delta$ so that 1-phonon processes should be energetically forbidden. Preliminary experiments using this technique [18] indicate that for monolayer coverage the velocity spectrum of atoms desorbed from a superconductor has a maxwellian character.

The work described here is the result of collaborations with M. Sinvani, D. Goodstein, M. Cole, and A. Ferdman.

References

1. K. Andres, R. C. Dynes, and V. Narayanamurti, Phys. Rev. A8, 2 501 (1977).
2. G. N. Crisp, R. A. Sherlock, and A. F. G. Wyatt, Low Temp Physics 14, eds. M. Krusius and M. Vuorio (North-Holland, Amsterdam, 1975) p. 455.
3. P. Taborek, M. Sinvani, M. Weimer, and D. Goodstein, J. Phys. (Paris) C6 pp. 825, 852, and 855 (1981).
4. P. Taborek, M. Sinvani, M. Weimer, and D. Goostein, Physica 107B, 247, (1981).
5. M. Sinvani, P. Taborek, and D. Goodstein, Phys. Rev. Lett., 48, 1259 (1982).
6. P. Taborek, Phys. Rev. Lett., 48, 1737 (1982).
7. M. Sinvani, P. Taborek, and D. Goodstein, Phys. Lett., 95A, 59 (1983).
8. M. Sinvani, D. L. Goodstein, M. W. Cole, and P. Taborek, Phys. Rev. B, to be published.
9. M. Weimer and D. Goodstein, Phys. Rev. Lett. 50, 193, (1983).
10. J. E. Lennard-Jones and C. Strachan, Proc. Roy. Soc. (London) A150, 442 (1935).
11. B. Bendow and S. C. Ying, Phys. Rev. B7, 622 (1973).
12. F. O. Goodman and I. Romero, J. Chem. Phys. 69, 1086 (1978).
13. Z. W. Gortel, H. J. Kreuzer, and D. Spaner, J. Chem. Phys., 72, 234 (1980).
14. Z. W. Gortel, H. J. Kreuzer, and R. Teshima, Phys. Rev. B 22, 5655, (1980).
15. F. O. Goodman, Phys. Rev. B27, 6478, (1983).
16. J. P. Cowin, D. J. Averbach, C. Becker, and L. Wharton, Surf. Sci. 78, 545 (1978).
17. V. Narayanamurti and R. C. Dynes, Phys. Rev. Lett. 27, 410, (1971).
18. P. Taborek and A. Ferdman, to be published.

Thermal Boundary Resistance Between Small Particles and Liquid He-3

T. Nakayama

Department of Engineering Science, Hokkaido University, Sapporo 060, Japan

Abstract

A review is presented of the current status of investigations on the anomalous thermal resistance of the boundary between small particles and liquid He-3 observed at temperatures below about 10 mK.

1. Introduction

The surprisingly small thermal resistance of the metal particle-liquid He-3 boundary at temperatures below about 10 mK has remained a puzzle since its first observation by AVENEL et al. [1] over the past ten years. The observed values of the resistance below 10 mK [1] are about several orders of magnitude smaller than the value predicted by the theory for a bulk solid-liquid He-3 boundary [2]. It is now well established by successive experiments [3-8] that the boundary resistance between small particles and liquid He-3 exhibits anomalous behavior below about 10 mK. This is shown in Fig.1 in which the boundary resistances or Kapitza resistances R_K measured to date are compared with the predictions of the theory for a bulk solid [2]. The observed resistances in Fig.1 deviate by several orders of magnitude from the T^{-3} law at temperatures below 10 mK. This suggests the presence of unknown processes transferring heat effectively and which has been a long-standing unsolved problem in low-temperature physics [9]. The aim of this article is to review current experimental and theoretical aspects of investigations of the above puzzling phenomenon.

2. Size effect of small particles

To make the boundary resistance sufficiently small at very low temperatures, it is necessary to increase the surface area between a solid and liquid He-3. For this reason, small metal particles of 1 μm diameter or less are used quite popularly in the heat exchanger [10]. BEKAREVICH and KHALATNIKOV [2] and GABORET [11] considered phonon energy from a solid to be transferred to the collective zero-sound modes in the Fermi liquid. TOOMBS et al. [12] justified these semi-phenomenological theories [2,11] from the microscopic viewpoint. When the phonon density of states proportional to ν^2 is taken into account, the Kapitza conductance $h_K(=1/R_K)$ should show a T^3 variation. This is, however, not the case for small particles as understood from the following arguments. Assuming that the sound velocity in the particle is given by v, the lowest frequency ν_o of the particle with diameter d is about v/2d. Figure 2 shows a schematic illustration of the lowest eigenmode of a particle with diameter d. For instance, the lowest ν_o becomes about 2 GHz≃0.1 K for the particle of 1 μm in diameter. If the heat exchange occurs from the excitation or absorption of zero sound

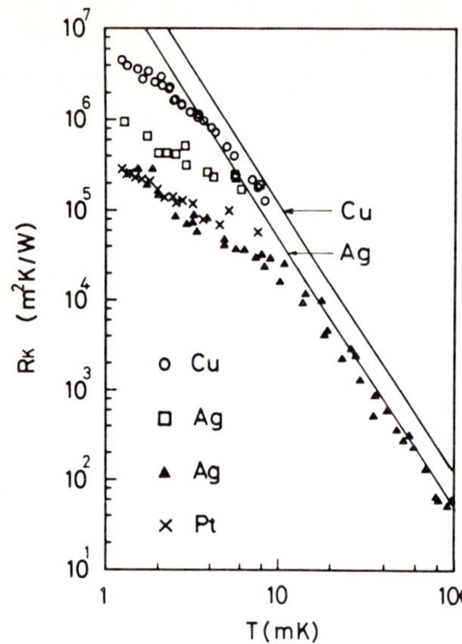

Fig.1 Observed boundary resistances R_K between small particles and liquid He-3 as a function of temperature. The symbols (▲,o,□,x) are taken from [4,5,6,16], respectively. The solid lines are the theoretical prediction [2]

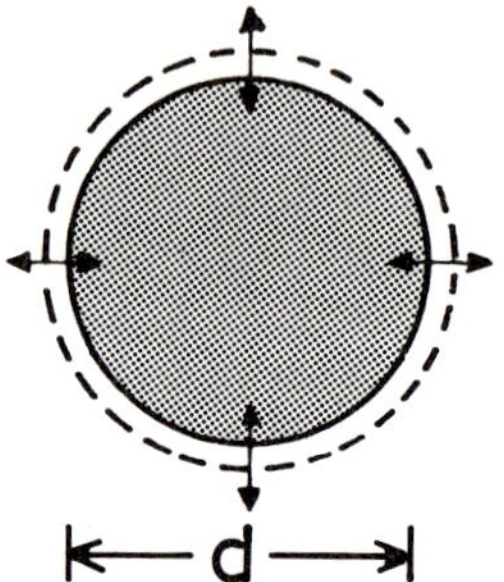

Fig.2 The lowest eigenmode of a particle with diameter d corresponding to the spheroidal mode

due to the surface vibrations of a particle as considered in [2,11], the resistance will increase exponentially at a temperature lower than that corresponding to ν_o.

The above expectation was demonstrated quantitatively in the case of isolated particles by NAKAYAMA and NISHIGUCHI [13] for the liquid He II and NISHIGUCHI and NAKAYAMA [14] for the normal liquid He-3. It should be emphasized that the theories [13,14] also predict the result for a bulk solid [2,11] as a limiting case of large d. The expression for the energy flux $\dot{W}(T)$ at finite temperature can be written down in terms of the surface displacement $\vec{u}$ of the particle as [14]

$$\dot{W}(T)=a_1\rho_L c_\ell \sum_{\substack{J\\ \alpha=s,t}} \int\langle|\dot{u}^J_{\alpha,r}(\vec{R})|^2\rangle_T d\Omega/4\pi + a_2\rho_L c_t \sum_{\substack{J,\alpha=s,t\\ \beta=\theta,\psi}} \int\langle|\dot{u}^J_{\alpha,\beta}(\vec{R})|^2\rangle_T d\Omega/4\pi. \quad (1)$$

The definition of the symbols in (1) is given in [14]. To obtain the Kapitza conductance, $h_K=d\dot{W}(T)/dT$, from (1), a detailed knowledge of the surface displacement $\vec{u}(R)$ of a particle is needed. The vibrational modes in a spherical particle consist of both the toroidal and spheroidal mode as obtained in [13]. The exponential behavior of the conductance at low temperatures is recovered by taking the first lowest eigenvalue in the expression (4.4) in [14]: the resistance R_p becomes $T^{-2}\exp(h\nu_o/k_BT)$, where ν_o corresponds to the frequency of the spheroidal mode given in Fig.2. Thus we have the exponentially increasing R_p originating from the finite-size effect of the small particle. At temperatures higher than that corresponding to the finite lowest eigenfrequency, the calculated resistance exhibits the T^{-3} dependence and the magnitude is in agreement with the bulk limit. The physical meaning of this result is clear since the dominant phonons contributing to the heat transfer in this temperature regime have much shorter wavelengths than the size of the particle, then the particle shape is irrelevant and the resistance approaches the bulk limit.

3. Magnetic coupling and adsorbed He-3 layer

As described in the previous Section, the theories based on the zero-sound excitations have failed to explain the experimental data at temperatures below 10 mK, indicating that for our system there should be another mechanism to transfer heat effectively. At first sight, a possible mechanism might be the magnetic coupling between the He-3 nuclear spins and electron spins. The first theoretical attempt along this line was made by MILLS and BEAL-MONOD [15], who studied the contributions from the dipolar and short-ranged coupling between conduction-electron spins and nuclear spins of quasi-particles. Their conclusion is that neither dipolar interactions nor short-ranged interaction of the exchange type can produce a significant transfer of energy. In addition, it has been shown experimentally that both magnetic impurities in sinter [4] and magnetic fields up to 0.6 kG [5] have no effect for the heat transfer.

Recently, however, PERRY et al. [16] reported that the boundary resistances between Pt powder and liquid He-3 depend on the applied magnetic fields. They used two different methods to determine the boundary resistance. In one method, pulses of eddy-current power were used to heat the Pt. The other method used a steady rf field to heat the Pt to a temperature greater than that of the He-3 by an amount of ΔT. The resistances obtained were insensitive to fields less than 2 kG, but decreased by a factor of three when the field was increased to 8 kG. These results are interesting because the magnetic coupling has been shown to be irrelevant to the anomalous boundary resistances. The analysis made in [15] is, however, not valid for the actual system, strictly speaking, since the surface is treated as an infinite potential barrier and as a result, both the He-3 and conduction-electron wave functions vanish at the boundary. As is well known, the wave function of the conduction electron spreads into the vacuum to a distance of the order of Fermi wavelength if the metal surface is clean. Thus there should occur an exchange interaction between the conduction-electrons and electrons of adsorbed He-3. The exchange interaction bears the Fermi contact coupling between conduction electrons and He-3 nuclear spin of the type of $H'=g_o \Sigma_\alpha \sigma_\alpha \cdot S_\alpha$, where σ and S are the Pauli spin operators for the conduction electrons and the localized He-3 nucleus, respectively. The theoretical analysis using this Hamiltonian was given in [16] by changing the roles of spins in the formula of the magnetic Kapitza resistance derived by LEGGET and VUORIO [17]: the localized electron spins and the nuclear spins of He-3 quasi-particles in [17] are replaced by the localized nuclear spins of He-3 adsorbed on Pt and the spins of conduction electrons in metal, respectively. The heat exchange between the spins of adsorbed He-3 and quasi-particles in the bulk liquid occurs through the exchange interaction and this is much faster than that between the adsorbed He-3 and the conduction electrons in Pt because of the much larger liquid He-3 density of states compared with the conduction-electron density of states.

A problem from the theoretical side is the estimation of the strength of magnetic coupling g_o between localized He-3 nuclear spins and conduction-electron spins. The observed dipole moments of inert-gas atoms (Ar, Kr, Xe) adsorbed on clean W [18] were used to estimate the dipole moment of adsorbed He-3 atoms on Pt of about 0.1 D [19]. Assuming a mixing of the 1s2p state with $1s^2$ ground state, PERRY et al. [16] obtained the coupling strength g_o of 20 mK which is sufficient to explain the observed magnitudes of the resistances. It should be noted, however, that a different approach to explain large dipole moments of inert-gas atoms adsorbed on W [18] was forwarded by FLYNN and CHEN [20]. They pointed out that the substantial change of bonding energy and work function associated with inert-gas atoms adsorbed on W [18] arises from an admixture of charge-transfer states. Figure 3 is the schematic diagram illustrating the suggested charge transfer state, from which we see that the bonding state depends on the sign of $\phi - I^*$, called optical switching. If the theory [20] is applied to the estimation of the coupling constant, g_o becomes of the order of 1 mK for Pt ($\phi - I^* \simeq 1$eV). Because the conductance h_K is proportional to g_o^2, this implies that a discrepancy of about two orders of magnitude exists in explaining the observed magnitude of the resistances from this analysis.

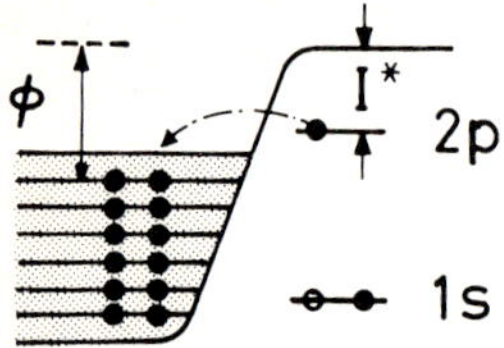

Fig.3 The schematic diagram showing the bonding state of He atom and the definition of the symbols ϕ and I*[20]

In addition, it should be emphasized that the large dipole moments of inert-gas atoms were observed for clean W surface [18]. It is not clear that the observations for clean surface are applicable to the surface exposed to air. Specifically, the surfaces of metal particles such as Ag or Cu react with the air molecules. Because it is unavoidable to expose the sinter to air for the construction of the heat exchanger, the surfaces of Ag sinters being used popularly seem not to be clean. As a result, the magnitude of g_o estimated for clean surface should be much reduced, implying the inefficiency of the Fermi contact interaction for these systems. In this connection, it is interesting to note that the boundary resistance between liquid He-3 and Pd particles increases when He-4 impurities are concentrated in liquid He-3 [8], which seems to indicate the importance of the Fermi contact interaction. Although these observations [8] involve the ambiguity as pointed out by HARRISON [9], it should be noted that the surfaces of Pt, Au or Pd do not react with oxygen.

4. Effect of sintering

The metal surfaces of Ag or Cu used in the heat exchanger have quite different characteristics compared with those of Pt, Au, and Pd. Ag or Cu is very reactive with oxygen and forms an oxide layer of about 100A thickness. In this connection, the electron micrograph of Al particle taken by BATSON [21] shows clearly the presence of the oxide layer. The oxide layer strongly prevents the exchange interaction between conduction electrons and He-3 electrons from producing the reasonable strength g_o to explain the observed resistance. This viewpoint is supported by the experiments [3,5], in which no effect of He-4 monolayers was observed. Recently, NISHIGUCHI and NAKAYAMA [22] developed a theory to explain the anomalous behavior of the resistance below about 10 mK taking into account the effect of sintering. This possibility was mentioned by FRISKEN et al. [23] without giving a quantitative explanation for the heat-transferring mechanism. I describe briefly the results given in [22].

In Sec.2, it was shown that the internal vibrations of the small particles are irrelevant to the heat transfer at low temperatures originating from the presence of the low-frequency cutoff caused by the finite-size effect. However, when the particles are bridged in the sintering process [24], there should be new low-energy vibrational modes associated with the 'large' mass of particles. Although the particles in the sinter are randomly connected to each other and the bridges are not uniform as seen from Fig.4, the disordered character of the sinter is irrelevant to the description of its thermal properties at around 10 mK. These bridges are approximated as rods with averaged cross section πa^2 and length ℓ in order to estimate the strength of the particle-particle coupling. Provided that the elastic deformation of the rod produces particle vibrations, we can derive the maximum frequency corresponding to the Debye cutoff frequency ω_c from the expression of the equilibrium conditions of force as $\omega_c=(\pi a^2 E/M\ell)^{\frac{1}{2}}$, where M and E denote the mass of the particle and Young's modulus of the rod, respectively. Then the particles execute periodic vibrations around their equilibrium positions at finite temperature and these constitute a perturbation to liquid He-3. Under these conditions, the thermal conductance h_e(= $1/R_e$) is expressed in the following simple form by applying (1) directly,

$$h_e = \frac{3k_B}{32\pi R^3}\,\frac{\rho_L}{\rho_S}\,(a_1 c_\ell + 2a_2 c_t)\left(\frac{\theta}{T}\right)^2 \mathrm{cosech}^2\left(\frac{\theta}{2T}\right), \tag{2}$$

where θ is the characteristic temperature of the sinter defined by $\theta = \hbar\omega_c/k_B$. Since the factors a_1 and a_2 in (2) are proportional to the inverse of the density of quasi-particle states and the zero-sound velocity, this theory explains the weak dependence on the pressure of the observed resistance [3,5]. KINGERY and BERG [25] have shown that the characteristics of sinters depend on certain conditions such as annealing temperature, heating time and packing ratio. For instance, in the case of the sintering process normally used in heat-transfer experiments such as the one-minute heating at 525°C in [4], the average length ℓ and radius a of the bridges are estimated to be of the order of one-tenth of the particle radius R using the relation derived in [25]. Taking a and ℓ to be one-tenth of R, the characteristic temperature θ is estimated to be equal to 10 mK for an Ag particle of 1 μm in diameter. Thus, as far as we are concerned with the temperature region around 10 mK, the Einstein model for oscillators adopted for obtaining (2) describes well the thermal properties of the system. This is the reason for the irrelevance of the disordered nature of the sinter to the heat transfer at the temperature region of interest here.

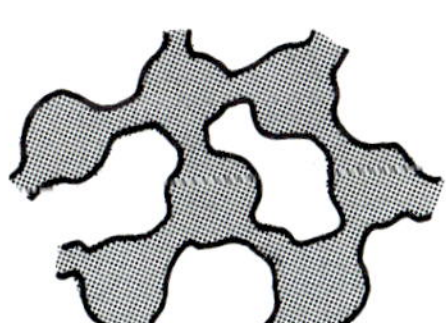

Fig.4 The illustration of sinter [24]

The total conductance h_K between sintered particles and liquid He-3 is obtained by the sum of two contributions (1) and (2); $h_K = h_p + h_e$, where h_p denotes the conductance due to the internal vibrations of particles derived from (1). Figure 5 shows a comparison between the experimental data [4] and the calculated resistance R_K for Ag sinter of radius 0.5 μm with the characteristic temperature θ=10 mK. The total resistance exhibits a T^{-3} variation above 10 mK as well as the case of a bulk solid, and below 10 mK, a shoulder appears when compared with the calculated resistance due to the internal vibrations. In order to deter-

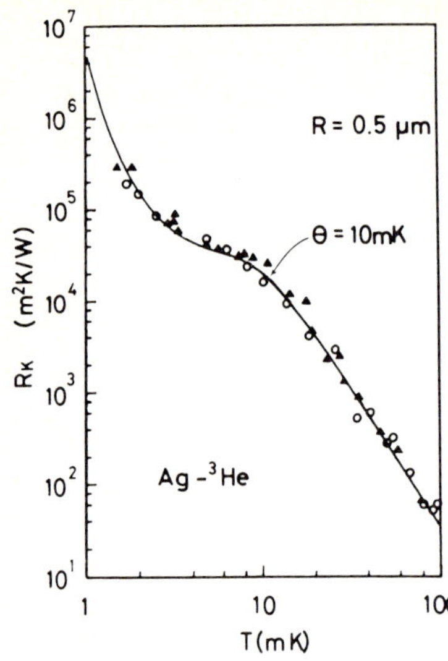

Fig.5 The boundary resistance R_K for silver particles of 0.5 μm radius versus temperature. The solid curve shows the calculated resistance with θ=10 mK. The observed data from [4] are given by (▲) for dirty A_g particles with R=0.44 μm in radius and by (o) for clearn A_g particles with R=0.55 μm, respectively

mine the physical origin of the shoulder, let us consider the low- and high-temperature limits of the conductances of h_p and h_e. The magnitude of h_e at sufficiently high-temperatures is obtained from (2) by replacing cosech^2 $(\theta/2T)$ by $4(T/\theta)^2$. For instance, h_e yields $3.73\times10^{-5}(W/m^2K)$ for 1 micron size Ag particle. On the other hand, h_p below 5 mK is negligible compared with h_e, while h_p above 10 mK increases with increasing temperature and the temperature dependence shows a T^3 variation. The magnitude of h_p becomes larger than that of h_e above 10 mK. Consequently, the total conductance h_K is mainly attributable to h_e at low temperatures, and to h_p in the high-temperature regime, respectively. The suggestion to be emphasized here is to make the characteristic frequency ω_c as small as possible in order to have an efficient heat exchanger.

5. Boundary resistance between Ag sinter and solid He-3

The thermal boundary resistance between solid He-3 and sintered Ag particles of size 700 A was observed at NAGOYA [26] at temperatures between 0.5 mK and 10 mK. MAMIYA et al. [26] obtained the same order of magnitude and similar temperature dependence of the resistances as those for liquid He-3, which is expressed approximately as $R_KT\simeq1.2\times10^3m^2K^2/W$. It is interesting to apply the mechanisms described in Sec.3 and 4 to the interpretation of the above results. The magnetic coupling in [16] does not give the temperature dependence of the resistance of $R_K\sim T^{-1}$ when liquid He-3 is solidified, because the exchange interaction of the Korringa type between localized He-3 and He-3 quasi-particles is crucial in reproducing the temperature dependence $R_K\sim T^{-1}$. Also we see that if the magnetic coupling is dominant, the R_K between solid He-3 and sinter should show quite different T dependence above and below the magnetic ordering temperature $T_c\simeq1$mK of solid He-3. But the experiments [26] do not show this tendency.

Now we try to explain the observations [26] using the mechanism described in Sec.4. In the case of solid He-3, the heat should be transferred from the interface by normal sound due to this mechanism.

The formalism of [22] applies directly to this situation by replacing the zero-sound velocities in (2) with those of normal sound. Because the velocity of zero sound ($c \simeq 2\times10^4$cm/sec) is nearly equal to that of normal sound in solid He-3 ($v\simeq(2-5)\times10^4$cm/sec), we see that R_K should have the same order of magnitude as well as a similar temperature dependence in the case of liquid He-3.

6. Cooling of He-3 and He-4 mixture

The cooling of He-3 and He-4 mixture down to 100 μK or less has been of great interest in connection with the possible discovery of the superfluid phase of the mixture. This is, however, not so easy to achieve due to the problem of the Kapitza resistance between the refrigerand and the helium sample. Until recently, the lowest temperature achieved in the system of Ag sinter and the mixture is 235 μK [27]. The R_K between the mixture and Ag sinter has been observed by RADEBAUCH et al. [28], FROSSATI [10] and OSHEROFF and CORRUCCINI [29]. The results are expressed as $R_K T^2 \simeq 16.2$ m^2K^3/W between 0.8 mK and 4 mK.

To explain these results, the magnetic coupling described in Sec.3 should be ruled out because all surfaces are covered with the He-4 layer. In addition, we have no zero-sound modes in the mixture and also other collective modes in the mixture are ineffective in the coupling with the low-energy vibrational modes of the sinter. These imply that the mechanism due to the excitation of the collective modes given in Sec.4 is also invalid for the cooling of He-3 quasi-particles in He-4. One of the possible direct interactions between He-3 quasi-particles in the mixture and the sinter is through the low-energy vibrational modes of the sinter described in Sec.4.

7. Concluding remarks

I have reviewed the current understanding of the puzzling phenomenon of the low thermal boundary resistance between small particles and liquid He-3 observed at temperatures below 10 mK. Although the problem has become clearer in recent years due to the accumulation of experiments and theories, we have to admit that there is still detailed work to be done to clarify the heat-transferring mechanism both theoretically and experimentally. This would be related to the various sub-fields of condensed matter physics: phonon, surface, powder, metal, magnetism, etc. The effort to explore the mechanism is of great significance in low-temperature physics.

The author would like to thank S. Ishi for some stimulating remarks. Y. Masuda is acknowledged for informing the author about the result for the thermal boundary resistance between solid He-3 and Ag sinter.

References

1. O. Avenel, M. P. Berglund, R. G. Gylling, N. E. Phillips, A. Vetsleseter, M. Vuorio: Phys. Rev. Lett. 31, 76 (1973)
2. I. L. Bekarevich, I. M. Khalatnikov: Sov. Phys. JETP 12, 1187 (1961)
3. J. M. Dundon, D. L. Stolfa, J. M. Goodkind: Phys. Rev. Lett. 30, 843 (1973) and D. L. Stolfa: Thesis, Univ. Calif. San Diego 1973
4. K. Andres, W. O. Sprenger: Proc. 14th Inter. Conf. on Low Temp. Physics (Notrh-Holland, Amsterdam 1975) Vol.1, p.123

5. A. I. Ahonen, P. M. Berglund, M. T. Haikala, M. Krusius, O. V. Lounasmaa, M. A. Paalanen: Cryogenics 16, 521 (1976).
6. A. I. Ahonen, O. V. Lounasmaa, M. C. Veuro: J. Phys. (Paris) 39, C6-265 (1978)
7. E. Varoquaux: J. Phys. (Paris) 38, C6-1605 (1978)
8. D. O. Edwards, J. D. Feder, W. J. Gully, G. G. Ihas, J. Landau, K. A. Muelhing, J. Phys. (Paris) 39, C6-260 (1978)
9. J. P. Harrison: J. Low Temp. Phys. 37, 467 (1979)
10. G. Frossati: J. Physics (Paris) 39 C6-1578 (1978)
11. J. Gaboret: Phys. Rev. 137, A721 (1965)
12. G. A. Toombs, F. W. Sheard, and M. J. Rice: J. Low Temp. Phys. 39, 273 (1980)
13. T. Nakayama and N. Nishiguchi: Phys. Rev. B24, 6421 (1981)
14. N. Nishiguchi and T. Nakayama: Phys. Rev. B25, 5720 (1982)
15. D. L. Mills, M. T. Beal-Monod: Phys. Rev. A10, 343 (1974) and ibid, A10, 2473 (1974).
16. T. Perry, K. DeConde, J. A. Sauls, D. L. Stein: Phys. Rev. Lett. 48, 1831 (1982)
17. A. J. Legget, M. Vuorio: J. Low Temp. Phys. 3, 359 (1970)
18. T. Engel and R. Gomer: J. Chem. Phys. 52, 5572 (1970)
19. This estimation of dipole moment (0.1D) of He atom adsorbed on clean W seems to be large. Using polarizability χ_{ob} and the observed dipole moment μ_{ob} for Ar, Kr, and Xe, we have at most $\mu_{He} \simeq \chi_{He}\ \mu_{ob}/\chi_{ob} \simeq 0.02\text{–}0.05$ D for clean W.
20. C. P. Flynn, Y. C. Chen: Phys. Rev. Lett. 46, 447 (1981), and C. P. Flynn, J. E. Gunningham: J. Phys. C15, L1169 (1982)
21. P. E. Batson: Solid State Commun. 34, 477 (1980)
22. N. Nishiguchi, T. Nakayama: Solid State Commun. 45, 877 (1983)
23. B.Frisken, F. Guillon, J. P. Harrison, J. H. Page: J. Phys. (Paris) 42, C6-858 (1981)
24. M. C. Veuro: Acta Polytech. Scand. Ph. 122, 1 (1978), see electron micrographs of Ag sinter given in Fig.16.
25. W. D. Kingery and M. Berg: J. Appl. Phys. 26, 1205 (1955)
26. T. Mamiya, Y. Sawada, H. Fukuyama, Y. Masuda: private communication
27. H. Chocholacs, R. M. Mueller, J. R. Owers-Bradlay, CH. Buchal, M. Kubota, F. Pobell: private communication
28. Radebaugh, J. D. Siegwarth, J. C. Holste, Proc. 5th Inter. Cryogenic Engineering Conf. (Kyoto, 1974) p.232
29. D. D. Osheroff, L. R. Corruccini: Phys. Lett. 82A, 38 (1981)

Scattering and Absorption of Ballistic Phonons by the Electron Inversion Layer in Silicon: Theory and Experiment

J.C. Hensel, R.C. Dynes, B.I. Halperin*, and D.C. Tsui†

Bell Laboratories, 600 Mountain Avenue
Murray Hill, NJ 07974, USA

1. Introduction

This paper is a brief review of our investigations[1,2,3] using ballistic phonons to probe the two-dimensional electron gas (2DEG) in the inversion layer of Si. Our goal in the beginning was to elucidate the nature of the electron-phonon interaction for a 2DEG, an important problem physically just as in 3D but one rather less amenable to the conventional experimental procedures. In fact, in the absence of substantive results suspicions have arisen (mostly from interpretation of electrical transport) that the electron-phonon coupling in 2D might be anomalously strong compared to 3D[4]. Our ballistic phonon experiments are able to answer unequivocally that this is not the case.

On a more general plane, evidence has only recently come to light[3] that ballistic phonons are also a highly effective way of probing certain dynamical properties of the 2DEG as represented by the complex linear response function $\chi(\vec{q},\omega)$. Until now probes of the 2DEG have been limited mostly to optical and electrical transport techniques (or some combination thereof), both largely ineffectual in this regard. Representing a new approach we shall see that scattering and absorption of ballistic phonons can sample $\chi(\vec{q},\omega)$ of the 2DEG throughout a substantial region of energy $\hbar\omega$ and momentum transfer $\hbar\vec{q}$ in the plane. To a first approximation this function is a universal property of the 2DEG, whether it be in the Si inversion layer, in a GaAs-(AℓGa)As heterostructure or on the surface of liquid He.

2. Experiments

Figure 1 depicts the experiment[1,2]. The Si sample is prism shaped with a large MOSFET (metal-oxide-semiconductor field-effect transistor) device fabricated on the (001) basal plane. Ballistic phonons generated by pulsed optical excitation propagate from source (evaporated constantan film heater) to detector (superconducting granular Aℓ bolometer) via reflection from the Si-SiO_2 interface thereby passing twice through the intervening inversion layer. A collimating slot cut in the median plane blocks direct propagation from source to detector. The inset in Fig. 1 displays a typical time-resolved spectrum consisting of longitudinal (LA), mode converted (MC), and transverse acoustic (TA) phonons. A square-wave potential (from threshold voltage V_T to some specified voltage V_g) applied to the gate of the MOSFET modulates the electron (areal) density from zero to a value $n_s = Ce^{-1}(V_g-V_T)$ where C is the oxide (8500 Å thickness) capacitance per unit

* At Physics Department, Harvard University, Cambridge, Massachusetts 02138.
† Present address: Princeton University, Princeton, New Jersey 08544.

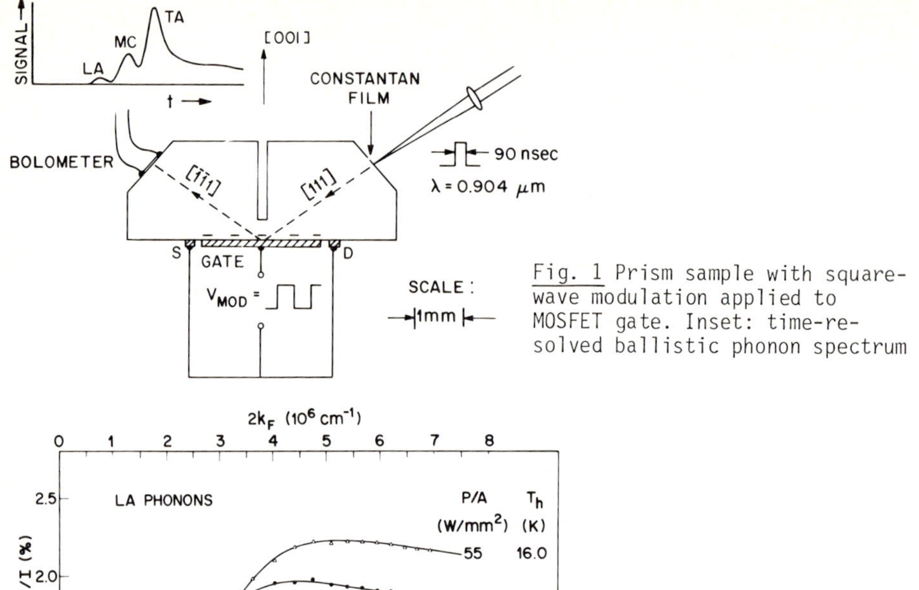

Fig. 1 Prism sample with square-wave modulation applied to MOSFET gate. Inset: time-resolved ballistic phonon spectrum

Fig. 2 Profiles of LA phonon attenuation: T = 2.15 K

area. The concomitant change ΔI in phonon intensity I is synchronously detected. Figure 2 shows ΔI/I for LA modes plotted versus $(V_g - V_T)^{1/2}$ which is directly proportional to the Fermi wavevector $k_F=(\pi n_s)^{1/2}$ (see upper scale). The experiment in effect provides a measurement of ΔI/I as the Fermi surface (a circle of diameter $2k_F$) is expanded in size. Four profiles of ΔI/I were taken each representing a specific heater excitation power density P/A. The corresponding heater temperature T_h was estimated from the black-body radiation law[5], $P/A = \sigma(T_h^4 - T_0^4)$, where T_0 is the ambient temperature and σ is a Stefan-Boltzmann constant. (See remarks on this assumption in a subsequent section.) Isotope scattering modifies the emitted Planckian spectrum producing a cutoff for q's > 4 x 10^6 cm^{-1} (for LA modes).

Each point in Fig. 2 can be regarded as the convolution of this phonon distribution $U(q,T_h)$ with some relative-intensity function ΔI(q)/I for the 2DEG. Crudely speaking, one can interpret the profiles in Fig. 2 by taking cognizance of the fact that the processes contributing to ΔI(q)/I tend to be maximal for q's in the vicinity of $2k_F$ by phase-space considerations, the result being that the convolution is some rough facsimile of $U(q,T_h)$. We do see in fact in Fig. 2 a certain resemblance to Planckian functions, the peaks shifting to larger $2k_F$'s with increasing T_h (Wien shift). In subsequent sections we shall consider the function $\Delta I(q)/I$ in more detail.

Profiles of $\Delta I/I$ for TA phonons have also been measured. While the shapes are similar to LA profiles, there is an overall shift to higher $2k_F$ owing to their larger q's for a given T_h.

3. Absorption Model

Initially[1,2] we attempted to interpret the attenuation $\Delta I/I$ in Fig. 2 in terms of absorption of ballistic phonons passing twice through the inversion layer. It is instructive to examine this model, despite its inadequacies, because it provides a stepping stone to a more sophisticated and realistic treatment.

The absorption of a ballistic phonon of wavevector $\vec{q}$ in the plane by a 2DEG at zero Kelvin scatters an electron from an occupied state $|\vec{k}| < k_F$ to a unoccupied state $|\vec{k}+\vec{q}| > k_F$, subject to conservation of crystal momentum (in the plane) and energy. The transition rate for this process is given by Fermi's Golden Rule,

$$\Gamma(q) = \frac{2\pi}{\hbar} |ME|^2 g_s g_v \frac{A}{(2\pi)^2} \int d^2k\ \delta(E(\vec{k}+\vec{q})-E(\vec{k})-\hbar\omega) \tag{1}$$

where A is a normalization area and $g_s g_v$ represents the spin degeneracy $g_s = 2$ and valley degeneracy $g_v = 2$. The electron-phonon matrix element (ME) and the phase-space integral in (1) are readily evaluated; and one obtains the following formula for absorption of phonons (mode j) propagating at an angle θ with respect to the normal to the plane,

$$\frac{\Delta I(q)}{I} = - \frac{D_j^2(\theta)}{\rho c_j^2 \cos\theta \sin\theta} \pi G(q) F(k_j) . \tag{2}$$

Here, ρ is the crystal density; c_j is the phase velocity, and $F(k_j)$ is a form factor (for phonon wavevector k_j normal to the plane) of order unity for all relevant phonons. The results consist of two parts: the electron-phonon coupling represented by a deformation potential constant $D_j(\theta)$ (in Herring's notation $D_\ell = \Xi_d + \Xi_u \cos^2\theta$ for LA modes) and a structural factor G(q) expressing the entire q dependence of the absorption. G(q), shown in the inset of Fig. 3, exhibits a cusp-like behavior in the neighborhood of $2k_F$ representing the absorption spectrum (or, according to detailed balance, the low-temperature emission spectrum as the case may be).

To make the comparison with experiment we must convolve $\Delta I(q)/I$ in (2) (doubled for 2 passes) with the phonon spectrum $U(q,T_h)$. The results for LA phonons shown in Fig. 3 are disappointing. The shapes of the profiles are at variance with the data in Fig. 2; but even more seriously, the theory (based on the "bulk" deformation potentials for Si, $\Xi_d = -6.0$eV, $\Xi_u = +9.0$eV) underestimates the strength of $\Delta I/I$ by an order of magnitude. Clearly the first and most obvious approach is inadequate. The problem is not that the absorption model is wrong, it is simply incomplete.

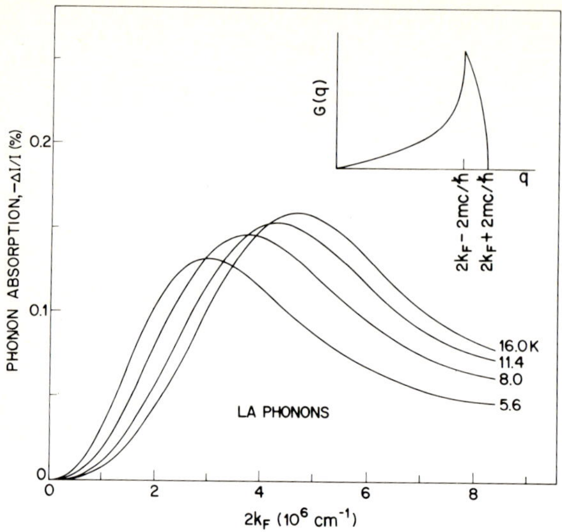

Fig. 3 Calculated profiles of absorption of LA phonons by a 2DEG
Inset: $G(q)$ for the case $mc/\hbar < k_F$

4. Dynamical Model

A clue to the dilemma was the realization that the reflectance of the phonons from the interface was very small, $r \sim 4\%$, as the angle of incidence $\theta = 54.7°$ is close to the Brewster angle. Had r been close to 100% the absorption model might have been a reasonable approximation. This not being the case, however, one must consider all other contributions, the most important one being the phonon amplitudes backscattered from the inversion layer itself. This amplitude can interfere with the amplitude of the specularly reflected primary beam. The more nearly comparable the two amplitudes are the stronger the interference effect, the enhancement factor being roughly $r^{-1/2} \sim 5$.

Our approach[3] is to calculate the linear response of the 2DEG to perturbations by the phonons. The radiation fields thus set up are contributed by both real and imaginary parts of the density-density response function $\chi(\vec{q},\omega)$ at wavevector $\vec{q}$ in the plane and frequency ω. This is a dynamical theory whereas the former approach was "static" in the sense that it dealt exclusively with $\mathrm{Im}\chi(\vec{q},\omega)$ as an absorption mechanism.

We first formulate the coupling between the phonons and the 2DEG. We shall regard the phonons classically as elastic waves and assume that both Si and SiO_2 are elastically isotropic. We adopt a coordinate system wherein z is normal to the Si-SiO_2 interface and the oxide occupies the half space $z < 0$. The coupling between phonons and electrons in the [001] occupied valleys is given by an interaction Hamiltonian,

$$H' = \int d^3r\, n(\vec{r})[a\varepsilon_{xx} + b\varepsilon_{zz}] . \tag{3}$$

$n(\vec{r})$ is the electron density, a and b are deformation potential constants ($a = \Xi_d$ and $b = \Xi_d + \Xi_u$), and ε_{ij} is the elastic strain tensor. The equations of motion for the displacement $\vec{u}$ are then

$$\rho\omega^2\vec{u} + \nabla\cdot\overleftrightarrow{\sigma} = \begin{cases} 0 \; , \; z > 0 \\ \dfrac{\delta H'}{\delta\vec{u}} \; , \; z < 0 \end{cases} \tag{4}$$

where $\overleftrightarrow{\sigma}$ is the stress tensor (elastic moduli relating $\overleftrightarrow{\sigma}$ and $\overleftrightarrow{\varepsilon}$ and density ρ are different in the two regions $z > 0$ and $z < 0$). Solutions to (4) in the oxide (source free) half-space are trivial (i.e., plane waves propagating in the +z direction), but in the half-space $z < 0$ occupied by electrons we have a source term to represent the driving force due to the perturbed electron density $n(\vec{r})$ making the problem more formidable. To proceed we approximate the profile of the electron probability density $f(z)$ normal to the plane by a rectangular profile of thickness s spaced a distance h from the interface located at $z = 0$; i.e., $f(z) = s^{-1}[\theta(z+h)-\theta(z+h-s)]$, where $\theta(z)$ is the unit step function. By making this simplification and taking advantage of the fact that the perturbations are weak (they are only a few percent) we are able to obtain an essentially "exact" analytical expression to first order in $\chi(\vec{q},\omega)$ for the total, far-field LA amplitude in reflection from which follows our final results (quoted for sake of brevity in the limit $s \to 0$),

$$\begin{aligned}\frac{\Delta I(q)}{I} = &- \frac{1}{R_{\ell\ell}} \frac{B^2}{\rho c_\ell^2 \cos\theta\sin\theta} [q\mathrm{Im}\chi\cos 2k_\ell h - q\mathrm{Re}\chi\sin 2k_\ell h] \\ &- \frac{2B^2}{\rho c_\ell^2\cos\theta\sin\theta} q\mathrm{Im}\chi \\ &+ \frac{R_{t\ell}}{R_{\ell\ell}} \frac{BC}{\rho c_t^2\cos\phi\sin\theta} [q\mathrm{Im}\chi\cos(k_t-k_\ell)h + q\mathrm{Re}\chi\sin(k_t-k_\ell)h] \\ &- \frac{R_{\ell t}}{R_{\ell\ell}} \frac{BC}{\rho c_\ell^2\cos\theta\sin\phi} [q\mathrm{Im}\chi\cos(k_t-k_\ell)h + q\mathrm{Re}\chi\sin(k_t-k_\ell)h] \\ &+ 4 \text{ more terms} .\end{aligned} \tag{5}$$

Here, $R_{\ell\ell}$, $R_{\ell t}$ and $R_{t\ell}$ are reflection coefficients, i.e., ratio of reflected to incident amplitudes, in an obvious notation; c_ℓ, c_t are the LA and TA phase velocities, respectively; and $k_\ell = q\cot\theta$ and $k_t = q\cot\phi$ are the LA and TA wavevectors, respectively, normal to the plane. (Angle ϕ is the angle of reflection of TA modes.) The electron-phonon coupling is represented by $B = a\sin^2\theta + b\cos^2\theta$ and $C = (a-b)\sin\phi\cos\phi$. A plot of (5) (in its entirety) is shown in Fig. 4(a) for $\chi(\vec{q},\omega)$ calculated for a noninteracting electron gas[6] at zero Kelvin [see Fig. 4(b)], revealing an appreciably different shape from the simple absorption result (cf. insert in Fig. 3).

Equation (5) although lengthy is quite transparent. A prominent feature is its oscillatory nature evidencing the anticipated interference between waves radiated from the 2DEG and waves reflected from the interface, the phase difference being essentially the ratio of path difference 2h to a phonon wavelength, $2\pi k^{-1}$. (In the experiment k's are limited to values $\lesssim 2\pi/2h$, so the oscillation is one cycle or less.) We also note that unlike the absorption model where only $\mathrm{Im}\chi$ appears (note that $\pi G(q) = q\mathrm{Im}\chi$) to represent dissipation, now both $\mathrm{Re}\chi$ and $\mathrm{Im}\chi$ are present in a mix depending upon the phase difference. The origin of the several terms in (5) is illustrated

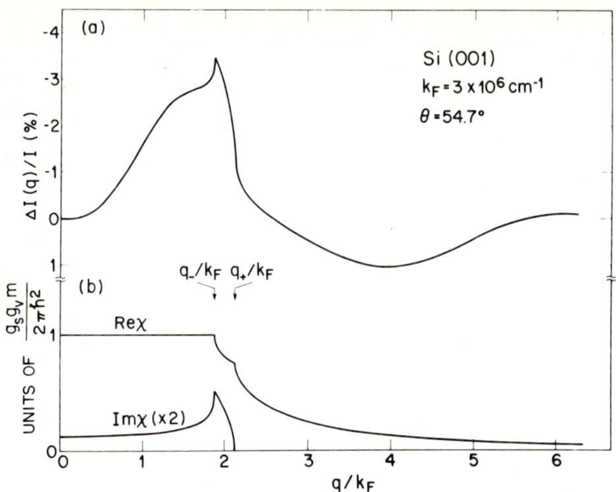

Fig. 4 (a) Predicted fractional change in reflected intensity $\Delta I(q)/I$ for LA phonons. Note: $q_{\pm} = 2k_F \pm 2mc/\hbar$. (b) $\chi(\vec{q},\omega)$ for noninteracting electrons

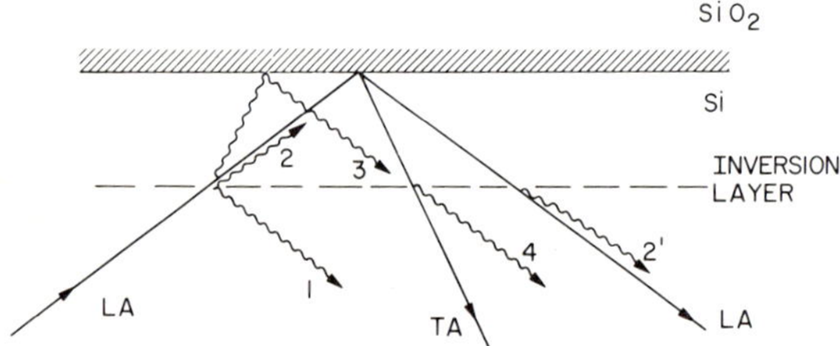

Fig. 5 Identification of individual contributions to $\Delta I(q)/I$

schematically in Fig. 5. The leading term in (5), enhanced by $R_{\ell\ell}^{-1}$ ($\approx$ -5.0), arises from direct radiation (wave 1 in Fig. 5) excited by the incident LA wave. The second term (roughly 10x smaller) is bona-fide absorption suffered by incident and reflected primary waves in passing through the electron layer as represented in Fig. 5 by destructively interfering forward scattered waves 2 and 2′. Apart from a form factor this one term is essentially the whole of the earlier result [cf. Eq. (2)]. The third and fourth terms involve TA phonons mode-converted upon reflection at the interface (waves 3 and 4, respectively). Omitted terms are higher order in "reflection number."

The convolution $\Delta I/I$ of Eq. (5) (all 8 terms) with the phonon distribution $U(q,T_h)$ is shown in Fig. 6. One sees that vis à vis the earlier model the agreement with the data in Fig. 2 is markedly improved. Now the overall magnitudes are in accord, and the shapes, particularly as regards to the elevated tails, are more realistic. There is a discrepancy, however, in the positions of the profiles, the calculated ones being shifted to a too large $2k_F$ for reasons unknown. One might suspect our description of the phonon spectrum, in particular the rather simplistic determination of T_h by a black-body radiation model for the heater film. This may not be as bad as it first seems inasmuch as there are two countervailing effects at work: the acoustic mismatch[7] between film and substrate which tends to raise T_h (and <u>worsens</u> the discrepancy) and radiation into the He bath [8] which

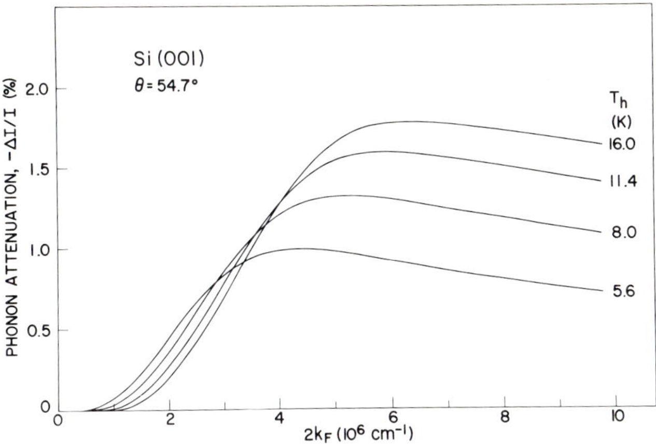

Fig. 6 Profiles of $\Delta I/I$ for LA phonons calculated according to the dynamical model

tends to lower T_h. Ultimately, frequency selective experiments exploiting the voltage tuneability of the MOSFET may sort out these issues.

Finally, it should be emphasized that the theory contains no free parameters except s and h which because of their somewhat artificial nature are not precisely definable physically. It is reassuring, however, that optimum agreement occurs when the rectangular profile midposition/interface separation, h-s/2, is adjusted to be ~ $3.5a_0$ (where a_0 is a variational thickness parameter for the lowest subband), rather close to the variational first moment of z for the charge density $z_0 = 3a_0$.

Analysis for the TA modes proceeds in basically the same way with two extra complications: firstly, the two orthogonal TA modes are degenerate in the [111] direction and exhibit internal conical refraction, but only one polarization couples with the electrons. Secondly, $\phi = 54.7°$ is beyond the critical angle for reflection of LA phonons (1st critical angle) making the reflection coefficients R complex.

5. Conclusions

Theory and experiment are now in substantial accord on the question of the strong attenuation of ballistic phonons by the inversion layer in Si leaving little doubt that the electron-phonon interaction for a 2DEG does not differ from that in 3D. Further, we see that it is feasible to do direct phonon spectroscopy with monochromatic phonons on the linear response function of a 2DEG, an important physical entity quite inaccessible by other means. This would be particularly interesting in the presence of a strong quantizing magnetic field. The theory can be extended to the problem of transmission and reflection of ballistic phonons in thin metal films (not necessarily 2D), either single or multiple, with interfaces — a subject of continuing interest.

Acknowledgement

We express our thanks to T. M. Rice and M. Lax for numerous invaluable discussions, to G. Kaminsky and J. P. Garno for sample preparation, and especially to F. G. Unterwald for his crucial role in the execution of the experiments. One of the authors (BIH) is grateful to the hospitality of the Institute of Theoretical Physics, Santa Barbara and for support by the NSF through Grant DMR 82-07431.

References

1. J. C. Hensel, R. C. Dynes, and D. C. Tsui: Surf. Sci. 113, 249 (1982); J. Phys. (Paris) Colloq. 42, C6-308 (1981).
2. J. C. Hensel, R. C. Dynes, and D. C. Tsui: Phys. Rev. B28, 1124 (1983).
3. J. C. Hensel, B. I. Halperin, and R. C. Dynes (to be published).
4. See, for example, T. Ando, A. B. Fowler, and F. Stern: Rev. Mod. Phys. 54 437 (1982).
5. O. Weis: Z. Angew. Phys. 26, 325 (1969).
6. F. Stern: Phys. Rev. Letters 18, 546 (1967).
7. W. A. Little: Can. J. Phys. 37, 334 (1959).
8. T. Ishiguro and T. A. Fjeldy: Phys. Lett. 45A, 127 (1983).

Low Wavevector Phonons in the 2-Dimensional Electron Solid on Liquid Helium

F.I.B. Williams

Service de Physique du Solide et Résonance Magnétique
Institut de Recherche Fondamentale
C.E.N. Saclay, 91191 Gif sur Yvette, Cedex, France

1. Introduction

Electrons trapped at the liquid-vapour (essentially vacuum) interface of helium at low temperatures make an ideally simple system of particles in interaction : two-dimensional motion with very small non-uniformity and exactly known monotonic power law interaction. From the experimental standpoint, the charge permits a strong, direct interaction with an electromagnetic probe field, while the low mass and scattering rate make possible quite direct experiments on the phonons despite the seemingly impossibly small number of particles (10^8-10^9).

I shall not try to give a complete review (see [1,2]), but only to summarize what the experiments have been able to say about the phonons. So far, the principal motivatation has been to learn about the phase diagram and the nature of the phase transition (s), and the phonon properties have been investigated mostly as a manifestation of these. Only the small and very small (to be defined below) wavevector regimes have been studied.

2. The System

A single electron is drawn from the vacuum towards the surface of the dielectric liquid helium by the polarization force, but it cannot penetrate unless it can supply the necessary 1eV or so to enter the conduction band. The lowest energy state (c.-8K) is one localized to some 75 Å above the surface with free motion in the plane (X-Y).

Adding further electrons to surface density n changes only the X-Y motion due to their repulsive interaction, $V = \sum_{ij} e^2/|\underline{r}_i-\underline{r}_j|$, at least if the Fermi energy, $E_F \approx \frac{\hbar^2}{m} n$, is much less than the 8K ($n \ll 10^{12}$ cm^{-2}). $\underline{r}_i$ is the 2-dimensional (X-Y) position vector of electron i, m and e the free electron mass and charge. At a finite temperature T there are 3 characteristic energies for the X-Y motion ; T, E_F and $V_{coul.} = e^2/a$ (a is defined by $\pi a^2 = n^{-1}$). All the experiments done to date pertain to the classical limit $T \gg E_F$ where we may specify the state of the system by a single dimensionless parameter

$$\Gamma \equiv V_{coul}/T \, . \qquad (1)$$

In principle, the results of experiments done at the same values of Γ are related by simple dimensional scaling factor.

3. Phases

There are at least two and possibly three thermodynamically distinct phases : solid and liquid with a possible subdivision into oriented liquid("hexatic") [3] and isotropic liquid. There is no distinction between liquid and gas. The solid-liquid transition takes place at $\Gamma=\Gamma_m \approx 135 \pm 5$ [4,5,6,7] (numerically $T_m[K] \approx 0.23\, n^{1/2}$ [10^8 cm^{-2}]) and is probably not of first order. No experimental evidence has been presented for or against the oriented liquid phase.

4. Phonons I

In the solid phase, one expects the usual 2 phonon branches ; their frequencies are given by

$$\omega_\ell^2 = \frac{2\pi e^2 n}{m} k \quad \text{and} \quad \omega_t^2 = \frac{\mu}{mn} k^2 \tag{2}$$

where $\mu=\mu_o f(\Gamma)$ is the shear modulus with $\mu_o=0.448\, e^2/a^3$ [8] being its value at $T=0$ ($\Gamma=\infty$) if we choose to define $f(\infty)=1$; $f(o) = 0$ says that the high temperature phase has no shear elasticity and presumably $f(\Gamma)$ has some sort of singularity at $\Gamma=\Gamma_m$.

5. Realization

Figure 1 illustrates schematically a real experimental configuration where the electron sheet is sandwiched between two planar control electrodes, of typical diameter 20-30 mm separated by 2-3 mm and confined by a guard ring at or near the (negative) potential of the upper electrode. The densities attainable in this configuration are limited by a macroscopic surface instability to $n \lesssim 2\ 10^9$ cm^{-2} situating the melting transition at $T_m \lesssim 1$K ($\sim n^{1/2}$), the transverse sound velocity at $v_t \lesssim 10^6$ cm s^{-1} ($\sim n^{1/4} f^{1/2}$).

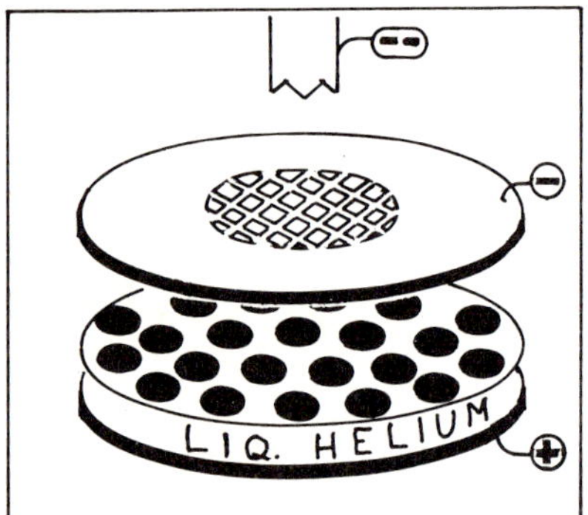

Fig.1. Schematic experimental configuration. The liquid helium level is set at height d above the lower ⊕ electrode at potential V with respect to the upper ⊖ electrode. A filament ⊖⊖, more negative than ⊖ , is heated briefly to emit electrons until the accumulated surface charge cancels out the field between the surface and the upper electrode. The surface density is then given by $n = V/4\pi de$. The guard ring needed to prevent them expanding is not shown

6. Phonons II

The electrodes screen the Coulomb interaction over long distances, turning the divergence of v_ℓ as $k \to 0$ in the plasmon branch into a more familiar linear dispersion with

$$v_\ell = \frac{4\pi e^2 n}{m} \frac{hd}{h+d} \lesssim 4.10^8 \text{ cm s}^{-1}$$

if h and d are the distances of the electrons from the upper and lower electrodes. Thus, apart from the large ratio of $v_\ell | v_t \approx 10^2$, the phonons are not very different from those of "ordinary" solids. This is even true in a sense for their damping. They are damped due to anharmonicity and to collisions with defects, these being helium atoms in the vapour for $T \gtrsim 0.7K$ and thermally generated rugosity of the helium surface ($\sim$ 1-2 Å) for lower temperatures [9,10,11]. This latter mechanism gives typically $10^{+6} < 1/\tau < 10^{+10}$ sec^{-1} ($\sim n^2$ for $n \gtrsim 10^8$ cm^{-2}.

7. Experiments

The first experiment was reported by GRIMES and ADAMS [4] ; they looked at longitudinal phonons of very small wavenumber (c. 2 cm^{-1} for $k_{Debye} \approx 10^5$ cm^{-1}) and frequencies up to c. 50 MHz. They did this by coupling the electrons to a radial electrostatic longitudinally polarized exciting field. On the ideas presented above, they should have seen the response from a longitudinal phonon of frequency $\omega = v_\ell k \approx 100$ MHz with $\omega\tau \approx 1$, whereas in fact they saw a series of sharp resonances starting around 8 MHz with widths of c. 2 MHz ! Clearly something had been forgotten.

8. Phonons III

Some years before, SHIKIN [12] had investigated the possibility that an isolated electron could trap itself in the deformation it induces by being pushed onto the helium surface. He found that this would happen only below c. 1 mK. But, stimulated by the surprising experiment just described, FISHER, HALPERIN and PLATZMAN [13] pointed out that for a many-electron system trapping occurs at much higher temperatures if long-range correlation sets in (solid phase).

Once in the solid phase, the time average pressure exerted by the electrons has the periodicity of the lattice and the surface is able to respond by deforming into a commensurate lattice of dimples some 1/10 Å deep into which the electrons trap. One must now take into account the displacement of the coupled deformation lattice when working out the electron dynamics and this alters dramatically the very low wavevector spectrum.

Displacing the deformation lattice in the X-Y plane involves setting the liquid into motion and, because it is superfluid, there is an inertia associated with this of mass per electron

$$M_g = \frac{1}{n} \rho/g \; g^2 \bar{\zeta}_g^{-2} \frac{1}{2} N_g \qquad (3)$$

where $\bar{\zeta}_g = \frac{eE}{\alpha} \frac{n}{g^2} \exp - \frac{1}{2} g^2 \langle u^2 \rangle$ is the time average amplitude of the gth Fourier component of the deformation. E is the pressing electric field ($\approx 2\pi ne$, usually), α is the surface tension and $\langle u^2 \rangle$ the time average mean square displacement of an electron about its lattice site. There is one such mass for each star of N_g reciprocal lattice vectors $|G|=g$ and it is coupled to the electron by a harmonic force of constant $M_g \Omega_g^2$ which represents the potential of the electron in the deformation. $\Omega_g^2 = \rho/\alpha \; g^3$ gives the frequency of capillary waves. These considerations lead to the very simple mechanical model for each $\underline{k}$ vector shown on Fig.2.

This model yields all the modes of the coupled system if all the $\{M_g\}$ are attached to the electron in parallel. Usually however the fundamental g_o dominates. Typically $M_{g_o}/m \approx 10$ for $n = 4.10^8$ cm^{-2} at $T \approx \frac{2}{3}$ Tm. Intui-

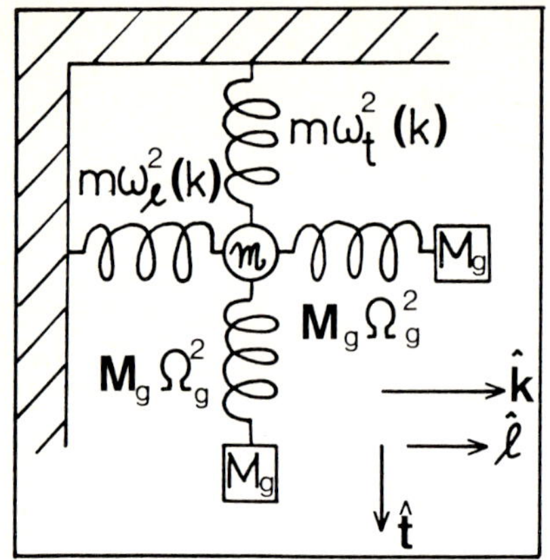

Fig.2. Model for dynamics of coupled electron-dimple lattices. Wavevector k(modulo g) and polarization are conserved, but each Fourier component g of the dimple lattice couples a mass $M_g \sim \exp -g^2\langle u^2\rangle$ to the electron mass m via a spring of force constant $M_g\ \Omega_g^2$, where Ω_g is the capillary wave frequency. $\omega_\ell(k)$, $\omega_t(k)$ are unperturbated longitudinal and transverse electronic phonon frequencies. Electromagnetic field couples by electronic charge e in the obvious way

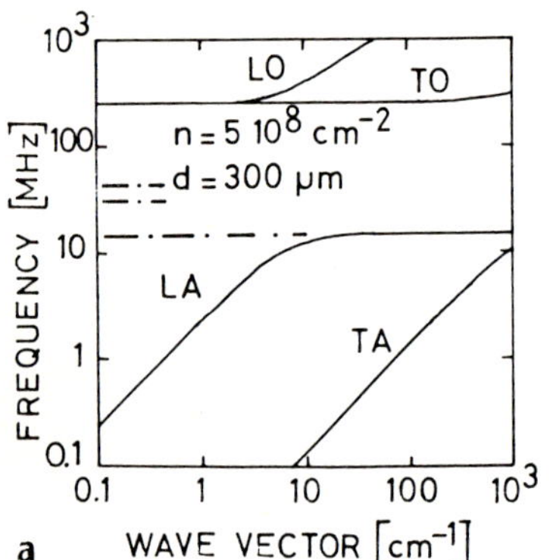

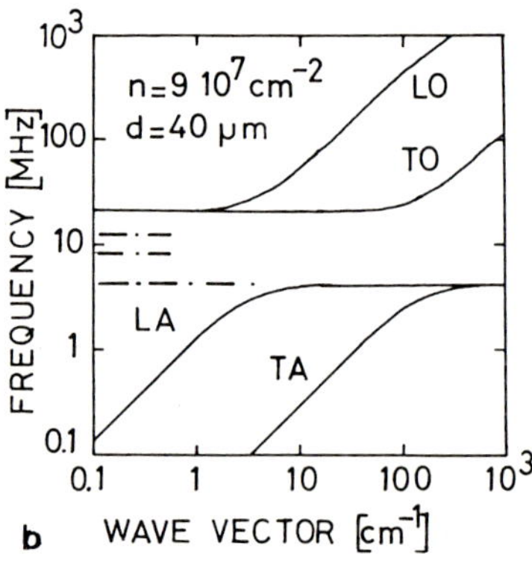

Fig.3. Log-Log plot of the low k dispersion of the coupled electron-dimple lattices for $n = 5\ 10^8\ cm^{-2}$ (a) and $n = 9\ 10^7\ cm^{-2}$ (b). The unperturbed electron modes are the dashed curves, the unperturbed capillary modes the dash-dotted curves. The longitudinal and transverse branches decouple for wave numbers beyond $k_p^o \approx \omega/v_p$, $p = \ell, t$

tion or elementary algebra then gives the low k dispersion curve shown on Fig.3. For each polarization the upper branch corresponds to antiphase ("optical") motion of the electron and its dimple and the lower branch to in phase ("acoustic") motion. The deviation from Ω_{g_o} of the k=0 optical frequency ω_o is a measure of the interaction. The two lattices decouple for $k \gg k_p^o \equiv \omega_o/v_p$, $p = \ell, t$ when the optical branches revert to the unperturbed phonon frequencies.

9. Experiments II

9.1 Very Low-k LA

GRIMES and ADAMS were of course seeing the longitudinal acoustic branch. In addition they saw a number of modes (Ω_1, Ω_2 on figure 3) at nearly unshifted capillary wave frequencies at higher g. The series of g so deduced corresponds to that for a triangular lattice.

9.2 Very Low-k TO

GALLET et al. [14] observed the transverse optical mode in the $k \ll k_t^o$ limit and were able to exploit it to deduce the temperature dependence of the shear modulus [15,16] through the Debye-Waller type factor $\exp -\frac{1}{4} g^2\langle u^2\rangle$ entering into its frequency in the $k\to 0$ limit

$$\omega_o^2 \approx \frac{(eE)^2 n}{m\alpha} \quad \exp -\frac{g_o^2}{4}\frac{T}{\mu}\int_{\omega_o}^{\omega_D}\frac{d\omega}{\omega} \quad . \tag{4}$$

Although the same general technique of radiofrequency spectroscopy was used in this experiment too, the coupling was achieved rather differently. An inductive longitudinal electric field

$$\underline{E} \approx \hat{x}\, Vk\, \cos(kx-\omega t), \quad k = \omega/c_e$$

was created at the electrons by a slow-wave meander line on the lower electrode linking the input and output transmission lines. It excites the longitudinal mode non-resonantly to electron velocity $v \approx \hat{x}\,\frac{e/m\,\omega E}{\omega_\ell^2(k)}$ which generates a transverse Lorentz force $F_y = \frac{e}{c} v_x B_z = \frac{\omega\omega_c}{\omega_\ell^2} eE_x$ in the pressure of a magnetic field B normal to the surface ($\omega_c = \frac{e}{mc} B$ is the electron cyclotron frequency). This force is then used to resonantly excite the transverse mode when the electromagnetic dispersion $\omega = c_e k$ crosses the electron dispersion $\omega_{t_o}(k) \approx \omega_o$ and it is detected by the additional loss from the transmission line. The experimental principle is made clear upon superposing the dispersion curve of the electromagnetic structure (Fig.4) on that of the electrons shown on Fig.3. Intersections indicate resonances and these are searched for by sweeping the frequency and monitoring the transmitted power.

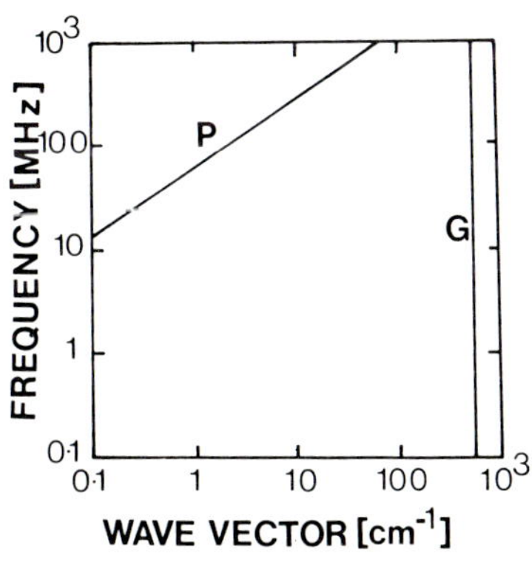

Fig.4. Log-log plot of the principle (P) and the first geometric (G) branches of the dispersion of the meander line electromagnetic structure used to detect the $k\to 0$. TO coupled mode and the $k \approx 500\ cm^{-1}$ decoupled T mode. This figure may be superposed on Fig.3

9.3 Low k Transverse

Very recently DEVILLE et al. [17] have observed the transverse mode in the region $k > k_t^o$ where the electrons are decoupled from the substrate and the dispersion is linear. They employed the same meander line as in the preceding experiment, but this time utilizing the electrostatic longitudinal field at the geometrical wavelength $k_G \approx 500\ cm^{-1}$ represented by the branch

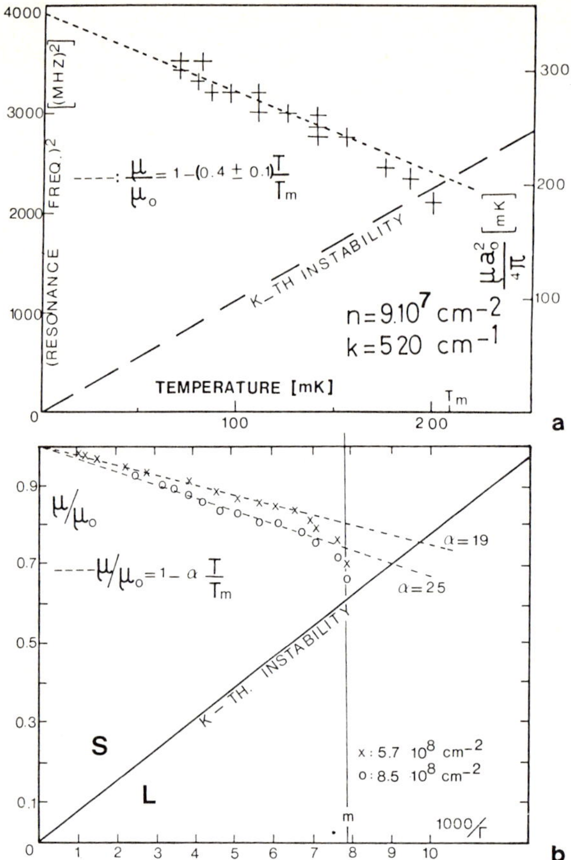

Fig.5 (a). The square of the k=520 cm^{-1} transverse phonon frequency vs temperature [17]. The right-hand scale gives the quantity $\mu a_o^2/4\pi$ relevant to the Kosterlitz-Thouless stability criterion for a solid. The criterion is represented by the line labelled K-Th . Solid phase should be unstable on the right of it. (b) The shear modulus normalized to its zero temperature value vs $\Gamma^{-1}(\sim T)$. The values are derived from experiments measuring the k=0 TO frequency ω_o [16]. The line k-Th has the same significance

"G" on Fig.4. To satisfy $k_t^o \ll k_G$ necessitates lowering the density to $n < 10^8$ cm^{-2} in order to lower ω_o ($\sim n^{3/2}$). To feel the effect of the 500 cm^{-1} field requires reducing the depth d of the helium to less than 50 μm. Once again, the Lorentz force is used to create a transverse excitation field.

The preliminary results of this experiment are shown on Fig.5a where the square of the transverse k = 520 cm^{-1} phonon frequency is plotted against temperature. The linewidth of the resonance was found to be considerably more than that expected from the electron-substrate scattering and presumably reflects the phonon-phonon and phonon-dislocation pair scattering. It varies roughly linearly in temperature at $T \ll T_m$, but more rapidly within about 10% of T_m, broadening out of existence at T_m.

10. Shear Modulus

It is straightforward to extract the shear modulus from the results of this last experiment from the relation $\frac{\mu}{mn} = v_t = \omega_t/k_G$ and that in how the right hand scale of Fig.5a is calculated.

On scaling these results as indicated in formula (2), they are found to be in essential agreement with the shear modulus deduced from the TO coupled branch at higher densities deduced by using formula (4) and shown on Fig.5b. The linear temperature dependence for $T/T_m < 0.8$ is slightly larger than both that "observed" by a molecular dynamics calculation [18] and that calculated analytically from the anharmonicity [19,20]. Both sets of results indicate a more rapid fall off near T_m, but neither is yet sufficiently precise to check the detailed form against the calculations [3] of the dislocation melting mechanism of KOSTERLITZ and THOULESS. However the value at $T=T_m$ is within 10% of the stability criterion for the solid phase

$$\frac{\mu a_o^2}{4\pi} \geq T$$

illustrated by the lines labelled K-Th of Fig.5. a_o is the triangular lattice constant and the limit $v_\ell >> v_t$ is taken.

11. Conclusion

The "simple" 2-D electron system shows an interesting low-k mode structure due to coupling with the deformable helium surface. Although this is an unwelcome complication as regards the study of the electrons, it did at least make it unexpectedly easy to see the onset of long-range correlation. It has finally proved possible to go beyond the region where this coupling is important to look directly at the pure phonon modes, albeit still in the small k-vector region.

Although this brief survey has been restricted to electrons on a "free" liquid helium surface, it should be mentioned that work is going on to put the electrons on a stiffer substrate, e.g. directly on solid neon or on a thin film of He on a solid substrate [21].

12. Acknowledgements

I should like to acknowledge particularly the help of past and present Saclay colleagues : Gérard Deville, François Gallet, Christian Glattli, Didier Marty, Jacqueline Poitrenaud, Aliro Valdes and, for their technical help, Claudine Heyer, Maurice Eisenkremer and Patrick Pari.

13. References

1. Yu. P.Monarkha and V.B. Shikin : Fiz. Nizk. Temp. *8*, 563 (1982) [Soviet J. Low Temp. Phys. *8*, 279 (1982)]
2. F.I.B. Williams : Surf. Sci. *113*, 371 (1982)
3. D.R. Nelson and B.I. Halperin : Phys. Rev. B *19*, 2457 (1979)
4. C.C. Grimes and G. Adams : Phys. Rev. Lett. *42*, 795 (1979)
5. A.S. Rybalko, B.N. Esel'son and Yu. Z. Kovdrya : Fiz. Nizk. Temp. *5*, 947 (1979)
6. D. Marty, J. Poitrenaud and F.I.B. Williams : J. Physique Lettres *41*, L311 (1980)
7. R. Mehrotra, B.M. Guenin and A.J. Dahm : Phys. Rev. Lett. *48*, 641 (1982)
8. L. Bonsall and A.A. Maradudin : Phys. Rev. B *15*, 1959 (1977)

9. V.B. Shikin and Yu. P. Monarkha : J. Low Temp. Phys. *16*, 193 (1974)
10. M. Saitoh : J. Phys. Soc. Japan *42*, 201 (1977)
11. A.J. Dahm and R. Mehrotra : J. Low Temp. Phys. *50*, 201 (1983)
12. V.B. Shikin : Zh. Eksprim. Teor. Fiz. *60*, 713 (1971) [Soviet Phys.-JETP *33*, 387 (1971)]
13. D. Fisher, B.I. Halperin and P.M. Platzman : Phys. Rev. Lett. *42*, 798 (1979)
14. G. Deville, F. Gallet, D. Marty, J. Poitrenaud, A. Valdes and F.I.B. Williams, in : *Ordering in Two Dimensions*, Ed. S.K. Sinka (North Holland, Amsterdam, 1980) p.309
15. G. Gallet, G. Deville, A. Valdes and F.I.B. Williams : Phys. Rev. Lett. *49*, 212 (1982)
16. A. Valdes : Thèse de Troisième Cycle, Université de Paris Sud, Orsay, France 1982
17. G. Deville, A. Valdes and F.I.B. Williams : to be published
18. R. Morf : Phys. Rev. Lett. *43*, 931 (1979)
19. D.S. Fisher : Phys. Rev. B *26*, 5009 (1982)
20. M. Chang and K. Maki : Phys. Rev. B *27*, 1646 (1983)
21. K. Kajita, J. Phys. Soc. Jpn. *52*, 372 (1983)

Reciprocity Theorem for Phonon Transitions at Ideal Interfaces Within the Acoustic Mismatch Model

O. Weis

Abteilung Festkörperphysik, Universität Ulm, Oberer Eselsberg
D-7900 Ulm, Fed. Rep. of Germany

The reflection, refraction and mode conversion of an incident plane sound wave at a plane lossless interface between two dissimilar media (L) and (K) can be calculated by applying continuum acoustics. If this treatment is transferred in an appropriate manner to the description of phonon transitions at the same interface, we use the name 'acoustic mismatch model'.

In this paper only the results of a detailed theoretical investigation [1] can be given, in which phonon transitions and the associated reciprocity are treated in full length for the most general case of an interface between two anisotropic solids. This general treatment includes of course all other material combinations as special cases. Figure 1 gives an example where half-space (L) is a germanium monocrystal and medium (K) is vacuum. The interface reduces to a free surface and is chosen here as the $(00\bar{1})$ crystal face.

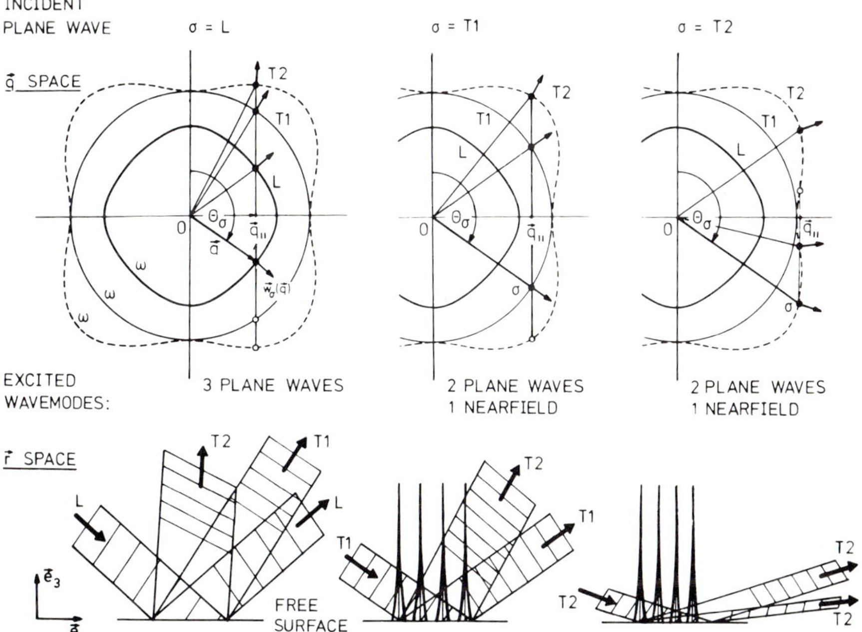

Figure 1: Plane wave reflection at a free interface

We consider now an incident plane wave of polarization σ, frequency ω and wave vector $\vec{q}$. This plane wave and the corresponding phonons can be represented in the usual way in $\vec{q}$ space by a point at the corresponding frequency surface $\omega_\sigma(\vec{q}) = \omega$ (see Fig. 1). The wavenormal $\vec{e}_{n\sigma}$ is the unit vector in $\vec{q}$ direction and gives also the direction of the phase velocity $\vec{c}_\sigma(\vec{e}_n)$:

$$\vec{q} = \vec{e}_{n\sigma} \cdot |\vec{q}| = \vec{e}_{n\sigma} \cdot \omega / c_\sigma(\vec{e}_n), \quad \vec{c}_\sigma(\vec{e}_n) = \vec{e}_{n\sigma} \cdot c_\sigma(\vec{e}_n). \tag{1}$$

In anisotropic media the energy velocity $\vec{w}_\sigma(\vec{q})$ (= group velocity of the corresponding phonon $\vec{q},\sigma$) deviates from the phase velocity, since it points always perpendicular to the frequency surface at the $\vec{q}$ end point:

$$\vec{w}_\sigma(\vec{q}) = \vec{\nabla}_{\vec{q}}\, \omega_\sigma(\vec{q}). \tag{2}$$

In our figures 1 to 3 the direction of energy flow is indicated by an arrow for all plane-wave states involved. In Fig. 2 an interface between two halfspaces of germanium is considered which corresponds in practice to a plane grain boundary. For convenience, we have chosen a mirror plane as plane of incidence in order to simplify the $\vec{r}$-space illustration since all group velocities lie within the mirror plane.

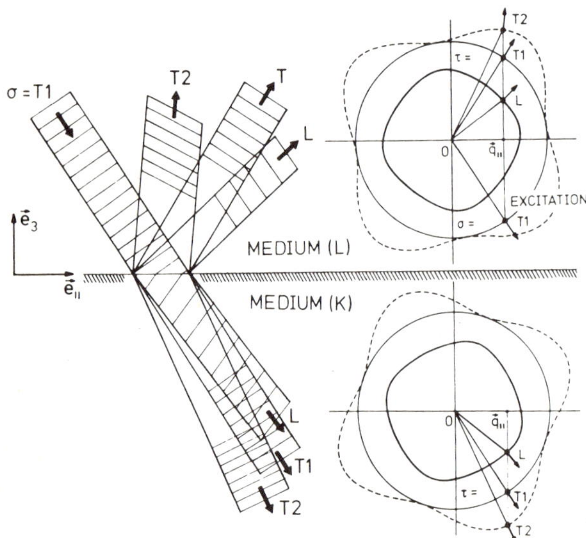

Figure 2: The incident plane wave σ = T1 excites 3 plane waves τ = L, T1, T2 in each medium (L) and (K)

The incident plane wave produces at the interface a wavefield with a trace component $\vec{q}_{\parallel}$ which lies in the interface and in the plane of incidence and is common to all excited wavemodes. We denote the excited wavemodes by the polarization index τ in order to distinguish them from the incident wave of polarization σ. In the general anisotropic case we will find 3 excited wavemodes τ in each halfspace (L) and (K): plane waves with energy flux away from the interface and/or nearfields showing elliptic polarization and an exponential decay (see Fig. 1). Whereas the plane waves can be determined from the cuts in $\vec{q}$ space, the nearfields have to be computed from the wave equation [2]. Applying boundary conditions

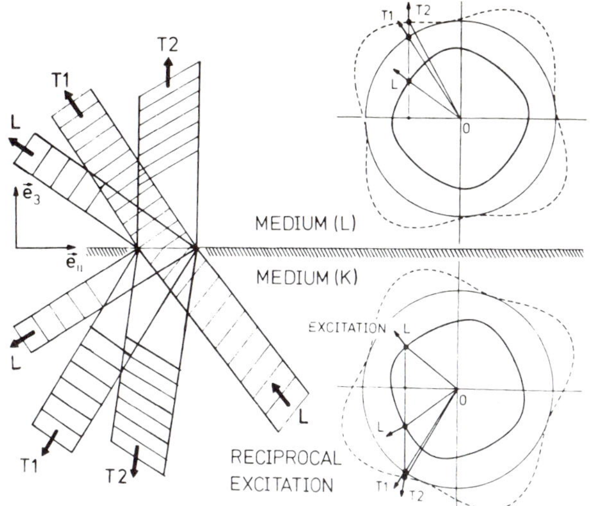

Figure 3: Example of a reciprocal excitation. The excited wave L in medium (K) of Fig. 2 is reversed and excites 3 plane waves L, T1, T2 in each medium (L) and (K). The exciting wave T1 in (L) of Fig. 2 appears in reversed form

(= continuity of particle displacement and of stress vector) the power transition coefficients $t_{\sigma\tau}^{(LK)}(\vec{e}_{n\sigma})$ from the incident plane wave σ (with wavenormal $\vec{e}_{n\sigma}$ in (L)) into a wavemode τ in medium (K) can be computed. In the case of a reflection the medium index (K) is equal (L). These calculated power transition coefficients have to be identified as the phonon transition probabilities, especially:

$$t_{\sigma\tau}^{(LL)}(\vec{e}_{n\sigma}) \equiv r_{\sigma\tau}^{(LL)}(\vec{e}_{n\sigma}) \quad \text{phonon-reflection probability and} \tag{3a}$$

$$t_{\sigma\tau}^{(LK)}(\vec{e}_{n\sigma}) \quad \text{phonon-transmission probability (K)} \neq \text{(L)}. \tag{3b}$$

One can prove that these quantities vanish for nearfields and that in any case the sum over all possible transition probabilities is unity -as expected. One can further prove [1] that the following reciprocity holds

$$t_{\tau\sigma}^{(KL)}(-\vec{e}_{n\tau}) = t_{\sigma\tau}^{(LK)}(\vec{e}_{n\sigma}) \quad \underline{\text{reciprocity theorem}}. \tag{4}$$

This theorem states that the phonon transition probability is the same for the direct transition and for the reversed where generator and detector are interchanged as well as polarization. Figure 3 serves as an illustration of this theorem. Theorem (4) is very useful in shortening numerical calculations of phonon irradiation of detector areas if phonon transitions at interfaces are involved. One can further prove [1] that detailed balancing exists at interfaces as a consequence of (4), provided the occupation number of phonons is the same in both media and depends only on frequency, i.e., the outgoing spectral phonon intensity of each polarization in any differential solid angle is the same as the incident intensity in reversed direction. In the special case of thermal equilibrium this is also a demand of the second law of thermodynamics.

1 O. Weis: in 'Nonequilibrium Phonons in Nonmetallic Crystals' (Ed.: A.A. Kaplyanskii, W. Eisenmenger, North-Holland Publishing Company) in Press.
2 O. Weis: Z.Phys. B 34, 55 (1979)

Phonon Scattering by Twin Planes

J.W. Vandersande, P.N. Chopra, and R.O. Pohl

Department of Physics, Clark Hall Cornell University
Ithaca, NY 14853, USA

In our ongoing research on the thermal conductivity of rocks, we noticed that the effect of grain boundaries commonly appeared to be insignificant relative to that of lamellar structures, such as twin planes, exsolution lamellae, and inclusions, within the grains. It has been found that these lamellae scatter phonons with a wavelength-independent rate [1]. In marble (recrystallized calcite), these scatterers appeared to be twin planes, since the twin spacing was approximately equal to the average phonon mean free path. To confirm this, we have measured the thermal conductivity of natural and plastically deformed single-crystal calcite ($CaCO_3$) and marble. Calcite samples with their long direction parallel to the [001] direction were cut from a large single crystal. These samples were deformed by applying a differential stress along this direction in a gas medium deformation apparatus operating at a confining pressure of 300 MPs at room temperature. This deformation geometry results in pronounced twinning [2]; and very little or no intracrystalline slip. The spacing of the twin planes in each specimen was determined from thin sections using an optical microscope. The Carrara marble specimens were deformed in a gas-medium deformation apparatus at 300 MPa confining pressure and temperatures of 500° C (heavily twinned) and 700° C (moderately twinned) in a similar manner to that described in [3]. As can be seen in Fig. 1, the low-temperature conductivity, Λ, of the twinned calcite is very much smaller than that of the undeformed material. A similar effect is observed in the twinned marbles, although to a lesser extent.

During the deformation, several visible cracks developed in the heavily deformed calcite sample. Cracks will change the path of the heat flow, and thus will change the apparent conductivity by a temperature-independent factor. The upper limit of the error introduced by these cracks can be estimated from the difference between the conductivities of the undeformed and the deformed sample at high temperatures. Assuming that this decrease, by a factor of two, is caused exclusively by these cracks, it would follow that the low-temperature data in a crack-free sample would actually be twice as large. For the moderately deformed sample, the error would be even smaller. Since we believe that the lowering of Λ at high temperatures is only in part due to the cracks, we ignore their effects entirely in the following discussion.

The average phonon mean free paths were determined from the gas kinetic formula $\ell=3\,\Lambda/C_V v$, and are plotted in Fig. 2. The Debye specific heat C_V and sound velocity v, $v=4.19\times10^5$ cm/s, were determined from transverse and acoustic speeds of sound measured at 160K [4]. Below 1K, ℓ is approaching constant values in all cases, indicating a wavelength-independent phonon mean free path. In Fig. 3, these low-temperature limits, ℓ_0, for calcite are compared with the average spacings d of the twin boundaries; also shown are data obtained on Bi [5], and the experimental and theoretical (Casimir) mean free paths in the undeformed samples. If each phonon were scattered diffusely at every interface (or

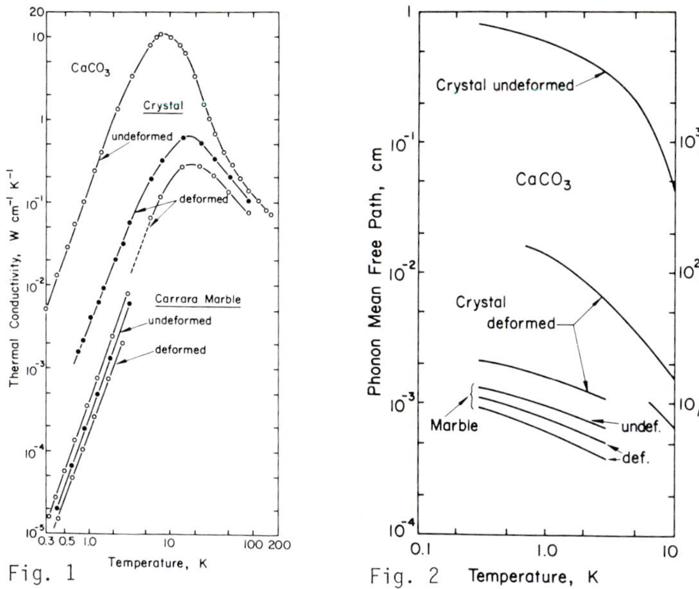

Fig. 1   Fig. 2

Fig. 1 Thermal conductivity of single crystal $CaCO_3$ and of Carrara marble after various deformations. Top curve: undeformed, 5x5 mm^2 cross section. The deformed samples were cylindrical, 7 mm diameter. All samples had rough surfaces. Note that the conductivity of the more heavily deformed sample appears to show a discontinuity near 5K; the sample had several cracks which may have changed between the low- and the high-temperature runs, thus changing the heat flow pattern through the sample. The conductivity of a moderately deformed marble, which lay between the deformed and undeformed sample, was omitted for clarity

Fig. 2 Average phonon mean free path, ℓ. For marble, the intermediate curve was for a moderately, the lower curve for a heavily deformed sample

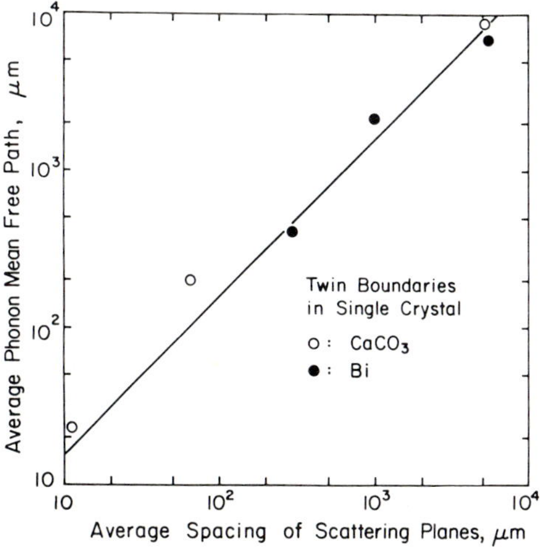

Fig. 3 $CaCO_3$ and Bi. Low-temperature limit of the average phonon mean free path, ℓ_0, vs. the average spacing, d, of the twin planes (and of the sample boundaries for the untwinned samples). Data for Bi after [5]. For the marble, with grain size averages between 100 and 150 μm, ℓ_0 and d were 15 μm and 56 μm; 12 μm and 19 μm; 10 μm and 15 μm for the undeformed, the lightly deformed, and the heavily deformed sample, respectively. In the undeformed sample, the distribution of twin boundaries was highly nonuniform, which may explain the poor agreement between ℓ_0 and d. Also, if the grain boundary spacing were included (112 μ for the undeformed sample), the average spacing would change from 56 μm to 37 μm

surface) we would expect $\ell_0 \approx d$. A good fit to the data is achieved with ℓ_0=1.6d (Fig. 3). The discrepancy, ≈ 60%, may indicate that the phonons are not diffusely scattered, but are partly reflected and partly refracted at the interface in the elastically anisotropic media. Matsuo and Suzuki [5] explained their results on twinned Bi with this mechanism, but performed the calculation of reflected vs. refracted power only for one angle of incidence. Calculation of the conductivity for random phonon incidences is very complicated and has not been attempted. On the other hand, the determination of ℓ_0 from Λ is rather uncertain, since the appropriate speed of sound to be used is uncertain, mainly because of the elastic anisotropy. The fact that the boundary scattering mean free path, 9 mm, in the undeformed calcite sample was also determined to be 60% larger than the theoretical Casimir value, which is 1.12d=5.5 mm, suggests that there may be problems of this kind, and that the true ℓ_0 may indeed be much closer to the spacing of the twin planes.

Whatever the reason for the strong phonon scattering, the fact that twin boundaries can have a large influence on the thermal conductivity of solids is now well established.

This work was supported by the National Science Foundation, Grant Nos. DMR-82-07079 and 81-15692, and through a contribution by the Exxon Production Research Company.

References

1. J. W. Vandersande and R. O. Pohl, Geophys. Res. Lett. 9, 820 (1983).
2. A. Nicolas and J. P. Poirier, "Crystalline Plasticity and Solid State Flow in Metamorphic Rocks", John Wiley Sons, New York, 1976, p. 216.
3. Schmid S. M., Paterson M. S. and Boland J. N., Tectonophysics, 65, 245-280, (1980).
4. D. P. Dandekar and A. L. Ruoff, J. Appl. Phys. 39, 6004 (1968).
5. T. Matsuo and H. Suzuki, J. Phys. Soc. Japan 43, 1974 (1977).

Boundary and Dislocation Scattering of Phonons in Lead Single Crystals

W. Odoni*+, P. Fuchs+, and H.R. Ott

Laboratorium für Festkörperphysik, ETH-Hönggerberg
CH-8093 Zürich, Switzerland

Abstract: Thermal conductivity measurements on various single crystals of lead between 0.05 and 1.2 K are analyzed considering phonon scattering by sample boundaries and by dislocation lines. It is demonstrated that the boundary-scattering length is indeed limited by sample dimension.

The gap in the electronic excitation spectrum of a superconductor provides a means to study the thermal transport properties of the phonon system of a metal because at temperatures well below the superconducting transition temperature T_c ($T < 0.15\ T_c$) the thermal conductivity λ of the considered material is entirely due to phonons and scattering processes other than phonon-electron scattering.

Because of its relatively high T_c of 7.2 K, lead has been chosen as a suitable material for such investigations earlier [1-5]. The first results in ref. [1] indicated an approximate T^3 dependence for λ as expected from phonon scattering at the sample boundaries [6] but the mean free paths as calculated from the experimental results were considerably smaller than the relevant sample dimensions. This discrepancy could not be removed by subsequent efforts [2,3] and it was conjectured that other scattering processes invoking grain boundaries or dislocation lines might be the reason for it. In ref. [4], a quantitative comparison between dislocation-line density and mean free path using a theoretical model due to Granato [7] gave a satisfactory agreement between theory and experiment and more recent measurements on high-quality single crystals [5] of lead revealed phonon mean free paths of the order of the sample diameter. In this work we present a systematic study of the size-limited phonon thermal conductivity of a metal, using lead as an example, by considering the importance of dislocation-line scattering of phonons at very low temperatures.

The thermal conductivities of cylindrical lead single crystals with various diameters and grown from a nominally 99.999% pure melt were measured between 0.05 and 1.2 K. A selection of experimental results is shown in Fig. 1. They clearly demonstrate that considerable deviations from a simple T^3 dependence of λ are observed below about 0.6 K, indicating the increasing effectiveness of an additional scattering mechanism other than boundary scattering with decreasing temperature. At the same time it should be noted that the mean free paths Λ_C calculated from λ/T^3 above 0.6 K are again smaller than

+ supported financially by the Schweizerische Nationalfonds

* now at Hilti AG, FL-9494 Schaan

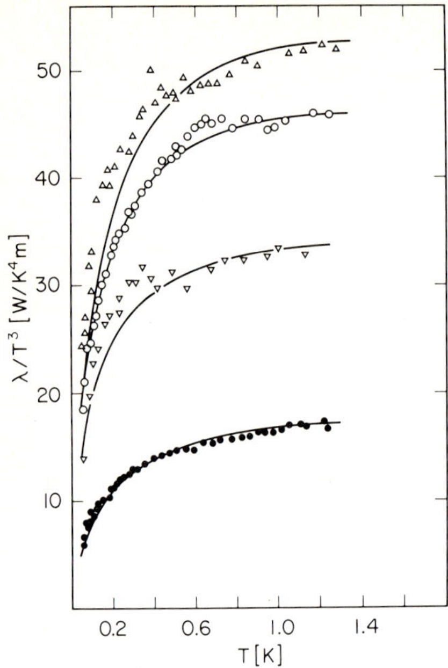

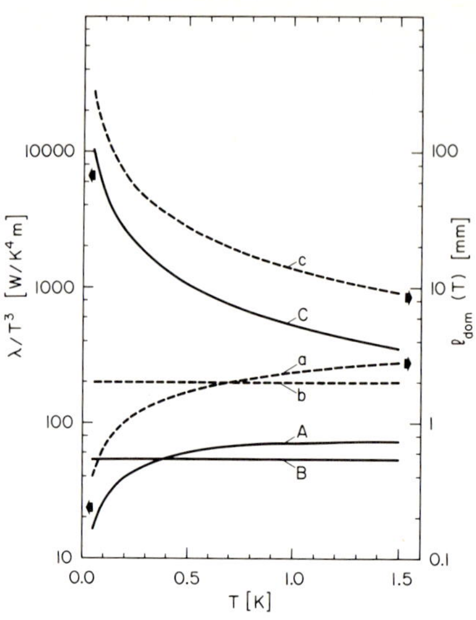

Fig.1: λ/T^3 for Pb single crystals with different diameters. Solid lines are fits considering phonon scattering at boundaries and dislocations

Fig.2: λ/T^3 and $\ell_{dom}(T)$ for the individual scattering mechanisms as explained in the text. A,B and C, see left-hand scale; a,b and c, see right-hand scale

the sample diameters. We analyzed our data assuming that λ can be described in the Debye approximation [8]

$$\lambda = \frac{k}{2\pi^2 v} \left(\frac{k}{\hbar}\right)^3 T^3 \int_0^{\theta/T} \tau(x) \frac{x^4 e^x}{(e^x-1)^2} dx \tag{1}$$

where $x=(\hbar\omega/kT)$, θ is the Debye temperature and v the sound velocity. For $\tau(x)$ we apply Matthiessen's rule

$$\tau^{-1}(x) = \tau_b^{-1}(x) + \tau_d^{-1}(x,N) \tag{2}$$

where $\tau_b = \ell_b/v$ describes the boundary scattering. $\tau_d(x,N)$ reflects the scattering of phonons at dislocation lines of density N and has been given by SUZUKI and SUZUKI [9], based on a more general formalism due to NINOMIYA [10], considering the possible fluttering of dislocation lines. The fit parameters entering our calculations are ℓ_b, the boundary-limited mean free path and N, the dislocation-line density. We neglect the static strain-field scattering, also proportional to N (see,e.g. [11]), because, as shown in Fig.2, the related mean free path is an order of magnitude larger than the sample diameters (curve c in Fig.2). In Fig.2 we show the calculated temperature dependences of both λ and ℓ_{dom} as they arise from dislocation-line (A,a), boundary (B,b) and strain-field (C,c) scattering, respectively. ℓ_{dom} is to be understood as $\ell_{dom} = v\,\tau(x_{max})$, where x_{max} denotes the maximum value of $x^4e^x/(e^x-1)^2$.

A selected collection of parameters used in the analysis of our thermal conductivity data is given in table I. The solid lines in Fig.1 were calculated by fully integrating (1) and inserting τ as given in (2). We note that

Table I: Parameters used in the analysis of a part of our thermal conductivity results

Sample	orient.	diam. mm	$\Lambda_C(1\ K)$ mm	N m^{-2}	ℓ_b mm	remark
13a	100	1.95	1.57	$3.1x10^{11}$	2.12	
13b	100	1.60	0.56	$1.45x10^{12}$	1.55	
13c	100	1.60	1.12	$4.25x10^{11}$	1.55	sample 13b, annealed at 80°C for 3 days

Λ_C, calculated from $\lambda = c_p v \Lambda_C$, where c_p is the lattice specific heat, is always less than the sample diameter and lacking a systematic variation in the sense that boundary scattering alone clearly cannot explain our experimental results. Including both scattering mechanisms mentioned above reveals that ℓ_b is essentially the sample diameter, as we intended to demonstrate. For N we obtain similar numbers as were found by O'HARA and ANDERSON by direct etch-pit counts. Moreover we find that annealing a sample results in a considerable decrease of N, leaving ℓ_b unchanged, as expected.

In conclusion, these experiments show that in high-quality single crystals of superconducting metals and well below the superconducting transition temperature T_c, phonon mean free paths of the order of millimeters can be obtained. In real crystals, the effective mean free path is determined by sample dimension and by scattering phonons at vibrating dislocation lines. A more extended account of this work will be published elsewhere [11].

References

1 J.L. Olsen and C.A. Renton, Phil. Mag. 43, 946 (1952)
2 H. Montgomery, Proc. Roy. Soc. A244, 85 (1958)
3 N.V. Zavaritskii, Zh. Eksp. Teor. Fiz. 38, 1673 (1960) {Sov. Phys. JETP 11, 1207 (1960)}
4 S.G. O'Hara and A.C. Anderson, Phys. Rev. B10, 574 (1974)
5 L.P. Mezahov-Deglin, Zh. Eksp. Teor. Fiz. 77, 733 (1979) {Sov. Phys. JETP 50, 369 (1979)}
6 H.B.G. Casimir, Physica 5, 495 (1938)
7 A. Granato, Phys. Rev. 111, 740 (1958)
8 see e.g. R. Berman, Thermal Conduction in Solids, (Clarendon Press, Oxford 1960)
9 T. Suzuki and H. Suzuki, J. Phys. Soc. Japan 32, 164 (1972)
10 T. Ninomiya, J. Phys. Soc. Japan 25, 830 (1968)
11 W. Odoni, P. Fuchs and H.R. Ott, to appear in Phys. Rev. B, Aug.(1983)

Diffuse Scattering of Thermal Phonons at Crystal Surfaces

T. Klitsner and R.O. Pohl

Laboratory of Atomic and Solid State Physics, Cornell University
Ithaca, NY 14853, USA

When an elastic wave encounters a smooth surface at low temperatures, it should be specularly reflected. Similarly, at a smooth interface, it should be reflected and refracted. In this paper, we report on our search for reflected thermal phonons, using thermal conductivity measurements, in the temperature regime in which the bulk phonon mean free path is long relative to the sample dimensions. The thermal conductivity will then depend on whether the phonons are specularly reflected or diffusely scattered at the sample surface.

The 5x5x50 mm samples were cut along the [111] direction with side faces pointing in [110] and [211], from a boule of Wacker Chemie floating zone semiconductor grade silicon of 650 Ohm cm resistivity at room temperature. Samples were kindly provided by Dr. V. Narayanamurti from Bell Labs. Final polish was by a colloidal silica suspension with particle size 300 Å. Before depositing the films, the surfaces were cleaned in a boiling organic solvent followed by a mixture of equal volumes H_2O_2 and NH_4OH. This treatment should not materially affect the native oxide layer. The evaporation was done in an electron beam evaporator evacuated with an ion pump. Pressure never exceeded 5×10^{-7} torr during evaporation at a deposition rate of 1 Å/s. To deposit the film uniformly over all four faces, the crystal was rotated around its axis at a constant rate of 20 rpm. Average film thickness was monitored during evaporation with a crystal monitor, and subsequently measured by Rutherford backscattering. Film structure was determined using a standard scanning electron microscope (SEM).

At low temperatures, the thermal conductivity Λ of a polished silicon crystal was found to be an order of magnitude larger than that of a crystal with surfaces roughened by sandblasting, see Fig. 1. Rough surfaces are known to scatter all thermal phonons diffusely. Thus, at the polished surface, the probability for diffuse scattering is reduced tenfold, i.e., each phonon will, on average, be specularly reflected ten times before it is scattered diffusely. The probability for specular reflection is also greatly reduced when thin metal or dielectric films are deposited onto the polished silicon surfaces. In these cases, however, the onset of the diffuse scattering occurs at a threshold temperature, see Fig. 1. Removal of the film restores the original conductivity of the polished sample. The polished, clean surface also appears to have such a threshold temperature. Whether this is resulting from some kind of residual surface defects (the silicon dioxide layer?), or from scattering centers in the bulk of the sample, cannot be decided at this time. Figure 2 shows the average inverse phonon mean free path, $\ell^{-1} = (\tau v)^{-1}$, determined from the gas kinetic formula, $\Lambda = (1/3)\, C_V\, v\, \ell$, where the experimental low-temperature specific heat [1], $C_V = 6.02\, T^3$ erg cm^{-3} K^{-4}, has been used to calculate the Debye speed of sound, $v = 5.94\times10^5$ cm s^{-1}. For the sandblasted crystal, ℓ^{-1} is independent of temperature to within 20% below 2K, and somewhat smaller than the Casimir prediction shown as the dashed line (inclusion of phonon focussing effects would increase the theoretical value of ℓ^{-1} by only 3% for heat flow along

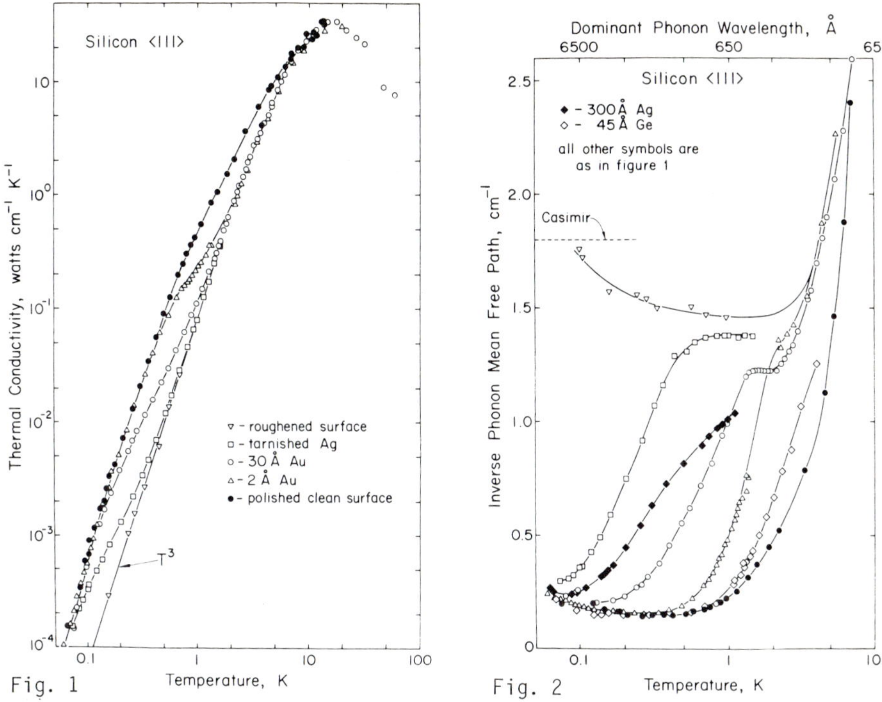

Fig. 1 Thermal conductivity of silicon with clean and polished surfaces, with roughened surfaces, and with polished surfaces carrying thin metal films. Heat flow along [111]. The tarnished Ag film (45 Å) contained sulfur and chlorine

Fig. 2 Inverse average phonon mean free path, ℓ^{-1}, for the samples shown in Fig. 1, and also for samples carrying a Ge film, and a thick Ag film. Note the scale on the top of the graph indicating the wavelength of the phonons carrying most of the heat at a given temperature (the dominant phonon wavelength.)

[111] [2]). The same temperature dependence of ℓ^{-1} has been observed previously in Al_2O_3 (Fig. 2 of ref. [3]). It will be ignored in the following. All films we studied enhance the diffuse scattering. The temperature range in which the diffuse scattering begins depends on the material and also on the film thickness. The question whether the enhancement of the diffuse scattering is resonant, i.e., whether the enhancement of ℓ^{-1} (i.e., of the scattering rate) peaks at a certain temperature cannot yet be answered. Thus, we will characterize the diffuse scattering through its threshold temperature T_{th}, defined as the temperature at which diffuse scattering begins to set in. For films of equal thickness (t ≈ 45 Å), we observed the following T_{th}: Ag, 60mk (see, however, the discussion below); Au, 150mK; Pt-15 wt% W, 200mK; Pd, 300mK; Pt, 400mK; and Si and Ge, 800mK. With Au films, we found T_{th} to decrease with thickness, varying from 400mK for t=2Å to 80mK for t=300 Å. For Ag, the onset of diffuse scattering was found to be shifted to a higher temperature for a thicker film (t=300 Å; T_{th}=80mK). The thin (45 Å) Ag film, however, had been allowed to sit in air for several days prior to the measurement. Rutherford backscattering later revealed considerable amounts of sulfur and chlorine. Hence, the data on this film may not be characteristic for clean Ag. The struc-

ture of thin metal films depends greatly on their mode of preparation. Under the conditions employed here, it may be assumed that all films of t<100 Å were discontinuous, in the form of droplets, with the possible exception of Ge and Si which may go down as a continuous amorphous layer. In the case of Au, we verified the droplet structure through SEM, and found particle sizes in agreement with earlier studies [4,5], e.g., droplet diameters of ~ 100 Å for t=40 Å. It might, therefore, be suspected that motional degrees of freedom of the rigid drops bonded to the surface, or internal degrees of freedom within the drops may cause the diffuse scattering. However, even thick films cause diffuse scattering: Au (t=300 Å) has a threshold temperature T_{th}=80mK (data identical to those obtained on 300 Å Ag, Fig. 2), and for Ti (t=530 Å), T_{th}=120mK. We found with the SEM that the 300 Å Au film was continuous, except for narrow cracks with average separation of ~ 1000 Å. Titanium films of thickness exceeding 500 Å are known to have the same electrical resistivity as bulk Ti [6], and hence should also be fairly uniform. It follows that the excitations causing the diffuse scattering are not restricted to isolated structures, and may even exist in bulk material. In Ti, for example, our film thickness is greater than the dominant phonon wavelength (370 Å at 1K, based on a Debye temperature of 380K and a Debye speed of sound of 3.34×10^5 cm/s). Electronic effects are unlikely as cause of the scattering, since the Ge and Si films are electrical insulators. It has previously been reported that not only thin metal films, but even thin layers damaged by ion implanation cause diffuse phonon scattering in sapphire crystals with polished surfaces [3]. A model to unify these two phenomena is based on the fact that surface waves (Rayleigh and evanescent waves) are strongly scattered by surface structures with dimensions comparable to their wavelength [7]. Conceivably, these structures will also provide a coupling between bulk phonons and surface waves. The variation we have seen with thickness and material could possibly be explained by the accompanying variations in island size and tightness of bonding to the surface. For the thick films it is possible that nonperfect binding to the surface could also provide a roughly periodic structure for coupling to the surface modes. It is interesting to note that Koos et al. have recently reported enhanced phonon transmission across a Cu film (2000 Å) to sapphire interface, which they have shown to be mediated by evanescent L modes at the interface [8]. The crucial question is whether this scattering is strong enough to explain our observations, considering the small density of states of surface phonons.

Since the diffuse scattering also occurs at films which are thick relative to the phonon wavelengths, the same scattering may also occur at interfaces between two media and, therefore, affect the heat flow across them. Work supported by the Materials Science Center, Cornell University.

References

1. P. Flubacher, A. Leadbetter, and J. Morrison, Phil. Mag. 4, 273 (1959).
2. A. K. McCurdy, H. J. Maris, and C. Elbaum, Phys. Rev. B 2, 4077 (1970).
3. R. O. Pohl and B. Stritzker, Phys. Rev. B 25, 3608 (1982).
4. L. Bachmann and L. Hilbrand, in Grundprobleme der Physik Dunner Schicten, Vandenhoeck and Rupprecht, Gottingen, 1966, p. 77.
5. T. Andersson and C. G. Granquist, J. Appl. Phys. 48, 1673 (1977).
6. D. Hacman, ref. [4], p. 561.
7. I. A. Viktorov, Rayleigh and Lamb waves, Plenum Press, New York, 1967.
8. G. L. Koos, A. G. Every, G. A. Northrop, and J. P. Wolfe, Phys. Rev. Lett. 51, 276 (1983), and in these conference proceedings.

Imaging of Specularly Reflected Phonons from a Crystal Boundary

G.A. Northrop

Physics Department and Materials Research Laboratory,
University of Illinois at Urbana-Champaign, 1110 W. Green Street
Urbana, IL 61801, USA

The ballistic heat pulse technique has been used by several authors to study the reflection of phonons at crystal boundaries [1,2]. These experiments examine the temporal evolution of a heat pulse scattered internally from a crystal surface for fixed generator and detector positions. By use of a mobile heat pulse source, we present in this paper a measurement of the spatial, or angular, character of heat pulse reflection in sapphire.

The experimental apparatus is identical to a previously detailed phonon imaging system [3], but the sample is configured for laser heat pulse scanning and detection on the same surface (Fig. 1a). A highly polished disk of $[1\bar{1}02]$ oriented sapphire 3.2 mm thick and 30 mm in diameter has a 50 × 50 μm^2 Al bolometer placed in the center of one face. The remainder of the surface is covered with a 2000 Å Cu film, which serves both as a heater film and bolometer contacts. The opposite surface, from which the heat pulses reflect, is left uncoated and the entire sample is immersed in superfluid He at 1.6 K. To form the reflection image a pulsed laser (a cavity-dumped Ar^+ ion laser with a 10 W, 15 ns pulse repeated at 100 KHz) is focused to 30 μm and raster scanned over the detector side of the sample. At each point in this scan the bolometer signal is sampled after a delay proportional to the minimum distance for reflection from the back surface of the sample. Thus, a fixed "effective" velocity is sampled throughout the scan.

The reflection images obtained for three different velocities are shown in Fig. 1. In these images the raster scan covers an area 12 mm square. The detector is the dark patch in the center, and the dark diagonal lines are insulating gaps in the Cu film. The high intensity near the detector in all three images is bulk scattering. The most striking features of these images are spatially sharp intensity structures, which are reminiscent of regular phonon focusing patterns, only more complex. We will show that these are due to specular reflection of phonons from the back surface of the sample, and that individual patterns may be associated with particular pairs of modes.

The first image, Fig. 1b, is for a velocity of 8.6×10^5 cm/sec, which is the average of the L (longitudinal) and ST (slow transverse) velocities. The diamond structure in the top half results from mode conversion from L to ST. The equivalent structure in the bottom results from ST → L. Fig. 1c, for a velocity of 7.0×10^5 cm/sec, shows two different structures. The large X results from FT → FT, the upper diamond shape from FT → ST, and the lower diamond from ST → FT. The third image, Fig. 1d, shows ST → ST at a velocity of 6.4×10^5 cm/sec. The sharp increase in intensity beyond a radius of 2.7 mm is due to the loss of L → ST and ST → L beyond the critical angle for those processes.

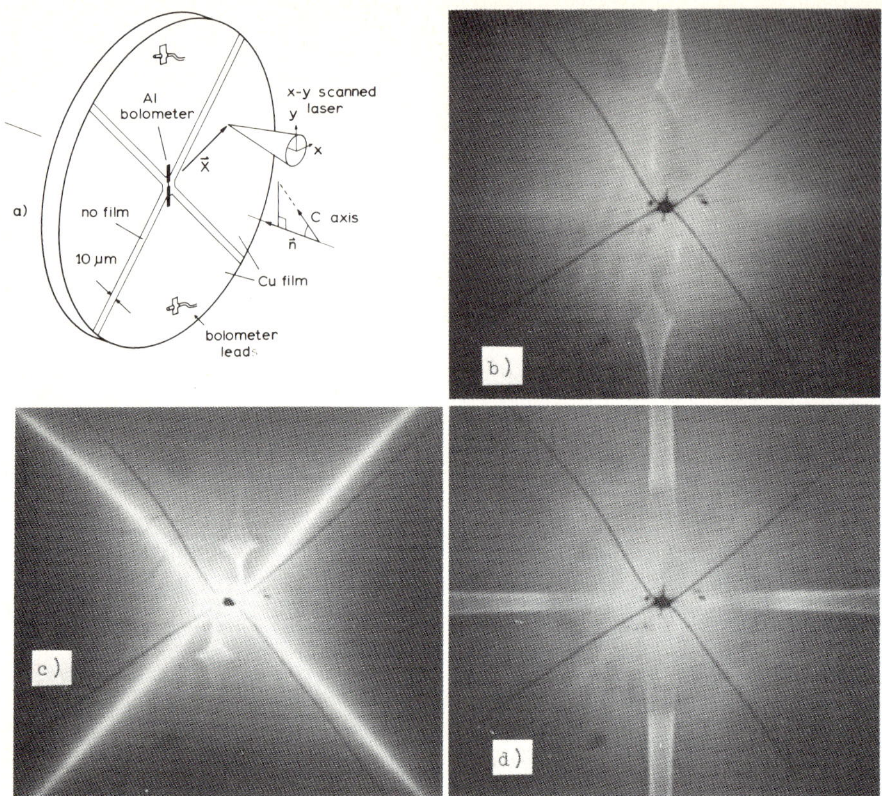

Fig. 1(a) Sample geometry showing laser raster scan. Reflection images (b) $V = 8.6 \times 10^5$ cm/sec, (c) 7.0×10^5 cm/sec, (d) 6.4×10^5 cm/sec

The condition for specular reflection is conservation of the wavevector component parallel to the surface. For specified incident and reflected modes this determines a unique destination ($\vec{X}$ in Fig. 1a) for a source phonon given by

$$\vec{X} = D\left[\frac{\vec{V}_{\alpha in}(\vec{k}_{in})}{|\vec{n}\cdot\vec{V}_{\alpha in}(\vec{k}_{in})|} + \frac{\vec{V}_{\alpha ref}(\vec{k}_{ref})}{|\vec{n}\cdot\vec{V}_{\alpha ref}(\vec{k}_{ref})|}\right], \tag{1}$$

where D is the sample thickness, $\vec{n}$ is the surface normal, $\vec{k}_{in}$ is the incident wavevector, $\vec{k}_{ref}$ is the reflected wavevector, $\alpha = L, ST, FT$ is the mode, and $\vec{V}_\alpha(\vec{k})$ is the group velocity. This may be expressed simply as

$$\vec{X} = \vec{X}\,(\vec{k}_{in}, \alpha_{in}, \alpha_{ref}) = \vec{X}_{\alpha_{in}\alpha_{ref}}(\vec{k}_{in}). \tag{2}$$

In the non-dispersive limit $\vec{X}$ depends only on the direction of $\vec{k}_{in}$, and $\vec{X}$ is restricted to the generation/detection surface. Thus (2) is a mapping from one two-dimensional space into another two-dimensional space, which makes the problem quite similar to that of conventional phonon focusing, but with the additional dependence on the reflection surface

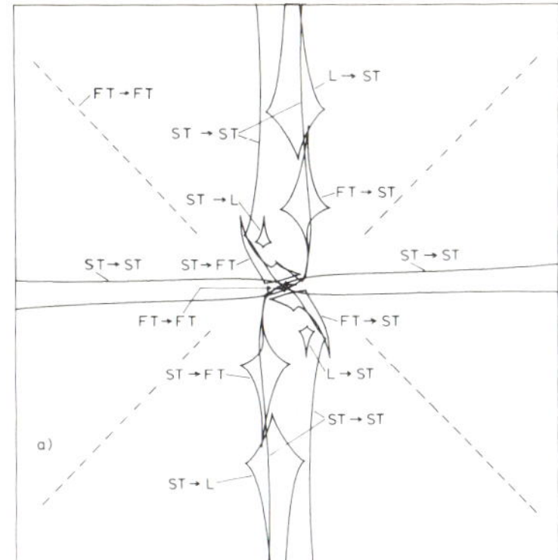

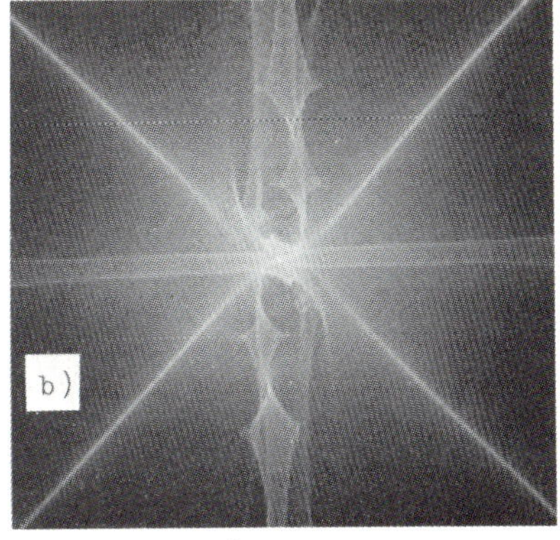

Fig. 2(a) Locations of singularities on the $\vec{X}$ surface. (b) Calculated (Monte Carlo) image for the singular mode pairs. The FT → FT reflection is only singular near the center, but is very intense along the dashed lines in (a)

orientation. As with phonon focusing, singularities in phonon flux will occur along lines where the Jacobian of this mapping goes to zero [3]. Thus specular reflection may result in singularity structures and associated intensity peaks and discontinuities.

Calculations for the [1$\bar{1}$02] orientation of sapphire indicate that of nine possible mode pairs only six show singular behavior. Fig. 2a shows lines of singular phonon flux plotted in the experimental geometry. Figure 2b is a Monte Carlo simulation of the specular reflections L $\rightleftarrows$ ST, ST $\rightleftarrows$ FT, ST → ST, and FT → FT, with all six represented in roughly equal numbers. The lack of horizontal mirror symmetry is due to the crystal surface normal being about 2.5° from the [1$\bar{1}$02]. Adjusting that angle in the calculation resulted in an excellent agreement between the calculated and experimentally observed singularity structure.

The imaging method unequivocally demonstrates the existence of the specular reflection of thermal phonons from a crystalline sapphire boundary. Furthermore, it allows a clear determination of modes and polarizations in the case of specular reflections. The imaging method should greatly enhance heat-pulse reflection as a method for studying the interactions of phonons with crystal surfaces.

This work was supported by the National Science Foundation under the MRL Grant DMR-80-20250 and equipment support from NSF DMR-80-24000.

References

1. P. Taborek and D. L. Goodstein, Phys. Rev. B 22, 1550 (1980).
2. D. Marx and W. Eisenmenger, Z. Phys. B 48, 277 (1982), and references.
3. G. A. Northrop and J. P. Wolfe, Phys. Rev. B 22, 6196 (1980).

Critical-Cone Channeling of Thermal Phonons from Solid/Solid Interfaces

A.G. Every*, G.L. Koos, G.A. Northrop, and J.P. Wolfe

Physics Department and Materials Research Laboratory,
University of Illinois at Urbana-Champaign, 1110 W. Green Street
Urbana, IL 61801, USA

Phonon imaging provides a powerful means of studying the anisotropy of ballistic phonon flux in crystals at low temperatures [1]. In the typical phonon-imaging experiment a point source of heat (focused laser [2] or electron beam [3]) is scanned across the metalized face of a crystal, and ballistic phonons emanating from this heated spot are detected by a bolometer mounted on the opposite face. Figure 1a shows an image obtained from one of the highly polished sapphire (Al_2O_3) crystals we have examined. Most of the structure in this image is due to bulk phonon focusing. Fig. 1b shows the corresponding theoretical image that has been generated by assuming a uniform distribution of phonon $\vec{k}$ vectors and pro-

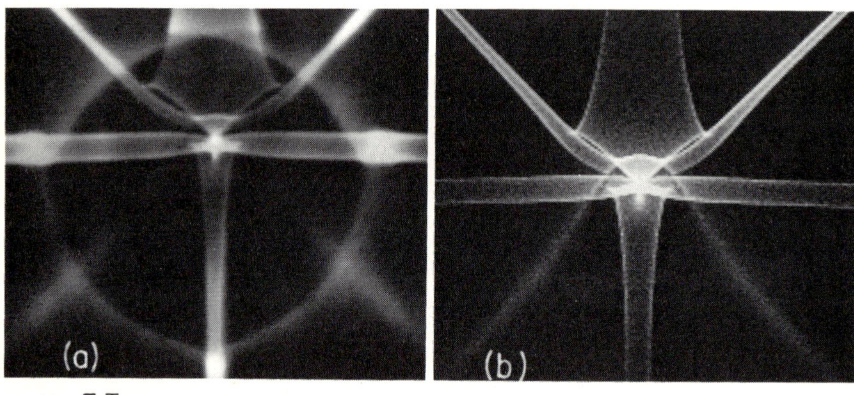

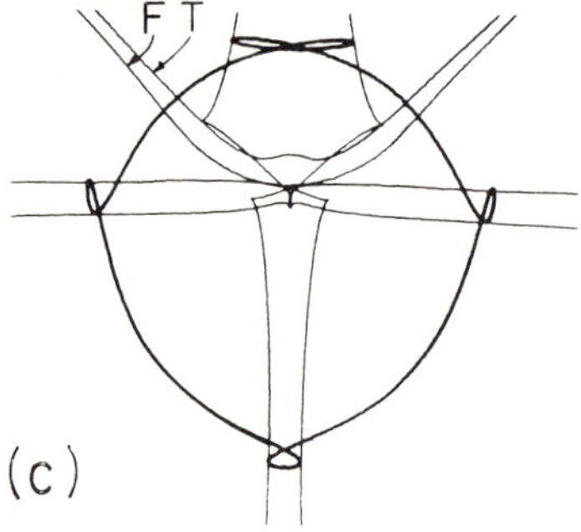

Fig. 1 (a) Ballistic phonon image for sapphire with faces cut normal to the [1102] direction. (b) Corresponding theoretical phonon image. (c) ST critical-cone contour (heavy line) superimposed on a map of the caustics for sapphire. The FT caustics are identified and all the remaining caustics are ST

*Permanent Address: Physics Department, Witwatersrand University, Johannesburg, South Africa

jecting the group velocity vectors for these phonons onto the viewing surface. Both images display essentially the same pattern of caustics, i.e., lines along which the focusing factor is mathematically singular, and they also both exhibit two ridges along which the focusing is large but non-singular.

The bright oval-shaped halo in the experimental image has no counterpart in the theoretical image, and is therefore not due to focusing. We have established that it corresponds to a concentration of slow transverse (ST) phonons with $\vec{k}$ vectors lying close to the critical cone for mode conversion of ST to longitudinal (L) waves at the sapphire surface. The heavy line in Fig. 1c shows the contour, formed by the ray vectors for these $\vec{k}$'s, in relation to the caustics. The corresponding channeling structure for fast transverse (FT) phonons is comparatively faint due to the fact that, for this particular crystallographic surface orientation, the critical cone FT phonons all happen to have polarization vectors directed very nearly perpendicular to the saggital plane of the $\vec{k}$'s. Such phonons are expected to participate fairly weakly in mode conversion and so show little sign of critical-cone effects. Other sapphire crystals with different surface orientations display both ST $\rightarrow$ L and FT $\rightarrow$ L channeling structures. These structures are accurately predicted by the critical cone condition, and their relative intensities are in accord with the polarization requirement.

Surface condition plays an important role. When one of the faces of the sapphire crystal is roughened the intensity of the channeling structure is reduced by a factor of about 2, while roughening both faces eliminates this structure almost completely. We attribute this critical-cone channeling to enhanced transmission across the metal/sapphire interface, mediated by longitudinal pseudo surface waves at the sapphire surface. The signature for these pseudo surface waves is a resonance in the mode conversion of bulk T waves, polarized in the plane of incidence, to evanescent L waves at the free sapphire surface. Figure 2 shows the theoretical phonon spectral emissivity, ε, into sapphire (assumed isotropic) for a number of different situations. Assuming conventional boundary conditions (i.e., continuity of displacement and traction force at the interface), the pseudo surface wave is almost completely suppressed when the adjacent metal film is well-matched acoustically to sapphire (as, e.g., in the case of copper). As a consequence there is a barely perceptible feature in the emissivity near the critical angle. In contrast, when the acoustic impedance of the metal film is much less than that of sapphire, the pseudo surface wave resonance shows up prominently in the emissivity. The emissivity for Al, an intermediate case, is shown in Fig. 2.

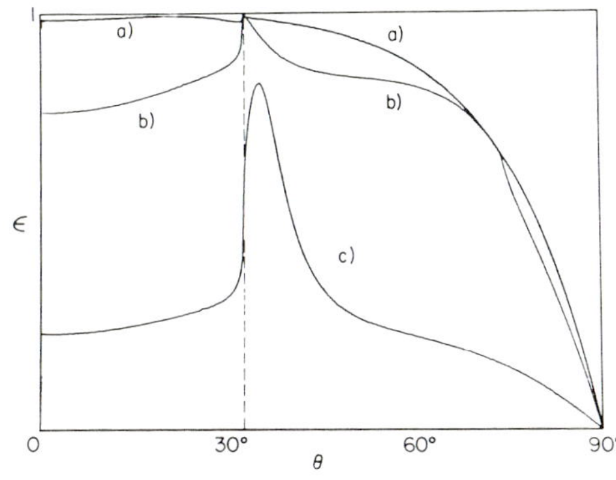

Fig. 2 Spectral emissivitiy, ε, into sapphire. a) Cu heater film, ideal bonding. b) Al heater film, ideal bonding. c) Cu heater film, "weak" bonding

Surprisingly, we find that experimentally the intensity of the critical-cone structure is relatively insensitive to the type of metal heater film employed. We believe that the reason for this is that the mechanical bonding between the metal film and sapphire is "weak", allowing relative motion of the two media at the contact surface. Figure 2 shows a curve of the emissivity calculated on the basis of modified boundary conditions that treat the interface as a distributed compliance. This curve shows reduced emissivity occurring overall, but mostly for angles other than the critical angle. It agrees qualitatively with the experimentally observed intensity. The process is reminiscent of the phenomenon of ultrasonic critical angle reflectivity [4], but there are a number of novel and distinctive features in the present case. It is the nature of the coupling between the two media which appears to be crucial rather than bulk attenuation, and it is a longitudinal pseudo surface wave and not the Rayleigh wave that mediates the process.

We anticipate that the study of critical cone channeling will provide important information on phonon-boundary interactions. The weak-coupling model developed here has a bearing on Kapitza resistance and phonon reflection effects at interfaces.

This work was supported by the National Science Foundation under the MRL Grant DMR-80-20250. One of us, A.G.E., would like to thank the University of the Witwatersrand and the CSIR for financial support.

References

1. For reviews, see W. Eisenmenger, J. de Phys. Colloque C6, C6-201 (1981); J. P. Wolfe, Physics Today 33, 44 (Dec. 1980).
2. G. A. Northrop and J. P. Wolfe, Phys. Rev. B 22, 6196 (1980).
3. R. Eichele, R. P. Huebener and H. Seifert, Z. Phys. B 48, 89 (1982).
4. R. E. Peterson and A. C. Anderson, J. Low Temp. Phys. 11, 639 (1973).

Anomalous Low Temperature Kapitza Resistance of a Paramagnetic Salt

G.J. Batey and P.C. Main

Department of Physics, University of Nottingham, University Park
Nottingham, NG7 2RD, England

1. INTRODUCTION

Historically the study of the boundary resistance between the paramagnetic salt cerium magnesium nitrate (CMN) and liquid ^{3}He has been important because of its influence on thinking about thermal coupling at very low temperatures. Bishop et al. [1] and Black et al. [2] used a spin-relaxation technique to determine the total thermal resistance between the Ce spins and ^{3}He. For single crystals they found $R \propto 1/T^2$ which they attributed to an internal resistance within the sample. In contrast, with powdered specimens, the total resistance actually fell for T < 20 mk and from these data the influential idea of a direct magnetic coupling between the ions and the ^{3}He nuclei was born and was later developed by Leggett and Vuorio [3]. The idea has been criticised strongly in a review paper by Harrison [4].

The CMN/^{3}He system can be considered as a set of interacting excitations, as shown in Fig 1. In the salt we have the spins on the cerium ions and the various phonon modes of the lattice, and in the liquid we have zero-sound phonons and the ^{3}He quasi-particles. In the diagram R_{S-L} is the spin-lattice resistance, R_K is the boundary or Kapitza resistance, and R_H is the resistance between the zero-sound phonons and the quasi-particles. The previous experiments had measured the total resistance between the spins and the quasi-particles, i.e. $R_{S-L} + R_K + R_H$, though R_H is expected to be entirely negligible in the particular regime of size and temperature.

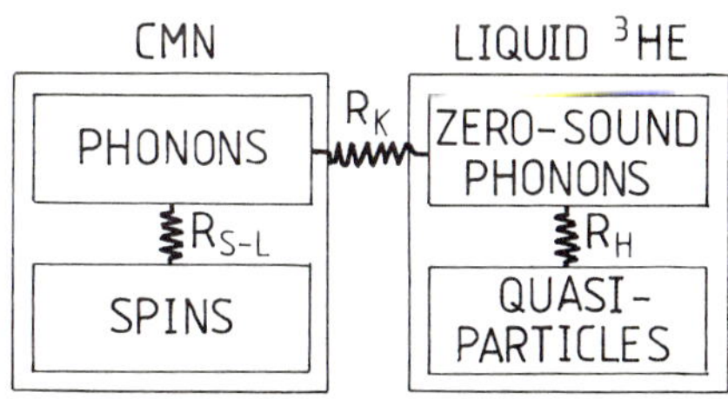

Fig. 1 The excitations and thermal resistances in the CMN/^{3}He system. See the text for the explanation of the symbols

2. EXPERIMENT

In the current experiment we tried to measure separately R_K and R_{S-L}. A small electrical heater was potted with epoxy resin into a hole drilled in the centre of a single crystal (approximately 12 mm x 3 mm x 3 mm) parallel to the long direction of the crystal. The crystal was then mounted on a graphite post and immersed in ^{3}He. The spin temperature was monitored using an S.H.E. low level impedance bridge incorporating a SQUID as a null detector. Referring to Fig. 1, heat is passed directly into the phonons of the crystal and then flows partly into the helium and partly into the spins until eventually the spins reach the same temperature as the phonons, T_P, given by

$$T_P - T_H = R_K Q \tag{1}$$

where T_H is the helium temperature and Q the continuous heater power. The spin temperature should tend exponentially to this value with a time constant τ given by

$$\tau = (R_K + R_{S-L})C_S \tag{2}$$

where C_S is the heat capacity of the spins (we are assuming $C_S >> C_L$, the lattice heat capacity), so that we are able to determine R_K and R_{S-L}.

3. RESULTS

Measured values of R_K are shown in Fig. 2, and values of $R_S + R_K$ determined as described above are illustrated by the closed circles in Fig. 3. The straight line in Fig. 2 corresponds to $AR_KT^3 = 8.5\ (\pm 1) \times 10^{-3}\ m^2KW^{-1}$ where A is the area of contact between the solid and the helium which compares with $AR_KT^3 = 5.5\ (\pm 1) \times 10^{-3}\ m^2KW^{-1}$ observed between 0.1K and 0.5K by Harrison and Pendrys [5]. The most prominent feature of the data is in Fig. 2 the departure from this line, noticeable below 70 mK but prominent for $T < 40$ mK. Originally we believed this to be evidence for an alternate heat path between the phonons and the quasi-particles, but we noticed that the time variation of the spin temperature after switching on the heater was slightly non-exponential. To test this further we carried out some normal spin-relaxation measurements. The thermal resistances derived from these measurements using the averaged specific heat capacity of CMN quoted by Harrison [4] are shown by the open circles in Fig. 3.

It is clear from Fig. 3 that the non-exponential variation of the spin temperature in the first experiment is an experimental artefact probably due to trapped ^{3}He. However, what is clear is that for $T < 40$ mK the open circles,

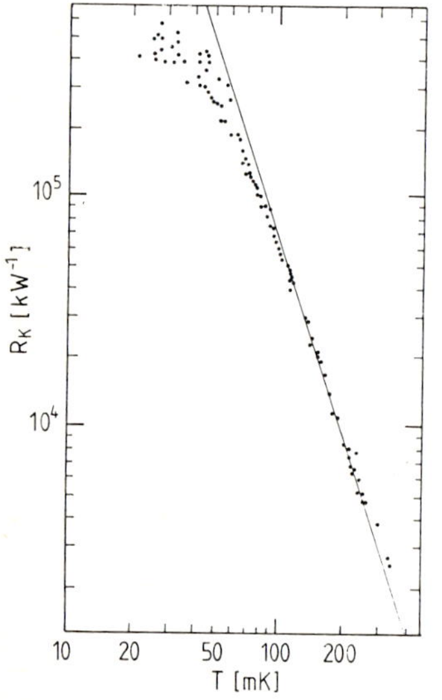

◀ Fig. 2 (left) The measured Kapitza resistance plotted against temperature. The straight line corresponds to $R_K \propto T^{-3}$

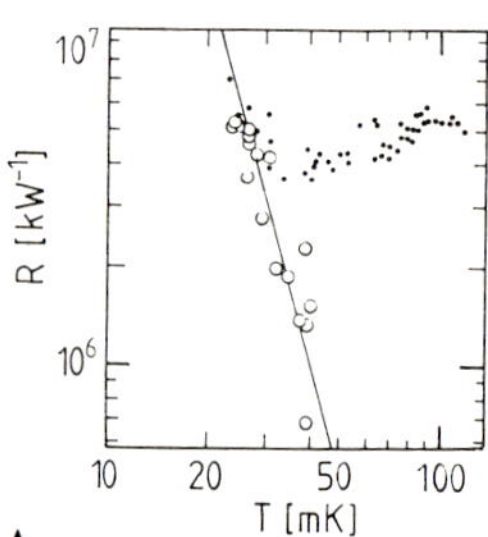

▲ Fig. 3 Thermal resistances derived from thermal time constants (closed circles) and spin-relaxation (open circles) plotted against temperature. The straight line is the same as in Fig. 2

and for $T < 30$ mK the closed circles, lie along the line which is drawn in Fig. 3, which is the same line as drawn in Fig. 2. In other words, below 40 mK the relaxation time given by (2) is dominated by the Kapitza resistance if we assume that the true resistance follows a $1/T^3$ variation over the whole temperature range. The departure in Fig. 2 must be due to an alternative heat path from the heater to the helium probably associated with a tortuous column of ^{3}He. This hypothesis is currently being tested using a crystal which has been grown around a heater to prevent the penetration of ^{3}He. In any case we are confident that we have measured $R_K \propto 1/T^3$ between a paramagnetic insulator and ^{3}He from 350 mK to 20 mK.

Although the results presented in Figs. 2 and 3 are for ^{3}He with a 1% ^{4}He impurity we have seen identical behaviour for pure ^{3}He. Since we are sure that we are measuring the Kapitza resistance and not some internal resistance in the crystal we would expect that if there was an alternative heat path due to a magnet coupling it ought to be apparent in this data as well as, for that matter, the data of Bishop et al. [1]. The fact that it isn't, and the absence of a change when ^{4}He is added to the ^{3}He is conclusive evidence for the absence of a magnetic coupling between CMN and liquid ^{3}He. The apparently anomalous results with powders have been explained by De Bruyn and Harrison [6] as a size effect.

4. ACKNOWLEDGEMENTS

One of us (GJB) wishes to thank SERC for financial support and the work as a whole was supported by SERC grant GR/B/00879.

5. REFERENCES

1. J.H. Bishop, D.W. Cutter, A.C. Mota and J.C. Wheatley: Journ. Low Temp. Phys. 10, 379 (1973)
2. W.C. Black, A.C. Mota, J.C. Wheatley, J.H. Bishop and P.M. Brewster: Journ. Low Temp. Phys. 4, 391 (1971)
3. A.J. Leggett and M. Vuorio: Journ. Low Temp. Phys. 3, 359 (1970)
4. J.P. Harrison: Journ. Low Temp. Phys. 37, 467 (1979)
5. J.P. Harrison and J.P. Pendrys: Phys. Rev. B8, 5940 (1973)
6. J.R. de Bruyn and J.P. Harrison: Physica 108 B+0, 915 (1981)

A Size Effect in the Kapitza Resistance to Dilute ^3He-^4He Mixtures

F. Guillon[1], J.P. Harrison, and A. Sachrajda

Department of Physics, Queen's University
Kingston, Ontario K7L 3N6, Canada

Abstract
The thermal resistance between ^{3}He quasiparticles and phonons in three dilute mixtures (0.03%, 0.1% and 0.3% ^{3}He) in sintered submicron copper powder has been studied. Analysis shows that the resistance was modified by a size effect arising from boundary scattering of the ^{3}He quasiparticles

1 Introduction

The transfer of heat between dilute ^{3}He-^{4}He mixtures and solids has been of interest primarily for the design of dilution refrigerator heat exchangers. These have been constructed with sintered copper or silver powder in order to increase the interface area between helium and solid and hence decrease the Kapitza thermal resistance [1]. The transfer of heat involves more than the Kapitza resistance however. In series with the Kapitza resistance, R_K, between helium phonons and solid phonons there are the thermal resistances between between ^{3}He quasiparticles (q-ps) and phonons in the helium (R_{p-q}) and between the phonons and electrons in the metal sinter, R_{p-e}. Only R_{p-q} will be considered here. As discussed elsewhere [2], in many experiments below 20 mK the measured total resistance was far smaller than the predicted R_{p-q} for the ^{3}He-^{4}He mixtures. It was speculated [3] that this was a size effect. Energy transfer between the ^{3}He q-p system and the phonon system in dilute ^{3}He-^{4}He mixtures requires an interaction between one phonon and two ^{3}He q-ps to conserve energy and momentum. If a boundary were to replace the second q-p in the interaction then the energy transfer would be increased by the ratio of the ^{3}He q-p mean free path for scattering by other q-ps, ℓ_{3-3}, to the m.f.p. for boundary scattering, ℓ_{3-b}. This is significant because in a 5% mixture ℓ_{3-3} is 5 μm at 10 mK and is proportional to T^{-2} below that temperature; for comparison typical heat exchangers have 0.5 μm pores.

In order to test this model it was necessary first to confirm the theoretical thermal resistance between ^{3}He q-ps and phonons in bulk mixtures [4,5] and then to repeat the experiment with sintered copper or silver powder in the experimental cell. A test of the bulk theory, made for a 0.1% mixture, showed the correct temperature dependence and a factor 2 disagreement in magnitude [6]. This experiment has since been repeated for 0.03% and 0.3% mixtures with the same conclusion [7].

2 Experiment

For the test of the bulk theory the experimental cell was fitted with three thermometers to measure the ^{3}He q-p, phonon and cell wall temperatures. In

---

1. Present address: Northwestern Univ., Evanston, Ill. U.S.A.

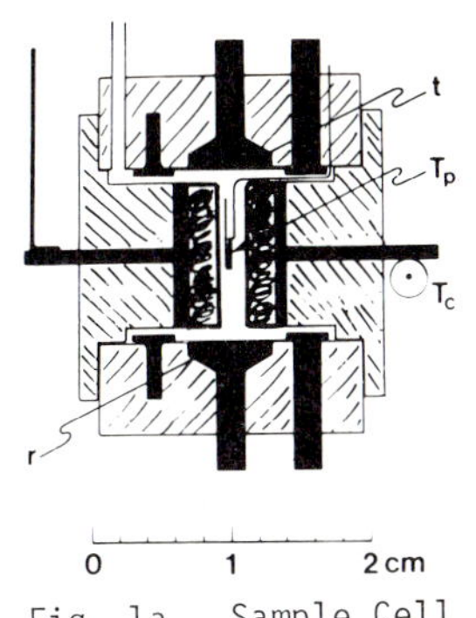

Fig. 1a Sample Cell

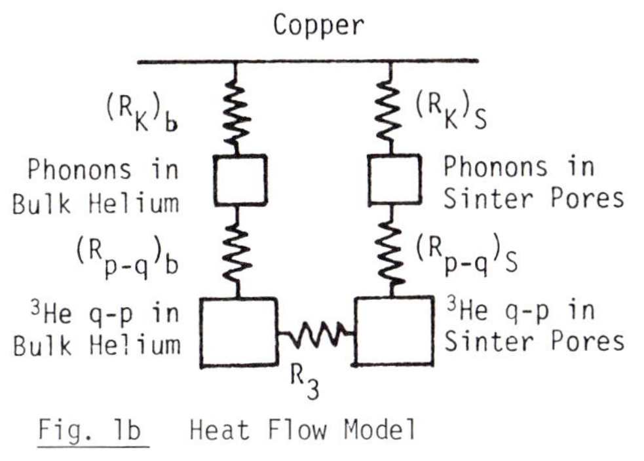

Fig. 1b Heat Flow Model

this way R_{p-q} and R_K were measured separately [6]. For the sinter experiment described here this was neither possible nor so necessary. Thermometers must be placed in bulk helium within the cell and therefore cannot directly measure the temperature of phonons or ^{3}He q-ps in the pores of the sinter. The cell is shown schematically in Fig. 1a. The phonon and cell wall temperatures (Tp and Tc) were measured with the same carbon resistors as used previously; the q-p temperature (Tq) was deduced from second sound velocity. The basic heat flow analysis model (Fig. 1b) treats the phonons in the bulk and in the sinter and the ^{3}He q-ps in the bulk and in the sinter as separate heat baths. The thermal resistances then are: R_{Kb}, the Kapitza resistance to the bulk phonons; R_{p-q} in the bulk helium; R_{Ks}, the Kapitza resistance to the phonons in the pores (negligible); R_{p-q} in the sinter (the quantity of interest) and the thermal resistance of the mixture in the pores which transfers heat into or out of the sinter. During the experiment Tc, Tp and Tq were measured as a function of time following a step change in the cell wall temperature, ΔTc, until thermal equilibrium was established. This was repeated over the temperature range 10 mK to 50 mK for ^{4}He containing 0.03%, 0.1% and 0.3% ^{3}He. It was observed that immediately after the step change there was a smaller step change in Tp (but not in Tq) and that Tp and Tq then approached equilibrium with the same time constant. The time constants are shown in Fig. 2. The results for the ratio ΔTp/ΔTc together with a more complete description of the time constant results and analysis will be published elsewhere [7].

3 Analysis of Results

The value of R_{p-q} in the sinter cannot be deduced directly from the measured quantities. Instead a numerical simulation of temperature versus time was made for the two cases: first that R_{p-q} in the sinter is equal to the previously measured bulk value (no size effect) and secondly that R_{p-q} is reduced by the ratio ℓ_{3-3}/ℓ_{3-b} (size effect). This simulation therefore required that the heat capacities and all other resistances be known. It was also necessary to know ℓ_{3-3} and ℓ_{3-b}, the pore diameter. The pore diameter deduced from the surface area of the sinter was 0.25 μm. This, together with tabulated heat capacity [8] and calculated q-p velocities was used to estimate the thermal resistance of the mixture in the pores. The q-p m.f.p., ℓ_{3-3}, was deduced from the temperature and concentration dependence of the viscosity of dilute mixtures [9]. For the range of the present experiments, ℓ_{3-3} was larger than the pore diameter by up to an order of magnitude.

The calculated time constants are shown in Fig. 2 as solid lines (no effect) and dashed lines (size effect).

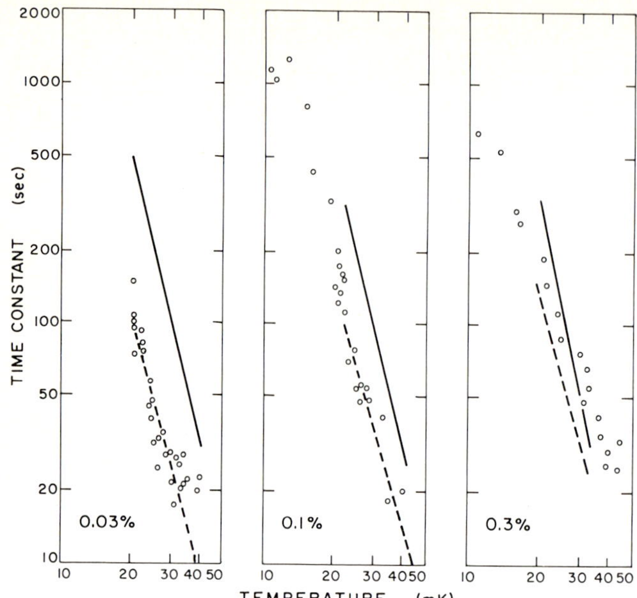

Fig. 2 Measured and Calculated Time Constants

4 Conclusion

The 0.03% and 0.1% mixtures clearly show the size effect. For the 0.3% mixture there is no agreement with either calculation. This may be because the bulk R_{p-q} was small for this mixture and was difficult to measure [7].

5 Acknowledgements

We wish to acknowledge the support of NSERC, Queen's University School of Graduate Studies and Research and the technical staff in the Physics Dept. The computer simulation was done by Drew Atkins.

1 O.V. Lounasmaa: Experimental Principles and Methods below 1K (A P 1974)
2 J.P. Harrison: J. Low Temp. Phys. 37, 467 (1979)
3 F. Guillon and J.P. Harrison: Phonon Scattering in Solids (Ed. H. Maris, Plenum 1980) p. 157
4 J.C. Wheatley, O.E. Vilches and W.R. Abel: Physics 4, 1 (1968)
5 T. McMullen: J. Low Temp. Phys. 51, 33 (1983)
6 F. Guillon, J.P. Harrison, T. McMullen and A. Tyler: Phys. Rev. Lett. 47, 435 (1981)
7 F. Guillon, J.P. Harrison and A. Sachrajda: to be published
8 R. Radebaugh: N.B.S. Monograph
9 K.A. Kuenhold, D.B. Crum and R.E. Sarwinski: Low Temperature Physics LT-13 (Plenum, 1974)

Kapitza Resistance Near 1 mK – The Shaking Box Model

A.R. Rutherford, J.P. Harrison, and M.J. Stott

Department of Physics, Queen's University
Kingston, Ontario K7L 3N6, Canada

Abstract
It is postulated that the dominant vibrational modes of a sintered heat exchanger in the temperature range 1 to 20 mK are localised oscillations of the powder particles. The energy transfer between ^{3}He in the pores of the sinter and these sinter oscillations is then calculated and yields good agreement with experimental results.

1 Introduction

The Kapitza resistance between liquid ^{3}He and sintered metal powder heat exchangers below 20 mK is significantly smaller than expected from acoustic mismatch theory. This is of inherent interest but has also been most important for the cooling of liquid ^{3}He into the millikelvin range.

There are two possible explanations: either there is an alternate heat path or acoustic mismatch theory becomes inapplicable below 20 mK. The early work focussed on a magnetic coupling between the liquid ^{3}He, a Fermi liquid, and magnetic impurities in the copper or silver powder, shown as path (a) in Fig. 1 [1]. However the dependence upon ^{3}He pressure and in particular magnetic field was not as strong as expected from the theory [2,3]. A revised model for magnetic coupling by way of the monolayer of solid ^{3}He formed in the surface of the metal powder particles, path (b), was successful in describing the temperature and field dependence of the Kapitza resistance between liquid ^{3}He and Pt powder[3].

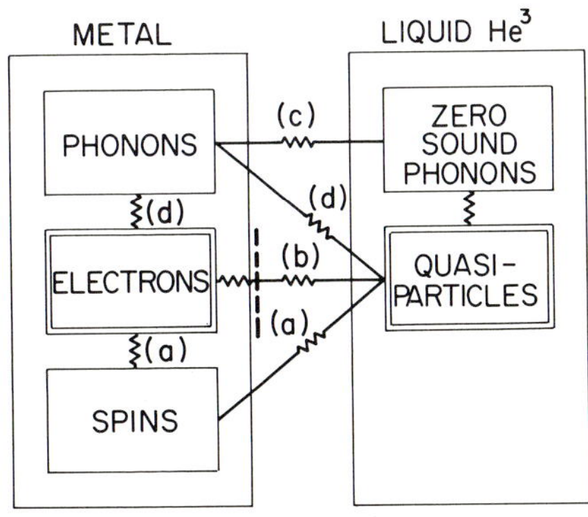

Fig. 1 Excitations of the Metal and of the ^{3}He

The alternative approach, the breakdown of acoustic mismatch theory, is based upon the small size of the metal powder particles in comparison to the wavelength of the Debye phonons excited in bulk metal at1 mK, about 0.5 μm and 30 μm respectively. Measurements of thermal conductivity [4], elastic constants, and sound velocity [5] have shown that in the long wavelength limit ($\lambda >> d$, where d is the powder particle diameter) effective medium, or sinter phonons propagate. Nishiguchi and Nakayama [6] have treated these sinter phonons with an Einstein oscillation model and have calculated the energy transfer to zero sound modes in the liquid ^{3}He (path (c)); by adjusting the Einstein temperature they were able to get good agreement with the experimental result of Andres and Sprenger. This present work describes another approach to the vibrational modes of a sinter and a calculation of the direct coupling of single quasiparticle excitation in the liquid ^{3}He to these vibrational modes (path (d)). Further explanation of the calculation and a more detailed comparison with experimental results will be published elsewhere [7].

2 Vibrational Modes of a Sinter

The largest wavelength of a bulk copper Debye phonon is ~ 1 μm, the size of the copper or silver powder particles. This corresponds to a minimum phonon frequency of about 5000 MHz, a frequency that dominates the excited phonon spectrum at about 100 mK. For $T < 20$ mK no bulk copper Debye phonons are excited. For very long wavelengths ($\lambda > 20$ μm) the sinter will appear to be a homogeneous effective medium [4,5] and phonons will propagate. This corresponds to a frequency range below 50 MHz or equivalent temperature range below 1 mK.

On the scale of 1 to 20 μm, electron microscope pictures show that sinters are very inhomogeneous and propagating phonons with wavelengths in this range are therefore unlikely. It is proposed that in the frequency range 50 to 5000 MHz the vibrational modes are localised oscillations of the powder particles, that the total number of these modes is 3N and that they are distributed over frequency with a constant density of states, in analogy with amorphous solids, up to the low frequency cut-off of the Debye spectrum of bulk phonons. N is the number of powder particles in the sinter and the number 3N is approximately the number of bulk phonon modes eliminated by the low frequency cut-off.

These inhomogeneity modes of the sinter, illustrated in Fig. 2, should give rise to a linear specific heat. For the metal sinters used in heat exchangers the term is only 10^{-4} times the electronic term and hence cannot

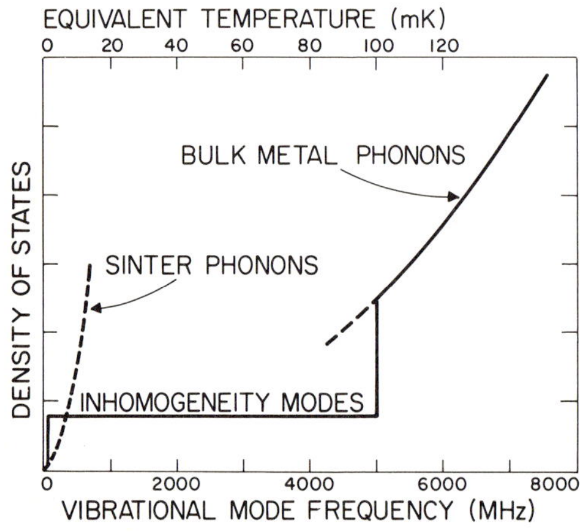

Fig. 2 Vibrational Modes of a 1 μm Sinter

be measured. However, Pohl and Tait [4] have measured the specific heat of four packed insulating powders and observed a linear term larger by an order of magnitude than is usual for an amorphous material but within a factor 2 or so with that predicted for inhomogeneity modes. Of course there could be other explanations [4].

3 Heat Transfer Calculation

For the purposes of modelling the heat transfer between liquid ^{3}He and the inhomogeneity modes of the sinter the system was modelled as a set of N cubic boxes, each the size of a pore, containing liquid ^{3}He. Each box is a harmonic oscillator of mass M with the frequencies distributed with the constant density of states. The Hamiltonian of the system is

$$H = \sum_j \left(H_{box} + \sum_i \frac{p_i^2}{2m^*} + V(x_i - X_j) \right) \quad (1)$$

$$= \sum_j \left(H_{box} + \sum_i \frac{p_i^2}{2m^*} + V(x_i) - X_j \sum_i \nabla V(x_i) \right) \quad (2)$$

in the small displacement limit. In Eq. 2 the first term represents the jth box, the second term the unperturbed ^{3}He quasiparticles and the third term describes the coupling. By means of the Fermi Golden Rule the energy transfer from ^{3}He to sinter modes was calculated [7] to give

$$\dot{Q}/\Delta T = \frac{2.42\, k_F^4 k_B^2\, a^2 DT}{\pi^2 M}$$

where k_F is the ^{3}He Fermi wave vector and D is the density of states. By taking the box edge a and mass M equal to the diameter (d) and mass of a powder particle this gives good agreement with experiment.

Acknowledgements

Research supported by NSERC and NSERC Summer Awards to A.R.R.

1, 2 See J.P. Harrison: J. Low Temp. Phys. 37, 467 (1979)
3 T. Perry et al.: Phys. Rev. Lett. 48, 1831 (1982)
4 R. Tait: Thesis, Cornell Univ. (1974)
5 B. Frisken et al.: J. de Phys. C-6 (Supp. 12) 858 (1981)
6 N. Nishiguchi and T. Nakayama: Solid St. Comm. 45, 877 (1983)
7 A.R. Rutherford, J.P. Harrison and M.J. Stott: (to be published)

Phonon and Roton-Induced Evaporation

A.F.G. Wyatt, M.J. Baird, and F.R. Hope

University of Exeter, Department of Physics, Stocker Road
Exeter EX4 4QL, England

We present results which show that the excitation-surface atom interaction is a single quantum process and that both energy[1] and the component of momentum parallel to the liquid surface are conserved. The essence of our experiments is to create a beam of excitations in the bulk liquid ^{4}He and let them propagate to the free surface. A bolometer in the space above the liquid detects atoms which leave the surface. The excitations in the liquid ^{4}He are not scattered by ambient excitations at 0.1 [K] and the saturated vapour pressure is so low that the evaporated atoms can travel ballistically to the detector. Using time of flight measurements we can estimate the energy of the excitation and the evaporated atom and hence analyse the interaction at the surface.

Consider an excitation in liquid ^{4}He travelling towards the free surface at an angle θ to the normal to the surface. Let the energy of the excitation be $\hbar\omega$, its momentum be $\hbar\underline{q}$ and group velocity $\underline{v}$. At the surface we assume the excitation is either reflected or gives all its energy to a surface atom. The atom is bound to the liquid with a binding energy E_B. If $\hbar\omega > E_B$ then the atom has sufficient energy to escape into free space and if energy is conserved in the excitation-atom system the atom will have kinetic energy $\hbar^2k^2/2m = \hbar\omega - E_B$, where $\hbar\underline{k}$ and m are the momentum and mass of the free atom respectively.

The direction of $\underline{k}$ will depend on the boundary conditions. If we approximate the wave packet of the excitation by a plane wave vector $\underline{q}$ and similarly the atom by a plane wave with wave vector $\underline{k}$ and we recognise the translational invariance of the surface it seems reasonable that there can be no change in momentum parallel to the surface, so $|\underline{q}|\sin\theta = |\underline{k}|\sin\phi$, where ϕ is the angle between $\underline{k}$ and the normal. The refracted angle is given by $\sin\phi = h|\underline{q}|\sin\theta/[2m(\hbar\omega - E_B)]^{1/2}$, where ω and q are related by the dispersion curve for the liquid ^{4}He at the S.V.P. For an excitation of given wave vector $\underline{q}$, ω can be found from the dispersion curve and using the latent heat per atom for E_B we can calculate $\sin\phi$. If the distance through the liquid is ℓ_1 and through the vacuum ℓ_2 then the total time for a signal starting with an excitation with wave vector q is $t(q) = \ell_1 v(q)^{-1} + \ell_2 v_a(k)^{-1}$, where $v(q) = d\omega/dq$ and $v_a(k) = [2(\hbar\omega - E_B)/m]^{1/2}$.

In the first set of experiments we use normal incidence so that $\theta = o = \phi$, and we vary ℓ_1 such that $\ell_1 + \ell_2 = 7.6$ mm. The arrangement is shown schematically inset in fig. 1. Phonons with energy greater than the 3-phonon cut off[2,3] ($\omega_c \sim 9$ K) have enough energy to evaporate an atom as $\hbar\omega_c > E_B$. For a particular ω and given values of ℓ_1 and ℓ_2, the total time has a minimum value.

The experiments were first done with purified ^{4}He[4] so that there was a negligible amount of ^{3}He on the surface. A typical received atom signal is shown as the upper curve in fig. 1. The value of ℓ_1 is varied by a

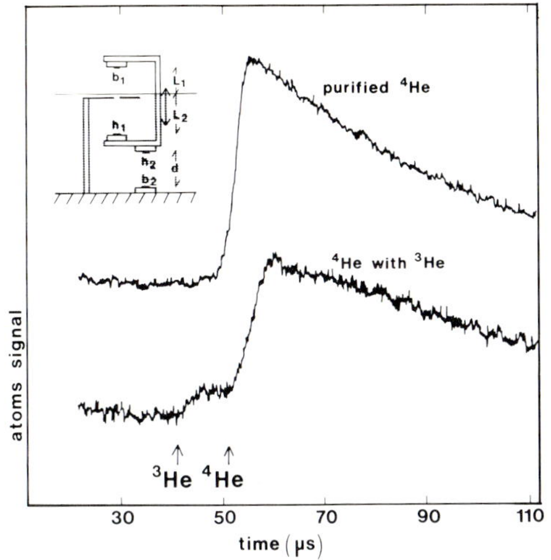

Figure 1. *The atom signals as a function of time due to phonons in the liquid ^{4}He striking the free liquid surface. The arrows indicate the calculated minimum total times*

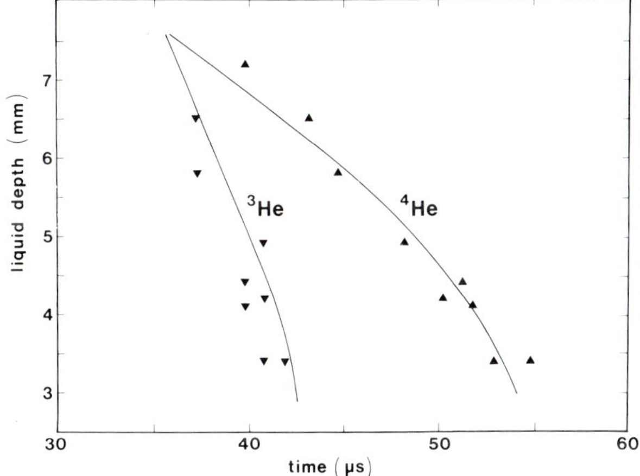

Figure 2. *The minimum total time for the signal is shown as a function of path length through the liquid ^{4}He for evaporated ^{3}He and ^{4}He atoms. The solid lines are calculated*

superconducting stepping motor and the minimum total times measured. The results for pure ^{4}He are shown in fig. 2 together with the solid line which is calculated using the dispersion curve[5,6] and taking E_B = 7.15 [K] [7]. The good agreement indicates that the assumption of a single excitation interacting with a single ^{4}He atom is valid.

To confirm our idea that the evaporated atom must come from the surface we added a little ^{3}He to the experimental cell. At the temperature of the experiment all the ^{3}He atoms will go to the free surface[8,9] and we estimate that there was a 23% monolayer. As the ^{3}He is bound to the surface with a lower energy than ^{4}He and as the ^{3}He atomic mass is less we expect the kinetic energy of the ^{3}He atom to be higher and the minimum total time to be less than for ^{4}He. That this is so can be seen in the lower trace in fig. 1 where the fastest signals for ^{3}He and ^{4}He are easily distinguished. The minimum total times are shown in fig. 2 together with the curve calculated using E_B = 5.0 [K]. The good agreement not only confirms the model but also the value of the ^{3}He binding energy found from quite different measurements and analysis.[9]

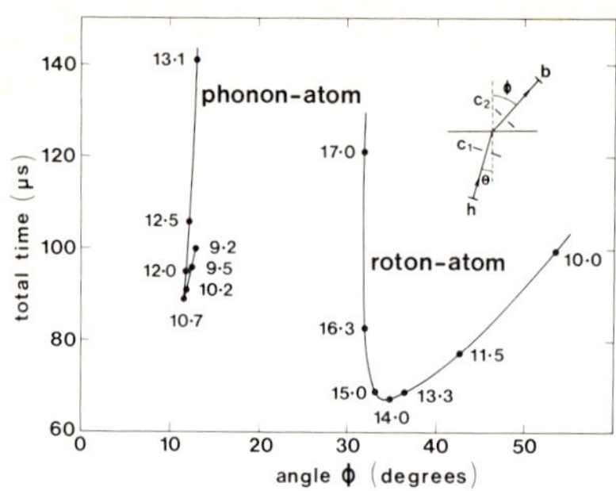

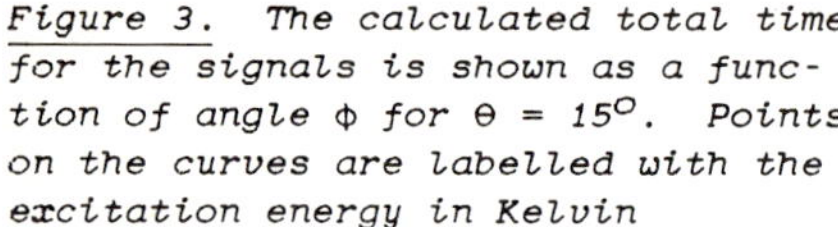

Figure 3. The calculated total time for the signals is shown as a function of angle ϕ for $\theta = 15^{o}$. Points on the curves are labelled with the excitation energy in Kelvin

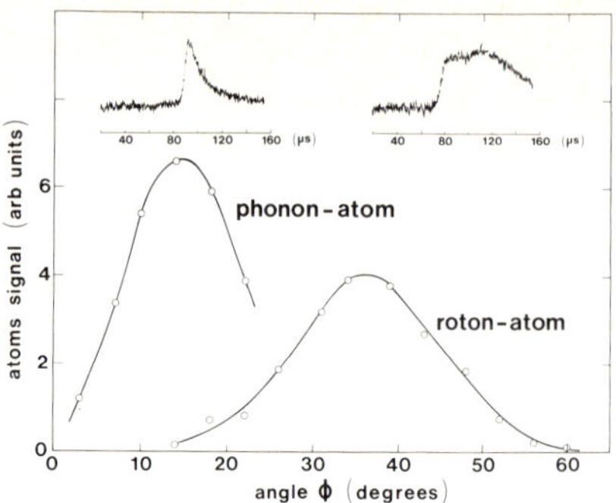

Figure 4. The angular distribution of the atom signal for $\theta = 13^{o}$ integrated over the first 17 μs. Inset are typical phonon-atom and roton-atom signals

To study roton-atom interactions we use the predicted large refraction angle for small angles of incidence which is quite distinguishable from the small angle expected for the phonon-atom interaction. The bolometer angle ϕ is varied and θ is constant at 13°, and $\ell_1 = \ell_2 = 6.5$ mm. Angles are measured to ~ $\pm 2^{o}$. The arrangement is shown schematically in fig. 3.

The calculated total time for different energies of phonons and rotons is shown as a function of ϕ in fig. 3. Both excitations show a minimum time of 89 μs and 67 μs respectively for phonons and rotons but we see that the angular spread is much larger for rotons than phonons.

As the detector is moved around angle ϕ we expect the minimum signal time to decrease as we move from phonon-atom to roton-atom signals. This can be seen in the measured traces shown inset in fig. 4. The measured times are 87 μs and 70 μs, which are in good agreement with the predicted values.

To demonstrate the angular separation of the two atom beams produced by the colinear phonon and roton beams we show in fig. 4 the signals integrated over the first 17 μs of the detected signal. It can be seen that the two effects are well resolved and the peaks occur within a few degrees of the angles predicted on the basis of the postulated boundary conditions.

From these experiments we can see that single phonons and rotons can desorb single atoms from the surface of liquid He with conservation of energy and parallel component of momentum at the liquid surface.

1. M. J. Baird, F. R. Hope and A. F. G. Wyatt. Nature 304, 325, (1983).
2. A. F. G. Wyatt, N. A. Lockerbie and R. A. Sherlock. Phys. Rev. Letters, 33, 1425, (1974).
3. R. C. Dynes and V. Narayanamurti. Phys. Rev. Letters, 33, 1195, (1974).
4. P. V. E. McClintock. Cryogenics, 18, 201, (1978). The sample used was kindly prepared for us by Dr. McClintock.
5. D. G. Henshaw and A. D. B. Woods. Phys. Rev. 121, 1226, (1961).
6. R. A. Cowley and A. D. B. Woods. Can. J. Phys. 49, 177, (1971).
7. J. Wilks. Liquid and Solid Helium, Oxford, (1967).
8. A. F. Andreev. J. Exptl. Theoret. Phys. (USSR), 50, 1415, (1966) and J. E. T. P., 23, 939, (1966).
9. D. O. Edwards and W. F. Saam. Prog. in Low Temp. Phys. VII A, 284, Ed. D. F. Brewer, North Holland, (1978).

Spectral Dependence of the Kapitza Resistance Between 0.5 K and 2.3 K

O. Koblinger, E. Dittrich, U. Heim, M. Welte, and W. Eisenmenger

Universität Stuttgart, Physikalisches Institut, Pfaffenwaldring 57
D-7000 Stuttgart 80, Fed. Rep. of Germany

The frequency dependence of the thermal boundary resistance between a solid and helium has been tested in earlier experiments [1] indicating a smooth and structureless transition between regular und anomalous Kapitza resistance. In our experiments, carried out with tunable monochromatic phonons, we find a sharp threshold in the decrease of reflectivity at about 85 GHz with additional structures at higher frequencies for a solid surface in contact with liquid or gaseous ^{4}He.

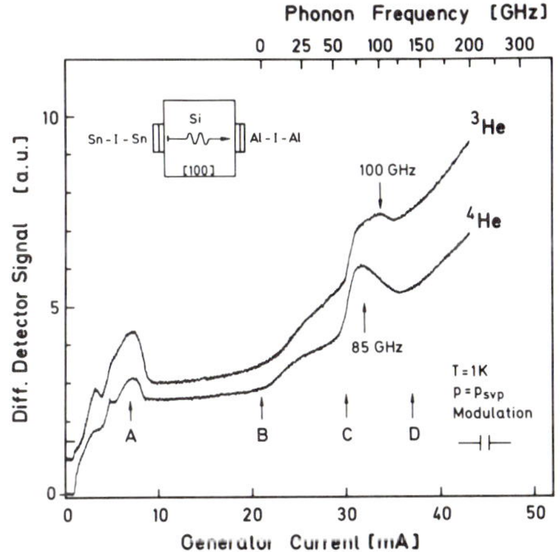

Figure 1: Detector signal of an Al detector vs. the current of a Sn generator. The sample is in contact with ^{3}He or ^{4}He gas respectively

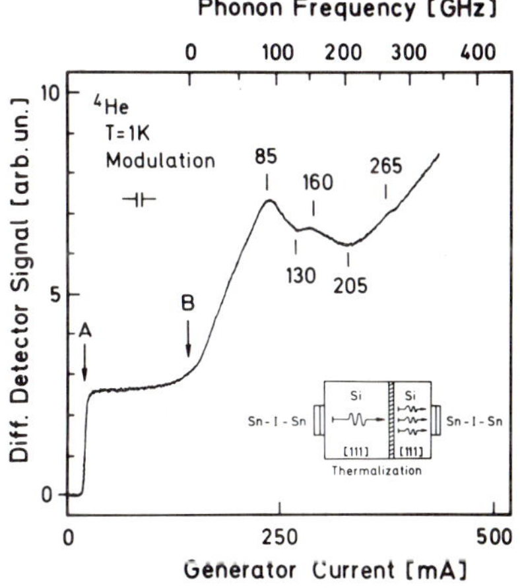

Figure 2: Bolometer signal vs. the current of a Sn generator obtained by the thermalizing arrangement which was immersed into liquid ^{4}He

Our experimental arrangements are shown as insets in Fig. 1 - 3. The Sn generator emits monochromatic phonons [2,3] both into the substrate and into helium. Phonons radiated into the substrate propagate ballistically to an Al junction working as a quantum detector [Fig. 1]. The sample is in contact with ^{3}He or ^{4}He gas respectively ($p = p_{svp}$, $T = 1.0$ K). It is important to realize that phonons which are reflected at the generator/helium interface also contribute to the detector signal. At small currents some peaks (A) due to nonlinearities in the I-V characteristics of the generator are observed. The following plateau is caused by a constant number of recombination phonons. Begin-

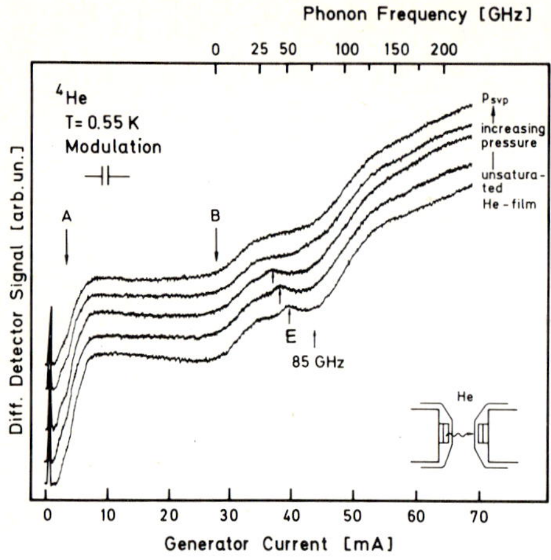

Figure 3: Differential detector sigal vs. the current of a Sn generator obtained by gas sound experiments in a volume between two tunnel junctions

ning with 0 GHz (B) monochromatic phonons are generated. If the phonon energy exceeds $2\Delta_{Det}$ (C) pairbreaking in the Al junction leads to a steplike sensitivity increase. At higher frequencies a significant difference in the detected phonon signal between ^{3}He and ^{4}He occurs. If ^{4}He gas is in contact with the sample a strong decrease of the phonon signal starting at 85 GHz is observed. After reaching a minimum at about 130 GHz the signal increases again, obviously at frequencies lower than $4\Delta_{Det}$ (D). If the sample was in contact with ^{3}He gas a small decrease of the phonon signal at about 100 GHz is followed by a minimum at about 120 GHz. In contrast we observed a complete disappearance of the frequency threshold phenomena when the generator first was covered with condensed nitrogen or a thin oil film and then was immersed into liquid ^{4}He or by keeping the sample under vacuum [4].

We therefore conclude that enhanced phonon transport from the generator into ^{4}He starts at 85 GHz. A corresponding threshold of increased phonon transport into ^{3}He is found at 100 GHz.

With a new kind of bolometer [5] we could extend both the observable frequency range from 0 GHz to 285 GHz and the temperature range from 1.0 to 2.3 K [inset of Fig. 2]. Between 0 GHz (B) and 85 GHz the detector signal increases with increasing phonon frequency and is almost structureless, whereas at 85 GHz a breakdown in the reflectivity is observed followed by additional structures at about 130, 160, 205 and 265 GHz. No significant influence of the helium temperature on the threshold frequency was observed in the temperature range from 0.95 K to the Λ point of ^{4}He, whereas at higher temperatures (normal fluid ^{4}He) only a small change in the signal slope at about 85 GHz occurs.

To study the phonon transport into helium we used two tunnel junctions which are mounted opposite each other [inset of Fig. 3]. The volume inbetween can be filled with gaseous or liquid helium respectively [6]. The He temperature is adjustable between 0.5 K and 2.3 K. Fig. 3 shows the detector signal vs. generator current for different ^{4}He pressure. Considering only the upper curve of Fig. 3. ($p = p_{svp}$, $T = 0.55$ K) at small currents (A) a plateau occurs. With increasing current (B) monochromatic phonons are generated. At 85 GHz a strong increase of the detector singal is observed reaching its maximum at about 130 GHz. At higher frequencies the signal increases again at 160 GHz and 265 GHz. The onset freqency of 85 GHz for enhanced phonon transport into helium did not depend on the pressure of the ^{4}He gas in the range of 10^{-4} Torr up to liquid helium. With ^{3}He gas between the junctions the threshold frequency shifts to 100 GHz. Both thresholds show a small temperature shift (roughly 20 % from 2.3 K to 0.5 K) to lower frequencies with decreasing temperature.

Below 1.0 K we observe at reduced ^{4}He gas pressure an additional structure (E) at frequencies below 85 GHz [lower traces of Fig. 3.]. With increasing He pressure the structure shifts to lower frequencies and disappears before reaching saturated vapour pressure. The additional structure might result from film thickness resonances in the helium film above the generator surface, whose thickness increases with increasing pressure [7]. The threshold at 85 GHz is not affected by the observed film thickness resonances. With ^{3}He we have not observed similar structures as yet.

Summarizing, we observed strongly enhanced phonon transport from a real solid into ^{4}He starting at about 85 GHz which is independent of the He gas pressure up to liquid helium. With the isotope ^{3}He the onset frequency for enhanced phonon transport shifts to 100 GHz. The threshold frequencies are almost independent of the He temperature in the range between 0.5 K and 2.3 K. Only slight shifts to lower frequencies with decreasing He temperature have been observed. No significant influence of the generator material (Sn, Al, SiO) or the generator preparation on the onset frequency has been observed. Moreover at low temperatures and thin He films we observed an additional structure which might be interpreted as film thickness resonances in the He film layer on the generator.

Our results may be interpreted in terms of strong phonon coupling by resonant excitations [8] or two-level transitions [9] of adatoms or surface imperfection in the van der Waals potential of the surface. This is also indicated by the strong influence of the atomic weight of the helium isotope ^{3}He and ^{4}He on the Kapitza threshold frequency. One finds that the ratio of the onset frequencies is about the square root of the mass ratio. Our results may also be compared with those of inelastic neutron scattering on grafoil with ^{4}He layers revealing dispersionless surface excitations, roughly at the same frequencies as we observed [10].

[1] E.S. Sabisky and C.H. Anderson: Sol. State Com. 17, 1095 (1975)
[2] W. Eisenmenger and A.H. Dayem: Phys. Rev. Lett. 18, 125 (1967)
[3] H. Kinder: Phys. Rev. Lett. 28, 1564 (1972)
[4] O. Koblinger: Thesis (1983), unpublished
[5] O. Koblinger, U. Heim, M. Welte, and W. Eisenmenger: Phys. Rev. Lett. 51, 284 (1983)
[6] U. Heim, R.J. Schweizer, O. Koblinger, M. Welte, W. Eisenmenger: J. Low Temp. Phys. 50, 143 (1983)
[7] C.H.Anderson and E.S. Sabisky: Phys. Rev. Lett. 24, 1049 (1970)
[8] M. Vuorio: J. Phys. C. 5, 1216 (1972)
[9] T. Nakayama: J. Phys. C. 10, 3273 (1977)
[10] H.J. Lauter, H. Godfrin, C. Tiby, H. Wiechert, P.E. Obermayer: Surface Science. 125, 265 (1983)

Kapitza Resistance of Laser-Annealed Surfaces

H.C. Basso, W. Dietsche, and H. Kinder

Physik Department, Technische Universität München
Institut für Festkörperphysik, D-8046 Garching, Fed. Rep. of Germany

P. Leiderer

Institut für Physik, Johannes-Gutenberg-Universität
D-6500 Mainz, Fed. Rep. of Germany

Our understanding of the phonon processes at the interfaces between two media is still quite rudimentary. Particularly notorious is the helium-solid interface, where the accoustic impedances differ by a large amount. It is well known that the Kapitza resistance, i.e. the thermal boundary resistance between helium and a solid, is usually up to 100 times smaller than predicted by the acoustic theory as formulated by Khalatnikov [1]. However, it was shown by J. WEBER et al. [2] that there was no discrepancy at surfaces of freshly cleaved crystals, i.e. at surfaces of exceptional good quality. Thus it is clear that surface irregularities must be responsible for the anomalous Kapitza resistance. Little progress has been made, however, in the understanding of how these irregularities mediate the anomalously strong phonon transmission.

In this paper we describe a technique which allows the in situ modification of a surface. Simultaneously the phonon loss into helium can be measured. From the effect of the surface modifications on the phonon losses, we can gain insight into the nature and the role of the surface irregularities.

We use the method of laser annealing which was introduced recently [3]. This technique uses high-power light pulses from a ruby laser to irradiate a Si surface. The energy of the pulse is sufficient to melt momentarily a thin (less than 5000 Å thickness) surface layer. It is known from several studies [3] that the molten layer recrystallizes in the same structure as the bulk and that the surface is free of impurities.

Our experimental set up is shown in Fig. 1.a. We used a vacuum chamber which was immersed in liquid helium at 1 K. The sample, a Si crystal, (100) oriented, of 4 mm thickness, was sealed to one side of the chamber. The test surface of the sample was at the vacuum side. It was polished by the supplier and without visible scratches. The phonon generator and the detector were placed about 0.5 mm apart on the outer side of the cystal. As generator we used a Sn film which was heated by irradiation with light pulses from an HeNe laser. We assume that the emitted phonon spectrum has a broad frequency band with a maximum at 2Δ , the energy gap of the Sn. We used phonon pulses of 100 ns duration. These pulses were, after reflection from the test surface, detected with an Al tunnel junction. The intensity of the reflected phonons depends , of course, on the losses at the test surface. The relative signal change during filling the chamber with helium could be measured with about 1 % precision. The reproducibility of the total sensitivity after a laser annealing of the test surface was about 30 %.

The ruby laser which we used for the annealing provided pulses of 40 ns duration (TEM_{oo} mode). From experiments with room-temperature Si, we concluded that an area of 0.12 mm^2 could be annealed with one laser pulse, i.e.

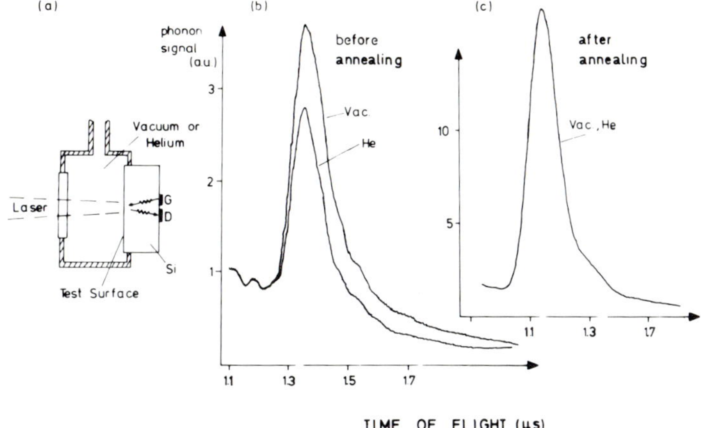

Fig.1 (a) Sample setup. Phonon pulses before laser annealing (b) and after annealing (c)

over this area was the energy density above the annealing threshold of 1 J/cm^2 [3]. Thus the energy of our laser was too small to anneal the whole test surface with one laser pulse. Therefore we used an optical scanning apparatus to move the laser beam across the test surface. With this method many annealed spots could be placed next to each other which barely overlapped.

The experimental results are shown in Figs. 1.b and 1.c. The intensities of the reflected phonons are plotted vs their respective times of flight. Due to phonon focusing transverse phonons only were detected. The two traces in Fig. 1.b were obtained before the annealing. The loss of intensity between the chamber being evacuated (vac) and filled with helium (He) was as expected for a "standard" Si surface [4]. The trace in Fig. 1.c was measured after annealing an 8 mm^2 area in the center of the test surface. Strikingly the reflected phonon intensity increased enormously (about fivefold). Note the different scales on the y axis. Probably this increase was due to the transition from diffusive to specular reflection of the phonons. Similar strikingly no difference between the traces was discernible if the chamber was evacuated or filled with helium. That means that the anomalous Kapitza transmission disappeared. From this result we conclude that laser annealed surfaces are as good as the cleaved ones of J. WEBER et al. [2].

In the next step we contaminated the test surface by irradiating the In seal between the Si crystal and the vacuum chamber several times with the ruby laser. The result is shown in Fig. 2.a. The absolute height of the phonon pulse was now considerably decreased,indicating that the phonons were now again nonspecularly reflected. The pulse shape, however, differed from that of the virgin state. It exhibited now a pronounced tail. Similarly to the virgin state of the test surface, the anomalous Kapitza transmission returned. Particularly the phonons in the tail were lost if the chamber was filled with helium.

With a repeated laser annealing we could decontaminate the surface again. This time we annealed an area of only 1mm^2. Nevertheless both the tail and the phonon loss into the Helium disappeared almost completely. This shows that the tail was not due to phonons travelling along long path lengths

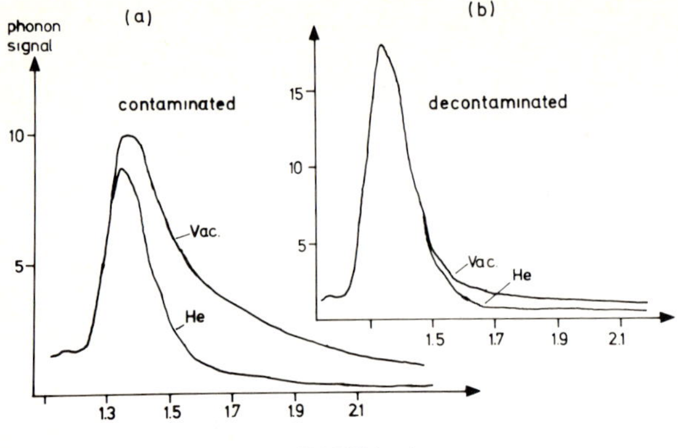

Fig.2 Phonon echoes after contaminating (a) and consequent decontamination by laser annealing (b)

under oblique angles which would lead to long times of flight. The tail must rather have been due to a time delay in the contaminant.

At the moment the exact chemical nature of the contamination is not known. It could be evaporated and recondensed water or In metal itself. The different phonon pulse shapes obtained with the virgin and with the contaminated surfaces indicate different origins of the anamolous Kapitza transmission in the two cases.

In conclusion we have shown that phonon-ideal surfaces can be prepared with laser annealing. The possibility of repeated contamination and annealing will very likely start a completely new branch of surface studies with phonons. With this technique we have demonstrated for the first time that the anomalous Kapitza transmission can be induced by condensing impurities on the interface between the solid and the helium. It is noteworthy that the existence of such impurities is a prerequisite for a recent model [5] of the anomalous Kapitza transmission.

1. For a review see: A. F. G. Wyatt: in Nonequilibrium Superconductivity, Phonons, and Kapitza Boundaries (K.E. Gray, Editor, Plenum New York-London, 1981) p.p. 31-72
2. J. Weber, W. Sandmann, W. Dietsche, and H. Kinder: Phys. Rev. Lett. 40, 1469 (1978)
3. For a review see: J. M. Poate and J. W. Mayer (Editors): Laser Annealing of Semiconductors (Academic, New York, 1982)
4. W. Dietsche and H. Kinder: J. Low Temp. Phys. 23, 27 (1976)
5. H. Kinder: Physica 107B, 549 (1981)

Discussions

PHONON-INDUCED DESORPTION OF HELIUM page 148
P. Taborek

K. Dransfeld: What is the difference between metallic and nonmetallic substrates?

P. Taborek: I don't know.

A.C. Anderson: In the measurements of heat transfer between Cu and He gas, we could account for the data if $\alpha = 1$. Was it possible to obtain a value of α from your data, assuming the thermodynamic model to be valid?

P. Taborek: α cannot be measured reliably in this way. The Caltech group has recently measured α in a reflection experiment with $\alpha \sim 3/4$.

L.J. Challis: Did you vary the state of the sapphire surface (polish, anneal etc.) and, if so, did you observe any change in the percentage of atoms emitted by the 1-phonon process with the state.

P. Taborek: No. The surface was the best available commercial polish.

K. Weiss: It would be interesting to compare your experiments with Wyatt's earlier experiments. Who did observe a cosine background in a different experiment!

W. Eisenmenger: Did you use any alternative pressure measurement besides evaluation of the recovery time?

P. Taborek: The pressure measuring technique can be compared to measurements using standard methods for $P > 10^{-4}$ torr. See Sinvani and Goodstein, Surf. Sci. 125, 291 (1983). I know of no other way of independently measuring the in situ pressure for $P = 10^{-6}$.

THERMAL BOUNDARY RESISTANCE BETWEEN SMALL PARTICLES AND LIQUID He-3 page 155
T. Nakayama

P.C. Main: Comment: The situation in very fine particles is very complicated with many resistances in series, some of which seem to be much smaller than expected theoretically, for example,the electron-phonon resistance, so it may not be right to attribute the observed T dependence to the boundary effect alone. Question: If the low temperature $R_K \propto 1/T$ dependence is due to sinter modes how do you explain the magnetic field effect seen by Perry et al.?

T. Nakayama: Your comment on the series resistance is well known. In particular this effect is important for the magnetic Kapitza resistance such as the resistance between SMN and liquid ^{3}He due to the localized electron spin-phonon interaction. For metal sinters, however, free electrons respond very fast to the vibrational modes of the sinter through the coupling of deformation type. This viewpoint could be accepted from the good agreement between our theory and experiments above 10 mK, in which the theory does not involve any fitting parameters. So it is correct to attribute the anomalous temperature dependence (not exactly proportional to 1/T) to the boundary resistance. For your question, the magnetic field dependence is observed only at the surface of Pt particles. As I said, the magnetic coupling of the Fermi type is very sensitive to the surface characteristics of metal. It should be required to see whether such a magnetic field dependence is really present or not in the case of Ag particles.

J.P. Harrison: Can you comment on the existence of zero sound collective modes when a quasiparticle mean free path is 25 μm at 1 mK (compared to 0.5 μm pores) and a zero sound phonon wavelength is about 10 x the pores size in the sinter.

T. Nakayama: The condition for existence of zero sound is $\omega\tau > 1$. Provided that the mean free path 1 is of the order of 1 μm as you suppose, we can estimate $\omega\tau > 10$ at 1 mK, which indicates really the existence of zero sound in our system. The zero sound does not behave like plane waves in the sinter because it must satisfy the complicated boundary condition. Still we have zero-sound modes in the sinter interacting with vibrations of sinter. You should also remember that sinters are very porous as seen from the electron micrograph of Ag sinter taken by Veuro in Ref. 24.

A.C. Anderson: Prof. Harrison suggested a relationship between R_K and sinter size. Can your model account for this dependence?

T. Nakayama: If the size of metal particles becomes large or small, we have the different characteristic frequency of the sinter because $\omega_c = (\pi a^2 E/M1)^{1/2}$. Taking account of this, we could explain successfully the experimental data of Ref. 5 in which rather large Cu particles of diameter 1 ~ 30 μm were used.

SCATTERING AND ABSORPTION OF BALLISTIC PHONONS BY THE ELECTRON INVERSION LAYER IN SILICON: THEORY AND EXPERIMENT page 163
J.C. Hensel, R.C. Dynes, B.I. Halperin, D.C. Tsui

V. Narayanamurti: Is your TA absorption coming from the fact that you are coming at an angle and not in a symmetry direction? Have you tried to fit it like you have done for the LA?

J.C. Hensel: This is entirely correct; for phonon k's parallel or perpendicular to the major axis of the electron ellipsoid, i.e. (001), the coupling vanishes. For intermediate directions the TA coupling is comparable to the LA coupling. We have not yet tried to fit the TA attenuation data. This case is slightly more complicated than the LA case and, consequently, less conclusive.

W. Eisenmenger: Do you expect a more detailed agreement between experiment and theory in using a more realistic electron distribution? Can your systems be used as phonon detectors, e.g., eventually by the electron-phonon drag effect?

J.C. Hensel: In answer to your first question, it is not expected that the results will be appreciably changed by use, say, of a variational electron distribution. A preliminary study shows that in form the solution changes rather little, the main effect being that the spacing parameter h is no longer arbitrary but fixed at a value of $3a_0$ in the arguments of the oscillatory terms. But this is approximately the actual value usual in making the fit in the present mode. As to the second question, we have not observed any effect in transport in the MOSFET channel due to the influence of ballistic phonons. Therefore, I am not optimistic that this would make a viable detector.

G. Meissner: Could you use your phonon detector in the case where a magnetic field perpendicular to the two-dimensional electron system is applied?

J.C. Hensel: For the particular detector we use (a granular Al detector) the experiment could not be directly transferred to a high magnetic field. For another more suitable kind of detector, e.g. a high T_c material, the same system would in principle work.

K. Dransfeld: Would it be possible to repeat the experiment with an electronic drift velocity around the velocity of sound?

J.C. Hensel: Yes, this is possible and interesting. But it would be better to do this with monochromatic phonons.

LOW WAVEVECTOR PHONONS IN THE 2-DIMENSIONAL ELECTRON SOLID ON LIQUID HELIUM page 171
F.I.B. Williams

W. Eisenmenger: Do you think it possible to visualize your electron lattice since the electron distance is larger then 3000 A?

F.I.B. Williams: In principle diffraction of light would be possible, but in practice the scattering cross section, even from the microscopic dimple lattice, is too low.

P. Leiderer: (Regarding Prof. Eisenmenger's question to detect the e lattice by light scattering): It certainly would be very hard to do light scattering off the electrons or also the $\lesssim 1$ Å deep dimples underneath them. However, one might succeed by some decoration technique.

JJ. Kim: Could you substitute protons in your experiment, and obtain any more information especially on the LA (plasmon) branch?

F.I.B. Williams: Yes, but protons would dissolve into the liquid where they would form a positive ion consisting of a "snowball" of He atoms as for the He^+ ion. The plasmons have been observed for the He^+ ion system held against the surface by G.A. Williams and M. Ott at UCLA. The $\omega\tau \ll 1$, however, for the transverse branch.

RECIPROCITY THEOREM FOR PHONON TRANSITIONS AT IDEAL INTERFACES WITHIN THE ACOUSTIC MISMATCH MODEL page 179
O. Weis

K. Weiss: Did you include Rayleigh modes in your analysis?

O. Weis: No, because I treated the excitation by plane waves. Strictly speaking, Rayleigh modes or Love modes, respectively, are freely propagating waves and are not coupled to an incoming or outgoing plane wave at an infinite ideal interface. Of course, there are angles of incidence where the excited exponential nearfields have properties very close to those of the mentioned free modes, but they are not identical.

A.F.G. Wyatt: The theorem depends on the translation symmetry of the boundary and therefore the interface has to be infinite. I wonder how the theorem is modified if one has finite interfaces. This is relevant to Kurt Weiss' comment, as surface waves can transport energy away from a small interface area.

O. Weis: Free surface modes can be involved as well as other outgoing waves in the case of a finite area. Their amplitudes are essentially determined by a spatial expansion of the field of excitation at the area, in the same way as it is done in solving diffraction problems. I have not treated this case. I guess that reciprocity is also valid for finite areas.

J.K. Wigmore: Is the reciprocity theorem valid when the phonon density of states has a different frequency dependence on the two sides of the interface? I have in mind a very thin heater film deposited on a dielectric sample.

O. Weis: I have only derived the theorem for half-spaces. In this case, the density of states is not involved. However, by considering detailed balancing the density of states comes in and within the model used the same ω^2 dependence occurs in both half-spaces.

H. Kinder: To my understanding, the reciprocity theorem is a special case of time-reversal symmetry which is of course broken if magnetic effects are included.

O. Weis: From time-reversal symmetry follows that energy surfaces have inversion symmetry, i.e. the relation $\varepsilon_\sigma(-\vec{q})=\varepsilon_\sigma(\vec{q})$ holds. Indeed, this was assumed to be valid, but it is not enough information to derive the reciprocity theorem.

W. Rehwald: Is the acoustical reciprocity theorem hurt in the case of magnetic effects, such as magnetic-field dependent elastic properties?

O. Weis: I have confined my treatment to pure elastic equations. If magnetic and piezoelectric effects are included, one has to extend the set of equations and boundary conditions by the presence of magnetic and electric fields. I expect that one has only to reverse the magnetic field (= axial vector) in order to set reciprocity. This is not necessary if only quadratic effects are studied.

PHONON SCATTERING BY TWIN PLANES page 182
J.W. Vandersande, P.N. Chopra, R.O. Pohl

S.J. Rogers: What was the relationship between the direction of heat flow in the measurements and the direction of application of the stress? Were the twinning planes perpendicular to the heat flux?

R.O. Pohl: Heat flow was in the (001) direction, the same direction in which the pressure had been applied. As far as I remember, the twin planes are inclined to the heat flow by approximately 40^0. The twin plane spacings quoted were measured perpendicular to the planes. A correction for this angle was not made in Fig. 3. (Thank you for pointing out this mistake to me).

K. Weiss: How certain is it that in your experiments you did really observe the "Casimir regime"? It would be very nice if it were indeed the case because this regime is not too easy to observe.

R.O. Pohl: In the undeformed $CaCO_3$ sample, a Casimir length of ~9 mm was determined from the experiment, using a Debye average of the speeds of sound and a Debye specific heat computed from this average speed . The theoretical Casimir length is 5.5 mm. My guess is that this discrepancy is the result of the Debye approximation. Usually, we obtain a better agreement (e.g. Si, Al_2O_3, Li F).

BOUNDARY AND DISLOCATION SCATTERING OF PHONONS IN LEAD SINGLE CRYSTALS page 185
W. Odoni, P. Fuchs, H.R. Ott

H.M. Rosenberg: Have you actually measured the specific heat of your sample to show that it is proportional to T^3 down to 0.1K?

W. Odoni: No, but the thermal conductivity measurements in the intermediate state, where the mean free path is constant, show that the specific heat has a cubic temperature dependence.

L.J. Challis: Comment on Rosenberg's remark: I am not sure why additional contributions to the specific heat such as nuclear specific heats due to impurities, etc. should affect the analysis since presumably they are localized and so do not transport heat.

DIFFUSE SCATTERING OF THERMAL PHONONS AT CRYSTAL SURFACES page 188
T. Klitsner, R.O. Pohl

K. Weiss: Do you see a chance to reach the hydrodynamic regime with your X-tals, specifically with the one with the "cleanest" surfaces?

T. Klitsner: It is a good idea. Extremely specular surfaces could help reach this regime by increasing the effective time between collisions with the walls of the crystal and thus allowing more N processes to occur between wall collisions. However, these crystals are neither pure enough nor specular enough to reach that regime.

L.J. Challis: Some metal ions present as impurities in bulk Si can of course act as localized donors or acceptors which are often very strong phonon scattering centres. Do you think there is any possibility that some of your metal ions have diffused into a surface layer?

T. Klitsner: I don't think so. I have checked for diffusion during Rutherford backscattering studies and found nothing. In addition I can wipe off the metal film and the scattering goes away. Perhaps the oxide layer on the silicon surface inhibits diffusion into the bulk. I should also mention that these same effects have been seen on sapphire surfaces.

J.K. Wigmore: (1) Did you deliberately not cool your substrates to make a more uniform film? (2) In order to simulate a real suface, would it not be a good idea to use a silicon film on bulk silicon (to avoid acoustic mismatch and possibly reflections in the film)?

T. Klitsner: (1) Yes, it turns out to be quite difficult to keep all the parameters the same while varying only one, so I have kept the deposition conditions as constant as possible, so far. However, I will try to make both more and less uniform films for my experiments in the future. (2) A silicon film has a much smaller effect than a gold film. It apparently does not form island droplets which seem to be important for this effect. If one could form these structures, then you are right, it would be a good choice.

IMAGING OF SPECULARLY REFLECTED PHONONS FROM A CRYSTAL BOUNDARY page 191
G.A. Northrop

W. Eisenmenger: I think that in the paper of Marx and myself basically the same calculation procedure including phonon focusing has been used. Applying this to silicon surfaces calculation and experiment indicate diffuse scattering in agreement with an experiment with purposely rough surface.

G.A. Northrop: To my knowledge both calculations are based upon the same physical principles, but are difficult to compare, being for different materials and in different domains (time vs space). I believe that the different experimental interpretations (diffuse scattering in Si and specular reflection in sapphire) resulting from these calculations are both correct, and that the samples/surfaces are indeed different.

S.J. Rogers: If you have mainly specular reflection it seems that multiple reflections from the top and bottom surfaces might play an important role at large angles from the normal. Did you see any effects?

G.A. Northrop: No, since the scan range used here limited the reflection angle to a little more than 45° from normal.

H. Kinder: I am not surprised that you find strong specular reflection because your sample was submersed in liquid helium. In this case we know from Taborek and Goodstein's work that the diffuse phonons are lost into the liquid helium.

G.A. Northrop: Yes, this is what we expected. The obvious next step is to repeat this experiment with a vacuum to see if the diffuse scattering increases.

J.P. Harrison: Is it possible to reduce the hot phonon temperature to reduce the phonon scattering in the bulk?

G.A. Northrop: Yes, by defocusing the laser or reducing its total power.

W. Grill: Where does the halo-like structure come from, which you observe in some of the pictures?

G.A. Northrop: This occurs when the incident slow transverse wave reaches the critical angle for conversion to longitudinal waves. Beyond that angle ($\sim 34^{o}$) the ST $\rightarrow$ L channel is no longer open and the ST $\rightarrow$ ST intensity increases, as observed in the data.

J.C. Hensel: You come to the conclusion that the diffuse scattering is weak relative to specular reflection on the basis of an absence of observable structure in the reverse focusing images. Isn't it possible that the diffuse scattering can still be quite strong, yet not concentrated sufficiently spatially and temporally to show up by imaging? Have you attempted to estimate integrated intensities?

G.A. Northrop: Although the diffuse scattering has contrast ratios of only about three, compared to at least 10 for specular reflection, the more intense structures in the diffuse pattern result from a fairly narrow set of paths, and hence should be concentrated in time. In particular, the diagonal structure in the diffuse calculation is due to FT → ST, and should arrive around the time of the specular ST → ST. This experimental image shows the expected diffuse structure, but it is much weaker than comparable specular structures.

ANOMALOUS LOW TEMPERATURE KAPITZA RESISTANCE OF A PARAMAGNETIC SALT page 197
G.J. Batey, P.C. Main

J. van Miltenburg: It is well known from bottleneck experiments (for instance in ruby) that solely because of the bottleneck the temperature of the phonons can be very different from that of the bulk. You say that this difference $T_p - T_H$ is solely caused by the Kapitza resistance and you incorporate the bottleneck effects in the path from the spins to the lattice. In my opinion you have to take the bottleneck effects into account as a resistance in series with the Kapitza resistance, so between the phonons and the bath.

T. Nakayama: I think you cannot conclude definitely the unimportance of the magnetic coupling from your experiments, because you measured the resistances only down to around 20 mK. Since the magnetic coupling is so weak, this coupling becomes important below 10 mK compared with the zero-sound excitation mechanism.

P.C. Main: Yes, this is true. In any case it is dependent on which theory one takes. All we are saying is that the original motivation for magnetic coupling seems not to be correct.

A.C. Anderson: Please comment on the condition of the sample before and after the measurement.

P.C. Main: The crystals were clear without any obvious faults and the faces were water polished. As far as we could tell with an optical microscope, the crystals were in the same condition after cycling.

A SIZE EFFECT IN THE KAPITZA RESISTANCE TO DILUTE ^3He-^4He MIXTURES page 200
F. Guillon, J.P. Harrison, A. Sachrajda

T. Nakayama: Why don't you consider the plausible direct channel between quasi particles and vibrations of sinter?

J.P. Harrison: The "shaking box" model suggests that direct coupling of quasi particles to sinter modes will not be important for a dilute mixture unless the temperature is below a few mK for 5 % ^3He and well below 1 mK for 0.1 % ^3He.

A. Ikushima: How could you deduce the relaxation time in the mixture from the viscosity data you showed? Could you do that even when the viscosity does not show $1/T^2$ behaviour?

F. Guillon: The relaxation time was deduced by extracting τ_η for the different concentrations given in the viscosity data of Kuenhold et al., using $\tau_\eta = \eta/P(T)$. These relaxation times were well fitted, below 100 mK, by the equation $\tau_\eta T^2 = A\,(1 + B\,(T/T_F)^2)$ for all concentrations in the Kuenhold et al. data. The concentration dependence of the two fitted parameters, A and B, allowed us to obtain τ_η for .03, .1 and .3 % by extrapolation and interpolation.

P.C. Main: Is it possible to say what is the temperature dependence of the phonon-quasiparticle resistance actually within the sinter?

F. Guillon: The temperature dependence of the phonon-quasiparticle resistance is given by $(R_{p-q})_b/l_{3-3}$, where l_{3-3} is the mean free path estimated by the viscosity data of Ruenhold et al.

KAPITZA RESISTANCE NEAR 1 mK - THE SHAKING BOX MODEL page 203
A.R. Rutherford, J.P. Harrison, M. J. Stott

F.W. Sheard: In calculating the heat transfer from the vibrations which shake the box to the ^{3}He quasiparticles in the box, do you take the interactions between the quasiparticles into account since ^{3}He is a strongly interacting Fermi liquid?

J.P. Harrison: Whilst ^{3}He is indeed a strongly interacting system, at 1 mK a single ^{3}He quasiparticle makes about 50 boundary collisions before interacting with another quasiparticle. The quasiparticle - quasiparticle scattering m.f.p. is 25 m at 1 mK and is proportional to T^{-2}. The nature of zero sound, the collective excitation of the quasiparticle gas, also is unclear given that the wavelength of a thermally excited zero-sound phonon is equal to the pore size at 5 mK and is T^{-1} below that temperature.

T. Nakayama: I think you need the physically plausible explanation of the proposed constant density of states of the sinter, because the sinter is not one-dimensional.

J.P. Harrison: The constant density of states was postulated in analogy with the constant density of states reduced for all glassy and amorphous systems.

T. Nakayama: It is not clear what the box of your model is actually in the sinter. The electron-micrograph of sinter shows that the box model is not a good one. The boxes are connected with the canals.

J.P. Harrison: I agree that the pores have openings. The box is just a first approximation to a pore.

T. Nakayama: Your model is based on the Fermi-gas model and also works for the ^{3}He/^{4}He mixture, but the experiments for the mixture show that the resistances have the different temperature dependence proportional to T^{-2}. Can you explain this discrepancy from your model?

J.P. Harrison: When the Fermi momentum for a ^{3}He - ^{4}He mixture is put into our heat transfer coefficient equation then reasonable numerical agreement with the experimental results is attained. Of course, the temperature dependence is wrong. We believe that below 1 mK heat transfer is via the shaking box and that above 1 mK the heat transfer is via phonons. The T^{-2} result is then seen as a change over from T^{-1} to T^{-3}. We plan experiments to test this.

P.C. Main: How far can your theory be said to apply to a packed powder rather than a sinter? The Princeton experiment shows the same T dependence with powder rather than sinter and yet they see a strong magnetic field dependence. What is this due to?

J.P. Harrison: The theory should apply to a packed powder. Our estimate of the thermal boundary resistance in the Princeton experiment is 20 % of the measured resistance. We surmise that most of the resistance was between electrons and vibrational modes. Unfortunately we know almost nothing about the electron-phonon interaction in small particles at very low temperatures.

R.O. Pohl: Are specific heat measurements needed in determining the thermal resistance?

J.P. Harrison: No. However, the model predicts that pressed or sintered fine powders should have a linear specific heat over a limited temperature range.

A.C. Anderson: How important was the contribution from adsorbed gases to the heat capacities measured by Pohl and Tait?

R.O. Pohl: Some of the low temperature specific heat anomaly may indeed have been caused by adsorbed gases. For a discussion, see R.O. Pohl in Topics in Current Physics, Vol. 24, W.A. Phillips (ed.), Springer 1981

PHONON AND ROTON-INDUCED EVAPORATION page 206
A.F.G. Wyatt, M.J. Baird, F.R. Hope

P. Leiderer: Do you have an idea why rotons with negative group velocity are absent?

A.F.G. Wyatt: My feeling is that they cannot be produced by a plane heater because its geometry is essentially different from a neutron which we know can create them.

W. Eisenmenger: Congratulations for your beautiful results. With respect to vapour-liquid equilibrium I would like to ask whether the ripplon concept invoked from atomic beam reflective experiments by E.O. Edward et al. is needed in addition to the one-phonon processes you have given evidence of. From our estimate of the balance of phonons with high energy and the atoms the one-phonon processes do not occur at a sufficient rate.

A.F.G. Wyatt: Our experiments show that at least some atoms are evaporated without exciting ripplons. Ripplons were introduced in the theoretical model to explain Edward's et al.'s results but there is no direct evidence for them and it is possible that their results can be explained without them. To answer the question about the equilibrium between liquid and vapour we shall need to know the efficiency of the excitation - atom process as a function of frequency. If it is high enough then I think it can account for the saturated vapour pressure.

SPECTRAL DEPENDENCE OF THE KAPITZA RESISTANCE BETWEEN 0.5 K AND 2.3 K page 209
O. Koblinger, E. Dittrich, U. Heim, M. Welte, W. Eisenmenger

K. Weiss: Did you try to analyze your results in terms of a model in which the energy is first absorbed by an atom at the interface and then goes into the phonon system? Such a mechanism could explain the threshold (discrete bound state for the atom)!

O. Koblinger: The isotope effects point to the helium atom and the threshold may be understood by atoms in distinct positions of the real surface with reduced binding energy as has been suggested by yourself. Furthermore, the experiments (Fig. 2) show that the phonon losses into ^{4}He are much stronger than into ^{3}He, which certainly is an additional clue to the mechanism.

T. Nakayama: By neutron scattering experiments, Lauter and others observed recently the discrete dispersionless spectra of ^{4}He adsorbed layer on graphite in which the spectra are quite similar to yours. Do you have any comment to explain this similarity?

O. Koblinger: As I said before, the observed structures in the reflectivity do not depend on the material of our generators, even if there is an oxide layer on the surface of the generator or it is covered with an insulator material like SiO. Therefore, it is quite reasonable that both groups observed the same surface excitations.

KAPITZA RESISTANCE OF LASER-ANNEALED SURFACES page 212
H.C. Basso, W. Dietsche, H. Kinder, P. Leiderer

E. Pater: Did you control the amount of contamination by repeatedly shooting on the seal?

W. Dietsche: We have not yet done so because the experiments were done very recently. But we will do this in the near future.

J.P. Maneval: The phonon signals reflected back from your indium-contaminated surface last for microseconds. Is not this long sticking time contradictory to a strong probability of capture of the phonons by the surface states?

W. Dietsche: The relaxation time depends on the coupling constant and the density of states. If the density of states in the contaminant is very large, the strong coupling to them and the long tail (slow relaxation) can be explained.

Part V

Quantum Liquids and Crystals

Chairmen:
A. J. Ikushima I. M. Khalatnikov P. J. King W. Rehwald K. Weiss

Crystallization Waves in Helium

A.Ya. Parshin

Institute for Physical Problems, USSR Academy of Sciences, Kosygin Street 2
SU-117973 Moscow, USSR

Abstract

Both theoretical and experimental evidence for a new quantum state of the ^{4}He liquid-solid interface are presented. The existence of this state implies the possibility of exactly non-dissipative crystallization and melting. Weakly damped oscillations of the interface due to periodic melting and crystallization-crystallization waves - may propagate along the interface under these conditions. The spectrum of these waves is measured and the frequency and temperature dependences of their attenuation are determined.

1. Introduction

The interface between solid and liquid ^{4}He has been under intensive study in recent years. The unique property of such an interface is its very high mobility, originally proposed theoretically [1] and then observed in various laboratories. In the field of phenomena connected closely with that of superhigh mobility, such as crystallization waves [1,2], anomalous sound transmission and Kapitza resistance [3-8], roughening transitions [2,9-12], the first one seems to be quite unusual and at the same time very easy to observe.

The physical nature of the interphase superhigh mobility and crystallization waves can be understood by taking into account the quantum delocalization effects. In fact, strictly nondissipative crystallization appears to be possible in the presence of a special type of coherent motion in a two-phase system, in contrast to the classical situation when the crystallization process is a result of random transitions of individual particles from one phase to another. Of course, the current theory is not sufficient for the complete description of this phenomenon, it is in essence only a correct guess of its nature.

2. Quantum Rough State

The following simple idea is principle in understanding of the unique surface properties of helium crystals at low temperatures. Let us consider an atomically smooth surface of a crystal in equilibrium with liquid phase at absolute zero (Fig.1). There is a step on the surface and a kink on the step. The displacement of the kink by the elementary translation vector along the step involves the transfer of a particle (helium atom) from one phase

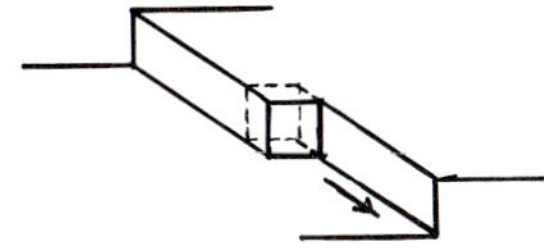

Fig.1 An elementary step with a kink

to another. The kink energy does not change as a result of this displacement since the chemical potentials of the phases in equilibrium are equal. Therefore, like other point defects in quantum crystals, such a kink behaves as a delocalized quasiparticle. Its energy is a function of its quasimomentum. So at absolute zero the state corresponding to the bottom of the energy band ($p = p_0$) is the ground stationary state of a solitary kink and the velocity of the kink is zero. Stationary states of close energies ($p \rightarrow p_0$) correspond to finite kink velocities and therefore to a continuous flux of matter from one phase to another. We conclude that the transfer of matter from one phase to another i.e., crystallization or melting, in this case is a coherent process occurring without energy dissipation. It is important to note also that corresponding stationary states are not separated from the ground state by any energy gap.

This example shows that two quite different states of a surface are possible at absolute zero. One of them is usual atomically smooth state without any surface defects. The other is a new quantum state including a number of delocalized kinks and steps of various configurations. This state could be called "quantum rough", because similarly to a classical atomically rough state the surface free energy α in this case is a smooth function of the surface orientation. The important property of that quantum rough surface is the existence of stationary states which are arbitrarily close in energy to the ground state and the continuous growth or melting of a crystal occurs via these states. The motion of an interphase boundary at T = 0 thus occurs without disturbing the phase equilibrium. In other words, the kinetic growth coefficient K (defined by the formula $V = K\,\delta\mu$, where V is the velocity of the boundary and $\delta\mu$ is the difference between the chemical potentials of the phases in contact) becomes infinite at T = 0 for a crystal with this type of surface. On the contrary, a smooth surface growth coefficient tends to zero with $\delta\mu$, even taking into account the possibility of quantum tunnelling [13].

In some respects a smooth surface could be compared with 2D insulator, and a quantum rough surface with 2D metal (the growth coefficient playes the role of the electrical conductivity). In these terms the transition between the two possible states of a surface is the metal-insulator transition. So the problem of quantum roughening appears to be connected closely with the problem of 2D localization.

Generally speaking, one crystal can have surfaces of both types, depending on their orientation. As for ^{4}He hcp crystals, it was known until this year that only two surfaces of the highest symmetry become smooth below 1,2 [K] and 0.9 [K], respectively, [2,9]. Recently a third roughening transition at 0.36 [K] was observed by WOLF, BALIBAR and GALLET [12] when investigating helium crystal growth down to 70 [mK]. Nobody knows what additional number of smooth surfaces appear at lower temperatures. FISHER and WEEKS [14] even argue that all surface orientations ought to be smooth at T = 0. However it is easy to see that this statement can not be correct concerning at least a class of surfaces close in orientation to any smooth one. More exactly, if the smooth surface is, say, (001) then all the surfaces of the type (1, M, MN + m), where M and N are sufficiently high numbers and $0 \leq m < M$, ought to be rough. In fact, each of those surfaces can be regarded as a stepped one with the distance between neighbouring steps (kinks) being equal to N(M), in the units of interatomic distances. Since the interactions between steps and kinks at long distances become negligible, at fixed width of the kink's energy band, kinks and steps remain delocalized, and we have a "metal-like" (quantum rough) state of the whole surface.

At finite temperatures the motion of a quantum rough boundary is accompanied by dissipation due to its interaction with bulk thermal excitations. Roughly speaking this dissipation should be proportional to the density of the normal component (including the solid phase phonons). It gives $K \sim T^{-4}$ at $T < 0{,}5$ [K] (the phonon region and $K \sim \exp(\Delta_r/T)$ at $T \gtrsim 0{,}5$ [K] (the roton region, Δ_r is the roton gap). Recently the quantitative theory of the growth kinetics has been advanced by ANDREEV and KNIZHNIK [15] and also by BOWLEY and EDWARDS [16], taking into account the space and time dispersion of K. Note that in the frames of the current theory all the essential dissipation mechanisms are due to the bulk thermal excitations but not the surface itself.

3. Crystallization Waves

As we have seen above, the quantum rough state of a surface gives rise to the possibility of strictly non-dissipative growth and melting at T = 0. If this is the case, then weakly damped oscillations, similar to normal capillary waves, may propagate along the surface at sufficiently low temperatures. Indeed, any deflection from the equilibrium crystal shape results in an increase in the surface energy. Therefore any disequilibrium crystal shape should change by crystallization or melting reducing its surface energy. On the other hand, since the densities of the two phases are different, growth and melting of a crystal give rise to a motion of a liquid, i.e., an increase in the kinetic energy of the system. As a result oscillations of an interphase boundary should occur if the total energy dissipation is small. The spectrum of these crystallization waves in the long-wave limit, where the compressibilities of the two phases are negligible, is given by

$$(\rho_1 - \rho_2)^2\omega^2 - \tilde{\alpha}\rho_2 k^3 - \rho_2(\rho_1 - \rho_2)gk + \rho_1\rho_2(mK)^{-1}i\omega k = 0 \quad (1)$$

where $\tilde{\alpha} = \alpha + \partial^2\alpha/\partial\varphi^2$, φ is the angular variable in the plane of $\vec{k}$, ρ_1 and ρ_2 the densities of the solid and liquid phases, respectively, and m the mass of an atom; here we take into account gravity g and damping due to the finite growth coefficient K.

Like the normal capillary waves, the oscillations described by (1) are unstable with respect to decay of one quantum into two of lower energy. Therefore, these oscillations are characterized by finite damping, even at T = 0. This damping, however, is very small for macroscopic wavelengths. At finite temperatures and for low-frequency oscillations the most important damping mechanism is due to the finite growth coefficient.

Non-dissipative crystallization significantly changes the spectrum of the elastic mode of surface oscillations, but does not short-circuit them, as one could think. In fact,ordinary Rayleigh waves are present on the quantum rough surfaces of a helium crystal, and their spectrum is determined only by the properties of the crystal, just as in the case of the interface with vacuum [5]. These waves do not attenuate by emission of phonons in the liquid, since the sound velocity in liquid helium exceeds the velocity of the transverse waves in the crystal and consequently, the velocity of the Rayleigh waves. Naturally, on smooth faces the velocity of the surface sound waves depends essentially on the properties of the liquid.

At very high frequencies, where the phase velocity of crystallization waves becomes of the order of sound velocities ($\omega \sim \sim 10^{11}$ [Hz] , $k \sim 10^{6} \div 10^{7}$ [cm^{-1}]) the dispersion relation (1) is not valid. High-frequency corrections to the spectrum of crystallization waves (and Rayleigh waves) have been discussed by UWAHA and BAYM [17] and PUECH and CASTAING [18] on the basis of macroscopic equations. Note in this connection that non-dissipative crystallization itself should take place only at frequencies lower than the probability of quantum tunnelling (per unit time) of "bare" kinks into the neighbouring position (this probability is of the same order of magnitude as the width of the kinks energy band). The experimental data on the anomalous Kapitza resistance [6-8] show that this quantity is at least 10^{10} [Hz] and perhaps even more. However, the effect of the crystal lattice in this case should be more serious than it seems at first sight. In fact, crystallization waves can exist at a surface of arbitrary Miller indices except for the lowest ones. It means that the elementary translation vectors of the surface can be as long as we please. Therefore the effect of zone boundaries could be neglected, strictly speaking, only in the limit of infinitely long waves. So the accuracy of the macroscopic approach in the high-frequency spectrum calculations [17,18] seems to be doubtful.

4. Experimental data

In practice the simplest means of exciting low-frequency crystallization waves is the mechanical vibration of the experimental cell, just as well as in the case of normal capillary waves on the surface of a liquid. Crystallization waves were first observed in this very way. Figures 2 a, b, c show motion-picture frames of excitation and damping of such oscillations. The first frame shows the convex meniscus, which is the calm boundary between the solid (lower) phase and the liquid. The succeeding frames demonstrate the behavior of the boundary following a knock on the outer wall of the cryostat. This manner of exciting surface oscillations becomes ineffective at temperatures higher than 0.7 [K] because of rapid increase of its damping with the temperature.

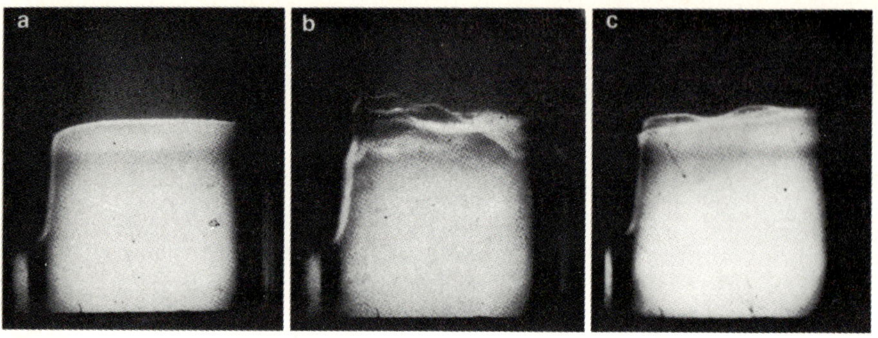

Fig.2. Crystallization waves at 0.5 [K]

In the case when the equilibrium meniscus contains a flat region which is in fact one of the atomically smooth surfaces, the oscillations can be observed only on the rounded sections of the miniscus, whereas the flat section remains perfectly immobile (of course, its outlines can oscillate due to oscillations of the rounded sections). Crystallization waves are observed not only under quasi-equilibrium conditions, when the crystal fills the entire lower part of the cell, but also, e.g., in situations when the crystal hangs from one of the side walls of the cell (Fig.3). In the latter case oscillations can be clearly seen on the rounded sections, and only on them. All these data demonstrate the difference between the behavior of quantum rough and classical atomically smooth surfaces.

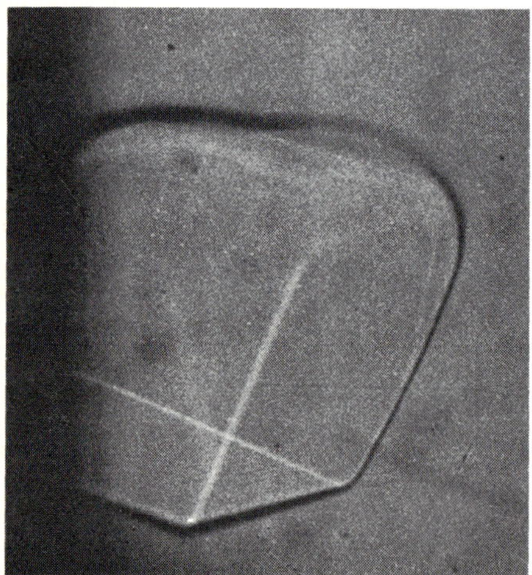

Fig.3. A crystal hanging from a side wall of the cell. The lower face is (0001) plane; the crystallization waves due to the vibration of the cryostat can be seen on the upper rounded surface

The measured crystallization wave spectrum for one of our samples at two temperatures, 0.360 and 0.505 [K], is shown in Fig.4. The dashed line corresponds to $\omega \sim k^{3/2}$, the solid line is the theoretical plot of $\omega(k)$ according to (1) with the gravitational term and $\tilde{\alpha} = 0.21$ [erg/cm^2] (the cor-

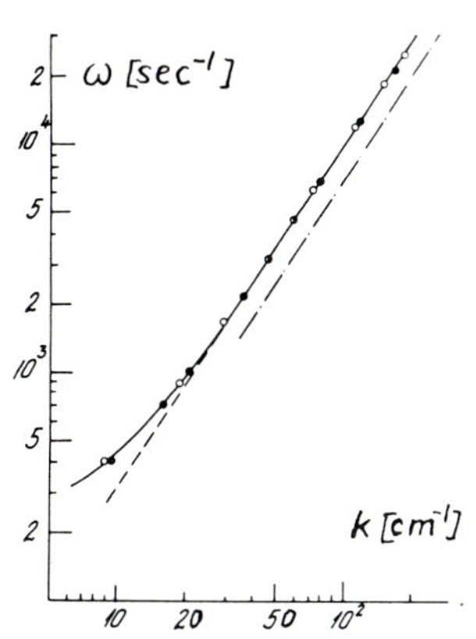

Fig.4. Spectrum of crystallization waves. ○ - T = 0.360 [K] , ● - T = 0,505 [K]

rection for damping can be neglected). The data for other samples agree essentially with the foregoing, except that entire curve is shifted somewhat in accord with the change of the value of $\tilde{\alpha}$; the $\tilde{\alpha}$ values from 0.097 [erg/cm^2] (the dash-dot line in Fig.4) to 0.23 [erg/cm^2] were registered. These data show that the surface energy α is temperature independent within the accuracy (5%), but significantly anisotropic. Now the systematic investigation of angular dependence of α is highly desirable.

As for the temperature dependence of α the following thing should be noted. The crystallization waves' quanta are elementary excitations of the ^{4}He quantum rough surface at low temperatures (in addition to Rayleigh waves). They are responsible for the temperature dependence of the surface energy. Since the frequency is proportional to $k^{3/2}$, the temperature-dependent component of surface energy is proportional to $T^{7/3}$, as in the case of capillary waves at the liquid-vapor interface. However, numerical estimates show that the current experimental accuracy is not sufficient to detect this temperature dependence.

Figure 5 shows the measured frequency dependence of the reciprocal damping length(i.e.,the imaginary part of the wave vector) $\varkappa(\omega)$ for the sample whose spectrum is shown in Fig.4. According to (1), $\varkappa \sim \omega^{1/3}$. Clearly this dependence is nearly satisfied, i.e., the observed damping is indeed due to the finite growth coefficient (the contribution of other possible damping mechanisms in this case probably is less than 1[cm^{-1}]). Fig.6 demonstrates the temperature dependence of the quantity $(mK)^{-1}$ calculated from the measured values of $\varkappa$ for three different samples.

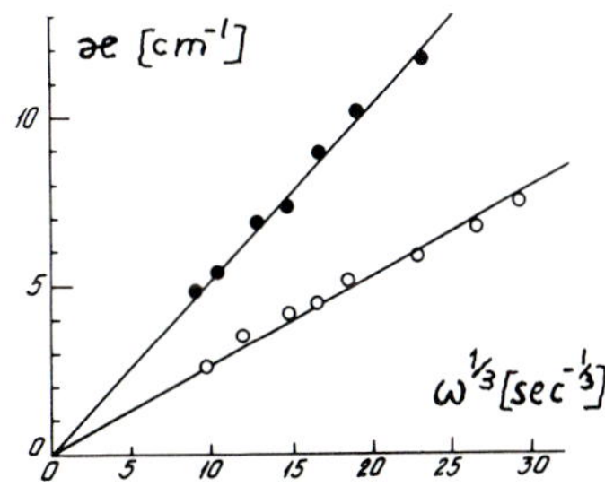

Fig.5. Frequency dependence of the damping of crystallization waves: ○-T = 0.360 K ; ●-T = 0.505 K . The sample is the same as that in Fig.4. The straight lines are $\varkappa \sim \omega^{1/3}$

It can be stated that at low temperatures the attenuation decreases more slowly than it would do in the case of pure phonon mechanism. This difference can be due to some additional temperature-independent mechanism of scattering of the crystallization waves (e.g. by crystal defects, mechanical vibrations, etc.). This "residual" attennuation can obviously vary greatly from sample to sample, and depending on the specific scattering mechanism, vary with frequency in accordance with some law. From this point of view it is easy to understand the systematic discrepancies between the data corresponding to two different frequencies for sample No.3 in Fig.6. Namely, in this case the residual damping value is $\varkappa_0 \approx 2\,[cm^{-1}]$ for both frequencies.

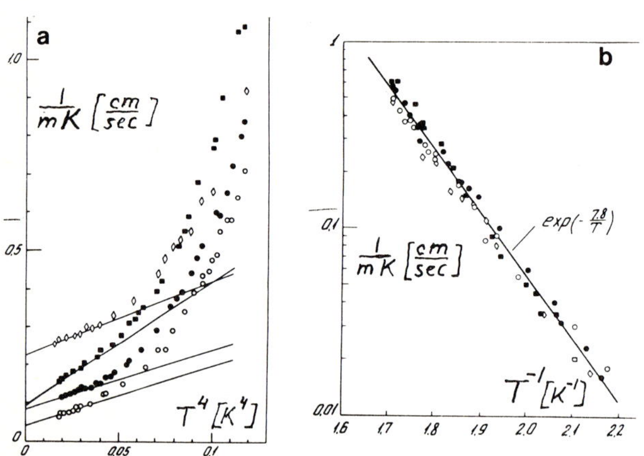

Fig.6. The temperature dependence of the crystallization wave damping: sample No.3, ○-1118 [Hz], ●-232 [Hz]; sample No.4, ■-827 [Hz], sample No.5, ◇-837 [Hz]. Figure 6a - the whole damping, Fig.6b - proposed roton contribution to the damping

The simplest interpretation of the data of Fig.6a is the following: we assume that the measured damping consists of three terms: residual, phonon ($\sim T^4$) and roton damping ($\sim exp(-\Delta_r/T)$). The straight lines in Fig.6a represent the contributions of the first two of these terms. The roton contribution determined in this manner is shown in Fig.6b. It is seen that all the data can be fitted in practice to a single exponential dependence. The obtained value Δ_r = 7,8 [K] indeed is close to that obtained by other methods. However, the numerical value of this roton damping is at least 20 times more than that predicted by quantitative theory [15] (measured phonon damping agrees in the order of magnitude with the theory). Moreover, the theory predicts another frequency dependence of the roton damping under these experimental conditions. BOWLEY and EDWARDS believe [16] that the agreement can be obtained by taking into account an inelastic process. Anyway, further experiments are needed now in wider temperature and frequency ranges.

Acknowledgements

It is a great pleasure to acknowledge the contribution of my colleagues A.F.Andreev, A.V.Babkin, K.O.Keshishev, V.I.Marchenko and A.I.Shalnikov to this research and to my understanding of the surface phenomena in helium crystals.

References

1. A.F.Andreev and A.Ya.Parshin: Sov.Phys. JETP 48, 763 (1978)
2. K.O.Keshishev, A.Ya.Parshin and A.V.Babkin: JETP Lett. 30, 56 (1979); Sov.Phys.JETP 53, 362 (1981)
3. B.Castaing and P.Nozieres: J.Physique 41, 701 (1980)
4. B.Castaing, S.Balibar and C.Laroche: J.Physique 41, 897 (1980)
5. V.Marchenko and A.Ya.Parshin: JETP Lett. 31, 724 (1980)
6. T.E.Huber and H.J.Maris: Phys. Rev. Lett. 47, 1907 (1981)
7. L.Puech, B.Hebral, D.Thoulouze and B.Castaing: J.Phys.Lett.43, L809 (1982)
8. P.E.Wolf, D.O.Edwards and S.Balibar: J.Low Temp.Phys.51, 489 (1983)
9. J.E.Avron, L.S.Balfour, C.G.Kuper, J.Landau, S.G.Lipson and L.S.Schulman: Phys.Rev.Lett. 45, 814 (1980)
10. S.Balibar and B.Castaing: J.Phys.Lett. 41, L329 (1980)
11. S.Ramesh and G.D.Maynard: Phys.Rev.Lett. 49, 47 (1982)
12. P.E.Wolf, S.Balibar and F.Gallet: Phys.Rev.Lett.to be published
13. A.F.Andreev: "Quantum Crystals", in Progress in Low Temperature Physics, 8, ed. by D.F.Brewer (North-Holland, Amsterdam)
14. D.S.Fisher and J.D.Weeks: Phys.Rev.Lett. 50, 1077 (1983)
15. A.F.Andreev and V.G.Knizhnik: Sov.Phys.JETP 56, 226 (1982)
16. R.M.Bowley and D.O.Edwards: J.Physique 44, 723 (1983)
17. M.Uwaha and G.Baym: Phys.Rev. B26, 4928 (1982)
18. L.Puech and B.Castaing: J.Phys.Lett.43, L-601(1982)

Phonon Transmission and the Kapitza Resistance Between Liquid and Solid Helium

H.J. Maris

Department of Physics, Brown University, Providence, RI 02912, USA

For many years the explanation of the Kapitza resistance R_K between liquid helium and ordinary solids has been an interesting, and unsolved, problem in low-temperature physics. There has been much discussion (for example, WYATT [1], and KINDER et al.[2]) about the influence of surface defects and dirt on the Kapitza resistance. For this reason, attention has been paid to the preparation of very ideal surfaces. One approach [2] has been to study the surfaces of crystals which have been cleaved at low temperatures. We became interested in another possible way to prepare an ideal surface. The idea was to grow a crystal of solid helium, and to measure the R_K between this crystal and liquid helium. It is possible to eliminate essentially all impurities from helium. In addition, the high quantum mobility of the helium atoms allows defects to anneal out, even at low temperatures. Thus, we thought that the problems of dirt and defects which plague ordinary Kapitza experiments could be avoided.

The major experimental difficulty that was anticipated was the very small value of the Kapitza resistance. The elastic properties of liquid and solid helium are well known, and it is straightforward to calculate R_K from Khalatnikov's acoustic-mismatch theory (AM) [3]. The result is

$$R_K = 0.15\ T^{-3} \quad \text{K. watt}^{-1}\ \text{cm}^2 . \tag{1}$$

At 1K this is several hundred times smaller than the experimentally measured R_K between liquid helium and copper. A way to measure this very small R_K is shown schematically in Fig. 1. Liquid and solid helium are contained in an experimental cell at 0.5K. Second sound pulses are generated in the solid, and are reflected from the liquid-solid interface. At each reflection some fraction of the phonons in the second sound pulse cross the interface, and then propagate as ballistic phonons in the liquid. Thus, a series of pulses of ballistic phonons is detected in the liquid. The ratio of the amplitudes of successive pulses is equal to the reflection coefficient for the second sound pulse at the interface, and this quantity can in turn be related to R_K. Measurements by this method [4] gave values which at 0.53K were equal to the Khalatnikov acoustic-mismatch value to within the experimental accuracy of ± 30%. Unfortunately, this method can only be used over a narrow range of temperature around 0.5K, where second sound propagates well in the solid.

While these experiments were being performed it became clear that the naive view of the liquid-solid helium interface as just another liquid-solid boundary was wrong. Andreev and Parshin (AP) [5] published their pioneering paper which predicted that the interface should have several special properties because of the large quantum delocalization of the surface atoms. These properties included the following. For classical crystals at low enough temperatures the equilibrium shape of the crystal is facetted, i.e.,the sur-

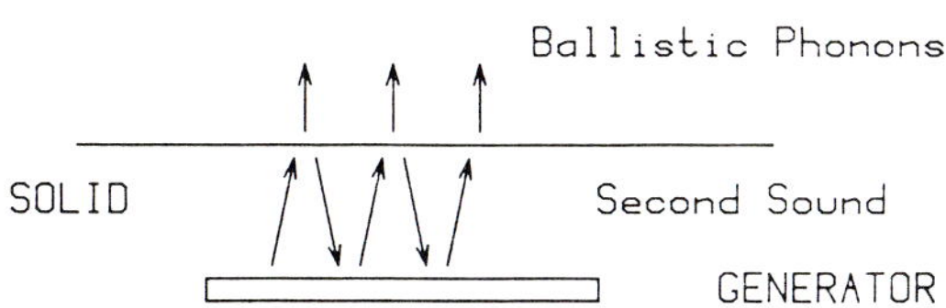

Fig. 1. Experiment to measure R_K at 0.5K

face consists of planar regions which are atomically flat. AP pointed out that it was possible that the equilibrium surface of helium might include rounded regions, which would be rough when viewed on the atomic scale. For these regions the growth rate of the solid phase from the liquid would be extremely large. Ordinary crystal growth is usually a thermally activated process, which varies with temperature according to an Arrhenius law. Thus, at low temperatures growth becomes extremely slow. On the other hand, AP proposed that for the rough surfaces of helium the growth rate increases with decreasing temperature, becomes infinite at T=OK, and is limited (rather than enhanced) by the thermal excitations which tend to damp the motion of the interface. Because of this high growth rate, they proposed that it should be possible to propagate melting-freezing waves on the interface, and these were soon observed by Keshishev, Parshin, and Babkin in [6] in the kHz range of frequency. Castaing and Nozieres [7] proposed that the transmission of sound across the interface would be a good method to study melting and freezing at the interface. Under ordinary circumstances when a sound wave is incident on a boundary between two phases, a pressure oscillation occurs at the interface, and it is this oscillation which generates the transmitted wave. However, if melting and freezing can occur very rapidly the liquid and solid will always be in a state of "phase equilibrium". This equilibrium corresponds to a certain definite constant pressure, and so no pressure oscillation can occur. Hence there is no sound transmission. Castaing, Balibar, and Laroche [8] found this to be true in experiments with a 1 MHz sound wave. Their experiment implies that equilibrium is reached by melting and freezing within less than 10^{-6} secs. They were able to confirm that the rapid melting and freezing only occurs on the rough surfaces, i.e.,it does not occur on the facets.

Because of these developments Marchenko and Parshin [9] considered how the Kapitza resistance would be modified if rapid melting and freezing (RMF) can occur in times less than the period of a typical phonon. In the first approximation, when RMF occurs there is no phonon transmission across the interface, for the same reason that there is no transmission of sound waves. Thus, this leads to an infinite Kapitza resistance. However, Marchenko and Parshin pointed out that there is in fact some transmission for thermal phonons because of capillary effects. The boundary conditions which exist between a liquid and a solid in phase equilibrium are

$$\sigma^S_{zz} + P^L + \alpha(1/R_1 + 1/R_2) = 0 \tag{2}$$

$$\sigma^S_{xz} = \sigma^S_{yz} = 0 \tag{3}$$

$$(\rho_S/\rho_L - 1)(P^L - P_o) - \alpha(1/R_1 + 1/R_2) = 0 \tag{4}$$

where $\sigma^S_{\alpha\beta}$ is the stress tensor in the solid, P^L is the pressure in the liquid, P_o is the equilibrium melting pressure, ρ_S and ρ_L are the densities of the solid and liquid, and R_1 and R_2 are the principal radii of curvature of the surface. α is the surface energy per unit area, and for simplicity it is assumed that this is equal to the surface tension. The surface is taken to be the plane z=0. If the surface is flat ($R_1, R_2=\infty$) we have $\sigma^S_{zz} = P^L = P_o$, and so there is no oscillating stress, and no transmission. However, a phonon which is incident on the interface at an oblique angle will cause the surface to be curved (Fig. 2). Then there will be a fluctuating stress in the solid and some phonon transmission. It is clear that the curvature of the interface increases with increasing phonon frequency Ω. It can be shown [9] that at low frequency the transmission is proportional to Ω^2. At higher frequency the transmission t_L from the liquid side varies more slowly with frequency [10], essentially because t_L is approaching its maximum value of 1. Theoretical results are shown in Fig. 3. This shows the value $< t_L(\Omega) >$ of t_L averaged over all angles of incidence. The dependence on angle is shown in Fig. 4, which also includes the results expected on the basis of the acoustic-mismatch (AM) model. From these results

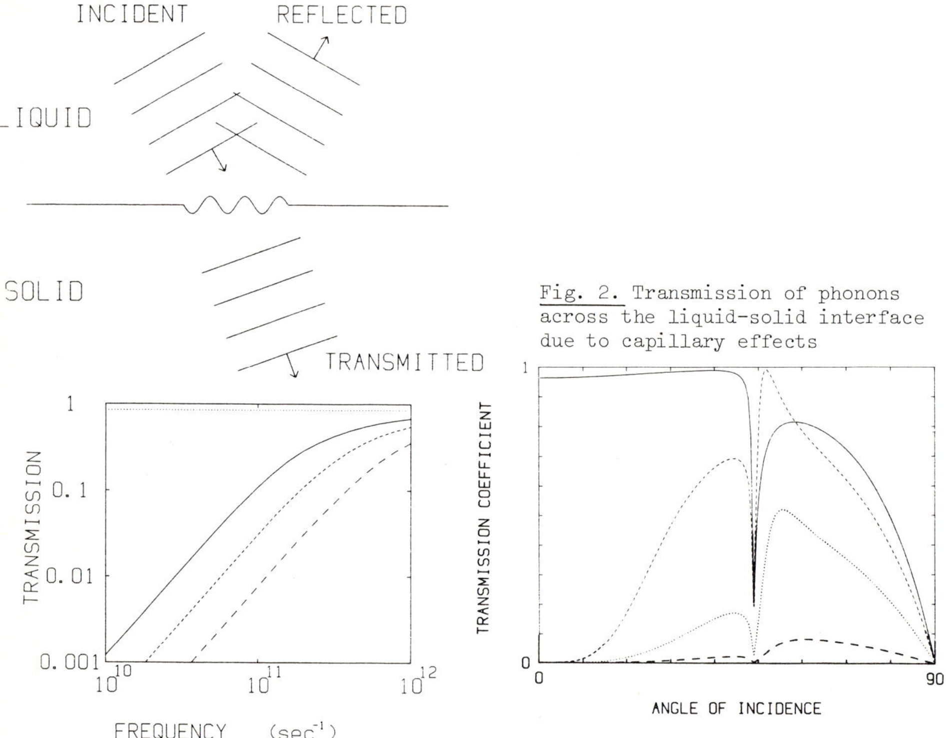

Fig. 2. Transmission of phonons across the liquid-solid interface due to capillary effects

Fig. 3. Transmission coefficient from liquid to solid helium as a function of frequency. ..... indicates the acoustic-mismatch result. ——, ----, and — — — are the predictions of the rapid melting-freezing theory assuming surface energies of 0.4, 0.2, and 0.1 c.g.s. units, respectively

Fig. 4. Transmission coefficient from liquid to solid helium as a function of angle. Solid line shows the acoustic-mismatch result. ----, ..., and — — — are the predictions of the rapid melting-freezing theory for frequencies of 10^{12}, 3×10^{11}, and 10^{11} sec^{-1}, respectively. The surface energy is taken to be 0.2 c.g.s. units

one can calculate the average transmission coefficient $< t_L(T) >$ for a Planck distribution of phonons with temperature T. The Kapitza resistance is related to $< t_L(T) >$ by

$$R_K = 0.13 \quad T^{-3}/< t_L(T) > \quad K.watt^{-1}.cm^2 . \tag{5}$$

At low temperature

$$< t_{L(T)} > = 37\ \alpha^2\ T^2 . \tag{6}$$

Thus, $R_K = 0.0036\ \alpha^{-2}\ T^{-5}$. R_K as a function of T is shown in Fig. 5. The theoretical results in Figs. (3-5) are given for several values of α. This is because this quantity is not accurately known, and because α is expected to vary with the crystallographic orientation of the surface.

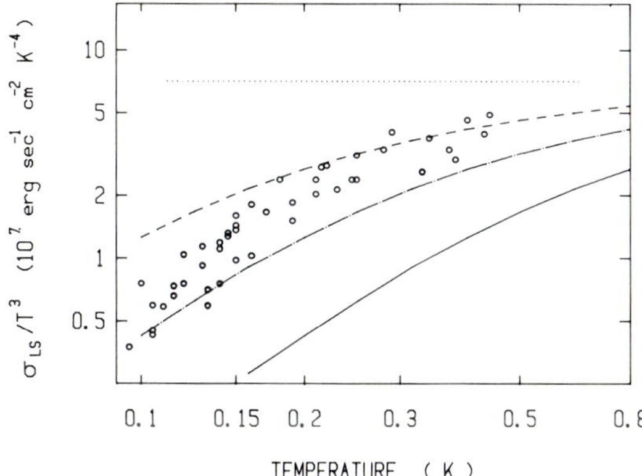

Fig. 5. Kapitza conductance σ_{LS} divided by T^3. ..... acoustic mismatch theory. ----, —.—.—. and —— are the predictions of the rapid melting-freezing theory for surface energies of 0.8, 0.4, 0.2 c.g.s. units respectively. Data points are from ref. [11]

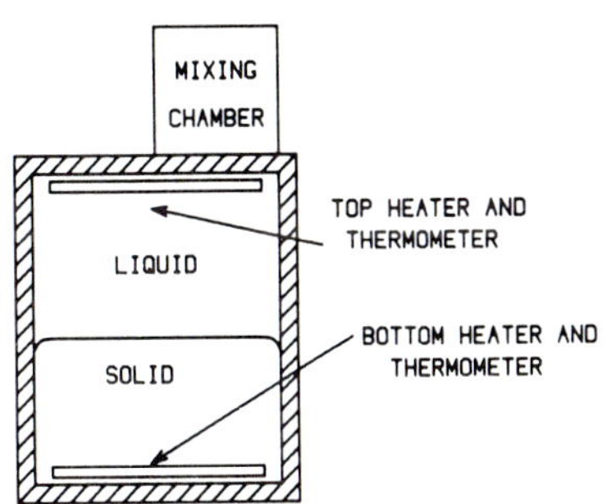

Fig. 6. Experimental apparatus

It is clear from the above discussion that the predictions of the RMF theory and the conventional AM theory are quite different, both as regards the temperature and the frequency dependence of the phonon transmission. However, the difference between the two theories is fairly small at 0.5K where we had measured R_K. To distinguish between the two theories we therefore made a measurement at lower temperatures [11] using the apparatus shown in Fig. 6. The cell contained two heaters and two thermometers, and was connected to the mixing chamber of a dilution refrigerator. A pulse of heat applied to the solid caused the temperature of the solid and liquid thermometers to vary with time as shown in Fig. 7. The R_K between liquid and solid helium could be determined from the time it took the liquid and solid temperatures to become equal. The R_K between the helium and the walls of the cell is much larger, and could be handled as a correction. The results are shown in Fig. 5. These data are accumulated from measurements on several different crystals, and part of the scatter is probably due to the different orientations of the crystalline surfaces. One can see that the results are in qualitative agreement with the RMF theory, and a fair quantitative fit can be obtained if α is allowed to be an adjustable parameter. (See, however, remarks to follow).

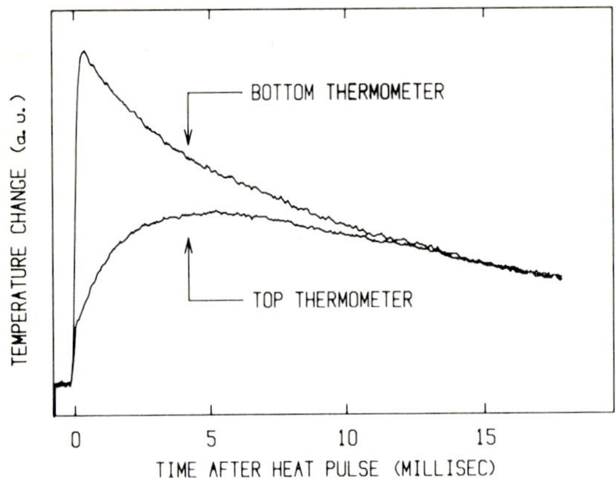

Fig. 7. Response of thermometers to applied heat pulse

Puech et al.[12] were able to make conventional static measurements of R_K. They used a very sensitive a.c. bridge together with closely matched carbon resistors to measure the small difference in temperature between the liquid and solid. From the work of Keshishev et al.[6] it was known that while the solid phase is growing the surface tends to be highly facetted. To study R_K for facetted surfaces, Puech et al. therefore made measurements of R_K while the solid was growing. The result was that R_K was approximately proportional to T^{-3}, as expected, for an interface at which rapid melting and freezing does not occur. The magnitude of R_K was in agreement with the acoustic-mismatch value (1) to within the experimental uncertainty.

For rough interfaces Puech et al. obtained data which below about 0.3K was well described by the law

$$R_K = AT^{-5} \tag{7}$$

where A was a constant. This is what is expected at low temperatures if RMF occurs. The coefficient A was not the same for all crystals, and ranged from 0.017 to 0.029. The orientations of some of the crystals was determined, and it was found that the phonon transmission was larger for crystal surfaces closer to the c-axis direction, then for other crystals. These results can be compared with the data of Huber and Maris [11] which gave A in the range 0.011 to 0.020 for crystals of unknown orientation.

Wolf, Edwards, and Balibar [13] have also performed static measurements of R_K. They used a very ingenious "superfluid melting-curve thermometer" [13], which measures a ΔT by converting it to a height difference via the fountain effect. The orientation of the crystals was determined from observation of the orientation of facets induced during growth. The results confirmed that the transmission of phonons across facetted surfaces was much larger than across rough surfaces. The values obtained for A for the rough surfaces were in reasonable agreement with the results of Puech et al., and Huber and Maris.

In addition to the measurements of the Kapitza resistance just described, there is one other experiment which strongly indicates that the RMF theory is qualitatively correct. In the R_K experiments of Huber and Maris [11] the heat pulse was applied to the solid, because in the solid phase phonon-phonon interactions are strong and the phonon distribution rapidly approaches a Planck distribution. Thus, the phonon flux incident on the interface is an equilibrium one described by an effective temperature. This is necessary if the experiment is to measure the Kapitza resistance. When a heat pulse was applied to the liquid some surprising results were obtained. A short time after the application of the pulse the temperature of the solid became larger than that of the liquid, even though the heat has been applied to the liquid. This effect could be explained in terms of a "phonon greenhouse". The heater emits high energy phonons which do not thermalize in the liquid. When these reach the interface they have a high probability of transmission because of their high frequency (see Fig. 3). Once in the solid they quickly decay into lower frequency phonons, which have a very small probability of transmission back into the liquid. Thus, the system acts in analogy to an ordinary photon greenhouse, and the solid, which plays the role of the inside of the house, becomes hotter than the liquid. The existence of this effect is strong evidence that the transmission increases rapidly with increasing frequency.

It is fair to say, therefore, that the experimental results support the RMF theory. This implies the remarkable result that melting and freezing can occur so rapidly that equilibrium is reached in times less than or equal to 10^{-11} secs. On a classical view of the interface this is very hard to understand. From a quantum mechanical viewpoint, however, it is reasonable. The idea is that conversion of liquid into solid occurs by the adiabatic deformation of a quantum mechanical state from a liquid to a solid configuration. From general quantum mechanical ideas a deformation of this type will be adiabatic and reversible if the time τ for the change satisfies

$$\hbar/\tau << \hbar\omega_{ex} \tag{8}$$

where $\hbar\omega_{ex}$ is the energy needed to raise the system to an excited state. For helium we may assume that $h\omega_{ex}$ is of the same order of magnitude as the roton energy Δ. Then (8) gives the condition

$$\tau >> 1.1 \times 10^{-12} \text{ sec} \tag{9}$$

which is well satisfied in the experiments.

An important problem which remains is that the experimental results are not in quantitative agreement with the theory. To get a reasonable fit to the data one has to assume that the surface energy α is in the range 0.4 to 0.6 cgs. More direct measurements of α (e.g.,[6]) give smaller values, and so the experimental phonon transmission is considerably larger than the theoretical value. Puech and Castaing [13] have proposed that this discrepancy may be due to a mass associated with the surface. It should be possible to test for this because this mass causes a finite transmission for phonons at normal incidence (cf. Fig. 4). It seems likely that there are other physical effects which need to be taken into account in a more accurate theory. The theory developed so far assumes the validity of the boundary conditions (2)-(4) derived by Gibbs for static equilibrium between phases, and it is possible that in time-dependent situations there may be several extra terms that occur. Phonon transmission then becomes a unique tool to study these effects.

Finally, we mention some experimental questions. It will be interesting to study the reflection coefficients for rotons incident on the liquid-solid boundary, and to investigate the polarization and angular distribution of the phonons produced in the solid. We are currently studying R_K between a dilute solution of ^{3}He in ^{4}He and solid ^{4}He. In this way we hope to measure the energy accommodation coefficient for a ^{3}He quasi particle at the interface.

This work was supported in part by the National Science Foundation through Grant No. DMR83-04224, and through support of the facilities of the Materials Research Laboratory at Brown University.

1. A.F.G. Wyatt: In Phonon Scattering in Condensed Matter, ed. by H.J. Maris (Plenum, New York, 1980) pp. 181-190.
2. H. Kinder, J. Weber, and W. Dietsche: In Phonon Scattering in Condensed Matter, ed. by H.J. Maris (Plenum, New York, 1980) pp. 173-180.
3. I.M. Khalatnikov: Zh. Eksp. Teor. Fiz. 22, 687 (1952).
4. T.E. Huber and H.J. Maris: unpublished work.
5. A.F. Andreev and A.Y. Parshin: Sov. Phys. JETP 48, 763-766 (1978).
6. K.O. Keshishev, A.Y. Parshin, and A.V. Babkin: JETP Lett. 30, 56-59 (1979); Sov. Phys. JETP 53, 362-369 (1981).
7. B. Castaing and P. Nozieres: J. Phys. 41, 701-706 (1980).
8. B. Castaing, S. Balibar, and C. Laroche: J. Phys. 41, 897-903 (1980).
9. V.I. Marchenko and A.Y. Parshin: JETP Lett. 31, 724-726 (1980).
10. H.J. Maris and T.E. Huber: J. Low Temp. Phys. 48, 99-109 (1982).
11. T.E. Huber and H.J. Maris: Phys. Rev. Lett. 47, 1907-1910 (1981); J. Low Temp. Phys. 48, 463-475 (1982).
12. L. Puech, B. Hebral, D. Thoulouze, and B. Castaing: J. Phys. Lett. 43, L809-814 (1982).
13. L. Puech and B. Castaing: J. Phys. Lett. 43, L601-608 (1982). See also A.M. Kosevich and Yu. A. Kosevich: Sov. J. Low Temp. Phys. 7, 394-396 (1981).

Transmission of Sound at the Solid-Liquid Interface of ^{4}He

B. Castaing, L. Puech, and G. Bonfait

Centre de Recherches sur les Três Basses Températures, CNRS, BP 166 X
F-38042 Grenoble Cedex, France

1. INTRODUCTION

It is well known (1) that the dynamics of growing a crystal from its melt depends on the nature of the interface. If it is rough, on the atomic scale, growing is easy, as the atoms of the liquid can adsorb about on any point of the interface, without energy barrier. One can define a mobility as the ratio of the flux J of mass across the interface to the difference in chemical potentials between the two phases : $\Delta\mu$.

On the contrary the interface can be smooth, like a cleaved surface. In this case growing is a slow process, controlled by the nucleation of the following layer or by the Franck and Reed sources (twin dislocations). BURTON, CABRERA and FRANCK (2) have been the first to note that an interface can go from the smooth state to the rough one as the temperature is raised. Some models of this roughening transition have been exactly solved and the resulting evolution of the crystal shape has been recently precised worked out (3).

The smooth interfaces correspond to the high density crystallographic planes. When growing the crystal, the rough interfaces, which progress more rapidly, tend to disappear and the crystal shape is limited by planes or facets which correspond to the smooth ones. On the contrary, on melting the facets do not progress and disappear. The crystal has a rounded shape.

It is all these old or new ideas which will be illustrated here by some experiments which have been performed on the helium 4 liquid-solid interface, at Paris and Grenoble, in the last three years. It must be pointed out that other experiments have been performed (4)-(6) on the same system, some of which being presented at this Conference.

2. TRANSMISSION OF SOUND

It is easy to understand how a high mobility of the interface can affect the sound transmission across it, specially in the case of ^{4}He at low temperatures where the latent heat is negligible (7). In the limit of infinite mobility and zero latent heat, the pressure at the interface remains equal to the equilibrium one, whatever the flux of atoms is across it. If the sound is emitted in the liquid there is no pressure oscillation in the solid and thus the transmission is zero.

In the real case, dissipative and reactive effects occur near the interface and the transmission is finite. In the case of ^{4}He it remains small, being thus a probe of the interface dynamic.

We shall discuss first the dissipative effects. There are two kinds of such effects. There are first the intrinsic ones of the interface: at a point where the interface is passing, on growing, there is a rapid disappearance of the superfluid order parameter and appearance of the solid one. These rapid changes give rise to retardation, or inertia effect. They can also a priori give rise to relaxation or dissipation effects.

Second, the movement of the interface perturbs the distribution of phonons on each side. This type of dissipative effects has been extensively discussed (8)-(10) and we shall refer to it as the Andreev-Parshin dissipation.

3. ONSAGER-LIKE APPROACH

3.1. General formalism

In the quasi-equilibrium regime, one can write Onsager-like equations for the interface (7). First it is to be noted that one cannot disregard the flux of energy across the interface, due to its convective part. In each phase :

$$J_E = JTs + Q$$

where s is the entropy per unit mass and Q the energy flux in the center of gravity frame. One can then write the Onsager equations :

$$\Delta\mu/T = aJ + bJ_E$$
$$\Delta T/T^2 = bJ + cJ_E$$

and the dissipated energy per unit surface is :

$$T\dot{S} = T(aJ^2 + 2bJJ_E + cJ_E^2) .$$

The thermodynamic stability requires the relation

$$b^2 < ac .$$

cT^2 is the usual Kapitza resistance, b can be called a Peltier coefficient and aT the growth resistance.

In the hydrodynamic regime, in both solid and superfluid, the excitations are decoupled from the ground state. The normal fluid moves at a velocity v_n different from the center of mass one. In the Andreev-Parshin picture, the dissipation vanishes when v_n is equal to the interface velocity v. As J_E is proportional to $(v_n - v)$, $J_E = 0$ implies $T\dot{S} = 0$. Thus a, and consequently b, are null. The dissipation is only due in this case to the Kapitza resistance. Finite values of a and b can only come from the intrinsic dissipation.

Note that this Onsager formalism can also be used in the case of a restricted geometry, say a tube of diameter d, in the ballistic regime. In this case, what is called interface is an extended region of length $\simeq$ d where strongly out of equilibrium phonon distributions occur. Then the effective a^* and b^* coefficients are not zero in the Andreev and Parshin picture. They have been calculated by BOWLEY and EDWARDS (10), and b^* has been recently measured (6).

3.2. Influence on the sound transmission

The influence of the dissipation on the sound transmission has been discussed in the quasi-equilibrium regime. The pressure transmission coefficient from the phase 1 to the phase 2 is :

$$T = \frac{2z_2}{z_1 + z_2 + \xi z_1 z_2}$$

where ξ is a kind of effective mobility of the interface, z_i the acoustic impedance of the phase i :

$$\xi^{-1} = +\left(\frac{\rho_1\rho_2}{\rho_1-\rho_2}\right)^2 \times$$

$$\times \left\{ aT + \frac{+ [T(s_1-s_2)^2] - \lambda_1\lambda_2 b^2T^3 + 2(\lambda_1 s_2 + \lambda_2 s_1)bT^2 + [(\lambda_1 s_2^2 + \lambda_2 s_1^2)cT^3]}{\lambda_1 + \lambda_2 + \lambda_1\lambda_2 cT^2} \right\}$$

where $\lambda_i = \rho_i C_i c_{IIi}$ is the inverse of the thermal impedance, ρ_i, C_i, c_{IIi} are the density, specific heat per unit mass and second sound velocity in phase i.

This complicated formula regroups different mechanisms. For example, the first term in brackets $T(s_1-s_2)^2$ corresponds to the influence of the latent heat, the oscillating temperature yielding a pressure oscillation and thus a transmitted wave, even for zero dissipation. The second term in brackets corresponds to the energy which is emitted as second sound by the oscillating interface. The contribution of this last term has the same shape as the growth coefficient found by Bowley and Edwards. As it is probably dominant, there should not be a strong difference between the hydrodynamic and ballistic regimes.

4. EXPERIMENT

The measure of the transmission of sound across the solid-liquid interface of ^{4}He has been performed at the E.N.S. (Paris) (11). The frequency used was 1 MHz. This experiment has shown the great difference of behavior between the smooth interfaces (facets) and the rough ones where the results were in good general agreement with the ideas of ANDREEV and PARSHIN (8). The crystallographic orientation of the interface was estimated from the value of the sound velocity in the solid.

4.1. Rough interfaces

The results are presented on the Fig. 1a. Note that, as a general rule, the transmission coefficient becomes less than 10^{-3} under .5 K. The mobility ξ obtained from this transmission coefficient is obviously well represented by a law $\xi=\xi_0 \exp T_0/T$ with a strong anisotropy for To.

The first possible interpretation is that T_0 is the activation energy of some excitation of the solid or of the interface. But there is no other indication of the existence of these excitations. Their range of energies recalls the Umklapp process ones but the way by which they would enter in the problem is not clear.

It is thus reasonable now to interpret the variation of ξ^{-1} as the superposition of a T^4 term due to the phonons and a term proportional to the number of rotons, that is to exp $-\Delta/k_BT$ with Δ/k_B = 7.2 K. Such a fit is presented as lines in the Fig. 1. The same value has been taken for the roton contribution whatever the crystallographic orientation. The phonon term supposed to be mainly due to the solid ones is anisotropic. Moreover the curves have been shifted by an arbitrary temperature-

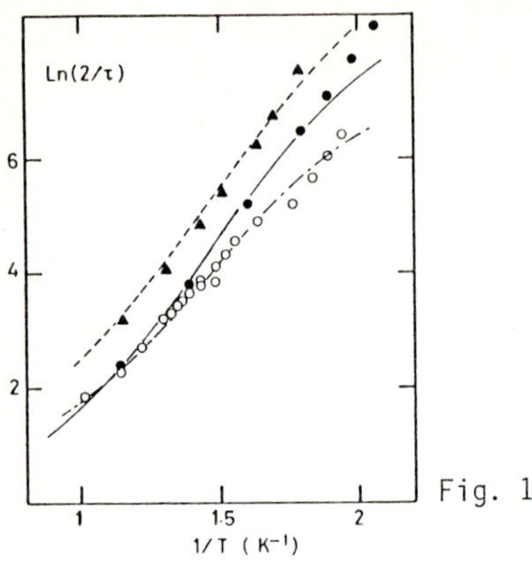

Fig. 1

Fig. 2

Fig. 1a): Some examples showing the dependence of the transmission factor versus the temperature. The lines represent theoretical fits. The angle θ between the c axis and the normal to the interface is :
▲,- - - - : θ=48° ; ●,———: θ=53° ;
0,——-: θ=70°

Fig. 1b): Hysteresis effect:
0, increasing the temperature;
●, decreasing the temperature ;
▲, after growing and melting the crystal

independent factor, which is justified by the observation of hysteresis effects for a same interface (Fig. 1b) when going to high temperature and then coming down. This is in agreement with the observation of KESHISHEV(4) et al. that going to temperatures larger than .9 K creates some defects on the interface, which can reduce the transmission of sound.

Two remarks must be made on this fit. First, the value obtained for the roton contribution is significantly higher than the term corresponding to second sound emission. One can argue that in the case of rotons, the mean squared thermal velocity, which might replace the second sound velocity c_{II}, in the ballistic regime is much higher than c_{II}. But it is then surprising that no transition is visible between the ballistic and the hydrodynamic regime.

Second the phonon term presents a strong anisotropy, by a factor 2. This has been also observed by KESHISHEV(4). While this is surprising, a detailed calculation taking into account the elastic anisotropy of the solid is necessary before ruling out this possibility.

4.2. Facets

For one of our samples, the basal plane was horizontal. In this case the behavior was completely different : the transmission was high even at low temperature. Thus the mobility is too low to be estimated by this technique. However, at a power level higher than some threshold , the transmission reduces suddenly.

This is coherent with the fact, discovered at the same time (12), that below 1 K such an interface is smooth. Growing occurs by nucleation of the following layer (which is a slow process), or via the existing steps which connect the emerging screw dislocations. But with a too small $\Delta\mu$, the step only bends, its radius of curvature R_c being inversely proportional to $\Delta\mu$. A larger $\Delta\mu$ can make R_c smaller than half the distance between dislocations. Then the crystal grows by twin spirals (Franck and Reed sources).

5. REACTIVE EFFECTS

What we call reactive or non dissipative effects are of two kinds : the surface tension effects and the inertia effects.

MARCHENKO and PARSHIN (13) have been the first to remark that the surface tension restores a sound transmission even for a perfect melting ($\Delta\mu \equiv 0$) but with an oblique incidence. In this case the interface is distorted in order to adapt to the velocity field and the capillary stresses due to this distortion create the transmitted wave. The shorter the wave length, the stronger is the transmission. MARCHENKO and PARSHIN found an energy transmission proportional to the frequency squared: $\omega 2$. They thus predicted an anomalous Kapitza resistance between solid and liquid : $R_K \propto T^{-5}$.

On the other hand, PUECH and CASTAING (14) have remarked that one must attribute an "inertia" to the interface. When going from the liquid state to the solid one, one must transform the short-range order to long-range order. This can be made by rearrangements of small clusters which occur in a finite time, the time it takes to the interface to pass on the considered point. The corresponding kinetic energy is proportional to J^2 and can be put in the symmetric form :

$$\frac{\sigma}{2\rho_S\rho_L} J^2$$

defining a surface inertia : σ.

This acts like a self-inductance in an electric circuit. In order to accelerate the interface, one must enhance this kinetic energy. This energy is taken from the difference in chemical potentials :

$$\Delta\mu = \mu_S - \mu_L = \sigma/\rho_S\rho_L \cdot \partial J/\partial T \quad .$$

The resulting pressure transmission is proportional to ω (for small ω) and the energy transmission is proportional to ω^2 like for the Parshin-Marchenko effect. At higher frequency, the usual transmission without melting freezing effect (acoustic mismatch) is restored.

6. EXPERIMENT

The first measurements of the Kapitza resistance which show the anomalous behaviour $R_K \propto T^{-5}$ are due to HUBER and MARIS (5) and are presented at this conference.

The method used by HUBER and MARIS was dynamical. They measured the time of recovering of thermal equilibrium between liquid and solid. We have performed high precision "static" measurements (15)-(16) whose results are presented on Fig. 2. One can see that they are consistent with a law

$$R_K T^3 = \gamma_3 + \gamma_5 T^{-2}$$

very close to the theoretical law. From the values of γ_5 one can deduce that the transmission is mainly due to the surface inertia and not to

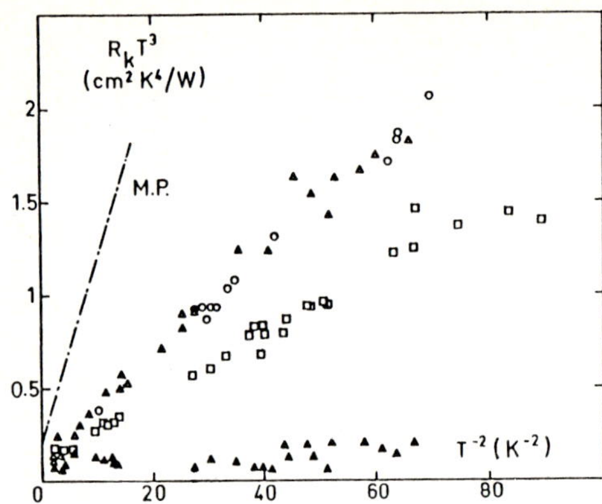

Fig. 2 : R_KT^3 as a function of T^{-2} (linear scales). ▲ : Facets ; △ : θ=45°; O : θ ≃ 80° ; □ : θ = 30°. θ is as in Fig. 1a

the Parshin-Marchenko surface tension effects. We moreover show evidence for the anisotropy of the surface inertia. Its order of magnitude is

$$\sigma \simeq 2 \times 10^{-9} \text{ kg/m}^2$$

which corresponds to 4 x 10^{-2} times, the mass of an atomic dense layer.

We have been also able to measure the Kapitza resistance of the facets. They exhibit a classical behaviour R_KT^3 = cste, which is an experimental proof that the anomalous behaviour of the rough interfaces is due to their mobility (17).

7. ROUGHENING TRANSITION

The strong difference at low temperature between the Kapitza resistance of facets and rough interfaces made us very sensitive to the presence of facets in the interface. With a given heat flux across the interface the temperature difference ΔT is smaller when the facet proportion is higher. In this way we have been able to observe the relaxation of a facet size after a disturbance (18).

In fact the interface is doubly out of equilibrium as we impose a continuous growing of the crystal in order to stabilize a rather large facet. The disturbance that we impose to this stationary state is to rapidly grow, then melt a small amount of crystal (Fig. 3a). After this the facet has nearly disappeared, and ΔT is large. Then ΔT comes back to the stationary smaller value. This relaxation is roughly exponential, the characteristic time of it we call the relaxation time τ.

Fig. 3b shows the evolution of τ^{-1} with the temperature. It is clear that τ diverges when approaching a temperature a little higher than 200 mK. We interpret this divergence as due to the approach of a roughening transition. This is strengthened by the fact that for a similar orientation P.E. WOLF et al. have recently observed such a transition (19), by a modification of the crystal shape. The divergence of τ would thus illustrate the critical slowing down associated with the roughening transition.

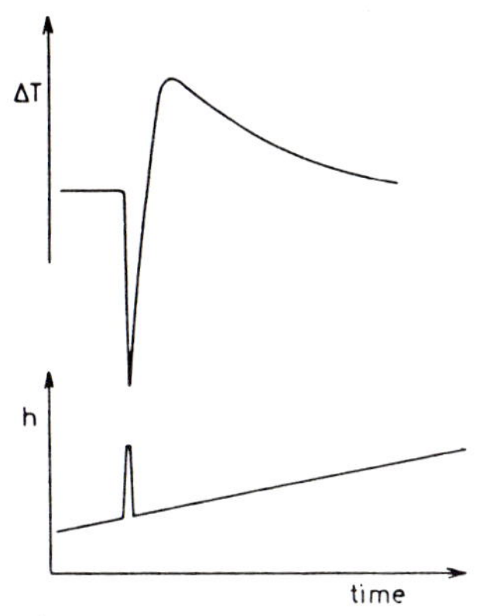

Fig. 3a) : Imposed evolution of the height h of the interface, and resulting evolution of the temperature jump ΔT across it, with a constant heat flux (schematic)

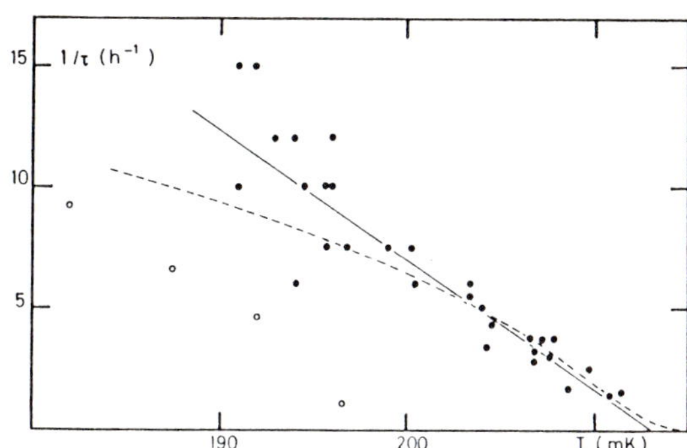

Fig. 3b) : τ^{-1} as a function of temperature. O before annealing up to 0.7 K during 3 hours ; O after annealing. The lines are two possible theoretical fits

8. CONCLUSION

In conclusion the set of experiments we have presented illustrates the high mobility of the rough liquid-solid interfaces of helium 4. Up to intermediate frequencies, this mobility is limited by phonons and rotons, and at higher frequency by the "inertia" of the interface. As for the facets, both the intermediate and high frequency measurements are coherents with a zero mobility. With higher driving forces the usual FRANCK and REED mechanism for facet growing has been also evidenced. Finally we have been able to contribute to further knowledge of the roughening transition by observing the critical slowing down which is associated with it.

REFERENCES

1. See for example D.P. Woodruff, The Solid-Liquid Interface, Cambridge University Press (1973).
2. W.K. Burton, N. Cabrera and F.C. Franck, Philos. Trans. Royal. Soc. 243, 299 (1951).
3. C. Jayaprakash, W.F. Saam, S. Teitel, Phys. Rev. Letters 50, 2017 (1983) and references therein.
4. K.O. Keshishev, A.Y. Parshin, A.B. Babkin, Sov. Phys. JETP 53, 362 (1981) (Zh. Eksp. Teor. Fiz. 80, 716 (1980)).
5. T.E. Huber and H.J. Maris, J. Low Temp. Phys. 48, 99 (1982).
6. P.E. Wolf, S. Balibar and D.O. Edwards, J. Low Temp. Phys. 51(1983).
7. B. Castaing and P. Nozières, J. de Physique 41, 701 (1980).
8. A.F. Andreev and A.Y. Parshin, Zh. Eksp. Teor. Fiz. 75, 1511 (1978) (Sov. Phys. J.E.T.P. 48, 763).
9. A.F. Andreev and V.G. Knizhnik, Zh. Eksp. Teor. Fiz. 83, 416 (1982) (Sov. Phys. J.E.T.P., 56, 226).
10. R.M. Bowley and D.O. Edwards, J. de Physique 44, 723 (1983).
11. B. Castaing, S. Balibar and C. Laroche, J. de Physique 41, 897 (1980).
12. S. Balibar and B. Castaing, J. de Physique Lettres 41, 329 (1980). J.E. Avron, L.S. Balfour, C.G. Kuper, S. Landau, S.G. Lipson and L.S. Schulman, Phys. Rev. Letters 45, 814 (1980).

13. V.I. Marchenko and A.Y. Parshin, Pis'ma Zh. Eksp. Teor. Fiz. 31, 767 (1980) (J.E.T.P. Letters 31, 724).
14. L. Puech and B. Castaing, J. de Physique Lettres 43, 601 (1982).
15. B. Castaing, G. Bonfait and D. Thoulouze, Physica B109 and B110, 2093 (1982).
16. L. Puech, B. Hebral, D. Thoulouze and B. Castaing, J. Physique Lettres 43, 809 (1982).
17. In the case of ^{3}He, preliminary measurements give $R_K T^3 \simeq 0.5$ cm^{2}K^4/W, at T $\simeq$ 60 mK, which is larger than the Khalatnikov result, but far smaller than in the ^{4}He case. The interface mobility, while unexpectedly large, is thus probably smaller than in the ^{4}He case.
18. L. Puech, B. Hebral, D. Thoulouze and B. Castaing, J. Physique Lettres 44, 159 (1983).
19. P.E. Wolf, S. Balibar and F. Gallet, to be published in Phys. Rev. Letters.

Propagation of High Frequency Phonons in Liquid He II

T. Haavasoja, V. Narayanamurti, and M.A. Chin

Bell Laboratories, Solid State Electronics, Research Laboratory
Murray Hill, NJ 07974, USA

The laws of conservation of momentum and energy strongly affect the propagation of high frequency phonons ($\hbar\omega \gg kT$) in He II [1]. The phonon dispersion curve exhibits an anomaly at $P \lesssim 20$ bar. For small wave vectors $q \lesssim 0.3$ Å^{-1} the phase velocity reaches a maximum above the velocity of sound thus allowing a spontaneous decay of a phonon below a certain threshold energy $E_c = \hbar\omega_c$. The presence of E_c is experimentally well established[2] by using superconducting Aℓ-oxide-Aℓ tunnel junction generators and detectors. We have observed a new threshold E_c'' of high frequency phonon propagation in He II at $T \leq 200$ mK by using Sn-oxide-Sn tunnel junction generators and Aℓ-oxide-Aℓ detectors. E_c'' occurs between the roton minimum and the phonon maximum energy of the dispersion curve suggestive of phonon-roton scattering.

The tunnel junctions were evaporated on two glass substrates placed parallel at a distance of 1.10 ± 0.05 mm. Maximum generator power was typically a few microwatts. Measurements were done by using the quasi-monochromatic modulation technique[3] at a frequency of ~ 500 Hz. The use of an Aℓ detector and a Sn generator allows us to study in great detail the signal structure at energies from $2\Delta_{A\ell} \approx 0.36$ meV to $2\Delta_{Sn} \approx 1.15$ meV.

In Fig. 1 we have plotted the signal derivative dS/dI_G with respect to generator current as a function of generator voltage at various pressures. At pressures $P \leq 12.5$ bar we observe a sharp rise in the signal level at E_c due to vanishing spontaneous decay. At $P > 12.5$ bar the signal rise is determined by the detection limit of $2\Delta_{A\ell}$. After the rise the signal levels off until a small bump occurs at E_c'. Finally, at E_c'' it starts slowly to diminish reaching almost the level just before E_c at $eV_G \lesssim 4\Delta_{Sn} \approx 2.3$ meV. This new threshold at E_c'' is not observable with an Aℓ generator and detector for two possible reasons. Firstly, with Aℓ junctions one must work at $eV_G > 4\Delta_{A\ell}$ where the sensitivity of the modulation technique becomes appreciably reduced. Secondly, the presence of the highly damped $2\Delta_{Sn}$ recombination phonons may be crucial to the appearance of E_c''.

The pressure dependence of the E_c, E_c', and E_c'' features is presented in Fig. 2. The data on E_c agrees with the results of DYNES and NARAYANAMURTI[2] as well as with our own measurements done with Aℓ junctions. The solid line in Fig. 2 represents an average of various determinations of the roton minimum energy $\Delta(P)$ taken from Ref. 4. It is obvious that the bump at E_c' originates from propagation of rotons. However, we have evidence that most of the signal observed at $\hbar\omega > E_c'$ is due to phonon propagation. Firstly, at SVP (see Fig. 3) the main part of the signal starts at $\hbar\omega > \Delta$. Secondly, we could not observe an expected change of phase of the signal due to different group velocities of phonons and rotons when passing $\hbar\omega = \Delta$.

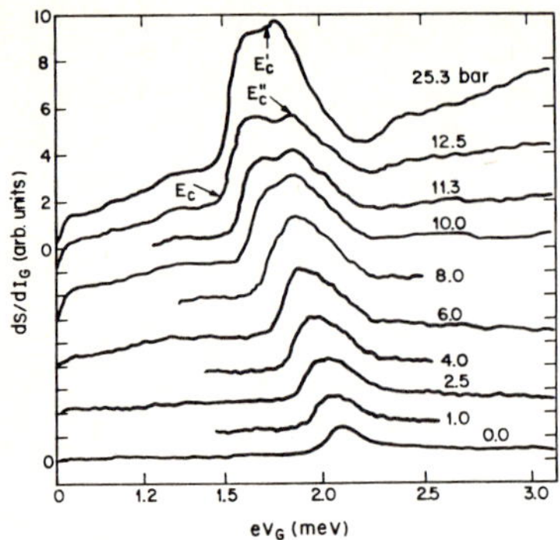

Fig. 1 The signal derivative vs the generator voltage

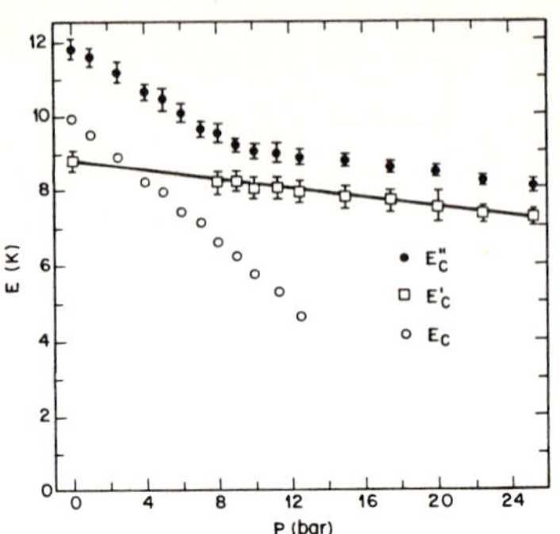

Fig. 2 The threshold energies vs pressure

As to the attenuation of phonon propogation at $\hbar\omega_c \geqslant E_c''$ it is obvious that it must originate from four-particle processes (4pp) or even higher order, since three-particle processes are not allowed due to conservation laws. A 4pp phonon-phonon scattering is not likely at $T \approx 0.1$ K and on the other hand, we do not expect it to exhibit strong energy dependence. Obviously the attenuation of the signal at $\hbar\omega \geqslant E_c''$ is due to phonon-roton scattering. Actually preliminary calculations[5] show that the phase space seems to open up for phonon-roton scattering at $\hbar\omega \geq E_c''$.

In Fig. 3 we have replotted dS/dI_G vs phonon energy E at SVP on an expanded scale. Following the discussion of PITAYEVSKI and LEVINSON[6] let us define $E_c^{(n)}(q_n)$ as the energy above which a spontaneous decay of a phonon into n phonons is not allowed. It can be shown that $E_c^{(\infty)}$ corresponds to q_∞ satisfying $v_p(q_\infty) = v_s$, i.e., the phase velocity v_p reaches the sound velocity v_s [6]. The value of $E_c^{(2)} = 7.89$ K shown in Fig. 3 is determined by the best available data of the dispersion curve at SVP according to DONNELLY et al.[7]. It is clear from Fig. 3 that the difference between $E_c^{(2)}$ and any reasonable

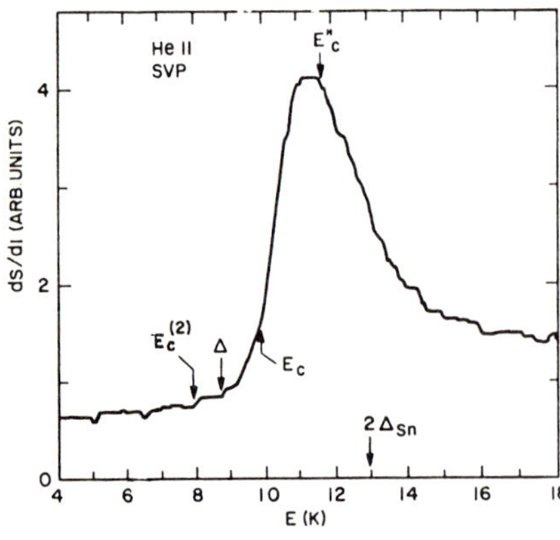

Fig. 3 The signal derivative vs phonon energy at SVP

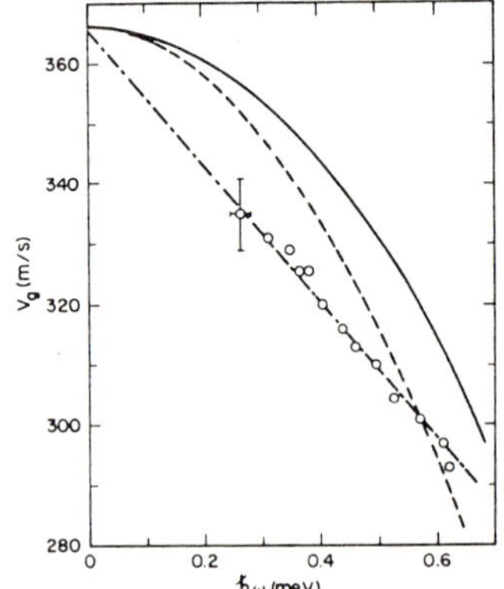

Fig. 4 The group velocity vs phonon energy at 25.3 bar

definition of E_c is more than the maximum experimental error $\lesssim$ 0.6 K. Actually, it turns out that our value of E_c = 9.85 ± 0.15 K as well as that of DYNES and NARAYANAMURTI[2] E_c = 9.5 ± 0.4 K agrees with $E_c^{(\infty)}$ = 9.85K of DONNELLY et al.[7]. The observed cut-off energy E_c must lie between $E_c^{(2)}$ and $E_c^{(\infty)}$ depending on the unknown decay rates $\Gamma^{(n)}$. The high measured value of E_c suggests that $\Gamma^{(n)}$ must be reasonably high for small values of n and the neutron scattering data analyzed by DONNELLY et al. yields a reasonable estimate of the dispersion curve for q < 0.6 Å^{-1}.

The rise of the dS/dI_G signal starts definitely before our value of E_c at SVP. This less steeply increasing part of the signal could originate from a finite lifetime of phonons against n-phonon decay or from rotons. However, if it were due to roton propagation, we would expect from time of flight measurements a group velocity $v_g(q_\infty) \approx$ 190 m/s[7] at the leading edge of the signal. The measured value is 220 ± 10 m/s which suggests propagation of phonons with wave vector q_n, n > 2. Thus the signal between Δ and E_c at SVP is more likely due to propagation of phonons than rotons. At higher pressures this kind of precursory signal was relatively much smaller if at all observable.

By tuning the energy gap of a granular Aℓ detector with magnetic field we obtained the group velocity as a function of phonon energy from time of flight measurements. The data is shown in Fig. 4 at 25.3 bar. The solid line approximates the data of SVENSSON et al.[8] and the dashed line corresponds to a fit of neutron scattering data to an expansion of the dispersion curve by BROOKS and DONNELLY.[9] Unfortunately our absolute accuracy of about 5% does not allow one to make definitive comparison to the other data. However, the precision of our data suggests that power series expansions of the dispersion curve for the whole phonon region 0 ≤ q < 1 Å^{-1} may not be adequate at q < 0.6 Å^{-1}. For example, the dash-dotted line would require a q^2 term in the expansion.

References

1. For a review, see H. J. Maris, Rev. Mod. Phys. 49, 341 (1977).
2. R. C. Dynes and V. Narayanamurti, Phys. Rev. Lett. 33, 1195 (1974), and Phys. Rev. B 12, 1720 (1975).
3. For a review, see W. Eisenmenger in Physical Acoustics, Vol. XII, p. 79, eds. W. P. Mason and R. W. Thurston (Academic Press, New York, 1976).
4. D. S. Greywall, Phys. Rev. B 18, 2127 (1978); ibid. 21 1329 (E) (1979).
5. R. N. Bhatt, private communication.
6. L. P. Pitayevski and Y. B. Levinson, Phys. Rev. B 14, 263 (1976).
7. R. J. Donnelly, J. A. Donnelly, and R. N. Hills, J. Low Temp. Phys. 44, 471 (1981).
8. E. C. Svensson, A. D. B. Woods, and P. Martel, Phys. Rev. Lett. 29, 1148 (1972).
9. J. S. Brooks and R. J. Donnelly, J. Phys. Chem. Ref. Data 6, 51 (1977).

Ultrasonic Attenuation in KCl:OH⁻ with High OH Concentration

M. Saint-Paul* and R. Nava

Centro de Fisica, Instituto Venezolano de Investigaciones Cientificas
Apartado 1827, Caracas 1010 A, Venezuela

J. Joffrin

Laboratoire de Physique des Solides, Université Paris-Sud, Bât. 510
F-91405 Orsay, France

The low temperature properties of KCl containing OH impurities have been studied in detail and the microscopic model is well known. The hydroxyl ion enters in substitution for Cl^- and has six equilibrium orientations among which it performs motions at low temperatures [1]. Besides its electric dipole moment, the substitutional OH^- ion, because of distortions in the host lattice, induces the existence of an "elastic dipole" which may interact with an applied stress field [2]. Ultrasonic measurements performed at low temperatures where the thermal energy is comparable with the energy of dipolar interactions among OH defects are presented here.

The measurements have been performed on crystals doped with OH concentrations of 1.5 and 3×10^{19} cm^{-3} corresponding to about 0.1 and 0.2 at.%. The OH concentration has been determined at Cornell and IVIC by measuring the optical absorption line at λ = 204.5 nm. The attenuation of longitudinal and shear acoustic waves propagating along a (100) direction has been measured with the standard pulse echo method between 15 and 150 MHz in the temperature range 1.5-30 K.

A large thermally activated anomaly is observed around 6 K for longitudinal waves. The excess attenuation, $\Delta\alpha = \alpha(T)-\alpha(30\ K)$, with respect to the value at 30 K, is reported only for the 1.5×10^{19} cm^{-3} OH concentration in Fig. 1. The anomaly observed at 3×10^{19} cm^{-3} has a maximum situated at the same temperature T_M but its low temperature side is slightly modified. Below 5 K $\Delta\alpha$ is an Arrhenius process characterized by an activation energy of about 41 K and a relaxation time of 3.5×10^{-12} s corresponding to a frequency of 4.6×10^{10} Hz (Fig. 2). This thermally activated relaxation has been observed first by Brugger et al . [3] in the same temperature and frequency range but discrepancies with our results exist around 2 K. Our results are in good agreement with recent Brillouin scattering experiments [4]. No inelastic anomaly is observed around 6 K at 100 MHz with shear waves propagating along a (100) direction in agreement with other published results [3,5]. From the Nowick-Heller selection rules the defect involved is the (100) elastic OH defect [3,5]. The results in Fig. 1 are analysed in terms of the classical theory developped by Nowick and Heller [2]. The experimental curve in Fig. 1 at a given frequency ω is broader than a Debye curve. A reasonable description is obtained by assuming a Gaussian distribution P(E) of the activation energies E centered around 41 K with a width of 20 K :

$$\Delta\alpha = \frac{ND^2}{6kT\rho v^3}\int_0^\infty \frac{\omega^2\tau}{1+(\omega\tau)^2}\,P(E)\,dE \qquad (1)$$

* Present address : Centre de Recherches sur les Très Basses Températures, CNRS, BP 166 X, 38042 Grenoble-Cédex, France

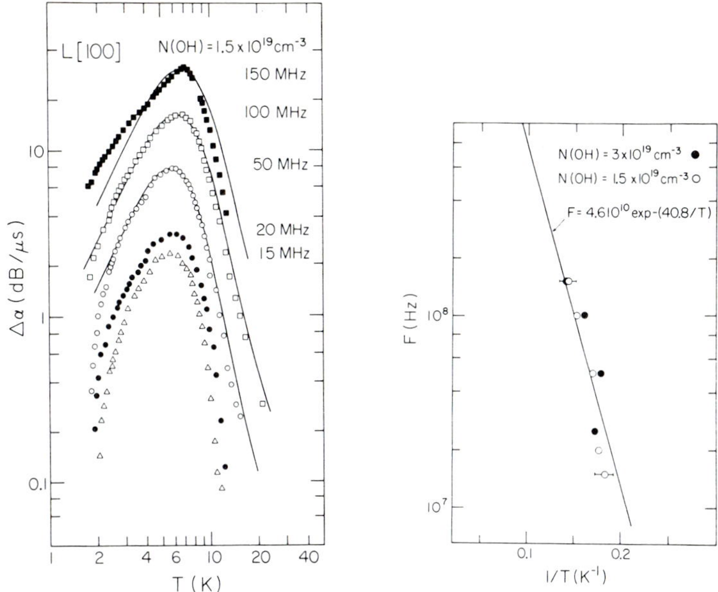

Fig. 1 : Ultrasonic excess attenuation Δα in dB/μ_s of longitudinal waves along (100) as a function of temperature. The curves have been calculated with expression 1

Fig. 2 : Frequency dependence of the inverse temperature T_M of the maximum of the ultrasonic longitudinal excess attenuation Δα

where $\tau = \tau_o \exp(E/kT)$, N is the OH concentration (1.5×10^{19} cm^{-3}), D is the coupling coefficient, ρ is the density (2 g/cm^3) and v is the sound velocity (4.5×10^5 cm/s). A coupling coefficient D ∿ 0.8 eV is deduced by fitting Eq. 1 to Fig. 1 . It corresponds to an elastic dipole of 3×10^{-24} cm^3 which is comparable with the reported value of 5.9×10^{-24} cm^3 [5]. The relaxation mechanism (Fig. 2) can be associated with the reorientations of the OH dipoles involving thermally activated transitions via the 32 cm^{-1} (46 K) excited state correlated with the optical absorptions and the decreases in the thermal conductivity [6]. Interaction effects among OH impurities are most probably responsible for the broadening of the ultrasonic relaxation with increasing OH concentration observed below 5 K.

In conclusion we want to outline that this study is part of an enterprise to explore the possibility of observing dipolar electric glass in this compound; further studies in different dynamic and temperature range are under way.

1 V. Narayanamurti and R.O. Pohl, Rev. Mod. Phys. 42, 201 (1970) and F. Bridges, Crit. Rev. Solids State Sci. 5, 1 (1975)
2 S. Nowick and W.R. Heller, Adv. Phys. 12, 251 (1963) and 14, 101 (1965)
3 K. Brugger, T.C. Fritz and D.A. Kleiman, J. Acous. Soc. Amer. 41, 1015 (1967)
4 R. Vacher, J. Pelous, J.F. Berret and M. Schmidt, J. Physique C-9, C9-517 (1982)
5 B. Chin Yap, Ph. D Thesis Cornell University (1973)
6 C.K. Chau, M.V. Klein and B. Wedding, Phys. Rev. Lett. 17, 521 (1966)

Localized Phonon Mode Associated with Dislocations in Alkali Halide Crystals

Y. Hiki and Y. Kogure

Tokyo Institute of Technology, Oh-okayama, Meguro-ku
Tokyo 152, Japan

F. Tsuruoka

Fakultät für Physik, Universität Konstanz
D-7750 Konstanz, Fed. Rep. of Germany

In a crystal containing dislocations, the state of phonons should be changed due to the dislocated lattice around the defects, and various dynamical properties of the crystal must be altered. Specific heat, thermal diffusivity and conductivity of metallic alloys /1/ and alkali halide crystals /2/, and ultrasonic attenuation in solid helium /3/ have been studied experimentally by the present authors, and they were confident that a kind of phonon mode was associated with crystal dislocations. It was called the quasilocal phonon mode, which was an in-band mode of phonons spatially localized around the defects. Study of scattering of thermal phonons by dislocations in alkali halide crystals seems to be especially adequate in investigating the existence of the quasilocal mode phonons, since one observes a large thermal resistance due to dislocation scattering which cannot be explained by the strain-field scattering /4/.

Firstly, our analyses of our experimental results on thermal properties of LiF will be shown /5/. We have simultaneously measured the thermal conductivity κ and thermal diffusivity D (= $\kappa/\rho C$; ρ is the material density, C is the specific heat) of dislocated LiF crystals by using the temperature-wave method /6/. At first, the specific heat data will be analyzed. Experimental values of specific heat C derived from κ and D are plotted against T^3 in Fig. 1. Data are identified by the percent reduction of the specimen by compression (Arabic numerals) and the specimen number (Roman numerals). Dashed lines represent Debye specific heat for the Debye temperature Θ_D = 740 K. Apparent increase of the specific heat and deviation from the T^3 law dependence can be seen in the data. When we anticipate extra phonons with

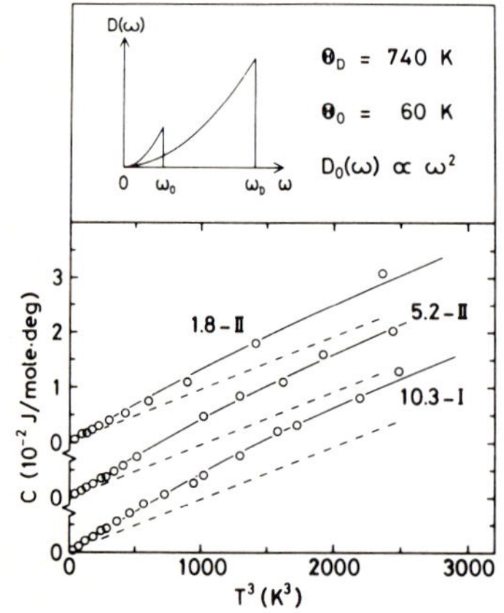

Fig. 1. Specific heat of dislocated LiF. Data are parameter-fitted by using the following formulas:

$$C=\left(\frac{\partial}{\partial T}\right)\int_0^\infty \frac{\hbar\omega D(\omega)}{e^{\hbar\omega/k_BT}-1}\,d\omega,$$

$$D(\omega)=D_D(\omega)+D_0(\omega)=P_D\omega^2+P_0\omega^m,$$

$$P_D=0 \quad \text{for} \quad \omega>\omega_D,$$

$$P_0=0 \quad \text{for} \quad \omega<\omega_1^0 \quad \text{and} \quad \omega>\omega_2^0.$$

Best fit is obtained for m = 2, ω_1^o = 0, ω_2^o = 60 K. Fitted values of P_D and P_o are used to determine the number of Debye phonons and that of quasilocal phonons. The dimension of the quasilocal phonon region is then obtained.

state density D_Q in addition to the Debye phonons with D_D, calculated values of specific heat represented by the solid curves in the figure are obtained. The extra density of state is considered to be that of the quasilocal mode phonons. The phonons are to be localized in a region around each dislocation line, and the dimension of the region is shown to be around 10^{-5} cm. Dislocation densities in specimens determined by an etch-pit method must be used in obtaining this result. Examples of the data of temperature dependences of thermal diffusivity D and conductivity κ are shown in Fig. 2 for a dislocated LiF crystal. Theoretical expressions (Callaway's formalism and Berman-Brock's relaxation rates) shown beside the figure are parameter-fitted to the diffusivity data, and here the relaxation rate for the resistive process τ_R^{-1} is the sum of the rates for boundary, isotope, and dislocation scattering. The normal-process relaxation rate τ_N^{-1} is also taken into account. Especially for the dislocation scattering, we choose the relaxation rate of the form $\tau^{-1} = \tau_F^{-1} + \tau_A^{-1} = F\omega^{-n} + A\omega^{m}$ and determine the parameters F, A, n, and m. The best fit is obtained when the values n = 1 and m = 2 are adopted. The term $F\omega^{-1}$ is dominant at lower temperatures and represents the phonon scattering by fluttering mechanism. The second term $A\omega^2$ is considered to be due to the scattering by quasilocal phonons, and is dominant at higher temperatures. Figure 2 (a) shows that only the two-term fitting leads to a good result. Thermal conductivity is also well fitted to the theoretical expression when the same method is used, as shown in Fig. 2 (b). The parameter A thus determined represents the strength of the scattering by quasilocal phonons, and is shown to be proportional to the density of dislocations in the specimen crystal. The reason why the relaxation rate for the scattering is proportional to ω^2 has already been discussed by considering an interaction process of thermal phonons and quasilocal phonons /5/.

Now thermal properties of dislocated LiF crystals have well been understood after taking into consideration the fluttering mechanism and the quasilocal-phonon scattering mechanism, some of the existing data on thermal conductivity of NaCl, KCl, NaF /7/ and LiF /8/ will be analyzed on the same basis. Data in the temperature range of 1.5 - 20 K, where the contribution of the dislocation scattering is dominant, are used for the analysis. At first, the relaxation rates for the boundary scattering and the isotope

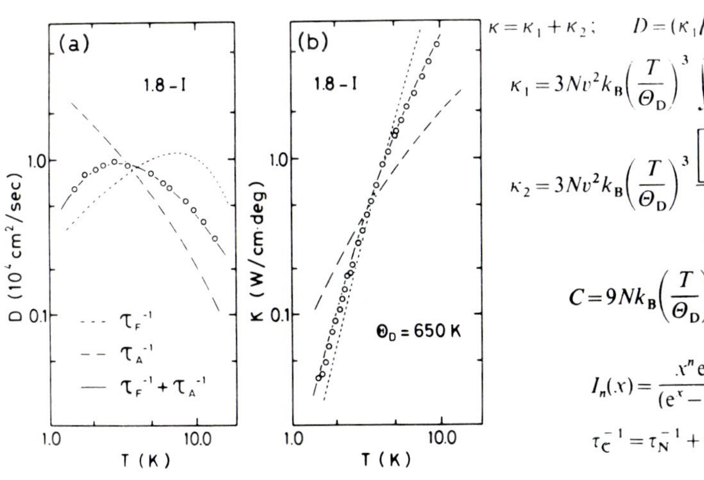

$$\kappa = \kappa_1 + \kappa_2; \qquad D = (\kappa_1/\rho C) + (\kappa_2/\rho C).$$

$$\kappa_1 = 3Nv^2 k_B \left(\frac{T}{\Theta_D}\right)^3 \int_0^{\Theta_D/T} \tau_C I_4(x)\,dx.$$

$$\kappa_2 = 3Nv^2 k_B \left(\frac{T}{\Theta_D}\right)^3 \frac{\left[\int_0^{\Theta_D/T} (\tau_C/\tau_N) I_4(x)\,dx\right]^2}{\int_0^{\Theta_D/T} (\tau_C/\tau_N\tau_R) I_4(x)\,dx}$$

$$C = 9Nk_B\left(\frac{T}{\Theta_D}\right)^3 \int_0^{\Theta_D/T} I_4(x)\,dx,$$

$$I_n(x) = \frac{x^n e^x}{(e^x - 1)^2}, \qquad x = \frac{\hbar\omega}{k_B T},$$

$$\tau_C^{-1} = \tau_N^{-1} + \tau_R^{-1}.$$

$$\tau_R^{-1} = v/L + 0.32x^4T^4 + (F\omega^{-n} + A\omega^m);$$

$$\tau_N^{-1} = 35xT^4.$$

Fig. 2. Thermal diffusivity and conductivity of LiF. Data are parameter-fitted by the formulas

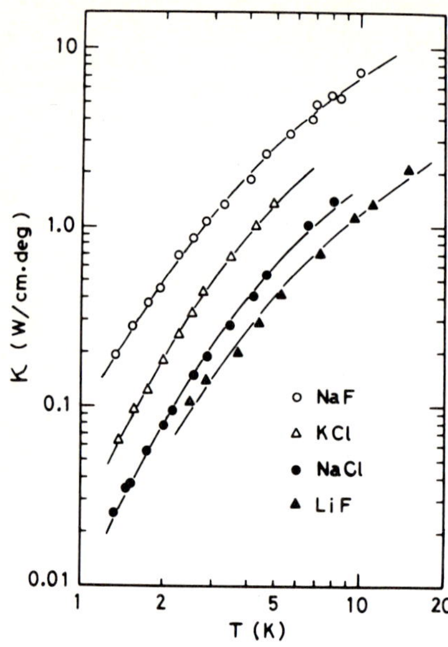

Fig. 3. Thermal conductivity of alkali halide crystals. Data by Taylor et al. (NaCl, KCl, NaF) and Sproull et al. (LiF) are analyzed by using the relaxation rate for the dislocation scattering:

$$\tau^{-1} = A\omega^{-1} + A\omega^{2}.$$

The values of determined A divided by dislocation density Λ:

	A/Λ (10^{-26} cgs)
NaCl	10.8
KCl	6.8
NaF	8.6
LiF	6.3
LiF*	5.2

* from thermal diffusivity

scattering are determined by using the conductivity data for an annealed crystal. The normal-process scattering is ignored in the present analysis. Then the dislocation relaxation rate $\tau^{-1} = F\omega^{-1} + A\omega^{2}$ is added to the determined boundary and isotope rates, and the parameters F and A are obtained by a data fitting for a crystal containing dislocations. For the Debye temperature required in determining the absolute values of F and A, the value calculated from elastic constants of the crystal is adopted. Figure 3 represents the results of the fit for four kinds of alkali halide crystals, and the fitting seems to be satisfactory. Values of the determined parameter A divided by the dislocation density in the specimen Λ are given in the table beside the figure. The adopted values of Λ are those given in the literatures /7,8/. The quantity A/Λ really represents the true strength of the scattering by quasilocal phonons. The value of A/Λ determined from the analysis of our thermal diffusivity data for LiF is also given in the table. The agreement of the two values for LiF is not unsatisfactory. It is finally concluded that the strength of the quasilocal-phonon scattering is not so much different in four kinds of alkali halide crystals.

References

1. Y. Kogure and Y. Hiki: J. Phys. Soc. Jpn. 39, 698 (1975)
2. H. Kaburaki, Y. Kogure and Y. Hiki: J. Phys. Soc. Jpn. 49, 1106 (1980)
3. F. Tsuruoka and Y. Hiki: Phys. Rev. B 20, 2702 (1979)
4. Y. Kogure and Y. Hiki: J. Phys. Soc. Jpn. 36, 1597 (1974); 38, 471 (1975)
5. Y. Hiki and F. Tsuruoka: J. Phys. Soc. Jpn. 50, 2355 (1981)
6. Y. Kogure and Y. Hiki: Jpn. J. Appl. Phys. 12, 814 (1973)
7. A. Taylor, H. R. Albers and R. O. Pohl: J. Appl. Phys. 36, 2270 (1965)
8. R. L. Sproull, M. Moss and H. Weinstock: J. Appl. Phys. 30, 334 (1959)

^3He-^3He Interaction in ^3He-^4He Liquid Mixtures Determined from Sound Attenuation

A.J. Ikushima, I. Fujii, M. Fukuhara, and K. Kaneko

The Institute for Solid State Physics, The University of Tokyo
Roppongi, Minato-ku, Tokyo 106, Japan

1 Abstract Interaction potential between ^3He quasiparticles in ^3He-^4He liquid mixtures is determined from the sound attenuation as a function of pressure in mixtures of ^3He mole fraction ranging from 0.0289 to 0.0803. Superfluid transition temperature of ^3He in the mixtures was estimated from the interaction potential.

2 Introduction A systematic experimental study on ^3He-^4He mixtures was made in 1966 by the Illinois group [1]. They made measurements of a number of properties under the saturated vapor pressure (S.V.P.) in two dilute mixtures of ^3He mole fraction X_3= 0.013 and 0.050. Since then, some number of measurements were made in the mixtures [2].

Recently, rapid developement of ultralow temperature techniques has stimulated very strongly an interest that we may see the superfluidity of ^3He in ^3He-^4He mixtures. The interaction between two ^3He quasiparticles is thought to be weak and attractive at least under relatively low pressures [3]. The superfluid phase of ^3He in the mixture will therefore be of s-type or BCS superfluidity. Furthermore, p-type superfluidity has been proposed to exist in the most concentrated mixture under a high enough pressure, 20 bars say[4]. Therefore, the superfluidity of ^3He in the dilute ^3He-^4He mixtures is very interesting in the sense that it can be the first example of the BCS superfluidity that ^3He ensemble shows, and that it will probably be the best candidate where one can observe both the BCS and p-type superfluidities if one changes pressure.

Unfortunately, the interaction between ^3He quasiparticles is so weak that the expected transition temperature to the superfluid state, T_c, is very low. T_c was once estimated in the paper by Bardeen, Baym and Pines(BBP) [5] to be 2μK under S.V.P. BBP assumed the ^3He-^3He interaction potential was of the form, $V(q) = V_0\cos(\beta q)$. The two parameters, V_0 and β , were determined from spin diffusion data for X_3 = 0.013 and 0.050. Since then, alternates of the BBP potential have then been proposed; dipolar and the polynomial types.

Recently, some groups have been trying to cool down mixtures using recent ultralow temperature techniques to find the ^3He superfluid phase [6]. However, because of a large Kapitza resistance between the mixtures and coolant, the lowest temperature ever obtained is still probably around 200 μK.

Instead of cooling mixtures by brute force, any measurement which allows us to deduce the ^3He-^3He interaction potential in the mixture should also be important. The interaction potential gives us a rather accurate estimate of T_c, and then the condition to get high T_c. Furthermore, the potential gives us much more information about other properties of the mixture. This paper is to report sound attenuation measurements in mixtures changing X_3 and pressure, one of the most relevant properties to the interaction.

3 Sound Attenuation and the Interaction Potential   Sound attenuation, and equivalent sound velocity dispersion, due to the interaction of ^{3}He quasiparticles in the mixture, gives us a characteristic relaxation time between two ^{3}He quasiparticles, which is the same relaxation time as that for the viscosity.

If the system is Fermi degenerated, the relaxation time τ_η is given by

$$(\tau_\eta T^2)^{-1} = \frac{9}{16}\cdot\frac{k_B^2\, n_3}{\hbar^3 v_F^3}\langle W(\theta,\phi)\cdot\sin^4(\theta/2)\sin^2\phi/\cos(\theta/2)\rangle_{AV} \qquad (1)$$

where n_3 and v_F are the number density of ^{3}He atoms and the Fermi velocity, respectively, and k_B and $\hbar$ have the usual meanings. And $W(\theta,\phi)$ is the transition probability from an initial state with two quasiparticles ($\vec{k}_1$, $\vec{k}_2$) to a final state ($\vec{k}_1'$, $\vec{k}_2'$), θ is the angle in k space between $\vec{k}_1$ and $\vec{k}_2$ (or equivalently that between $\vec{k}_1'$ and $\vec{k}_2'$), and ϕ is the angle between the plane determined by $\vec{k}_1$ and $\vec{k}_2$ and that determined by $\vec{k}_1'$ and $\vec{k}_2'$.

The interaction potential $V(\vec{q})$ is then related to $W(\theta,\phi)$ as

$$W(\theta,\phi) = \frac{2\pi}{h}\,[(3/4)|V(\vec{q})|^2 + (1/4)|V(\vec{q}\,')|^2 - (1/2)|V(\vec{q})V(\vec{q}\,')|], \qquad (2)$$

where $\vec{q} = \vec{k}_1'-\vec{k}_2 = \vec{k}_2'-\vec{k}_2$, and, $\vec{q}\,' = \vec{k}_2'-\vec{k}_1=\vec{k}_1'-\vec{k}_2$ are the scattering vectors.

4 Experimental Procedure   We measured the acoustic attenuation at around 20 mK and above. Especially at temperatures lower than 100 mK say, we had to be careful not to introduce large heat input. See Ref.[7] for the details.

Temperature in this experiment was essentially measured by using a CMN magentic thermometer. To calibrate the CMN thermometer, we used a Ge thermometer calibarated against ^{3}He gas pressure, and also NBS's temperature standards with several superconducting transitions.

5 Experimental Results and Discussions   Figure 1 (a) and (b) show typical results of the sound attenuation under S.V.P. and 10 bars, respectively. The peaks observed in the range of 20-200 mK are due to the collision between ^{3}He quasiparticles. The present peak temperatures and those in earlier work under S.V.P. are in disagreement [8]. The disagreement would be due to the difference in temperature scales among the groups. In the present experiment, the temperature scale at least from 20 to around 200 mK is quite sure because this range was very well calibrated against the superconducting transition temperatures of $AuAl_2$(162.58 mK), Ir (98.90 mK) and Be (23.05 mK).

To calculate the ^{3}He-^{3}He interaction potential under S.V.P., we took into account the data for X_3 = 0.0289, 0.0401, 0.0573 and/or X_3 = 0.0803. The last mixture can be homogeneous over ca. 3 bars, and therefore, we took the data in this mixture at 3.2 bars.

We then assumed for making the calculation easier that the interaction potential be expanded as

$$V(q) \cong a_0 + a_1 q^2 + a_2 q^4 \qquad (3)$$

and made the least squares fitting for a_0, a_1 and a_2. Figure 2 shows the potential thus deduced under S.V.P. and 10 bars. For a comparison, the famous BBP potential under S.V.P. is also shown in the figure.

We can then estimate the superfluid transition temperature T_c for s pairing of ^{3}He atoms in the mixture. T_c was found around 1μK at $X_3 \cong 1\%$ under both S.V.P. and 10 bars.

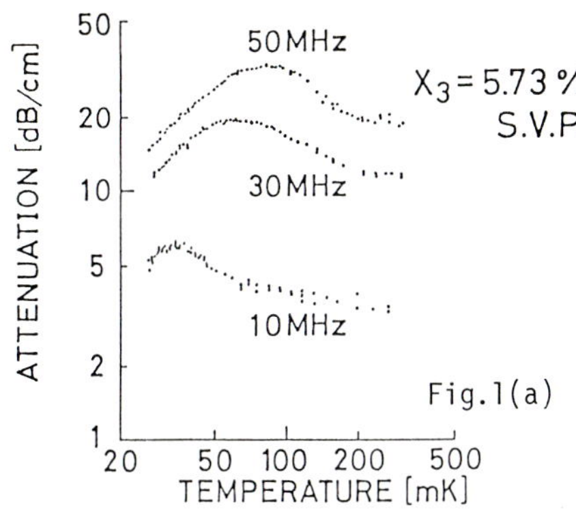

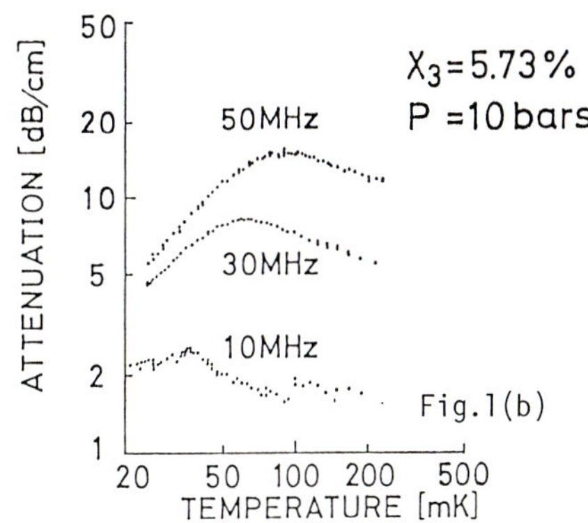

Fig.1 (upper). Sound attenuation in 5.7% ^3He-^4He mixture under (a) S.V. P. and (b) 10 bars

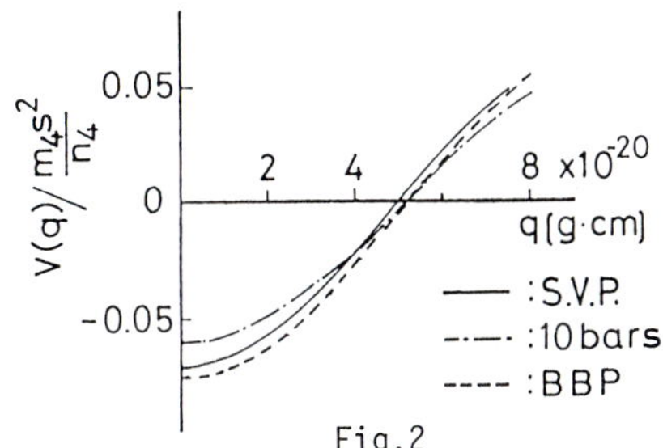

Fig.2 (left). ^3He-^3He interaction potential in ^3He-^4He mixtures with the pressure as a parameter

6 References

1. A.C.Anderson, W.R.Roach, R.E.Sarwinski and J.C.Wheatley: Phys. Rev. Letters 16, 263 (1966).
2. G.Baym: "The Physics of Liquid and Solid Helium. II", ed. K.H.Bennemann and J.B.Ketterson (John Wiley & Sons Inc., 1978) p.123; A.I.Ahonen, M.A.Paalanen, R.C.Richardson and Y.Takano: J.Low Temp. Phys. 25, 733 (1976); E.S.Murdock, K.R.Mountfield and L.R.Corruccini: ibid.31, 581(1978)
3. See, for example, C.Ebner and D.O.Edwards: Physics rep. 2, 77 (1970).
4. M.B.Hoffberg: Phys. Rev. A5, 1963 (1971).
5. J.Bardeen, G.Baym and D.Pines: Phys. Rev. Letters 17, 372 (1966), and Phys. Rev. 156, 230 (1967).
6. R.E.Mueller, H.Chocholacs, Ch.Buchal, M.Kubota, J.R.Owers-Bradley and F.Pobell: Proc. Symp. on "Quantum Fluids and Solids" (Sanibel Is., 1983), to be published.
7. A.J.Ikushima, I.Fujii and K.Nara: Jpn. J.Appl. Phys. 21, Suppl. 21-3. 20 (1982).
8. B.M.Abraham, Y.Eckstein, J.B.Ketterson and J.H.Vignos: Phys. Rev. Letters 17, 1254 (1966).

Ultrasonic-Attenuation and Pressure Measurements in Phase-Separated Solid ^{3}He-^{4}He Mixtures

I. Iwasa and H. Suzuki

Department of Physics, University of Tokyo, Bunkyo-ku
Tokyo 113, Japan

1. Introduction

The phase separation of solid ^{3}He-^{4}He mixtures is known to obey fairly well the regular solution theory, in which the phase separation temperature T_{PS} is given by

$$T_{PS} = 2T_c(1-2X_4)/\ln((1-X_4)/X_4). \tag{1}$$

Here X_4 is the atomic concentration of ^{4}He and $T_c = 0.38$ K is the critical temperature [1]. MULLIN [2] made a more detailed theory of phase separation based on the self-consistent phonon theory and found an excess volume on mixing ΔV given by

$$\Delta V = -0.4X_4(1-X_4) \quad [\mathrm{cm}^3/\mathrm{mole}]. \tag{2}$$

He predicted a pressure increase on phase separation at a constant volume, which was verified by PANCZYK et al. [3] by using a capacitive strain gauge.

Recently hystereses with respect to temperature have been observed in the thermal-conductivity [4], the ultrasonic-attenuation [5], and the X-ray [6] measurements on mixtures, which were related to the phase separation. In this paper we give the results of ultrasonic-attenuation and pressure measurements and compare them with each other and with other experiments.

2. Ultrasonic Attenuation

The relative change in the attenuation of a longitudinal wave at 10 MHz was measured in bcc ^{3}He crystals containing 480 or 1600 ppm ^{4}He at various stress amplitudes. Figure 1 shows the attenuation in a crystal with X_4 = 1600 ppm and the melting pressure P_m = 33 atm at high and low amplitudes denoted by I = 50 dB and 20 dB, respectively. The attenuation at 50 dB, α(50dB), began to increase at a phase separation temperature T_A = 98 mK on cooling and showed a large hysteresis due to the dislocations created by the phase separation process [5]. The attenuation at 20 dB, α(20dB), rose at a temperature about 20 mK lower than T_A and showed a smaller hysteresis. Both hystereses were reproducible when the crystal was warmed above 400 mK. Similar hystereses were observed in a crystal with X_4 = 480 ppm and P_m = 38 atm. In crystals with X_4 = 480 ppm and P_m = 33 atm, however, α(20dB) was independent of temperature and the hysteresis of α(50dB) differed from that in Fig.1 in that the decrease of attenuation around 100 mK on warming was absent.

3. Pressure

Using a capacitive strain gauge, we measured the pressure of solid mixtures whose concentration of ^{4}He was 4100 or 8700 ppm. Each crystal was grown at a constant pressure. Figure 2 shows a typical result of measurement on a crystal with X_4 = 8700 ppm. When the crystal was rapidly cooled to the lowest

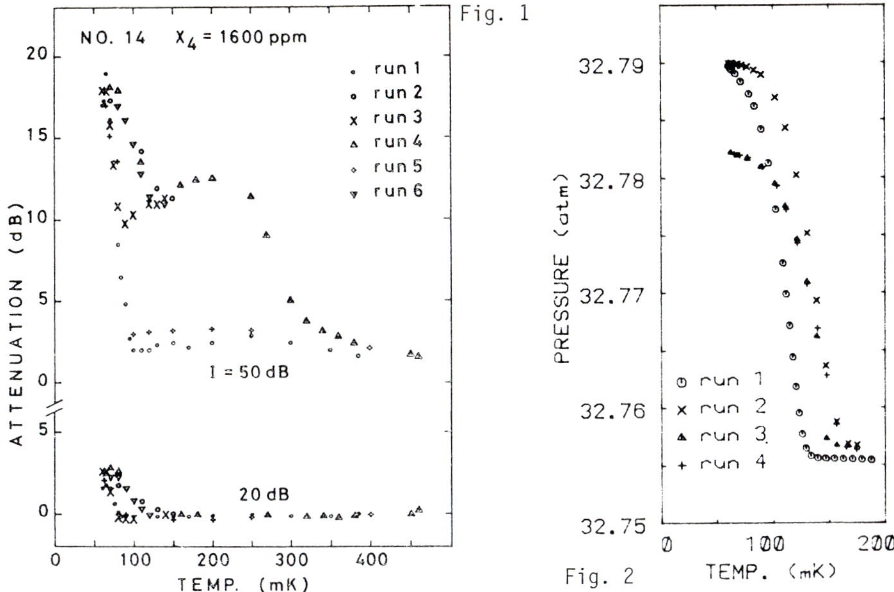

Fig. 1. Ultrasonic attenuation in a mixture crystal with $X_4 = 1600$ ppm grown at 33 atm. Stress amplitudes at I = 50 and 20 dB are 3×10^4 and 1×10^3 dyn/ cm^2, respectively. Runs 1, 3 and 5 are cooling and runs 2, 4 and 6 are warming

Fig. 2. Pressure in a mixture crystal with $X_4 = 8700$ ppm. Runs 1 and 3 are cooling and runs 2 and 4 are warming. In run 1 the temperature was rapidly lowered. In runs 2-4 each data point represents the pressure after the temperature was held constant at least 30 minutes

temperature (run 1), the pressure increased by 0.035 atm as a result of the phase separation, and the subsequent warming (run 2) presented a hysteresis. On the other hand, when the crystal was slowly cooled (run 3), the pressure increase was only 0.025 atm, and the pressure change was reproducible on the next slow warming (run 4) at temperatures below 130 mK. For runs 2-4 the temperature was held constant at least 30 minutes at each data point in Fig.2, and the relaxation of pressure was recorded.

In the case of a crystal with X_4 = 4100 ppm the hysteresis did not vanish even though the temperature was slowly varied. The pressure increase was 0.013 atm on rapid cooling and 0.011 atm on slow cooling.

4. Discussion

The phase separation of a dilute solution of ^{4}He in ^{3}He is regarded as precipitation of ^{4}He atoms in the almost pure ^{3}He matrix which has the bcc structure. In Mullin's theory the ^{4}He-rich phase is assumed to be bcc, but the equilibrium structure of pure ^{4}He is hcp. If the precipitates take the hcp structure, the excess volume on mixing is

$$\Delta V = -0.4X_4(1-X_4)+0.2X_4 \qquad [cm^3/mole] \qquad (3)$$

because the molar volume of bcc ^{4}He is 0.2 cm^3/mole larger than that of hcp ^{4}He. Equations (2) and (3) are the extreme cases of totally bcc and totally hcp precipitates, respectively. At a constant volume the pressure change due to phase separation is given by $\Delta P = -B\Delta V/V$, where B = 300 atm is the bulk modulus and V = 24 cm^3/mole is the molar volume of the mixture. Then, we can estimate from ΔP values in Fig.2 that 60% of the precipitates are bcc in the rapidly cooled mixture with X_4 = 8700 ppm and 17% of them are bcc if the mixture is slowly cooled. The relaxation time determined from run 2 of Fig.2 is shorter than that determined from run 4, and the average diameter D of the precipitates is estimated to be 7.6×10^{-5} cm for run 2 and 1.04×10^{-4} cm for run 4 at 120 mK.

The hysteresis of α(20dB) may be related to the amount of hcp precipitates which is expected to vary hysteretically. The attenuation due to Rayleigh scattering is very small, but it may be enhanced by the mode conversion at the surface of the hcp precipitates because the bcc ^{3}He crystal is highly anisotropic (the anisotropy factor is 10∼12). Otherwise resonant absorption in the hcp precipitates or at their surface may be responsible for the hysteresis of α(20dB).

The change in the thermal conductivity is caused by the scattering of thermal phonons at the precipitates. The hysteresis of the X-ray mosaic width as well as the hysteresis of α(50dB) is due to the dislocations created by the phase separation.

References

1. D.O. Edwards, A.S. McWilliams and J.G. Daunt: Phys.Rev.Lett 9, 195 (1962).
2. W.J. Mullin: Phys.Rev.Lett. 20, 254 (1968).
3. M.F. Panczyk, R.A. Scribner, J.R. Gonano and E.D. Adams: Phys.Rev.Lett. 21, 594 (1968).
4. A.S. Greenberg and A. Armstrong: Phys. Rev. B22, 4336 (1980).
5. I. Iwasa, N. Saito and H. Suzuki: J.Phys.Soc.Jpn. 52, 952 (1983).
6. B.A. Fraas and R.O. Simmons: Physica 107B, 277 (1981).

Effects on Anharmonicities and Broken Time-Reversal Invariance on Static and Dynamic Properties of Two-Dimensional Electron Solids

G. Meissner

Theoretische Physik, Universität des Saarlandes
D-6600 Saarbrücken, Fed. Rep. of Germany

Making extensive use of sum-rule techniques, also when time-reversal symmetry is broken we were able to study theoretically effects of anharmonicities and intense magnetic fields on two-dimensional electron solids. Collective excitations can be investigated by the present method both classically and quantum mechanically. By explicitly evaluating the shear modulus in the classical limit as well as in the extreme quantum limit we have analyzed dislocation-mediated melting of a triangular electron lattice.

Two-dimensional electron systems realized in Si metal-oxide semiconductors and GaAs-AlGaAs heterojunctions have rather high densities, i.e., $n \geq 10^{11}\mathrm{cm}^{-2}$, in contrast to electrons on the surface of liquid helium. Therefore, strong magnetic fields applied perpendicular to the two-dimensional plane are presumably necessary for these systems to form a two-dimensional electron solid. Due to localization of the electrons of charge $q \equiv -e$ within the order of the Larmor radius, $r_L = (c\hbar/eB)^{1/2}$, in the presence of magnetic fields of amplitude B, qualitatively one expects the formation of a lattice, if r_L is sufficiently smaller than the mean electron distance $a_o = (\pi n)^{1/2}$, in order to minimize the Coulomb interaction between the electrons. A filling factor $\nu \equiv 2\pi r_L{}^2 n < 1$ of the lowest Landau level, therefore, seems to be a plausible condition for the formation of a two-dimensional electron solid at T = OK. For such a filling factor $\nu < 1$ it seems furthermore reasonable that the solid phase exists for all finite temperatures T, where the ratio of the mean kinetic to the mean potential energy, i.e., $k_BT/(e^2/\varepsilon r_L)$, is less than some critical value $\Gamma_c(\nu)$, if ε denotes the background dielectric constant.

As an illustration of our method for studying static and dynamic properties of a two-dimensional electron solid we will indicate in this paper the application to a microscopic calculation of the critical ratio Γ_c and discuss some related questions.

Let us start from the rather general relation

$$\det \left|\delta(14)\delta(51') - K(121'3)G(34)G(52)\right| \equiv 0 \tag{1}$$

which holds due to the instability of the interacting electron system in a uniform positive background potential to the formation of charge density waves. In (1) the notation $1 \equiv \underset{\sim}{r}_1, t_1$ for the position $\underset{\sim}{r}_1$ in the x-y plane and for the time t_1,etc., is used in the one-electron Green's functions G and in the electron-hole interaction K. Evaluating the Hartree-Fock contribution of the shielded potential approximation of (1) under the condition $\nu = 2\pi r_L{}^2 n < 1$ we obtained $\Gamma_c(\nu) \simeq 0.557\ \nu(1-\nu)$, which corresponds to the mean-field result of FUKUYAMA, PLATZMAN, and ANDERSON [1]

$$\frac{k_B\ T_M}{e^2/(\varepsilon r_L)} = 0.557\ \nu(1-\nu) \tag{2}$$

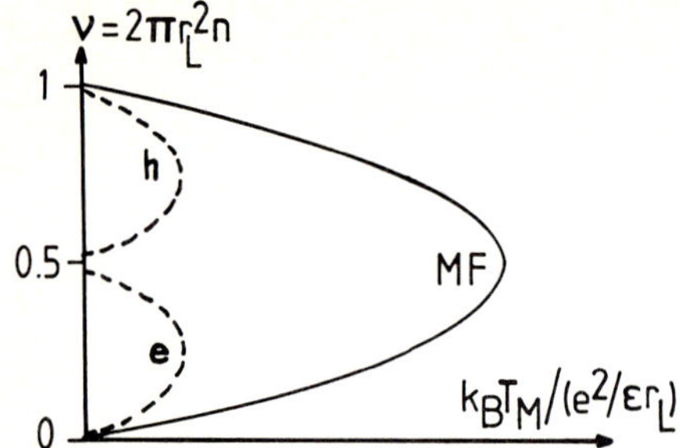

Fig. 1. Schematic phase separation curve: electron density n versus temperature T. The result of mean-field theory for T_M is indicated by MF, the result of the Kosterlitz-Thouless approach is indicated for the electron lattice (e) and for the hole lattice (h)

for the transition temperature T_M depicted in Fig. 1. Concerning reciprocal lattice vectors in deriving (2) the additional simplifying assumption of unidirectional charge density waves has been used.

Due to a gap $\Delta \sim 0.37e^2/(\varepsilon r_L)$ in the one-electron spectrum collective excitations are expected to play an important role in the charge-density wave state. The relation between frequencies $\tilde{\omega}_j$ and wave vectors $\underset{\sim}{k}$ of the associated modes j = 1,2 are to be determined from zeros of the determinant of the real part of the inverse displacement response function, i.e.,

$$|\mathrm{Re}\,\chi^{-1}_{\alpha\alpha'}(\underset{\sim}{k},\omega_j(\underset{\sim}{k}))| \equiv 0 \quad , \quad \alpha,\alpha' = 1,2 \ . \tag{3}$$

We have been able to obtain the following exact spectral representation [2]

$$\chi^{-1}_{\alpha\alpha'}(\underset{\sim}{k},z=\omega+i\eta) = m^*\{z^2\delta_{\alpha\alpha'} + iz\omega_c\varepsilon_{\alpha\alpha'} - M_{\alpha\alpha'}(\underset{\sim}{k},0) + iz\gamma_{\alpha\alpha'}(\underset{\sim}{k},z)\} \tag{4}$$

where m* denotes the effective electron mass, $\delta_{\alpha\alpha'}$ is the Kronecker delta, $\varepsilon_{\alpha\alpha'}$ the antisymmetric tensor, and $\omega_c = eB/m^*c$ the cyclotron frequency. Formally, the term in (4) with $\varepsilon_{\alpha\alpha'}$ results from the second frequency moment of the spectral function of displacements which would vanish identically in a time-reversal invariant system. For the static limit of the dynamical matrix the rigorous sum rule

$$M_{\alpha\alpha'}(\underset{\sim}{k},0) = M^{(\infty)}_{\alpha\alpha'}(\underset{\sim}{k}) - \int_{-\infty}^{+\infty} d\omega\gamma_{\alpha\alpha'}(\underset{\sim}{k},\omega)/\pi \tag{5}$$

may be derived [2] with the high-frequency dynamical matrix $M^{(\infty)}_{\alpha\alpha'}(\underset{\sim}{k})$ and the spectral width function $\gamma_{\alpha\alpha'}(\underset{\sim}{k},\omega)$. In the long-wave-length limit we find

$$M_{\alpha\alpha'}(\underset{\sim}{k},0) = \frac{2\pi e^2 n}{\varepsilon m^*}\frac{k_\alpha k_{\alpha'}}{k} + Z_{\alpha\alpha'\beta\beta'}k_\beta k_{\beta'} + O_{\alpha\alpha'}(k^4) \tag{6}$$

where the expansion coefficients

$$Z_{\alpha\alpha'\beta\beta'} = (2N)^{-1}\left\{\frac{\delta^2 F}{\delta u_{\alpha\beta}\delta u_{\alpha'\beta'}} + \frac{\delta^2 F}{\delta u_{\alpha\alpha'}\delta u_{\beta'\beta}}\right\} \tag{7}$$

are rigorously related to second derivatives of the free energy F with respect to deformations $u_{\alpha\beta}$ due to the elastic sum rule [3]. In a triangular lattice [4] $n = \nu/(2\pi r_L^2) = (\sqrt{3}\,a_o^2/2)^{-1}$ and the shear modulus

$$\mu(T,B) = m^* n\, Z_{1122} \ . \tag{8}$$

Assuming the melting of the electron lattice in general to be dictated by dissociation of dislocation pairs we may estimate the melting temperature T_M from the Kosterlitz-Thouless criterion [5] for a two-dimensional electron solid

$$4\pi\, k_B\, T_M/a_o^2 = \mu(\nu, T_M) \tag{9}$$

both in the classical limit and in the extreme quantum limit.

We have examined the temperature dependence of the shear modulus in the classical limit for vanishing magnetic field previously [2] using perturbation theory to evaluate the effect of anharmonicities on the static limit of the dynamical matrix in one-loop expansion. The resulting linear temperature decrease

$$\mu(T) = \mu_o\,[1 - 26.4\, k_B T\varepsilon/e^2(\pi n)^{1/2} + O(T^2)] \tag{10}$$

modifies the temperature-independent expression in harmonic approximation $\mu_o = m^* n Z^{(o)}_{1122} = 0.245065\, n^{3/2}\, e^2/\varepsilon$ and is consistent with dislocation-mediated melting.

Here we want to investigate the renormalization of the shear modulus μ_o or of the coefficient $Z^{(o)}_{1122} = 0.097767\, \nu^{1/2}\, e^2/(\varepsilon r_L m^*)$, respectively, by strong magnetic fields in the extreme quantum limit. To this end in an iterative procedure we start by disregarding dispersive anharmonicities in (5) taking $M_{\alpha\alpha'}(\underset{\sim}{k},0) \simeq M^{(\infty)}_{\alpha\alpha'}(k)$ where

$$M^{(\infty)}_{\alpha\alpha'}(\underset{\sim}{k}) = \frac{1}{m^*}\sum_{\ell\neq 0}\,[1 - e^{-i\underset{\sim}{k}\cdot\underset{\sim}{R}(\ell)}]\langle\nabla_\alpha\nabla_{\alpha'}e^2/\varepsilon|\underset{\sim}{x}(\ell) - \underset{\sim}{x}(0)|\rangle\ . \tag{11}$$

Instead of replacing position operators $\underset{\sim}{x}$ in (11) by their mean values $\underset{\sim}{R} \equiv \langle\underset{\sim}{x}\rangle = (R_1, R_2)$ which amounts to the harmonic approximation, however, we determine the effective force constant tensor $\langle\nabla_\alpha\nabla_{\alpha'}e^2/\varepsilon|\underset{\sim}{x}(\ell) - \underset{\sim}{x}(0)|\rangle$ by evaluating $\langle\ldots\rangle$ at zero temperature using for example the Slater determinant $\Psi(\underset{\sim}{r}_1 \ldots \underset{\sim}{r}_N) = (N!)^{-1/2}\det|\psi(\underset{\sim}{r}_k - \underset{\sim}{R}(\ell))|$ for the ground-state wave function of the electron solid with localized one-electron wave functions $\psi(\underset{\sim}{r}-\underset{\sim}{R}) = (2\pi r_L^2)^{-1/2}\exp\{-[(\underset{\sim}{r}-\underset{\sim}{R})^2 - 2i(x_1R_2 - x_2R_1)]/4r_L^2\}$. From an expansion of (11) analogous to (6) we can calculate the renormalized coefficient $Z^{(\infty)}_{1122}$. Using (8) and (9) we finally obtain the shear modulus $\mu^{(\infty)}$ and the melting temperature $T_M^{(\infty)}$, respectively, replacing the classical one, i.e. $4\pi k_B T_M^{(o)} = 0.112892\, \nu^{1/2}e^2/(\varepsilon r_L)$. As a remarkable feature, the $T_M^{(\infty)}$ versus ν curve exhibits two broad maxima on the left- and right-hand side of $\nu = 1/2$ where it vanishes (Fig. 1). This behavior found independently by MAKI and ZOTOS [6] can be interpreted as the electron lattice to be locally stable for $0 \leq \nu < 1/2$, while the hole lattice is locally stable for $1/2 < \nu \leq 1$.

I am grateful to Professor K. Maki for a copy of his work prior to publication.

References

1. H. Fukuyama, P.M. Platzman, and P.W. Anderson, Phys. Rev. B19, 5211 (1979)
2. G. Meissner, in: Lecture Notes in Physics 177, ed. by G. Landwehr (Springer, Heidelberg, 1983) p. 70
3. G. Meissner, Phys. Rev. B1, 1822 (1970)
4. G. Meissner, H. Namaizawa, and M. Voss, Phys. Rev. B13, 1370 (1976)
5. J.M. Kosterlitz and D.J. Thouless, J. Phys. C6, 1181 (1973)
6. K. Maki and X. Zotos, preprint.

Charge-Induced Deformation of the ^{4}He Solid-Superfluid Interface

J. Bodensohn and P. Leiderer

Fachbereich Physik, Johannes Gutenberg-Universität
D-6500 Mainz, Fed. Rep. of Germany

D. Savignac

Technische Universität München, D-8046 Garching, Fed. Rep. of Germany

The interface between solid and superfluid ^{4}He has revealed unusual dynamic properties. Characteristic of this quantum system are the very high thermal conductivity of the superfluid phase, and an extremely small heat of fusion below temperatures of about 1 K. As a result, equilibrium at the solid-superfluid ^{4}He interface is established quite rapidly, which gives rise, e.g., to melting - crystallization waves [1,2] - similar in appearance to surface waves on a free liquid surface - and to anomalous transmission of sound [3] .

When a perturbation is acting on the solid-liquid interface, the response of the system is governed by two coupled equations which describe material and thermal transport. The kinetic coefficients characterizing these processes have been calculated from the density of thermally excited phonons and rotons by BOWLEY and EDWARDS [4] . Experimentally the relevant parameter for the motion of the interface, the kinetic growth coefficient K, has been derived from the damping of melting-freezing waves by KESHISHEV et al. for $T < 0.6$ K [2] . At higher temperatures the relaxation times for crystal growth were too long to be measured with this technique. Here we describe a method which is applicable also in the range of extremely long relaxation times, so that temperatures above 1 K, where the density of thermal excitations is high, become accessible, and in addition to the hcp phase of solid He also the bcc phase can be studied. As a by-product, also the interfacial tension of the solid-liquid boundary is obtained.

Our method is based on melting a thin layer of the helium crystal by the application of a small electrostatic pressure, and then recording the relaxation of the interface back to its original equilibrium position as the pressure is removed. For this purpose, the interface is charged with negative ions, which are generated in the superfluid by a field emission tip and drawn toward the solid by an electric field of the order of 1 kV/cm. The ground-state energy of the ions in the crystal is about 18 meV higher than in the liquid [5] , the charges therefore encounter an energy barrier high enough that they are prevented from penetrating into the solid and thus accumulate at the interface.

The pressure p_{el} which the space charge layer exerts upon the solid gives rise to a difference in chemical potential per unit mass, $\Delta\mu = p_{el}/\rho_s$, between the solid and the liquid phase, whose densities are ρ_s and ρ_l, respectively. As a result, the crystal melts in the charged area to such a depth ξ_o that the electrostatic pressure is balanced by the gravitational pressure $p_g = (\rho_s - \rho_l)\, g\, \xi_o$. (Here we have assumed that the wave vector characterizing the deformation is small compared to the inverse capillary length a^{-1} [5] - 10 cm^{-1} for the hcp-superfluid interface - so that contributions from the interfacial tension are negligible; temperature differences are neglected as well.)

When the electric field is switched off, the crystal grows to its original shape with a velocity of the interface $v = m_4 K \Delta\mu(t)$, where m_4 is the ^{4}He atomic mass, and the chemical potential difference now is given by $\Delta\mu = \xi(t)\, g\, (\rho_S - \rho_l)/\rho_S$. Since $v = d\xi/dt$, the interfacial position varies in height as $\xi(t) = \xi_0 \exp(-t/\tau)$ with a relaxation time

$$\tau = \frac{\rho_S}{(\rho_S - \rho_l)\, g} \cdot (m_4 K)^{-1} \quad . \qquad (1)$$

An example for the growth of an hcp ^{4}He crystal at 1.35 K after turning off the electric field is shown in Fig. 1. As expected, the position of the interface changes exponentially. The time constant here is 11 sec, corresponding to a growth resistance $(m_4\ K)^{-1} = 9.6$ m/s. One might suspect that by repeated melting and growth the crystal quality deteriorates, which could influence the kinetic processes at the interface. We found, however, that the reproducibility of the relaxation times was not affected as long as the applied electric field did not exceed a critical value E_c where the interface becomes unstable (see below).

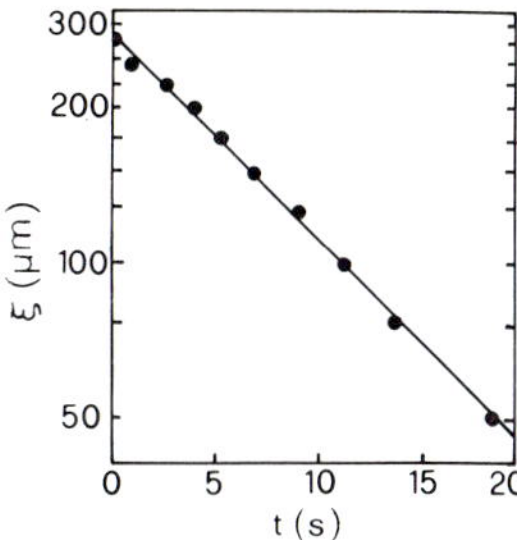

Fig. 1: Semilog plot of the relaxation of a ^{4}He hcp-superfluid interface at T = 1.35 K, after a crystal layer of ξ_0=270 μm had been melted by applying an electrostatic pressure $P_{el} = 4.5 \times 10^{-7}$ bar to the interface. The change in crystal thickness as P_{el} is turned off at t = 0 is measured with a resolution of 10 μm by means of an interferometric technique [5]

The temperature dependence of the relaxation time for crystal growth is plotted in Fig. 2. The data are compatible with a dependence

$$\tau \propto \exp(-\Delta/k_B T) \qquad (2)$$

where Δ is the roton energy, suggesting that - as supposed earlier [2] - the thermal excitations in the liquid phase dominate the crystallization rate. The crystal structure apparently is not relevant because the same relation is found for both the hcp- and the bcc-superfluid interface within the scatter observed for various crystal orientations. The data for τ , when extrapolated according to (2) to $T < 0.6$ K, are in remarkably good agreement with the results of KESHISHEV et al. [2] regarding that the extrapolation extends over 4 orders of magnitude.

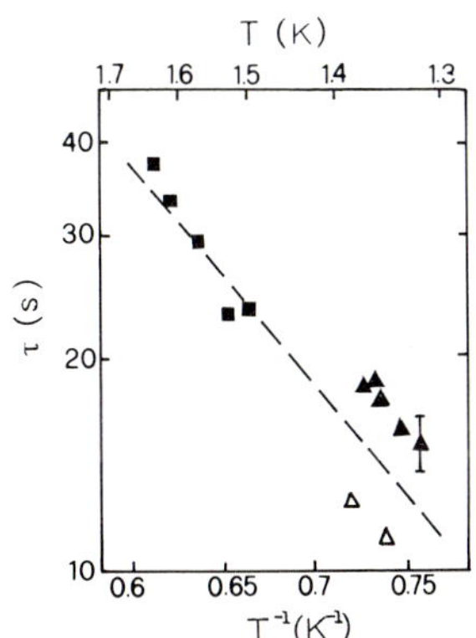

Fig. 2: Relaxation times for the growth of a bcc (squares) and two hcp (open and full triangles) crystals of ^{4}He in contact with the superfluid pase. The dashed line represents a temperature dependence $\tau \propto \exp(-\Delta/k_B T)$, with $\Delta/k_B = 7.2$ K. A typical error bar is indicated

As already indicated, the interface becomes unstable at high electric holding fields E, a phenomenon to be understood in terms of the dispersion relation for melting - freezing waves in the presence of charges [5]

$$\omega^2 = \frac{\rho_l}{\rho_s-\rho_l} gq + \frac{\rho_l}{(\rho_s-\rho_l)^2}\bar{\alpha}q^3 - \frac{i\omega q \rho_s \rho_l}{m_4 K(\rho_s-\rho_l)^2} - \frac{\rho_l}{4\pi(\rho_s-\rho_l)^2} q^2 \varepsilon E^2 \quad . \qquad (3)$$

Here again the subscripts s and l refer to the solid and the liquid phase, respectively; ε is the dielectric constant, and the effective interfacial tension in the form $\bar{\alpha} = \alpha + \partial^2\alpha / \partial\varphi^2$ includes the dependence of α on the orientation of the surface with respect to the crystal axes. Although in the temperature range $T > 1.3$ K investigated here these waves are normally overdamped (i.e., the third rhs term predominates), equation (3) predicts that for E above a critical field E_C a band of modes around a wave vector

$$q_c = [g(\rho_s - \rho_l)/\bar{\alpha}]^{1/2} \qquad (4)$$

will grow instead of decay in time and therefore lead to an instability of the interface. For electric fields near E_C this range of unstable modes is given by $|q - q_c| \leq 2q_c\, [(E - E_c)/E_c]^{1/2}$.

We have observed the growth of these modes by the same interferometric method as the stationary deformations at lower fields. As E is raised above E_C, corrugations spontaneously develop at the interface, whose preferential orientation is governed by the anisotropic surface properties of the helium crystal. For $(E-E_C)/E_C \ll 1$ the wave vector of these deformations approaches the critical wave vector q_C. Using (4) this provides a new method to obtain the interfacial tension between the liquid and the solid phase. Preliminary measurements for an hcp-superfluid interface (orientation not determined) at 1.35 K yield $\bar{\alpha} = 0.12 \pm 0.01$ erg/cm^2, in good agreement with earlier results [6] .

We appreciate valuable discussions with Prof. D.O. Edwards. This work was supported by the Deutsche Forschungsgemeinschaft.

References

1. A.F. Andreev and A.Y. Parshin, Zh.Eksp.Teor.Fiz. 75, 1511 (1978) [Sov.Phys. JETP 48, 763 (1978)]
2. K.O. Keshishev, A.Y. Parshin, and A.B. Babkin, Zh.Eksp.Teor.Fiz. 80, 716 (1981) [Sov.Phys. JETP 53, 362 (1981)]
3. B. Castaing, S. Balibar, and C. Laroche, J.Physique 41, 897 (1980)
4. R.M. Bowley and D.O. Edwards, J. Physique 44, 723 (1983)
5. D. Savignac and P. Leiderer, Phys.Rev.Lett. 49, 1869 (1982)
6. J. Landau, S.G. Lipson, L.M. Määtänen, L.S. Balfour, and D.O. Edwards, Phys.Rev.Lett. 45, 31 (1980)

Pinning of Dislocations in Helium by Large Ultrasonic Stresses

J.R. Beamish[1] and J.P. Franck

Department of Physics, University of Alberta
Edmonton, Alberta T6G 2J1, Canada

Measurements (1,2,3) on hcp ^{4}He have shown that below about half the melting temperature, the ultrasonic velocity (v) and attenuation (α) do not have the simple adiabatic behavior expected for dielectrics. The anomalous behavior, which for ^{4}He has been attributed to dislocations, was recently observed (4,5) in hcp and bcc ^{3}He. The Granato-Lucke model (6) of dislocation-sound wave interactions explained both the ^{3}He and ^{4}He results and permitted the dislocation density, loop length and damping to be determined.

A new feature observed in the experiments was the presence of hysteresis in the 3 MHz velocity and attenuation when the crystals were thermally cycled in the presence of ultrasonic pulses. The hysteresis was due to the pulses themselves and only appeared when the crystals were cooled below about 300 mK. The crystals recovered to their original state when they were warmed above about one-half of their melting temperature. Further study showed that the changes in v and α depended only on the total number of pulses applied and occurred faster for larger pulse amplitudes and for lower temperatures.

By using low-amplitude measuring pulses which did not cause any hysteresis and applying varying numbers of large pulses at low temperatures, the anomaly could be studied at different stages. Figure 1 shows the anomalous velocity (the high-temperature adiabatic velocity changes have been subtracted) and attenuation in a typical hcp ^{3}He crystal (18.6 cm^3/mole). When first cooled, this crystal showed the beginnings of a positive velocity anomaly but the attenuation increased so rapidly that measurements were impossible below 0.8K. By applying large 3 MHz pulses (stress amplitude 4000 dyne/cm^2) at 160 mK, the attenuation could be greatly reduced. The data labelled t=0.50 (upright triangles) in Fig. 1 were taken after just enough large pulses were applied to make measurements possible. After warming to near melting, then cooling again, the original large attenuation returned. The data shown as circles, squares and inverted triangles show the results of successively larger numbers of large 3 MHz pulses applied at 160 mK. The effect of the large pulses was thus to first change a positive velocity anomaly to a negative one and then to reduce the magnitude of the negative anomaly while decreasing the attenuation monotonically.

This behavior, typical of anomalies in all ^{3}He crystals, suggests a shortening of the dislocation loops since, in the Granato-Lucke model, loops longer than the resonant length l_c make a positive contribution to the velocity while loops shorter than l_c make a negative one (3,6). We attempted to fit the four sets of velocity data to the Granato-Lucke model by varying only the average loop length L. The results are shown as solid curves in Fig. 1a

[1]Present address: Physics Department, Brown University, Providence, R.I. 02912 U.S.A.

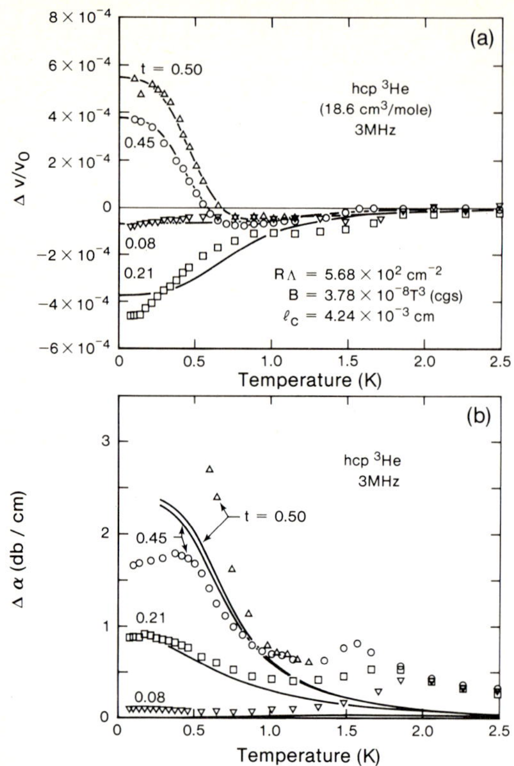

Figure 1:
a) Velocity anomaly at 3 MHz in an hcp ^{3}He crystal. The four sets of data were obtained after applying successively more and larger 3 MHz pulses near 160 mK. Solid lines represent fits to the Granto-Lucke model using parameters shown
b) Attenuation corresponding to the velocities of Fig. 1a. Solid lines are the predictions of the Granato-Lucke model using the dislocation parameters derived from the velocity data

where the dislocation density $R\Lambda$ and damping B were determined from the first set of data (upright triangles). The distribution of loop lengths was assumed to be of the form $N(l)=(\Lambda/L^2)\exp(-l/L)$ and only L was varied to fit the other three anomalies. The average loop length L is characterized in Fig. 1 by the parameter $t=L/l_c$.

The fits to the anomalies shown in Fig. 1a indicate that the large 3 MHz pulses reduce the average loop length (t decreases from 0.50 to 0.08). Using the dislocation parameters derived from the velocity fits, the attenuation predicted by the Granato-Lucke model can be calculated. Figure 1b compares the calculated attenuation to the measured values. The agreement between the two strongly supports the loop-shortening interpretation. Further confirmation comes from the data at higher frequencies. The largest pulses applied at 9 or 21 MHz (stress amplitudes 400 dyne/cm^2) did not affect the anomaly or cause any hysteresis but large 3 MHz pulses changed the 9 and 21 MHz anomalies in a manner also indicating a shortening of the dislocations.

The mechanism by which the 3 MHz pulses shorten the loops is not clear. We added ^{4}He impurities to the ^{3}He (originally 1.35 ppm ^{4}He) to see whether they were involved in the pinning. At low concentrations (< 430 ppm), there was little effect. With 0.53% ^{4}He, both the anomaly and the hysteresis were eliminated. Increasing the pulse amplitude in these crystals increased the attenuation in agreement with the Granato-Lucke (6) theory of breakaway from impurity pinning centers. It thus appears that the ^{4}He impurities are not directly involved in the pinning by large stresses. The most plausible mechanisms involve intersections between adjacent dislocations under the influence of the 3 MHz pulses. In a regular dislocation network, the spacing D between adjacent loops is comparable to the average loop length L. The maximum displacement of a resonant loop ($l=l_c$) can be written as (6)

$$\zeta = 4a\ \sigma/\pi\omega B \qquad (1)$$

where a is the Burger's vector, σ is the stress amplitude, ω is the sound frequency and B is the dislocation damping. For the crystal of Fig. 1, $a = 3 \times 10^{-8}$ cm and $B = 3.78 \times 10^{-8}\ T^3$ (cgs), so for the large 3 MHz pulses (σ = 4000 dyne/cm^2) the maximum displacement at 160 mK is $\zeta = 5 \times 10^{-3}$ cm which is larger than the average loop length L. Thus the large 3 MHz pulses can cause the loops to intersect and be pinned. From (1), the displacement is largest (and pinning thus possible) at large stresses, low frequencies and low damping (i.e., low temperatures), in agreement with our observations. A lower limit on the damping is given (7) by radiation damping $B_R = (1/8)\rho a^2\omega$ where ρ is the density (.22 g/cm^3). For our maximum stresses at 9 MHz (400 dyne/cm^2), the radiation damping limits the maximum displacements to about 2×10^{-4} cm, an order of magnitude less than the average loop length L, so intersection of adjacent loops is unlikely.

When two loops intersect, they may form a jog or they may interact in a more complicated manner. In a previous paper (4), we found that the dislocation network recovered to its original form around half the melting temperature with an activation energy equal to that of mobile vacancies in helium. Although this does not distinguish between the possible pinning mechanisms, it is consistent with, e.g., jog formation, since mobile vacancies are required to anneal out jogs.

Acknowledgements

We would like to thank Dr. I. Iwasa for valuable discussions. This work was supported in part by grants from the Natural Sciences and Engineering Research Council of Canada.

1. R. Wanner, I. Iwasa and S. Wales: Solid State Commun. 18, 853 (1976)
2. I.D. Calder and J.P. Franck: Phys. Rev. B15, 5262 (1977)
3. I. Iwasa, K. Araki and H. Suzuki: J. Phys. Soc. Jpn. 46, 1119 (1979)
4. J.R. Beamish and J.P. Franck: Phys. Rev. Lett. 47, 1736 (1981)
5. J.R. Beamish and J.P. Franck: Phys. Rev. B26, 6104 (1982)
6. A.V. Granato and K. Lucke: J. Appl. Phys. 27, 583 (1956)
7. J.A. Garber and A. V. Granato: J. Phys. Chem. Solids 31, 1863 (1970)

Discussions

CRYSTALLISATION WAVES IN HELIUM page 226
A.Ya. Parshin

P. Leiderer: We have recently obtained data on the growth resistance in the temperature range from 1.3 to 1.7 K. The results closely follow the law $(mK^{-1})\alpha\exp(-\Delta/kT)$ as suggested by your experiment, both for the hcp and bcc phase.

W. Eisenmenger: How can we understand this miracle of fast moving liquid-solid interfaces in terms of energy balance? Are there phonons emitted or absorbed in crystallisation and condensation respectively?

A.Ya. Parshin: At zero temperature there is no energy transfer between the phases because the chemical potentials of the phases are equal in equilibrium. At finite temperatures the problem is very complicated. It has been discussed by Andreev and Kniznic (15) and by Bowley and Edwards (16).

W. Eisenmenger: Have there been observations of ^{3}He crystallisation waves?

A.Ya. Parshin: We have studied ^{3}He crystallisation down to 0.3 K and did not observe any surface oscillations. According to the current theory oscillations of that type can exist only at temperatures below 1 mK, where liquid ^{3}He is becoming superfluid and ^{3}He crystals become magnetically ordered.

TRANSMISSION OF SOUND AT THE SOLID-LIQUID INTERFACE OF ^{4}He page 241
B. Castaing, L. Puech, G. Bonfait

K. Weiss: When looking for differences between ^{3}He and ^{4}He did you think about collective Fermi-liquid effects?

B. Castaing: The temperature at which we have made the experiment was probably too high to expect well-behaved Fermi-liquid effects. On the other hand,zero sound for example looks very much like ordinary sound in ^{4}He.

W. Eisenmenger: Is there a simple model for a limit of the crystal growing velocity, as perhaps that atoms can not arrange in a row faster than with sound velocity?

B. Casting: For ^{3}He there is probably a small critical velocity, due to the poor coupling between spins and phonon degrees of freedom, above which we cannot speak of quasiequilibrium. For both ^{4}He and ^{3}He the final limit would be, as you suggest, controlled by the exchange frequency between liquid and solid. It would be of order 1m/s as indicated by the steps velocity.

PROPAGATION OF HIGH FREQUENCY PHONONS IN LIQUID He II page 249
T. Haavasoja, V. Narayanamurti, M.A. Chin

H. Kinder: Was this a DC experiment? I am interested in the difference in the times of flight of rotons vs phonons. Can you comment on this?

V. Narayanamurti: We did both DC and pulsed experiments. Phonons near the anomalous decay threshold travel at 220 m/s at SVP. The rotons travel slower and most of our signal (95%) consists of phonons. Most of the rotons come down in energy to near the minimum and travel slowly as a driven roton second sound. We have studied scattering of such phonons with rotons.

K.F. Renk: I would like to ask a question concerning the threshold where linear dispersion sets in. Do you expect or do you see any anomalies? If you have exactly linear dispersion you may see an enhancement of the decay because a decaying phonon propagates parallel over a long distance with the phonons in which it decays.

V. Narayanamurti: In reality you always have either "normal" or anomalous dispersion. If you put a finite linewidth for almost "linear" dispersion you get attenuation as pointed out by Ter Haar. To get the correct attenuation you need anomalous dispersion as pointed out by Maris and Maszey. In our experiment we go well into the normal dispersion region for long mean free paths. We could try to enhance the scattering in a double-pulse type of experiment. We have done such experiments with phonons and rotons moving colinearly.

M. Lax: Show the E(k) vs k slide again and repeat your definitions E_c, E_c' and E_c''.

V. Narayanamurti: E_c turns out to be very close to the threshold for a phonon decay as predicted by Pitaecskii and Levinson and the known He dispersion curve. At SVP it is ~ 9.5 K. E_c' is close to the roton minimum as a function of pressure. E_c'' is the new threshold which depends markedly on pressure and we ascribe to phonon-roton scattering enhanced by the bending over of the dispersion curve.

ULTRASONIC ATTENUATION IN KCl:OH$^-$ WITH HIGH OH CONCENTRATION page 252
M. Saint-Paul, R. Nava, J. Joffrin

S. Hunklinger: Are the acoustic measurements and the dielectric measurements published previously consistent?

M. Saint-Paul: It is correct. An inspection of the elastic and dielectric anomalies in KCl:OH reveals apparent differences. The temperature T_F at the maximum of the dielectric constant scales with the OH concentration, see reference Känzig et al., while the temperature T_M at the maximum of the acoustic measurements does not depend on OH content.

W.K. Arnold: Your data are fitted with an Arrhenius relaxation process. Is there no contribution by other relaxation mechanism such as one-phonon or multi-phonon processes (Raman , Orbach processes)?

M. Saint-Paul: It is clear that the experimental results show clearly two regime behaviours: One characterized by a thermally activated activation via the known 32 cm^{-1} excited state independent of OH concentration,and another which appears below 2 K and where the interacting effects among OH are present

A complete answer about the nature of the latter relaxation mechanism will be certainly given by the experiments which we are planning to perform below 1 K.

R. Vacher: In relation to Dr. Arnold´s question, I would like to stress that in our Brillouin scattering, we observe a peek around 20 K that we interpret as a thermal process, an increase of attenuation with decreasing temperature, attributed to resonant absorption by tunneling systems, and nothing in between. Therefore, phonon-assisted tunneling relaxation seems to be absent in this material. The explanation could be that the distribution of tunneling energies does not extend to energies high enough to allow relaxation at temperatures around 5 K.

EFFECTS OF ANHARMONICITIES AND BROKEN TIME-REVERSAL INVARIANCE ON STATIC AND DYNAMIC PROPERTIES OF TWO-DIMENSIONAL ELECTRON SOLIDS page 263
G. Meissner

F. Williams: In the extreme quantum limit, what is the behaviour of melting for the electron lattice near the origin?

G. Meissner: Drastical differences between the behaviour of melting in the classical limit and in the extreme quantum limit are expected at the intersection of the corresponding phase separation curves rather than at the origin in the ν-T plane.

CHARGE-INDUCED DEFORMATION OF THE ^{4}He SOLID-SUPERFLUID INTERFACE 266
J. Bodensohn, P. Leiderer, D. Savignac

B. Castaing: Have you measured the anisotropy of the surface tension?

P. Leiderer: So far we have extracted from our data only the absolute value of the interfacial tension which appears to agree well with previous experiments. However, a more careful analysis of the interference patterns will yield the anisotropy of the interfacial tension.

A. Ikushima: 1.) You mentioned that the growth rate is governed by rotons. What is the mechanism, then? 2.) How accurate is the value, 7.6 Kelvin, in the temperature dependence of the growth rate? This question is because the value seems to be a little different from neutron data.

P. Leiderer: 2.) Our measurements, which extend over only a small temperature range, are also compatible with neutron data, which gave for the roton energy $\Delta E = 7.2$ K at the melting curve. The quoted value 7.6 was only used for the extrapolation to the data of Keshishev et al. and might be due , e.g., to a slight temperature dependence of ΔE. 1.) According to a recent theory of Boley and Edwards (ref. 4), the growth resistance is determined by the number of thermal excitations, at least in the ballistic regime, where the mean free path of the phonons and rotons is large. However, in the hydrodynamic regime (small mfp), as in our experiment, the theory predicts a vanishing growth resistance and hence is in contradiction to the measurements.

PINNING OF DISLOCATIONS IN HELIUM BY LARGE ULTRASONIC STRESSES page 269
J.R. Beamish , J.P. Franck

F. Williams: Might the frequency used to create the pinning points favorize a certain length?

J.P. Franck: We have been able to generate sufficient stress at the fundamental of the transducers (3 MHz). We believe that the same effect would also be obtained at other frequencies, provided large enough dislocation amplitudes can be generated.

A.J. Ikushima: How large is the dislocation displacement at your largest stress? This question is because you might have to worry about the Peierls stress, which would then make the apparent damping smaller.

J.P. Franck: The largest dislocation amplitude for "large" pulses (σ = 4000 dyne/cm^2) was about 5 x 10^{-3} cm of the order of the average dislocation length. This may be the cause of the observed pinning. Measurements, however, are also made at lower stresses (1000 dyne/cm^2 and below), where the dislocation amplitude is small compared to the average dislocation length.

I. Iwasa: I understand that you have determined the activation energy of vacancy from your experiment. There is still a controversy on the nature of the vacancy in solid helium,if it is a localized one or wavelike. I would like to know your view.

J.P. Franck: Our activation energies agree with those measured by the X-ray density method (R.O. Simmons and collaborators) in the bcc phase, but not in the hcp phase of ^{3}He. It is possible that vacancies move by tunneling in bcc ^{3}He, but not in hcp ^{3}He.

B. Castaing: What is the ^{4}He concentration in your samples and do you have a physical reason for taking an exponential form for the length distribution of dislocations?

J.P. Franck: The ^{4}He concentration, measured after the experiments were performed, was 1.35 ppm ^{4}He. experiments were also conducted with 47, 430 and 5300 ppm ^{4}He. The number of pulses required for pinning is smaller for a 430 ppm crystal, compared to the "pure" crystal. At 5300 ppm ^{4}He the dislocation lines are completely pinned and no anomalies are seen.

Part VI

Cooperative Phenomena

Ultrasonic Velocity and Attenuation Near the Cooperative Jahn-Teller Dilation in Cerium Ethyl Sulphate

J.T. Graham and J.H. Page

Department of Physics, Queen's University
Kingston, Ontario K7L 3N6, Canada

The anomalous enhancement of the Schottky specific heat peak [1] and the temperature dependence of the electric susceptibility [2], [3] in cerium ethyl sulphate (CeES) have been interpreted using a model [4] in which the separation between the two lowest Ce^{3+} Kramers doublets varies with temperature due to the coupling between these electronic states and the lattice phonons. Since the effect occurs in the absence of a phase transition, these experiments indicate that CeES is an unusual example of a cooperative Jahn-Teller system in which the dominant coupling is to the uniform A_{1g} strain modes, e_{zz} and/or $\frac{1}{2}(e_{xx}+e_{yy})$, which do not lower the crystal symmetry, but change the c/a ratio of the hexagonal unit cell. To investigate this effect, we have measured the temperature dependence of the velocity of longitudinal ultrasonic waves propagating along the c axis, and find that the corresponding elastic constant C_{33} softens at low temperatures (Fig. 1), providing direct evidence of Jahn-Teller coupling to this A_{1g} lattice mode in CeES.

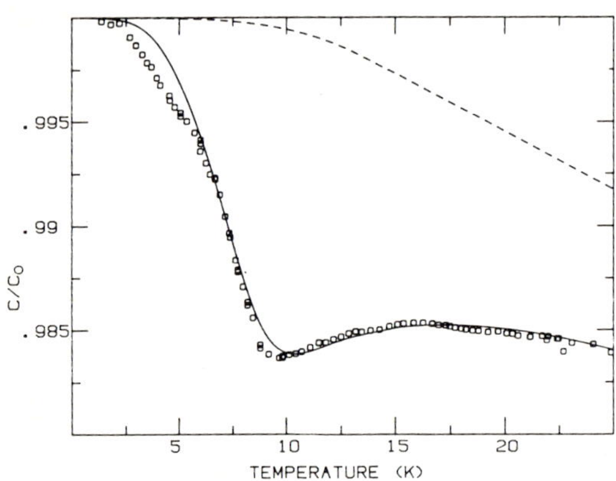

Fig. 1 Temperature dependence of the elastic constant C_{33} at 32 MHz. A small correction has been made for the measured change in sample size due to the temperature-dependent c/a ratio

We have also measured the ultrasonic attenuation for this mode (Fig. 2); strong relaxation peaks are observed, giving new data on spin-lattice relaxation in this material. Expressions for the temperature dependence of the elastic constants and ultrasonic attenuation in Jahn-Teller systems have been considered by several authors (e.g. see [5] [6]). A similar derivation for CeES gives

$$\frac{C_{33}}{C_{33}^{o}} = \frac{v_p^2}{v_o^2} = 1 - \frac{1}{1+\omega^2\tau^2}\,\frac{\mu_p g_\gamma}{1-\lambda_p g_\gamma} \tag{1}$$

$$\alpha = 4.34\,\frac{\omega^2\tau}{1+\omega^2\tau^2}\,\frac{v_o^2}{v_p^3}\,\frac{\mu_p g_\gamma}{1-\lambda_p g_\gamma}\;. \tag{2}$$

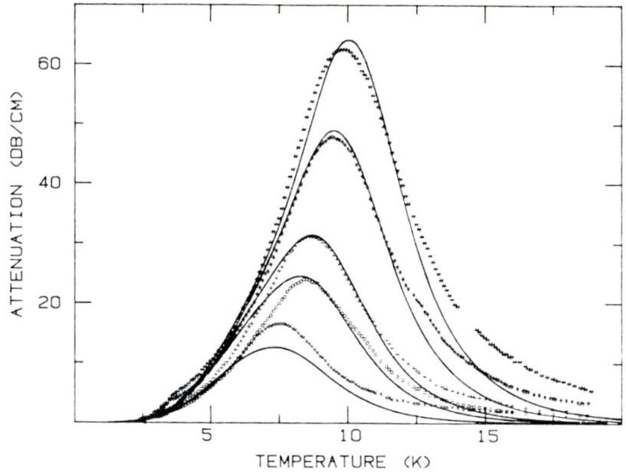

Fig. 2 Temperature dependence of the attenuation at 32(x), 58(□), 75(+), 120(*), and 160 MHz(#)

Here C_{33}^{o} and v_o are the unperturbed values of elastic constant C_{33} and ultrasonic velocity v_o, and μ_p and λ_p are the Jahn-Teller coupling constants for the phonon mode $\beta(=e_{zz})$. (Since there are two strain modes of A_{1g} symmetry, μ_p may be less than the total strain coupling μ_T, although $\lambda_p=\lambda_T$.) $\tau = \tau_s/(1-\lambda_p g_\gamma)$, where τ_s is the spin relaxation time between the two Kramers doublets and we have allowed for the possibility that τ^{-1}may be comparable with the ultrasonic frequency ω. g_γ is the static spin susceptibility; here it is evaluated in a thermodynamic regime γ=s', which is adiabatic in the sense that the entropy of the spin plus lattice system is constant [6]. In this case, $g_{s'}=(g_T C_L+g_s C_H)/(C_L+C_H)$. $g_T=(1/kT)(1-\tanh^2(\mathbf{w}/kT))$ is the isothermal susceptibility, $g_s=0$ is the "essentially isolated" susceptibility, 2w is the separation of the Kramers doublets, C_H is the spin specific heat at constant molecular field, and C_L is the appropriate lattice specific heat. The existence of peaks in the attenuation (Fig. 2) at frequencies between 30 and 160 MHz suggest that the magnitudes of relaxation times involved are consistent with spin-lattice relaxation. In this case we expect the relaxation rate τ_s^{-1} to be $A(w/k)^3 \coth(w/kT) + BT^7$, where the first term describes the direct process between the two Kramers doublets and the second term accounts for Raman processes, which dominate at high temperatures. With this form for the relaxation rate, we have fitted (2) to our attenuation data, giving the solid curves in Fig. 2. The overall agreement can be seen to be good, except at high temperatures, where deviations from the T^7 behaviour of the Raman processes are expected [7]. The coefficient B for the Raman term was determined from the temperatures at which the peaks in the attenuation occur, giving B=6.5±10 rad $s^{-1}K^{-7}$, in good agreement with the data of KRIGIN et al.[7]. With this value of B, the magnitude of the direct process was found by optimizing the fit below about 7K. The best fit gives A=2.5±.5x10^6 rad $s^{-1}K^{-3}$, in good agreement with a point charge calculation [8] but about an order of magnitude higher than measured values for dilute Ce^{3+} in other ethyl sulphates [9]. However, we do not observe the enormous enhancement of the relaxation rate measured in concentrated CeES by CLOVER [10]. In the fit the scaling parameter μ_p was found to be 0.12 K. w and λ_T were obtained from electric susceptibility data [2], [3], using the approximation $\mu_T=-3\lambda_T$ to obtain λ_T from the measured value $\lambda_T+\mu_T$=1.5K. In calculating the susceptibility $g_{s'}$, some uncertainty may arise in estimating the appropriate lattice specific heat C_L. It is interesting that our data can be well explained if the measured T^3 lattice contribution [1] is used, whereas if C_L is estimated using only those phonons within a resonant bandwidth of the Ce spins, the fit is much worse.

The elastic constant data (Fig. 1) was fitted using (1) with the same values of $g_{s'}$, λ_p, and τ as in the attenuation analysis, but with a somewhat larger value of μ_p=0.2 K. When the anharmonic background contribution (dashed curve) is also included, a good description of the experimental data is obtained as shown by the solid curve in Fig. 1. We conclude that although the effect is quite small, the softening of the elastic constant C_{33} provides the first direct evidence that the anomalous low-temperature properties of CeES are caused by a Jahn-Teller "distortion" in a uniform A_{1g} mode. It would be interesting to make velocity and attenuation measurements for the other A_{1g} mode $\frac{1}{2}(e_{xx}+e_{yy})$ for which the Jahn-Teller coupling may be larger.

Acknowledgements

Informative discussions with D.R. Taylor, and research assistance from NSERC are greatly appreciated.

References

1 Horst Meyer and P.L. Smith: J. Phys. Chem. Solids 9, 285 (1959)
2 D.R. Taylor, D.B. McColl, J.P. Harrison, R.J. Elliott, and L.L. Goncalves: J. Phys. C 10, L407 (1977)
3 J.P. Harrison and R.J. Stubbs: J. Low Temp. Phys. 51, 679 (1983)
4 J.R. Fletcher and F.W. Sheard: Solid State Commun. 9, 1403 (1971)
5 J.R. Sandercock, S.B. Palmer, R.J. Elliott, W. Hayes, S.R.P. Smith, and A.P. Young: J. Phys. C 5, 3126 (1972)
6 J.H. Page and S.R.P. Smith: J. Phys. C 16, 309 (1983)
7 I.M. Krigin, S.N. Lukin, G.N. Neilo, and A.D. Prokhorov: Phys. Stat. Sol. (b) 104 K21 (1981)
8 P.L. Scott and C.D. Jeffries: Phys. Rev. 127, 32 (1962)
9 G.H. Larson: Phys. Rev. 150 264 (1966)
10 R.B. Clover: Physica 68, 519 (1973)

Ultrasonic Attenuation at Megahertz Frequencies in the Cooperative Jahn-Teller System $TmVO_4$

J.H. Page

Physics Department, Queen's University, Kingston, Ontario K7L 3N6, Canada

S.R.P. Smith

Department of Physics, University of Essex, Colchester, Essex CO4 3SQ, England

Because of the simplicity of its energy levels and the dominance of long-range interactions, $TmVO_4$ may be considered an archetype amongst a series of rare earth compounds which undergo cooperative Jahn-Teller phase transitions at low temperatures (for a review, see GEHRING and GEHRING [1]). As such, the static properties of this structural phase transition have been extensively studied, although the dynamic behaviour, which gives rise to the ultrasonic attenuation, has received relatively little attention. In a recent paper on microwave ultrasonic attenuation in $TmVO_4$ [2], we showed that the attenuation is caused by the relaxation of the Tm^{3+} electronic excitations which are coupled by the Jahn-Teller interaction to the acoustic phonons. Here we report an extension of these attenuation measurements to lower frequencies (60 to 300 MHz) where, because of reduced attenuation, a more complete investigation is possible. The data are interpreted using the linear response function treatment of PAGE and SMITH [2] to obtain information on the damping of the spin excitations over a range of temperatures and magnetic fields in the undistorted phase ($T>T_D$ = 2.15 K).

The ultrasonic attenuation in $TmVO_4$ was measured for longitudinal waves propagating in both the [110] and [100] directions, enabling the properties of both the soft (B_1) and 'competing' (B_2) modes to be investigated [2]. Figure 1 shows the increase in the attenuation at 311 MHz as the temperature is lowered from 30 to 4 K. For the [110] data, a small relaxation peak is superposed near 20 K on the main contribution; however the peak was not observed with other samples, and is probably due to spin lattice relaxation of a magnetic impurity which is of no interest here. A magnetic field parallel to the tetragonal axis also affects the softening of the B_1 and B_2 modes (e.g.,see [3]), and causes the attenuation to decrease rapidly with field as shown in Fig. 2. Above 1 Tesla, the change in attenuation becomes vanishingly small at these frequencies, enabling the background value to be

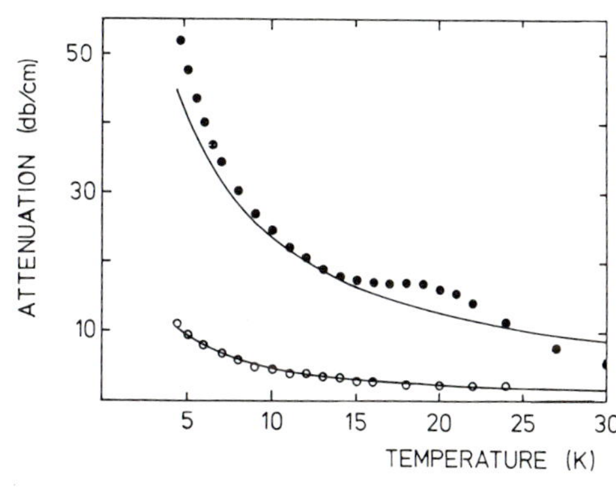

Fig. 1 Temperature dependence of attenuation for longitudinal ultrasonic waves at 311 MHz. Propagation directions: [110], ●; [100], ○

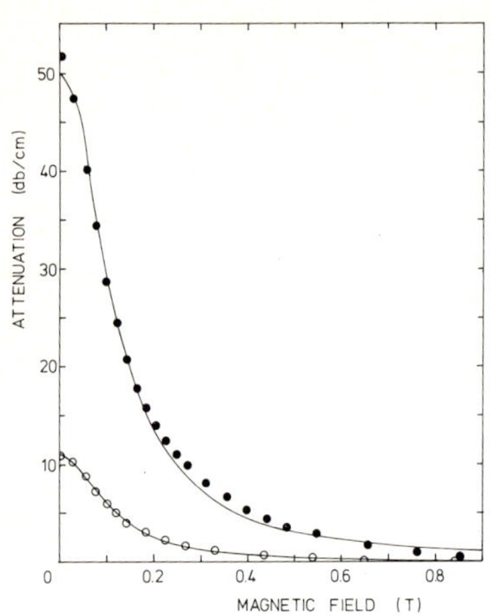

Fig. 2 Field dependence of ultrasonic attenuation at 311 MHz. T=4.4 K. ([110] ●; [100] ○)

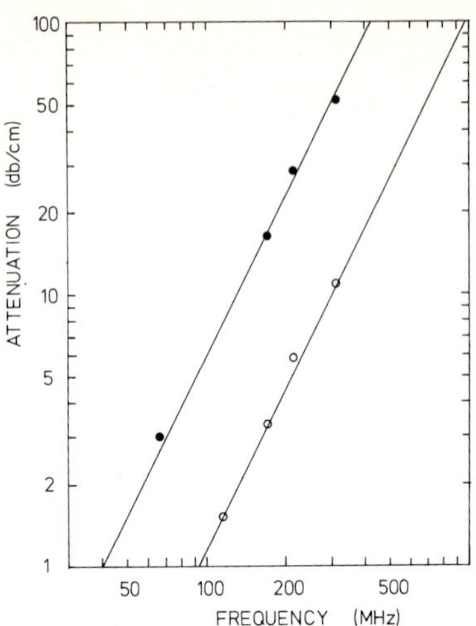

Fig. 3 Frequency dependence of attenuation of longitudinal waves at 4.4 K and zero magnetic field. ([110] ●; [100] ○)

readily established. The observed quadratic frequency dependence of the attenuation is shown in Fig. 3.

Following PAGE and SMITH [2], the ultrasonic attenuation is found to depend on two phenomenological damping parameters Γ_1 and Γ_2 which describe the relaxation of the electronic spin excitations of site symmetry B_1 and B_2 respectively. For [110] longitudinal ultrasonic waves, the attenuation (in dB per unit length) at ultrasonic frequencies such that $\omega^2 \ll \Gamma_1\Gamma_2$ is given by

$$\alpha = 4.43 \left[\frac{\Gamma 2}{\Gamma_1\Gamma_2 + 4H_M^2}\right] \frac{\omega^2}{v_0} \frac{\mu_p g_s}{(1-\lambda_p g_s)^{\frac{1}{2}}(1-(\lambda_p+\mu_p)g_s)^{3/2}} \tag{1}$$

where Γ_1, Γ_2, $H_M = \frac{1}{2}g\mu_B B/\hbar$, μ_p, λ_p and g_s^{-1} are expressed in rad s^{-1}. v_0 is the longitudinal ultrasonic velocity unmodified by the Jahn-Teller interactions and $g_s = (1/H_M)\tanh(H_M/kT)$ is the static spin susceptibility. λ_p and μ_p are the Jahn-Teller coupling constants for [110] longitudinal phonons, related to the soft mode coupling parameters λ and μ by $\lambda_p = \lambda_1$ and $\mu_p = \mu_1 c_{66}/[\frac{1}{2}(c_{11}+c_{12}) + c_{66}]$. When the magnetic field B is zero, the term in square brackets becomes Γ_1^{-1} (cf. [4]) and the dependence on Γ_2 is eliminated. The attenuation of [100] longitudinal waves is described by an expression similar to (1) but with Γ_1 and Γ_2 interchanged, and with $\lambda_p = \lambda_2$ and $\mu_p = \mu_2 \frac{1}{2}(c_{11}-c_{12})/c_{11}$, where λ_2 and μ_2 refer to the B_2 mode. Note that the coupling constants for the B_1 and B_2 modes are known from measurements of the elastic constants [3], [5] so that the only unknown parameters in (1) are the damping parameters Γ_1 and Γ_2.

With the basevalue determined experimentally by the high-field limit of the attenuation at 4 K, the solid curves in Fig. 1 show the comparison of theory and experiment in zero field assuming constant values of the damping parameters, $\Gamma_1/2\pi$ = 5.3 GHz and Γ_2 = 12 GHz for the [110] and [100] data respectively. It is clear that the temperature dependence of the attenuation is largely determined by the T^{-1} variation of the static susceptibility g_s; however it is also apparent, at least for the [110] results, that the curve departs systematically from the data, indicating that Γ_1 may decrease with temperature in this range. Between 4 and 12 K the observed variation in Γ_1 is roughly linear and may be as much as 20%, although there is a large uncertainty in this value since the analysis is quite sensitive to background errors. For the attenuation measured as a function of magnetic field, the curves in Fig. 2 represent the best fit of the theory with both Γ_1 and Γ_2 as field-independent adjustable parameters. Although a good fit is obtained, the values of the parameters are not in satisfactory agreement with the values obtained in zero field at the same temperature, there being a discrepancy of a factor of 5 in the product $\Gamma_1\Gamma_2$. This anomalous field dependence, along with the behaviour of the spin damping parameters below 4 K, will be the subject of further study.

Acknowledgements

Research support by NSERC is gratefully acknowledged.

References

1 G.A. Gehring and K. Gehring: Rep. Prog. Phys. 38, 1 (1975)
2 J.H. Page and S.R.P. Smith: J. Phys. C: Solid State Phys., 16, 309 (1983)
3 J.H. Page and H.M. Rosenberg: J. Phys. C: Solid State Phys., 10, 1817 (1977)
4 K.M. Leung, D.L. Huber and B. Lüthi: J. Appl. Phys., 50, 1831 (1979)
5 R.L. Melcher, E. Pytte and B.A. Scott: Phys. Rev. Lett. 31, 307 (1973)

Thermoelectric Power of $TiSe_{2-x}S_x$ Mixed Crystals

A.A. Lakhani and S. Jandl

J.P. Jay-Gerin

Université de Sherbrooke, C.H.U., Sherbrooke Québec J1H 5N4, Canada

C. Ayache

Centre d'Etudes Nucléaires de Grenoble, SBT/LCP, BP 85X
F-38041 Grenoble Cedex, France

We present thermoelectric power (TEP) measurements as a function of temperature in the range 7-450 K of $TiSe_{2-x}S_x$ mixed crystals for x = .1, .5, .75, 1 and 1.25. The semimetallic 1T - $TiSe_2$ is known to undergo an antiferrodistortive phase transition below 200 K which is accompanied by some peculiar features in transport properties. Especially, the TEP exhibits a pronounced negative dip around 150 K. Similarly to that $TiSe_2$ behaviour, mixed samples presenting the superlattice state are also observed to possess the dip. In contrast, the TEP of the non-transforming crystals varies monotonously with temperature. We interpret the occurrence of the dip in transforming crystals on the basis of a phonon-drag effect associated with the transition. A previously proposed antiferroelectric mechanism is thus concluded to be irrelevant to the $TiSe_2$ problem as it would imply strong intrinsic anharmonic effects and reduction of dragging effect. Instead, our analysis favours electron-phonon interpocket coupling as the driving mechanism. However, opposite to the transitions occurring in metallic dichalcogenides, the "nesting conditions" are not thought to be essential in the present case as the dip position does not shift upon mixing. Finally, an evaluation is provided for the non-renormalized energy of the softening phonon L_1^- (1).

An extended presentation of this work will appear in The Physical Review.

Ultrasonic Study of Melting of Crystalline Solids

Y. Hiki and J. Tamura

Tokyo Institute of Technology, Department of Applied Physics
Oh-okayama, Meguro-ku, Tokyo 152, Japan

The present authors are investigating the phenomenon of melting transitions by measuring the attenuation of ultrasound in melting crystalline solids. Ice of normal water was chosen as the specimen material for the convenience of the experiment. A single crystal of ice is grown by the Bridgman method in a fused-quartz sample cell. Pulse ultrasound waves are emitted into the specimen by exciting a quartz transducer bonded on the outer wall of the cell. The attenuation of sound is measured at frequencies of 15 - 85 MHz by a pulse reflection method using the Matec apparatus. The determined attenuation values were shown to be reliable enough and to represent the true vibrational energy loss in the specimen /1/. The measurements are made at temperatures from - 8°C up to the melting point.

It was concluded that the origin of the ultrasonic attenuation in ice crystals in the present ranges of sound frequency and temperature was due to the vibrating dislocations /2-4/. The bases were as follows:
(i) The internal friction Δ was determined as a function of sound frequency f. A broad peak was always observed and the position of the peak was different from those of known relaxation peaks (see Fig. 1). The shape of the peak is well fitted to the theoretical expression of the overdamped resonance of pinned dislocations /5/.

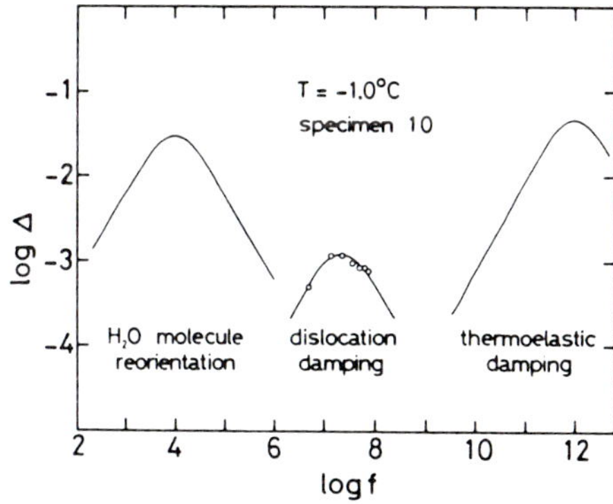

Fig. 1. Frequency spectrum of internal friction. Two calculated relaxation peaks and an experimental dislocation damping peak are shown. The dislocation peak is fitted to the overdamped resonance:

$$\Delta = \frac{\Delta_0 \omega\tau}{1 + \omega^2\tau^2}$$

Here $\Delta_0 \propto \Lambda L^2$, $\tau \propto BL^2$, where Λ, L and B are the dislocation density, the pinning length, and the damping constant

(ii) By increasing the temperature, the height of the resonance peak increased and the position of the peak shifted to the lower frequency. This was considered to be mainly due to the decrease of pinning length. The binding energy for the pinning was thus determined. The value was reasonable when an elastic interaction between dislocations and Na^+ ions, which were the main impurities, was considered.

(iii) The frequency of overdamped resonance was higher in crystals doped with impurities, being reasonable because the doping decreased the pinning length. The effect was most pronounced for Na^+-ion doping, which was understandable since the ionic radii of Na^+ and O^{--} ions were largely different.
(iv) The density of dislocations evaluated from the overdamped resonance model was $10^6 - 10^7\ cm^{-2}$. The values were acceptable when the method of preparing specimen crystals was considered.

The temperature dependence of the density of dislocations will now especially be noted. As shown in Fig. 2, the dislocation density Λ increases very rapidly near the melting temperature. The value decreases again to the lower level when the temperature is lowered. It is also found that the temperature dependence is very sensitive to the impurity doping, as shown in Fig. 3. Production of dislocations by thermal stresses in the crystal is not likely to be the origin of these phenomena, because thermal expansion of ice monotonously changes with temperature and is not sensitive to the impurity content. We are continuing to study precisely the effect of doping by Na^+ ions. The temperature dependence of dislocation density is more remarkable in highly doped crystals. It seems that, in the limit of the highest doping, the dislocation density is proportional to $\{1 - (T/T_{melt})\}^{-1}$, where T and T_{melt} are in absolute temperature. Our present interpretation for these result is as the following:
(i) Dislocations are produced in a crystal near the melting temperature due to a thermal fluctuation.
(ii) The produced dislocations can easily move and be annihilated with each other or go out from the crystal surfaces in the case of high purity crystals.
(iii) Only in the doped crystals, the produced dislocations can clearly be detected since the movement of dislocations is restricted.

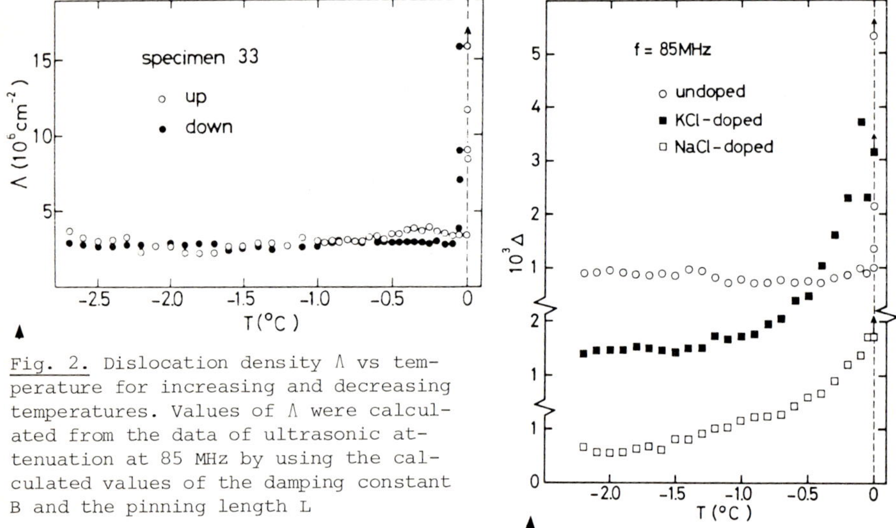

Fig. 2. Dislocation density Λ vs temperature for increasing and decreasing temperatures. Values of Λ were calculated from the data of ultrasonic attenuation at 85 MHz by using the calculated values of the damping constant B and the pinning length L

Fig. 3. Internal friction vs temperature at 85 MHz. Results for an undoped crystal and for crystals doped with KCl and NaCl are shown. The doping amounts to 2 ppm of K^+ or Na^+ ions. Because of the insensitivity of the overdamped resonance loss to pinning length at higher frequencies internal friction values shown can be regarded as being proportional to the dislocation density. Namely, the internal friction is proportional to Λ/B in the limit of high frequency and is independent of L

(iv) Even in the purest crystal, the enhanced dislocation density can be seen at temperatures very close to the melting point.
(v) The present interpretation is compatible with the molecular-dynamics computer simulation of melting crystals /6/, and the dislocation theory of liquid state /7/.

References

1. J. Tamura and Y. Hiki: Proc. 3rd Symp. on Ultrasonic Electronics, Jpn. J. Appl. Phys. 22, Suppl. 22-3, 208 (1983)
2. Y. Hiki and J. Tamura: Proc. 7th Intern. Conf. on Internal Friction and Ultrasonic Attenuation in Solids, J. Phys. (France) 42, C5-547 (1981)
3. J. Tamura and Y. Hiki: Proc. 2nd Symp. on Ultrasonic Electronics, Jpn. J. Appl. Phys. 21, Suppl. 21-3, 95 (1982)
4. Y. Hiki and J. Tamura: Proc. VI Intern. Symp. on the Physics and Chemistry of Ice, J. Phys. Chem. (1983), in press
5. A. V. Granato and K. Lücke: J. Appl. Phys. 27, 583,789 (1956)
6. R. M. J. Cotterill, W. D. Kristensen and E. J. Jensen: Philos. Mag. 30, 245 (1974)
7. S. Mizushima: J. Phys. Soc. Jpn. 15, 70 (1960)

Piezoacoustic Observation of Acoustic Soft Mode in KH_2PO_4 Crystal

J.Y. Koo, T.W.Yoo, and J.J. Kim[1]

Physics Department, Korea Advanced Institute of Science & Technology
P.O. Box 150, Chongyangni, Seoul, Korea

The critical dynamics of polarization in KDP crystal near the transition temperature T_c features a complex low frequency structure due to the soft mode coupling to other degrees of excitations (1, 2). Brillouin scattering and dielectric measurements (3, 4) were applied to study the coupling in the frequency ranges of 10^8 Hz and 10^6 Hz, respectively. We extend here the study of the soft mode coupling to the frequency range of 10^4 Hz by use of the time domain observation.

Approaching the transition temperature T_c the soft mode of $\omega_0^2 \propto (T-T_0)$ is coupled with the shear wave acoustic mode, which can be described by the effective Hamiltonian (3) :

$$H = \frac{1}{2} m^* \dot{P}_3^2 + \frac{1}{2}(\rho/q^2)\dot{S}_6^2 + \frac{1}{2} m^* \omega_0^2 P_3^2 + \frac{1}{2} C_{66} S_6^2 + a_{36} P_3 S_6 ,$$

where the last term represents the piezoelectric coupling (a_{36}) between the soft mode polarization P_3 and the acoustic phonon shear strain S_6, and m^* is the mass equivalent density (3). The piezoelectric oscillation encountered in the present work corresponds to the shear wave of the $q \simeq 0$ macroscopic limit, $q = \pi/L_x$, limited only by the sample dimension L_x, and ρ/q^2 will be substituted by $I^* = \rho L_x^2/\pi^2$, the moment of inertia equivalent density.

The equations of motion can thus be obtained as

$$m^*\ddot{P}_3 + m^*\omega_0^2 P_3 + a_{36} S_6 = 0, \qquad I^*\ddot{S}_6 + C_{66} S_6 + a_{36} P_3 = 0.$$

And, solving for the normal modes, we obtain, with $m^*\omega_0^2 = \alpha(T-T_0) = \alpha\,\Delta T$,

$$\omega_\pm^2 = \frac{1}{2}(\alpha\,\Delta T/m^* + C_{66}/I^*) \pm \{\frac{1}{4}(\alpha\,\Delta T/m^* + C_{66}/I^*)^2 + a_{36}^2/I^*m^* - \alpha C_{66}\,\Delta T/m^*I^*\}^{\frac{1}{2}}.$$

The ω_- coupled mode approaches zero frequency when $\Delta T = a_{36}^2/\alpha\, C_{66}$, that is, at $T = T_0 + a_{36}^2/\alpha\, C_{66}$, and is termed the acoustic soft mode. This acoustic soft mode dominates the low frequency polarization dynamics of KDP crystal near T_c. A phenomenological damping constant can be introduced (3) in the equation of motion to represent all the left over interactions with all the other excitations, when the $\tilde{\omega}_-$ normal mode becomes essentially a damped harmonic oscillator. The response function $\tilde{\chi}(t,t')$ and its Fourier transform $\chi(\omega)$ for the acoustic soft mode in the damped oscillator approximation can be written as

$$\chi(\omega) = 1/m^*(\omega_-^2 - \omega^2 - i\gamma\omega),$$

$$\tilde{\chi}(t,t') = \eta(t-t')\exp\{-\gamma(t-t')/2\}\sin\{\sqrt{\omega_-^2 - \gamma^2/4}\,(t-t')\}/m^*\sqrt{\omega_-^2 - \gamma^2/4} ,$$

where $\eta(t-t')$ represents a step function.

1) To whom correspondence should be addressed.

As shown in Fig.1 a c-cut KDP sample, silver-coated in vacuum for electrodes, is temperature controlled within ± 0.002 K across T_c, excited by a square pulse electric field, and the response signal is observed on the oscilloscope.

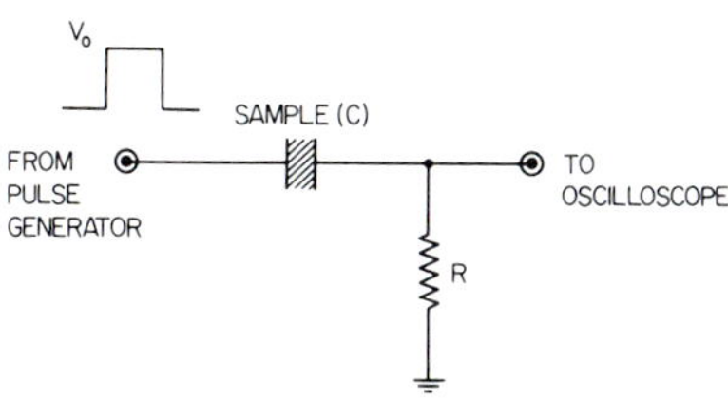

Fig.1 Circuit for measuring response function. Square pulses of 20 V and 10 msec are applied to the sample C of 8x6x0.6mm, response currents from C are fed into R(=30 Ω), and $V_R(t)$ is observed on the scope near T_c

The sample capacitor is not a simple capacitance but a damped harmonic oscillator element, which is usually represented by a LCR parallel or serial resonance element in the resonance experiments of frequency domain (5). However, in the present work the damped harmonic oscillator response itself is to be examined directly in the time domain as a result of the acoustic soft mode dynamics of KDP, which can be all incorporated into the time dependence of the KDP capacitor polarization:

$$C = \varepsilon C_0 = (\langle\varepsilon\rangle + \Delta\varepsilon)C_0 = \langle\varepsilon\rangle C_0 + 4\pi\Delta\vec{P}\cdot\vec{E}\, C_0/E^2 = \langle\varepsilon\rangle C_0 + 4\pi \Delta P_3\, C_0/E_0$$

$$= \langle\varepsilon\rangle C_0 + A\, \Delta P_3/V_0,$$

where $\Delta\varepsilon$ represents fluctuations from the thermal equilibrium average $\langle\varepsilon\rangle$, $C_0 = A/4\pi d$ capacitance for the plane parallel capacitor of area A and separation d, and $E_0 = V_0/d$.

Our signal of observation from the circuit of Fig.1 is $V_R(t) = Ri(t)$. Since we have $i(t) = V_0 dC/dt$ for $t \gg RC$ and $\Delta P_3 \simeq Q^* X_s(t)$, where $X_s(t)$ represents the acoustic soft mode displacement,

we obtain $i(t) = V_0 dC/dt = A\, d(\Delta P_3)/dt = A\, Q^* dX_s(t)/dt$.

On the other hand we know from the linear response theory,

$$X_s(t) = \int_{-\infty}^{t} dt'\, \tilde{\chi}(t-t')\, F^{ext}(t'),$$

and we find, for $F^{ext}(t) = E_0\eta(t)$,

$$dX_s(t)/dt = \frac{d}{dt}\left\{ \int_{-\infty}^{t} dt'\, \tilde{\chi}(t-t')\, E_0\eta(t')\right\} = E_0\tilde{\chi}(t).$$

Thus we see that $V_R(t) = Ri(t)$, as displayed on the oscilloscope with a step function excitation field $E_0\eta(t)$, represents a replica of the response function of the acoustic soft mode, that is, the KDP crystal near T_c.
Our field of excitation is not exactly a step function but a square pulse of 10 msec. However, all the signals are observed to decay completely within 0.5 msec at all temperatures near T_c, and the 10 msec (≫0.5 msec) square pulse is good enough to approximate the step function.

The signal i(t) observed at each temperature was best fitted to the damped oscillator $\tilde{\chi}(t)$ to obtain $\omega_-(T)$ and $\gamma(T)$ of the piezoacoustic soft mode. Representative signals observed at various temperatures are shown in Fig.2, from which the temperature dependence of frequency ω_-, damping constant γ, and amplitude I_0 are obtained as shown in Fig.3.

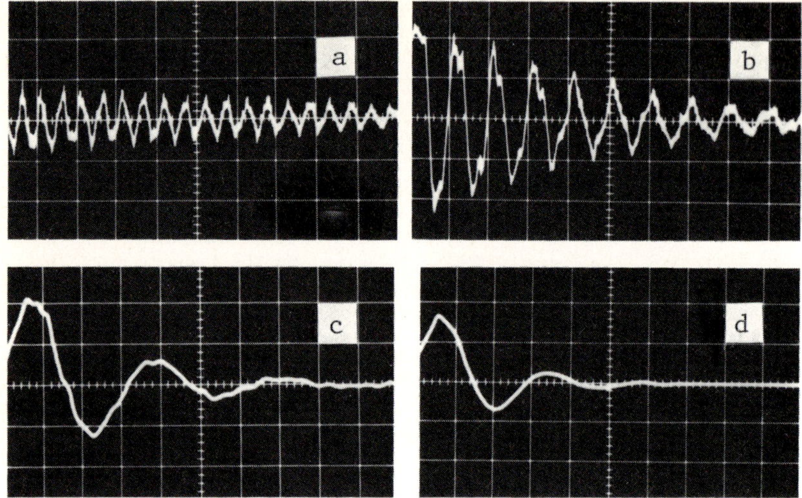

Fig.2 Representative response signals at various temperatures: (a) at T = 135.1 K, vertical and horizontal scales are (50 mV/div, 20 μsec/div), respectively; (b) T = 127.6 K, (100 mV/div, 20 μsec/div); (c) T = 122.5 K, (500 mV/div, 20 μsec/div); (d) T = 100.6 K, (500 mV/div, 20 μsec/div)

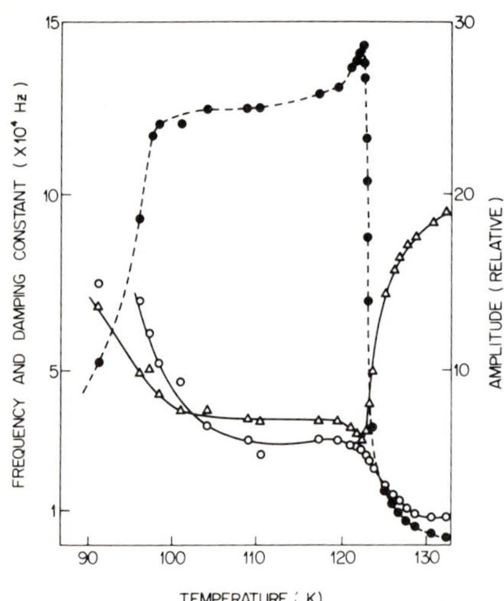

Fig.3 Temperature dependence of the piezoacoustic mode: frequency ω_- (Δ), damping constant γ(o) and amplitude I_0(•) as obtained from the damped oscillator fit

The resonance frequency ω_- is thus found to show the soft mode behaviour, decreasing form ~120 Hz (first observable with measurable amplitude) at ~135 K to ~26 KHz (as obtained from the damped oscillator fit, not the apparent frequency which is ~16 KHz) at very near T_c. The contribution of this ω_- mode to dielectric constant at low frequencies and a possible contribution to the $\omega \simeq 0$ spectral density of light scattering from the ω_- piezoacoustic mode of polarization fluctuations will be considered elsewhere.

Acknowledgements

This work was supported by the Korea Science and Engineering Foundation.

References

1 R. A. Cowley and G. J. Coombs: J. Phys. C6, 143 (1973): M. D. Mermelstein: Phys. Rev. B23, 3139 (1981): E. Courtens and R. W. Gammon: Phys. Rev. B24, 3890 (1981)
2 R. Blinc and B. Zeks: Soft Modes in Ferroelectrics and Antiferroelectrics (North-Holland, Amsterdam 1974)
3 E. M. Brody and H. Z. Cummins: Phys. Rev. Lett. 21, 1263 (1968): R. W. Gammon, E. Courtens and W. B. Daniels: Phys. Rev. B27, 4359 (1983)
4 N. E. Tornberg and R. P. Lowndes: J. Phys. C10, L549 (1977)
5 W. P. Mason: Phys. Rev. 69, 173 (1946)

Brillouin-Scattering Study of Sound Velocity of Quartz at α-β Transition

H. Unoki, H. Tokumoto, and T. Ishiguro

Electrotechnical Laboratory, 1-1-4 Umezono, Sakura-Mura, Niihari-Gun Ibaraki 305, Japan

It is well known that quartz (SiO_2) undergoes the first-order phase transition at 573°C from the lower temperature trigonal (α) phase to the higher-temperature hexagonal (β) phase accompanying an anomaly in the velocity of a particular acoustic phonon. But the mechanism of the transition has not been fully understood. In this paper we propose a mechanism of the transition based on the elastic constants derived from Brillouin scattering. The measurements were carried out at temperatures up to 700°C with emphasis on the propagation direction dependence in three principal planes as well as their temperature dependence.

There are six and five independent elastic constant tensor components in α and β phases, respectively. The temperature dependence of these constants has been measured by means of the mechanical resonance method[1-3]. The mechanism of the phase transition has been described[4] by softening of a zone-center optical phonon or the rotational oscillation of the rigid SiO_4 tetrahedra around the binary axes, with a temperature dependence proportional to $(T-T_0)^{1/2}$, where T_0 is about ten degrees below the transition temperature T_C. The Brillouin scattering measurements of the crystalline quartz have partly been performed[5,6]. Our major purpose of the experiment is to see if the temperature dependence of the sound velocity reduced by the mechanical resonance is reproduced at the measuring frequency of about 20 GHz and if some other softening of the particular transverse phonon propagating along a direction apart from the principal axes[7] can be observed.

The specimens were of a cylindrical form, to provide the angular-dependence measurement. The apparatus for measurement is the same as that reported before [7]. The result of the angular dependence in the X-Z plane at room temperature and at 580°C is shown in Fig.1. There seems no pronounced softening in the transverse phonon velocity in a particular direction within the X-Z plane. The temperature dependence of phonon velocity along the X direc-

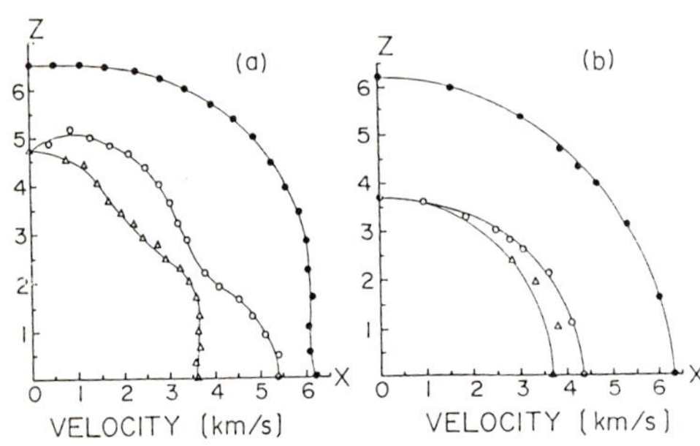

Fig. 1 Angular dependence of phonon velocities in X-Z plane at (a) room temperature and (b) 580°C

tion and the Z direction is shown in Fig.2, in the form of some elastic-constant-tensor components. Some of the components derived from transverse-phonon velocity were not separated because of line overlapping. The sharp hyperbolic temperature dependence of c_{11} and c_{33} around T_C derived from the longitudinal-sound velocity is exactly reproduced in the Brillouin-scattering measurement as shown in Fig.2. Thus it is suggested that a soft phonon mode should actually contribute to the longitudinal-acoustic-phonon branch in a significant frequency range of the reciprocal space around k = 0. We must interpret this phenomenon at any rate, since it has been believed that there should be no anomaly in the elastic constant at high temperature phase where there is no bilinear coupling between soft mode and the lattice strain[8].

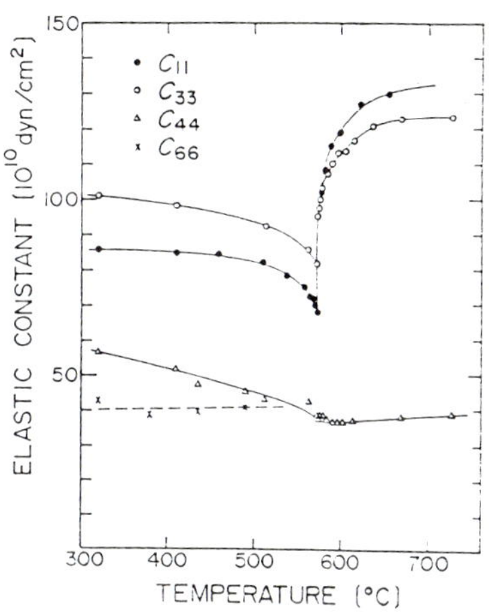

Fig. 2 Temperature dependence of elastic constant components

First, we take notice of the lattice contraction of quartz crystal below T_C[9]. Grimm and Dorner[10] have recently given a geometrical calculation on the α-phase structure of quartz, induced by the rotation of the SiO_4 tetrahedron and proved nearly equivalent contraction of lattice constants along both X and Z axes. Therefore, we can see that in the high temperature phase a breathing-mode oscillation of the unit cell is introduced by the tetrahedron oscillation at a frequency of twice of the latter, since the rotational shifts of both $+\phi$ and $-\phi$ directions yield the lattice strain of the same sign. This kind of interaction can be considered as though the longitudinal acoustic branch exhibits, in the k-ω space, an anticrossing behavior with a virtual excitation of an optical mode whose frequency is $2\omega_0(k)$, where $\omega_0(k)$ is the frequency of the tetrahedron oscillation. In such a case we know that the gradient of the coupled acoustic phonon branch at k = 0, the renormalized elastic constant $\bar{c}$, can be expressed[7] as $c - a^2/(T - T_0)$, where c is the elastic constant with no interaction and a is a constant proportional to the strength of the coupling. The potential-energy density U as a function of the order parameter ϕ and the compressive strain e, taking the coefficient $\bar{c}$ in the elastic energy into account, is given by

$$U(\phi,e) = \frac{1}{2}b\phi^2 + \frac{1}{4}d\phi^4 + \frac{1}{6}f\phi^6 + g\phi^2 e + \frac{1}{2}\bar{c}e^2 . \qquad (1)$$

Here, the coefficient b is proportional to $T - T_0$, d and f are relevant constants and g is strength of static quadratic coupling between ϕ^2 and the compressive strain e. When $U(\phi,e)$ is rewritten as a function of ϕ only, we have,

$$U'(\phi) = \frac{1}{2}b\phi^2 + \frac{1}{4}(d - 2g^2/\overline{c})\phi^4 + \frac{1}{6}f\phi^6. \tag{2}$$

The coefficient of the ϕ^4 term, $d - 2g^2/\overline{c}$, turns negative at a temperature higher than T_0, since $\overline{c}$ approaches zero, during cooling. The negative coefficient of the fourth-order term of the order parameter causes a discontinuous change of the phase. This will be the origin of the first-order phase transition of the quartz.

References

[1] E. W. Kammer, T. E. Pardue, and H. F. Frissel: J. Appl. Phys. 19, 265 (1948).

[2] V. G. Zubov and M. M. Firsova: Soviet Phys. Crystallography 7, 374 (1962).

[3] U. T. Hochli and J. F. Scott: Phys. Rev. Lett. 26, 1627 (1971).

[4] S. M. Shapiro, R. G. Gammon, and H. Z. Cummins: Appl. Phys. Lett. 9, 157 (1966).

[5] S. M. Shapiro and H. Z. Cummins: Phys. Rev. Lett. 21, 1578 (1968).

[6] J. D. Axe and G. Shirane: Phys. Rev. B1, 342 (1970).

[7] H. Tokumoto and H. Unoki: Phys. Rev. B27, 3748 (1983).

[8] J. C. Slonczewski and H. Thomas: Phys. Rev. B1, 3599 (1970).

[9] A. H. Jay: Proc. Roy. Soc. A142, 237 (1933).

[10] H. Grimm and B. Dorner: J. Phys. Chem. Solids 36, 407 (1975).

Phonon-Soliton Interaction in K_2SeO_4

W. Rehwald

Laboratories RCA Ltd., Badenerstrasse 569
CH-8048 Zürich, Switzerland

1. Discommensurations at the Lock-in Transition

An incommensurate phase is characterized by an order parameter that is modulated with a wavelength incommensurate with the lattice periodicity. Close to the transition temperature T_i this modulation is sinusoidal (plane-wave regime); the corresponding elementary excitations are fluctuations in amplitude and phase of the ordering coordinate (amplitons and phasons).

With increasing order parameter higher order interactions cause a change of the ordering wavevector with temperature that finally ends in a transition into a commensurate phase, where the ordering wavevector becomes a rational fraction of a reciprocal lattice vector, e.g. in K_2SeO_4 $q_i \rightarrow a^*/3$. The modulation is, close to the lock-in temperature T_c, no longer described by plane waves, but by an alternating sequence of commensurate regions and discommensurations (phase solitons), in which the order parameter phase changes rapidly with respect to the underlying lattice |1|. This soliton lattice generates a new periodicity that causes a splitting of the excitation spectrum at the corresponding Brillouin-zone boundary. Approaching the lock-in transition from above, the intersoliton distance increases towards infinity |2|. In the commensurate phase only the high frequency branch remains and behaves like a regular phonon (quasi phason). In potassium selenate the incommensurate phase extends between T_i = 127 K and T_c = 93 K.

2. Soliton - Phonon Interaction

Besides the primary order parameter, represented by a phonon coordinate Q, there exist two secondary order parameters, the shear strain e_5 and the electric polarization P_3, that couple to Q by fourth-order anharmonic interaction. The free coupling energy density is written:

$$F_c = \frac{1}{2} g_3 P_3 (Q^3 + Q^{*3}) + \frac{1}{2i} g_5 e_5 (Q^3 - Q^{*3}) . \tag{1}$$

Consequently there exists an alternating periodic array of static strain e_5 and polarization P_3 in the soliton phase. An external stress σ_5 perturbs the balance between regions with positive and negative strain |3|. This results in a decrease in the stiffness c_{55}:

$$c_{55}(T) = c_{55}^{\infty} - \frac{3g_5^2}{2b_6} \cdot \frac{K(k) - E(k)}{K(k)} . \qquad (2)$$

K(k) and E(k) are the complete elliptic integrals of modulus k, which is zero at T_i, increases in the plane-wave region, and approaches k = 1 in the soliton regime. Its value is determined by the minimum of the total free energy.

The dynamical soliton response should be observable in the imaginary part of c_{55}, i.e. in the ultrasonic attenuation.

3. Static and Dynamical Elastic Measurements

By measuring sound velocity and ultrasonic attenuation in a pulse-echo experiment both the real and imaginary parts of c_{55} were determined. In the incommensurate phase c_{55} softens, reaching a minimum around T_c (Fig. 1). No thermal hysteresis was observed and there is no difference between the two sound modes (a,c) and (c,a), which both produce the same strain e_5.

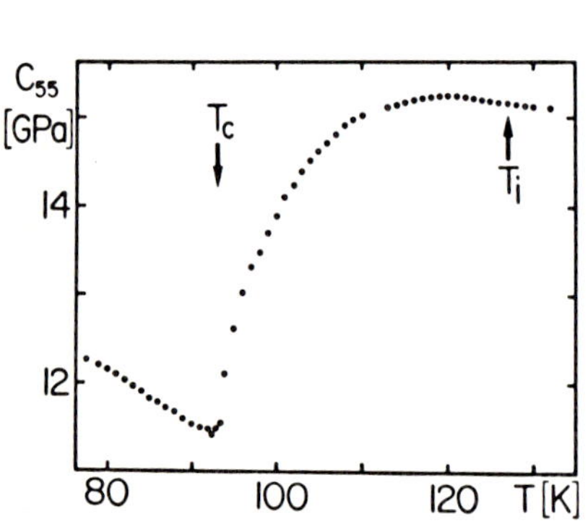

Fig. 1: Shear stiffness c_{55} in the incommensurate phase

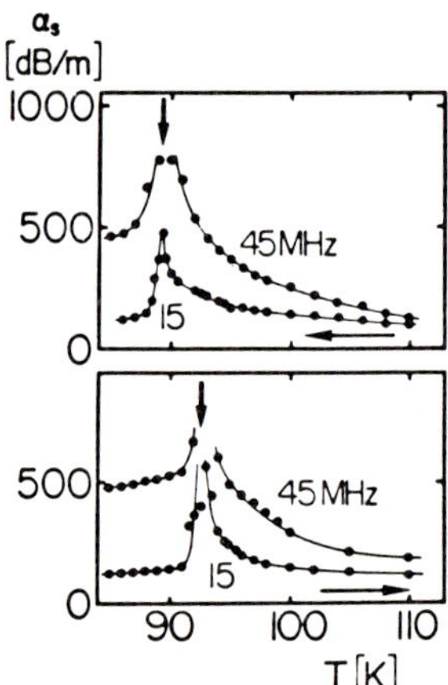

Fig. 2: Ultrasonic attenuation during cooling and warming up

The corresponding attenuation, however, shows a considerable thermal hysteresis (Fig. 2). The maxima measured during cooling and warming up differ by about 3.5 K.

4. Discussion

Applying (2) to the measured c_{55} values, the modulus k can be calculated as a function of temperature, and from that the soliton density n_s (Fig. 3). The soliton regime, where a power law in $(T - T_c)/T_c$ holds with an exponent 0.67, can be clearly distinguished from the plane-wave regime, where the interaction with the phason determines the softening, and a different description has to be used |4|.

The attenuation is strongly dependent on the history of the sample. This suggests that pinning of the soliton lattice |5| and perhaps also of the phason determines the measured dynamic behaviour.

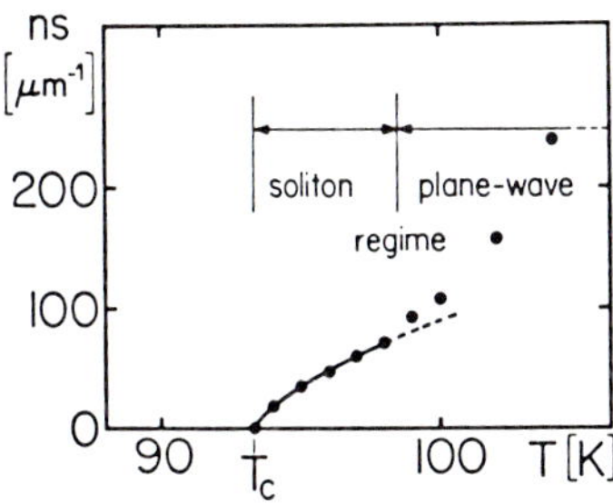

Fig. 3: Soliton density as a function of temperature

References

1. W.L. McMillan: Phys. Rev. B 14, 1496 (1976).
2. A.D. Bruce, R.A. Cowley: J. Phys. C 11, 3609 (1978).
3. V. Dvořák, O. Hudák: Ferroelectr. 46, 19 (1982).
4. W. Rehwald, A. Vonlanthen: Solid State Commun. 38, 209 (1981).
5. H.-G. Unruh: J. Phys. C 16, 3245 (1983).

Thermal Conductivity of Cooperative Jahn-Teller E-b_1,b_2 Systems

W. Mutscheller and M. Wagner

Institut für Theoretische Physik III, Universität Stuttgart
Pfaffenwaldring 57, D-7000 Stuttgart 80, Fed. Rep. of Germany

1. Introduction

Some rare earth compounds such as $TmVO_4$ undergo a structural phase transition at low temperatures (cooperative Jahn-Teller systems). As shown by measurements of DAUDIN and SALCE [1], this phase transition strongly influences the energy transport properties of these materials. Our aim is the theoretical analysis of this anomalous behaviour.

2. *Model*

The Hamiltonian of cooperative E-b_1, b_2 systems is given by [2], [3]

$$(1) \qquad H = H_p + \sum_{\vec{q}\lambda} (A^2_{\vec{q}\lambda} Q_{\vec{q}\lambda} \sigma^z_{-\vec{q}\lambda} + A^1_{\vec{q}\lambda} Q_{\vec{q}\lambda} \sigma^x_{-\vec{q}}) .$$

H_p is the phonon Hamiltonian. The second part of H represents the electron-phonon coupling. The electronic system is described by pseudo spin operators. Cooperativity arises by means of unitary transformations, which we write in the following general form:

$$(2) \qquad \tilde{H} = e^{-S} H e^{S} = H_p + H_c + H' .$$

In our concept the Ising-like H_c describes the cooperative behaviour of the "dressed" electrons, whereas H' defines the interaction between the two subsystems. Up to second order in the coupling it is given by

$$(3) \qquad H' = \sum_{\vec{k}\lambda} A^1_{\vec{k}\lambda} (b_{\vec{k}\lambda} + b^+_{-\vec{k}\lambda}) \sigma^x_{-\vec{k}} + \sum_{\substack{\vec{k}\vec{k}' \\ \lambda\lambda'}} \Gamma\binom{\vec{k}\ \vec{k}'}{\lambda\ \lambda'} (b_{\vec{k}\lambda} + b^+_{-\vec{k}\lambda})(b_{\vec{k}\lambda'} - b^+_{-\vec{k}'\lambda}) \cdot \sigma^y_{-\vec{k}-\vec{k}'} .$$

From our analysis we come to the conclusion that mainly the first (resonant) scattering term in H' is responsible for the anomalous behaviour of thermal conductivity.

3. *Transport theory*

The Kubo formula [4] for thermal conductivity is the starting point of our calculation,

$$(5) \quad \kappa = \frac{\beta}{VT} \int_0^\infty dt e^{-\varepsilon t} \int \frac{d\lambda}{\beta} < \vec{J}(0)\vec{J}(t+i\hbar\lambda)> = \frac{\beta}{VT} \tilde{C}\,(Z=i0) = \frac{\beta}{VT} \sum_{\vec{q}\lambda} \Gamma_{\vec{q}\lambda} \tilde{C}_{\vec{q}\lambda}(z=i0),$$

where $\tilde{C}(z)$ denotes the Laplace transform of the current-current correlation function

$$(6) \quad C(t) = \int \frac{d\lambda}{\beta} < \vec{J}(0)\ \vec{J}(t+i\hbar\lambda)> .$$

MORI and ZWANZIG [5], [6] have shown that autocorrelation functions of the form (6) can be expressed in terms of memory functions. The heat conductivity then finally reads:

$$(7) \quad \kappa = \lim_{z \to i0} C(t=0)\ (z + \tilde{K}(z))^{-1} \quad \text{or}$$

$$(7a) \quad \kappa = \lim_{z \to i0} \sum_{\vec{q}\lambda} \Gamma_{\vec{q}\lambda}\ C_{\vec{q}\lambda}(0)\ (z + \tilde{K}_{\vec{q}\lambda}(z))^{-1}$$

where the memory function is given by

$$(8) \quad K(t) = < L\vec{J} \mid e^{iLt} \mid L\vec{J}> < \vec{J} \mid \vec{J} >^{-1} .$$

Here L denotes the Liouvillian. A simple comparison with the usual heat conductivity integral [7] shows that K(z=i0) can be identified with the inverse relaxation time of the Callaway theory.

4. *Results*

As one can see from Hamiltonian (3) the memory function splits in a resonant and a nonresonant part. The exact form of the relaxation time depends on the particular approximation which is used to evaluate the memory functions. For this purpose we have tried the van Hove limit [8] and an exponential ansatz [9]. In the first case the resonant part of the inverse relaxation time is of the form [10]

$$(9) \quad \tau_R^{-1} = \Gamma\ \Omega^5\ (\sinh\ (\beta\ \Omega))^{-1}$$

where Ω denotes the meanfield splitting of the electronic two-level system. In the second case it is of the form

$$(10) \quad \tau_R^{-1}(\omega) = f_1(\omega,\ \Omega,\ T)\ \Omega^2\ f_2(\omega)\ ((\omega-\Omega)^2 + (\Omega^2\ f_2(\omega))^2)^{-1} .$$

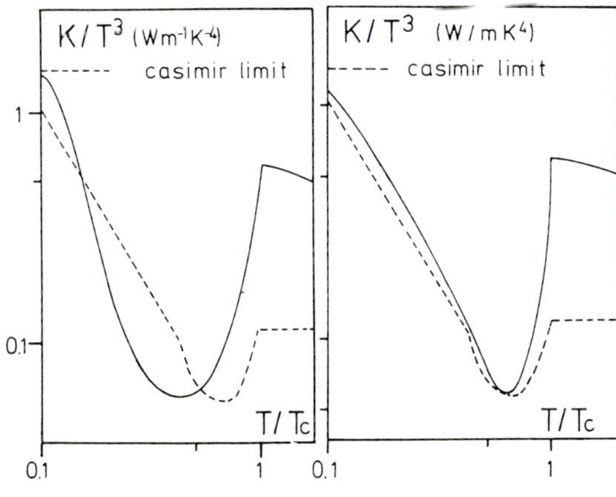

Thermal conductivity of $TmVO_4$
Dotted lines: Measured curve. The boundary scattering has been included by means of a Callaway superposition

Fig. 1. Curve obtained by using (9)

Fig. 2. Curve obtained by using the exponential ansatz

The result of the two approaches is drawn in Figs. 1, 2. In both cases the qualitative agreement is reasonable, whereas the quantitative behaviour is satisfactory only in the second case. The nonresonant part of the relaxation time shows no significant Ω dependence and is responsible for only a small but continuous decrease of $\kappa/_{T}3$.

References

1. Daudin, B., Salce, B., J. Phys. C 15, 463 (1982)
2. Gehring, G.A., Gehring, K.A., Rep. Progr. in Phys., 38, 1 (1975)
3. Melcher, R.L., Phys. Ac. XII, 1 (1976)
4. Kubo, R., J. Phys. Soc. Jap. 12, 570; 1203 (1957)
5. Mori, H., Progr. Theor. Phys. 33, 423; 34, 399 (1965)
6. Zwanzig, R., "Lectures in Theoretical Physics", vol. 3, 106 New York: Interscience (1961)
7. Callaway, J., Phys. Rev. 113, 73 (1959)
8. Van Hove, L., Physica 21, 517 (1955)
9. Berne, B.J., Harp, G.D., Adv. in chem. Phys. XVII, 63, (1970)
10. Daudin, B., Mutscheller, W., Wagner, M., to be published

Part VII

Free Carriers

Chairmen: **R.P. Huebener J.P. Maneval N. Perrin**

Magnetic Field Dependence of Ultrasonic Attenuation in Heavily Doped Ge:Sb

H. Sakurai and K. Suzuki

Department of Electrical Engineering, Waseda University, 3-4-1 Ohkubo Shinjuku-Ku, Tokyo 160, Japan

T. Miyasato

Institute of Scientific Industrial Research, Osaka University Suita, Osaka 565, Japan

The dependence of the ultrasonic attenuation in Sb-doped Ge on the temperature and the uniaxial stress has been investigated over a wide concentration (N) region by two of the authors (H.S. and K.S.).[1,2] The behavior of the attenuation coefficient α has been found to be complicated in the intermediate concentration region, where there exists the critical concentration N_c ($\simeq 1.5 \times 10^{17}$ cm^{-3} Sb impurities)[3] for the metal-nonmetal transition.

The magnetic field(B) dependence of the ultrasonic coefficient, $\alpha(B)$, has been reported in the lightly doped n-type Ge by one of the authors (T.M.) and his collaborators,[4,5,6] which seems not to be explained based on a simple model of the interaction between an acoustic wave and an isolated donor.

In this paper, the experimental results of $\alpha(B)$ in Sb-doped Ge have been reported over a wide concentration region from near the lower limit of the intermediate to the high region, in the temperature range 1.5 to 4.2 K and 380 MHz.

The Ge single crystals doped with Sb were grown by the Czochralski method. Samples were cut from the single crystals in the shape of a rectangular parallelepiped of $5 \times 6 \times 10 \sim 15$ mm. The impurity concentration N was determined from the room-temperature resistivity using the Irvin curve. The variation of impurity in each sample is less than 5 percent of the nominal concentration. The samples were not intentionally compensated. Measurements of $\alpha(B)$ for the [110] longitudinal wave were performed in seven samples (N = 0.31, 0.69, 0.83, 1.2, 1.8, 2.1 and 3.3×10^{17} cm^{-3}, respectively), where the magnetic fields up to 5.5 T were applied along the [111] axis. The acoustic wave was generated by means of ZnO thin film transducer which was grown onto one face of the sample by an rf-sputtering method.

The experimental results are summarized as follows. For 3.1×10^{16} cm^{-3}, $\Delta\alpha(B) = \alpha(B) - \alpha(0)$ monotonically decreases with increasing B, and the variation becomes larger with decreasing T. The behavior of $\Delta\alpha(B)$ for 6.9×10^{16} cm^{-3} is anomalous, that is, $\Delta\alpha(B)$ slightly increases with B at low B, and becomes negative at high B and low T. For 8.3×10^{16} cm^{-3}, $\Delta\alpha(B)$ gradually increases with B and exhibits a peak around 4 T. Figure 1 shows the behavior of $\Delta\alpha(B)$ for 1.2×10^{17} cm^{-3} which is just below N_c. For $N \geq 1.2 \times 10^{17}$ cm^{-3}, $\Delta\alpha(B)$ gradually increases with B. Assuming that $\Delta\alpha(B) \propto B^n$ above 4 T, it is found that $n < 1$ for 1.2×10^{17} cm^{-3}, $n \simeq 1$ for 1.8×10^{17} cm^{-3} and $n > 1$ for $N \gtrsim 2.1 \times 10^{17}$ cm^{-3}, but $n < 0$ for $N \lesssim 8.3 \times 10^{16}$ cm^{-3}. These are consistent with a classification of the concentration region which was tentatively made from the results of the ultrasonic attenuation on the temperature and the uniaxial stress.[1,2]

The concentration dependence of $\Delta\alpha(B)$ at 5 T is shown in Fig. 2. Although there is a lack of a result for a concentration between 1.2×10^{17} and 1.8

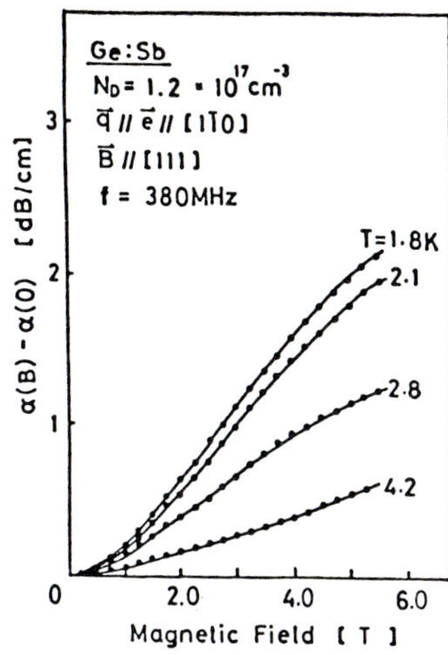

Fig. 1. The behavior of α(B) - α(0) in the sample with 1.2×10^{17} cm^{-3} Sb impurities

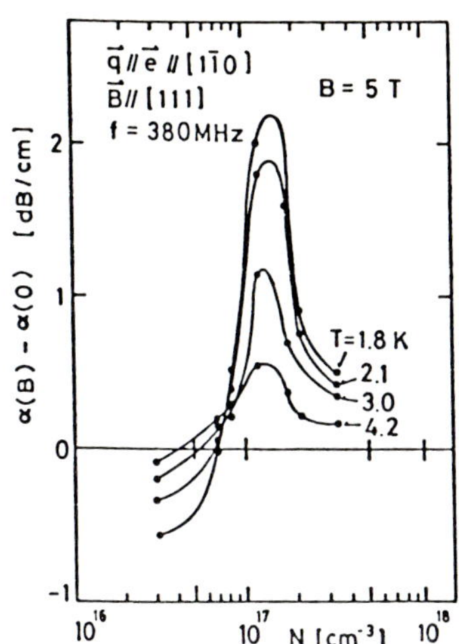

Fig. 2. The concentration dependence of α(B) - α(0) at 5 T and at 1.8, 2.1, 3.0 and 4.2 K

$\times 10^{17}$ cm^{-3}, Δα(B) seems to exhibit a peak near N_c as a function of the concentration. The similar results were also obtained at 3 and 4 T. The temperature dependence of Δα(B) below 4.2 K are as follows. For 3.1×10^{16} cm^{-3}, Δα(B) rapidly increases with T, and for N ≳ 2.1×10^{17} cm^{-3}, Δα(B) monotonically decreases with increasing T. The temperature dependence of Δα(B) is reversed at the concentration between 6.9×10^{16} and 8.3×10^{16} cm^{-3}. The frequency dependence of Δα(B) is much stronger in the concentration region around 7×10^{16} cm^{-3} than in other regions.

For N ≳ 2.1×10^{17} cm^{-3}, the behavior of Δα(B) can be qualitatively explained based on a simple free electron gas model which could reproduce the temperature and the uniaxial stress dependence of the ultrasonic attenuation in this region,[1,2] provided that we take the broadening of the Landau levels into consideration. However, we have no adequate model which can explain the results of Δα(B) in the intermediate concentration region, at present.

Acknowledgements

We are grateful to M.Tokumura of Osaka university for his advice on the rf-sputtering of ZnO thin films. We are also grateful to other members of our laboratory for their assistance in the experiment.

References

1. H.Sakurai and K.Suzuki: submitted to J.Phys.Soc.Jpn.
2. H.Sakurai and K.Suzuki: submitted to J.Phys.Soc.Jpn.
3. G.A.Thomas et al.: Phys.Rev. B26 (1982) 2113.
4. T.Miyasato et al.: J.Phys.Soc.Jpn. 35 (1973) 1668.
5. T.Miyasato and F.Akao: J.Phys.Soc.Jpn. 41 (1976) 502.
6. T.Miyasato et al.: J.Phys.Soc.Jpn. 49 (1980) 2219.

Phonon Attenuation in Heavily Doped p-Type Semiconductors

T. Sota and K. Suzuki

Department of Electrical Engineering, Waseda University, 3-4-1 Ohkubo Shinjuku-Ku, Tokyo 160, Japan

D. Fortier *

DPC/SSS, CEN Saclay, F-91190 Gif-sur-Yvette, France

The phonon relaxation rate in heavily doped n-type semiconductors with many-valley structure such as Ge and Si has been already derived in our previous papers [1, 2] where are included the effects of screening by conduction electrons and of the electrons having a finite relaxation time due to both the intravalley and the intervalley scattering by impurity atoms. In this paper we extend the method to calculate the phonon relaxation rate in heavily doped p-type semiconductors. Here the interband hole-phonon interaction is explicitly taken into account as well as the intraband one. We also report the experimental results of the low-temperature thermal conductivity κ in heavily doped p-type GaSb and Si and the analyses.

We consider only the J=3/2 valence bands with symmetry Γ_8 in semiconductors with the diamond and the zincblende structures where in the latter the linear $\mathbb{k}$ (the wave vector of a hole) terms are neglected [3]. The phonon relaxation rate is derived by solving the equations of motion for the single-particle density matrix within the self-consistent field and the relaxation time approximations. We consider both the deformation potential coupling and the piezoelectric one for the hole-phonon interaction. In order to make calculations feasible, the following approximations are made. (i) The interband scattering rates of the holes due to impurities are identical to each other. (ii) In calculating the matrix elements of the hole-phonon interaction Hamiltonian H'_{ex}, we put $\theta'=\theta$ and $\phi'=\phi$, where $\theta'(\theta)$ and $\phi'(\phi)$ denote the polar angle and the azimuthal one of $\mathbb{k}'(\mathbb{k})$, respectively. Here the angular dependence arises from the fact that the wave functions in the valence band at $\mathbb{k}\simeq 0$ have p-like characters. This approximation is good for $q<k_F$ where q is the wave number of the phonons and k_F is the Fermi wavenumber of the hole. However it leads to underestimation of the phonon attenuation in the large-q region. (iii) The squared matrix elements of H'_{ex} averaged over θ and ϕ are used. The final expressions for the relaxation rate through the deformation potential coupling and the piezoelectric one are given by, respectively,

$$\tau^{-1}(q,\lambda)=2\omega/(\rho_m\bar{v}_\lambda^{\,2})\,\mathrm{Im}(R_D D_D^{\,2}+R_S D_S^{\,2}+R_{IB} D_{IB}^{\,2}) \qquad (1)$$

and

$$\tau^{-1}(q,\lambda)=2\omega/(\rho_m\bar{v}_\lambda^{\,2})\,\mathrm{Im}(R_D D_p^{\,2})\,. \qquad (2)$$

Here ω is the angular frequency of the phonon and $\bar{v}_\lambda$ is the averaged sound velocity with the polarization direction λ, where $\lambda=1$ stands for longitudinal mode and 2,3 for transverse modes and ρ_m is the crystal density. R_D, R_S and R_{IB} represent the response function to the dilational(D_D), the shear (D_S) and the interband(D_{IB}) component of the hole-phonon interaction, respectively and D_p denotes the piezoelectric coupling constant.

---

* Present address: Laboratoire de Biophysique, Université René Descartes, Faculté de Médecine, Necker-Enfants-Malades 156, Rue de Vangirand, 75730 Paris Cedex 15, France

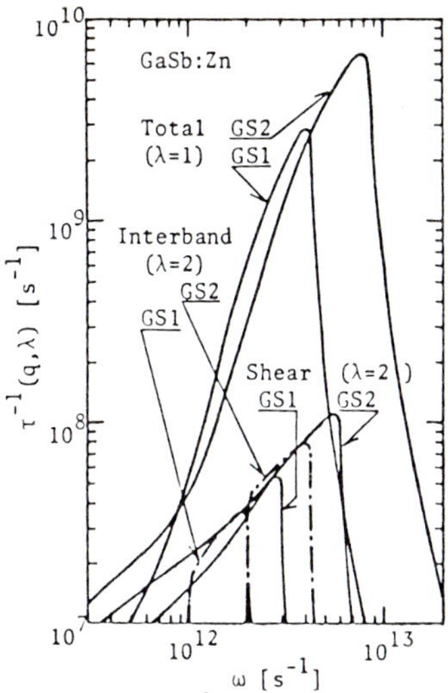

Figure 1: $\tau^{-1}(q,\lambda)$ versus frequency

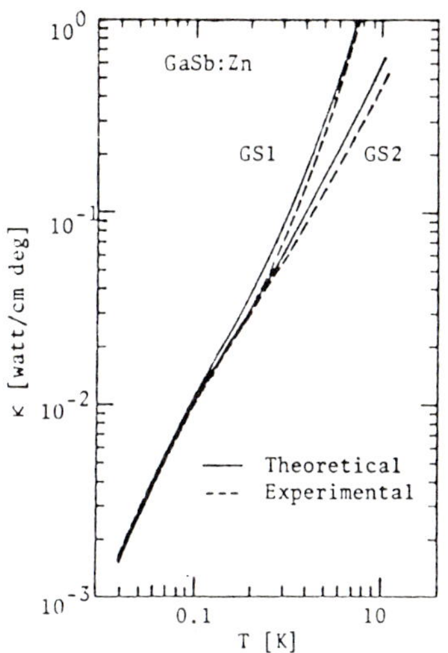

Figure 2: Thermal conductivity versus temperature

Figure 1 shows the frequency(ω) dependence of $\tau^{-1}(q,\lambda)$ for $\lambda=1$ and 2 in two samples of GaSb, GS1 and GS2 (see Table 1), which will be used in analysing the thermal conductivity below. In calculations we have used the following values of the physical parameters: $\bar{v}_1=3.97\times10^5$ cms^{-1}, $\bar{v}_2(=\bar{v}_3)=2.78\times10^5$ cms^{-1}, $D_d=-8$ eV, $D_u=3$ eV, $D_u'=4$ eV, $\tau_l\geq\tau_h$ (10^{-14} s)=6.7(GS1), 1.5(GS2), γ'(10^{12} s^{-1})= 8 (GS1), 10 (GS2) where τ_l and τ_h denote the relaxation time of the light hole and the heavy hole, respectively, and γ' is the interband scattering rate of the holes. The following results have been found. (i) The ω dependence of $\tau^{-1}(q,\lambda)$ is qualitatively different from that obtained previously [4, 5]. The difference arises from the fact that there exist the shear terms with R_S and the interband terms with R_{IB} in Eq. (1) but not in the previous expressions. (ii) The contribution of the interband hole-phonon interaction occurs only in a certain frequency region. Its relative importance strongly depends on the ratio of the light hole mass to the heavy hole one and it can become large for transverse modes. (iii) The thermal phonon scattering via the deformation potential coupling is much more effective in semiconductors such as GaSb and InSb than that via the piezoelectric one.

The thermal conductivity κ of GaSb and Si doped with Zn and B, respectively, was measured. The characteristics of our samples are given in Table 1 where N is the impurity concentration and L is Casimir's length and ρ is the electrical conductivity at the low temperature. The concentration dependence of κ in GaSb and Si is found to be similar to that in Ge, InSb and GaAs [6, 4, 5, 7].

Table 1 Characteristics of the samples

Material: Impurity		N (10^{19} cm^{-3})	L (cm)	ρ (10^{-3} Ωcm)
GaSb:Zn	GS1	0.57	0.32	3.5
	GS2	4.5	0.32	0.7
Si : B	S1	1.1	0.44	8.7
	S2	4.0	0.44	3.5

We have calculated κ of various p-type semiconductors using Eqs. (1) and (2). As an example, in Fig. 2 is shown the comparison between the theoretical and the experimental results of κ for GS1 and GS2. The physical parameters used in calculations of κ are those used in calculations of $\tau^{-1}(q,\lambda)$ and given in Table 1. As for A in $A\omega^4$ which expresses the relaxation rate by Rayleigh scattering due to static imperfections such as impurities, isotopes and strain fields, somewhat larger values of A, A($10^{-44}s^{-3}$)=8(GS1), 10(GS2), are used to obtain a better fit in the higher-temperature region as done in [8]. As can be seen from Fig. 2, semiquantitative agreement with the experimental results has been obtained. The concentration dependence of κ in p-type Ge, InSb and GaAs can be also explained by Eq. (1) derived herein. However Eq. (1) can not quantitatively explain κ of Si though qualitatively it works. Some discrepancy will arise from the fact that we ignore the J=1/2 valence bands of Si where the energy difference between the J=3/2 and J=1/2 valence bands is small.

Details of this work will be published elsewhere.

We wish to thank Y. Matsuyama for his help in numerical calculations.

References

1. T.Sota and K.Suzuki: J.Phys.C 15 6991 (1982)
2. T.Sota and K.Suzuki: J.Phys.C 16 4347 (1983)
3. G.L.Bir and G.E.Pikus: Symmetry and Strain-induced Effects in Semiconductors (John Willey and Sons, Ink., New York 1974)
4. C.R.Crosby and C.G.Grenier: Phys.Rev. B4 1258 (1971)
5. P.Fozooni, N.H.Zebouni and C.G.Grenier: J.Phys.C 13 4285 (1980)
6. J.A.Carruthers, J.F.Cohchran and K.Mendelssohn: Cryogenic 2 160 (1962)
7. R.O.Carlson, G.A.Slack and S.J.Silverman: J.Appl.Phys. 36 505 (1965)
8. V.V.Kosarev, P.V.Tamarin and S.S.Shalyt: Phys.Stat.Sol.(b) 44 525 (1971)

Ballistic Phonon Transport in Ge:P Under Magnetic Field

T. Miyasato and M. Tokumura

The Institute of Scientific and Industrial Research, Osaka University
Mihogaoka 8-1, Ibaraki, Osaka 567, Japan

K. Suzuki

Department of Electrical Engineering, Waseda University, 3-4-1 Ohkubo
Shinjuku-Ku, Tokyo 160, Japan

The magnetic-field dependence of the scattering of ballistic phonons in Ge doped with 5×10^{14} [P/cm³] was measured under [100] magnetic field up to 5.5 [T], and the direction of the phonon wave vector is along the $[1\bar{1}0]$ axis. The phonon scattering for FT and L modes showed the magnetic-field dependence. For FT mode the relative signal height of the transmitted heat pulse shows saturation at fields more than 4~5 [T], which is similar to that of Ge:Sb, but not to that of Ge:As in the similar concentration region. This is in contrast to the results of thermal conductivity where the behavior of Ge:P more resembles that of Ge:As. The experimental results were compared with the theoretical calculation which is based on the phonon scattering by the isolated donor and/or donor pairs, and it seems that both mechanisms can not explain the experimental results fully.

1. Introduction:

The donor impurity of V-group atoms in Ge has a shallow donor level and is expressed by an H-atom model, and whose ground state splits into the lowest singlet(A_1 level) and the upper triplet(T_2 level). This is called the valley-orbit splitting, which is due to the valley-orbit interaction and the central cell correction, and the energy separation is denoted by 4Δ whose value is strongly dependent on the species of donor atom, i.e., 3.71, 32.8 and 49.1 [K] for Sb, P and As doped Ge, respectively. The donor electron in the ground state of many-valley semiconductors such as Ge is strongly coupled with the lattice system (i.e., acoustic phonon or strain, etc.) through the deformation potential coupling. When the magnetic field is applied along the [100] axis, the degeneracy of the four pairs of conduction band minima is kept and there occurs a shrinkage of donor wave function (i.e., the effective Bohr radius). Then the energy separation 4Δ increases with the magnetic field, which decreases the scattering probability of the ballistic heat-pulse phonons. The shrinkage of donor wave function changes the cut-off function which gives the upper limit of wave number of phonons that couple with neutral donor electrons, which means that the shrinkage of the donor wave function increases the scattering probability of ballistic heat-pulse phonons.

The [100] magnetic field dependence of the ballistic heat-pulse transport in Sb- or As-doped Ge was reported by MIYASATO et al.[1][2][3], and it was clarified that in As-doped Ge, the relative transmitted heat-pulse power increases with magnetic field up to 6 [T], and the effect was far larger than the numerically estimated results based on the phonon scattering by neutral donor electrons under [100] magnetic field. And in Sb-doped Ge, the relative transmitted heat-pulse power increases with magnetic field up to about 4 [T] and shows a saturation tendency at higher fields, which is the opposite trend to the numerical estimation on the basis of the theory above mentioned.

Then in order to investigate these discrepancies, we carried out the experiment of the ballistic heat-pulse transmission in P-doped Ge under the same conditions as for As- or Sb-doped Ge, and the experimental results were compared with the theoretical calculations based on (a)the phonon scattering by neutral donor electron under [100] magnetic field [4], and (b)the phonon scattering by polar pairs on the basis of GORTEL's theory [5] taking into consideration the shrinkage effect under [100] magnetic field [6].

2. Experimental Technique and Procedure:

An Au thin film heater and the CdS thin film bolometer [7] were used in the present experiment in order to avoid the magnetic-field effect of the phonon generator and detector system, and the details were given in Ref.[1]. The magnetic field was applied along the [100] axis, and the ballistic heat-pulse propagated along the $[1\bar{1}0]$ axis.

3. Experimental Results:

The transmitted ballistic heat-pulse signals for L, FT and ST modes were observed, but the behavior of FT mode is shown here because L mode signal is very small on account of phonon defocussing phenomena and ST mode signal is on the ramp pulse. The relative signal height at various heater temperaturs T_h is shown as a function of the magnetic field, namely, the signal height at zero field is normalized to be unity.

The magnetic-field dependence of the relative signal height at T_o=4.2 [K] is shown in Fig.1, and that at T_o=1.8 [K] is also given in Fig.2. In case of T_o=4.2 [K], the relative signal height decreases at lower fields and then increases, and shows a saturation tendency at higher fields. In case of T_o=1.8 [K], the relative signal height increases monotonically at lower fields, and shows a saturation tendency at higher fields, and the effect of the magnetic field on the signal height is larger compared with that at T_o=4.2 [K].

4. Discussions and Conclusions:

(a)The phonon scattering by neutral donor electrons under [100] magnetic field was calculated for T_o=1.8 [K] and T_h=2.6 [K]. The theoretical formula for this mechanism was given in Ref.[1], and the magnetic field dependence of 4Δ for P-doped Ge was given experimentally by AGGARWAL et al.[8], and the magnetic field dependence of a^*(the effective Bohr radius) was calculated by MILLER [4] (the magnetic field dependence of a^* was also calculated by MIKOSHIBA [6], but it seems that this calculation over-estimates the shrinkage effect to explain the present experimental results). The numerically estimated results based on these theories were given in Fig.3 by solid curve, and the numerical estimation was also carried out by neglecting the shrinkage effect of a^* and the results were given by the dotted curve. The scale of the vertical axis is enlarged by ten times compared with that in Fig.1.

(b)The phonon scattering by donor pairs on the basis of the theory by GORTEL taking into account the shrinkage effect of a^* by the [100] magnetic field [6] was calculated for T_o=1.8 [K] and T_h=2.6 [K]. In his theory, the frequency was limited between 46 and 105 [GHz], so the numerical estimation was carried out within this frequency region, but it seems that it is not so unreasonable to employ his theory for the present experiment because higher frequency phonons will not play an important role in this mechanism. Another problem is that his calculation was for L mode, while the experiment was carried out for FT mode, but it seems that this difference is not so essential in the analysis of the experimental results and the calculation for FT mode is too complicated to carry out the numerical estimation. The numerically estimated results for L mode for T_o=1.8 [K] and T_h=2.6 [K] are shown in Fig.4. The scale of the vertical axis is enlarged by one thousand times compared with that in Fig.1.

As seen from these estimations, both mechanisms can not explain the experimental results fully.

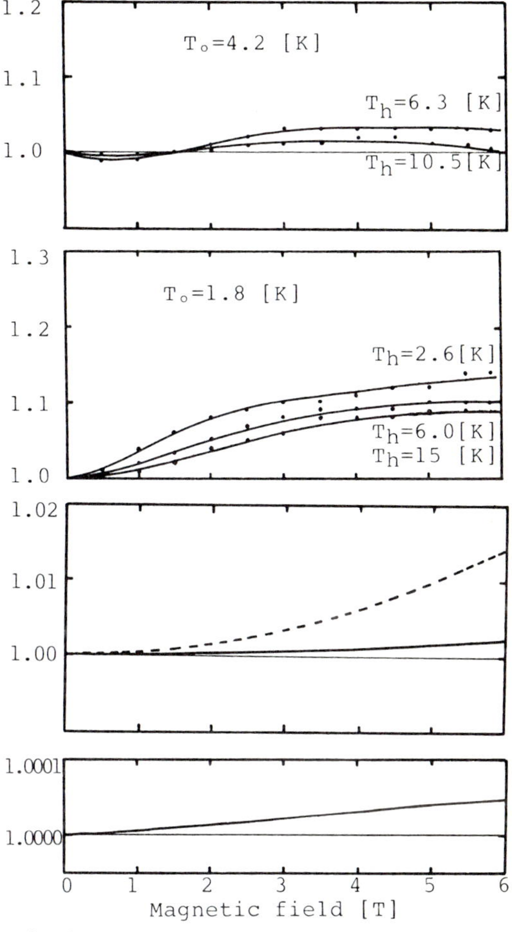

Fig.1. Magnetic-field dependence of ballistic heat-pulse transport in Ge:P. n=5x10^{14}P/cm^3,T_o=4.2 [K],T_h= 6.3 and 10.5 [K],H// [100],q//[1$\bar{1}$0], FT mode

Fig.2. Magnetic-field dependence of ballistic heat-pulse transport in Ge:P.n=5x10^{14}P/cm^3,T_o=1.8,T_h=2.6,6, 15 [K].H//[100],q//[1$\bar{1}$0],FT mode

Fig.3. Magnetic-field dependence of ballistic heat-pulse transport in Ge:P by neutral donor electrons (see text)

Fig.4. Magnetic-field dependence of ballistic heat-pulse transport in Ge:P by donor pairs (see text)

Acknowledgements

The authors would like to express their deep thanks to Prof.R.L.Aggarwal for his kindness in offering P-doped Ge.

References

1. T.Miyasato et al.:J.Phys. Soc.Jpn.50(1981)1986.
2. T.Miyasato et al.:J.Phys. (Paris)C6(1981)658.
3. T.Miyasato et al.:Phonon Scattering in Condensed Matter. ed.by H.J.Maris. p.425-428.(Plenum Press) (1980)
4. A.Miller:thesis, Rutgers Univ.(1960)(unpublished)
5. Z.W.Gortel:J.Phys.C9 (1975)693. ibid.707.
6. N.Mikoshiba:Phys.Rev. 127(1962)1954.
7. T.Ishiguro & S.Morita: Appl.Phys.Lett.25(1974) 533.
8. R.L.Aggarwal et al.:J.Phys. Soc.Jpn.49(1980)Suppl.A. 197.

Hole-Phonon Interaction in Wurtzite-Type Semiconductors

M. Singh* and J. Leotin

Service des Champs Magnétiques Intenses, Laboratoire de Physique des Solides INSA, Avenue de Rangueil, F-31077 Toulouse-Cedex, France

1. Introduction

Neutral shallow impurities in semiconductors are very effective scatterers of thermal phonons at low temperatures. Diamond [1] and zincblende crystals [2] have been widely investigated. In the present paper we calculated for the wurtzite-type semiconductors (CdS, CdSe, ZnO) the phonon relaxation rates due to elastic scattering by shallow acceptors. Taking also into account boundary and point defect scattering we then calculated the phonon conductivity of cadmium sulphide at low temperatures ($T < 10$ K). A large decrease due to the hole-phonon scattering is found.

2. Theory

The three two fold degenerated valence bands in wurtzite-type semiconductors [3] have the following angular momentum symmetries: $\Gamma_9 (J = 3/2, M_J = \pm 3/2)$; Γ_7 $(J = 3/2, M_J = \pm 1/2)$ and Γ_7 $(J = 1/2)$. The remote $\Gamma_7 (J = 1/2)$ band [3] is neglected and the ground acceptor states labelled by $n = 1, 2, 3, 4$ correspond to $M_J = 3/2, 1/2, -1/2, -3/2$ respectively. The ground-state envelope function is assumed to have only s-like contribution. R_B denotes its Bohr radius.

The coupling of the phonons to the acceptor holes is described by a strain hamiltonian derived by Pikus [4] from symmetry arguments:

$$H_s = (C_3 \varepsilon_{zz} + C_4 (\varepsilon_{xx} + \varepsilon_{yy}))\, J_z^2 + C_5 (J_-^2 \varepsilon_+ + J_+^2 \varepsilon_-) + C_6 ([J_z J_+] \varepsilon_{-z} + [J_z J_-] \varepsilon_{+z}) \tag{1}$$

where $[AB] = \frac{1}{2}(AB + BA)$, $\varepsilon_\pm = \varepsilon_{xx} - \varepsilon_{yy} + 2i\, \varepsilon_{xy}$,

$\varepsilon_{\pm z} = \varepsilon_{xz} \pm i\, \varepsilon_{yz}$, $J_\pm = J_x \pm i\, J_y$.

J_α is α component of $J = 3/2$ angular momentum operator, C_i are the deformation potential constants and $\varepsilon_{\alpha\beta}$ are the conventional strain components. Expanding the $\varepsilon_{\alpha\beta}$ into normal modes, the matrix element between two hole acceptor states takes the form :

$$\langle n | H_s | n' \rangle = \sum_{q,\, t} (\hbar\, \omega_{qt} / 2\, M\, v_t^2)^{1/2}\; f(q_t)\; (a_{qt} + a_{qt}^+)\; C_{qt}^{nn'} \,. \tag{2}$$

ω_{qt}, a_{qt}, a_{qt}^+, v_t are respectively angular frequency, destruction and crea-

* Department of Physics, McGill University, Montreal (Canada).

tion operators and velocity of phonon with wave vector q in the branch t. M is the crystal mass. The coefficient $C^{nn'}_{qt}$ referred to as coupling parameters have been thoroughly calculated [5] in the same way as reference [2]. Second Born approximation was used to derive the hole-phonon relaxation rates for elastic scattering. We obtained the following :

$$\tau_e^{-1}\,(qt) = A_t\, T^2\, x^4\, F(x)\, (H_1(x)\, W_t + H_2(x)\, v_t) \tag{3}$$

where :

$$A_t = N\, k_B^2\, f_t(x)/100\pi\, v_1^5\, v_t^2\, \hbar^4\, \rho^2 \;;\; x = \hbar\,\omega_{qt}/k_B T \;;\; x_{AB} = \Delta_{AB}/k_B T$$

$$f_t(x) = \left[1 + (R_B\, k_B T/2\hbar\, v_t)^2\, x^2\right]^{-4} \;;\; F(x) = f_1(x) + \tfrac{3}{2}\, f_2(x)\,.(v_1/v_2)^5$$

$$H_1(x) = (x^2 + x_{AB}^2)/(x^2 - x_{AB}^2)^2 \;;\; H_2(x) = 2\, x^2/(x^2 - x_{AB}^2)^2$$

$$W_1 = 8\, C_6^4 + 2048\, C_5^4 \;;\; W_2 = 7\, C_6^4 + 512\, C_5^4 \;;\; W_3 = 5\, C_6^4 + 2560\, C_5^4$$

$$V_1 = 256\, C_5^2\, C_6^2 \;;\; V_2 = 144\, C_5^2\, C_6^2 \;;\; V_3 = 240\, C_5^2\, C_6^2 .$$

N is the number of holes per unit volume. Δ_{AB} is the absolute energy difference between the valence bands $\Gamma_9(J = 3/2)$ and $\Gamma_7(J = 3/2)$. The subscripts 1, 2, 3 stand for the longitudinal and the two transverse modes respectively. Notice that C_3 and C_4 do not contribute to the relaxation rates.

3. Results and Discussion

The phonon conductivity of cadmium sulphide at low temperatures (T <10 K) has been calculated by :

$$K(T) = (k_B T^3/6\pi^2\,\hbar^3) \sum_{t=1}^{3} v_t^{-1} \int_0^{\theta_D/T} (x^4\, e^{-x}/(e^x - 1)^2\, \tau_{qt}^{-1}\,(x)\,)\,dx \tag{4}$$

$$\tau_{qt}^{-1} = \tau_B^{-1} + \tau_{pt}^{-1} + \tau_{pp}^{-1} + \tau_{e(qt)}^{-1} .$$

τ_B^{-1}, τ_{pt}^{-1}, τ_{pp}^{-1} and $\tau_e^{-1}(q,t)$ are the relaxation rates due to the boundary, the point defects and phonon-phonon and hole-phonon scattering respectively. We take $\tau_{pt}^{-1} = A(k_B/\hbar)^2\, x^4\, T^4$ and $\tau_{pp}^{-1} = B.(k_B/\hbar)^2\, x^2\, T^5$. The numerical values of parameters are listed below :

$\rho(g.cm^{-3}) = 4.8$	$v_1\ (10^5\ cm.sec^{-1}) = 4.42$	$v_2\ (10^5\ cm.sec^{-1}) = 1.94$
$C_5\ (eV) = -\ 1.5$	$C_6\ (eV) = 2.4$	$\Delta_{AB}\ (meV) = 16$
$A(10^{-22}\ sec.K^{-2}) = 16.5$	$B(10^{-22}\ sec.K^{-3}) = 2.74$	$\tau_B^{-1}\ (10^6\ sec^{-1}) = 1.0$
$N(10^{16}\ cm^{-3}) = 1.0$	$\theta_D(K) = 219.3$	$R_B(\mathring{A}) = 19$ and 7 .

The calculated values of K(T) as a function of temperature are drawn in fig. 1 for a pure and a doped sample with $10^{16}\ cm^{-3}$ acceptors. For the doped sample two different values previously reported for the Bohr radius were taken. The decrease of K(T) due to the hole-phonon interaction is shown clearly. Also K(T) is very sensitive to the value of the Bohr radius acceptor state. Fig. 2 shows a plot of the relaxation rates τ_B^{-1}, τ_{pt}^{-1}, $\tau_e^{-1}(q_1)$, $\tau_e^{-1}(q_3)$ against x at 2 Kelvins. The phonon-phonon relaxation rate τ_{pp}^{-1} is negligible at this temperature for any x value.

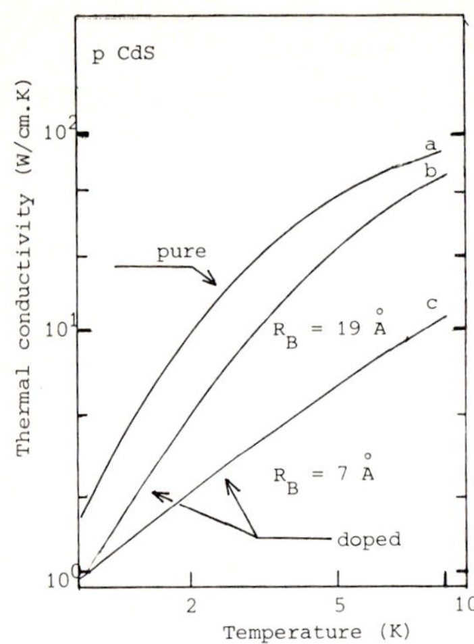

Fig. 1. Temperature variation of the thermal phonons' conductivity in CdS for a pure (Curve a) and a doped sample (Curves b and c) with $10^{16} cm^{-3}$ acceptors. The Bohr radius R_B has the value of 19 Å for Curve b and 7 Å for Curve c

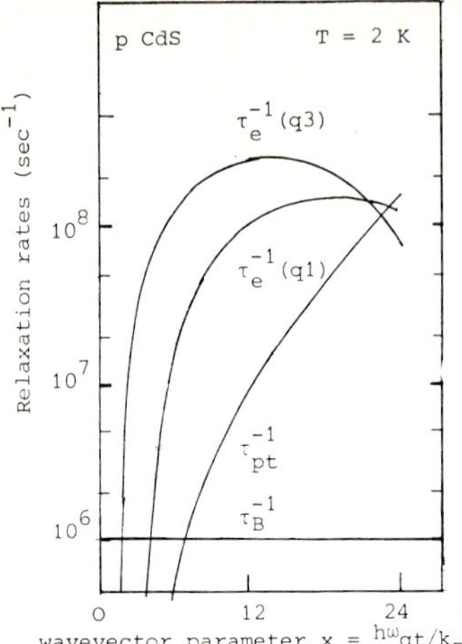

Fig. 2. Wave-vector dependence of phonon relaxation rates in CdS at T = 2 K for longitudinal ($\tau_e^{-1}(q1)$ and transverse phonon modes ($\tau_e^{-1}(q3)$ scattered by acceptor holes, for point defect τ_{pt}^{-1} (q) and boundary scattering τ_B^{-1}

The authors are not aware of experimental data on p-type doped CdS sample in the temperature range where the theoretical above calculations are performed.

ACKNOWLEDGMENT

Prof. P.R. Wallace made a critical reading of the manuscript and provided a grant from N.S.R.C. Canada.

REFERENCES

1. R.W. Keyes : Phys. Rev. 122, 1176 (1961).
   K. Suzuki and N. Mikoshiba: J. Phys. Soc.Jpn, 31, 186 (1971).
   M. Singh and G.S. Verma: Phys. Rev., B 18, 5625 (1978).
2. M. Singh: Phys. Rev., B25, 1214 (1982) and references therein.
3. D.W. Langer, R.N. Euwema, K. Era, T. Koda : Phys. Rev. B2, 4005 (1970).
4. G.E. Pikus, Sov. Phys. Solid State, 6, 261 (1964).
5. To be published.

The Observation of Strongly Coupled Magnetic Ions in Al_2O_3 by Low Temperature Thermal Expansion Measurements

I.J. Brown* and M.A. Brown

Department of Physics, Loughborough University of Technology
Loughborough, Leicestershire LE11 3TU, England

1. Introduction

Strongly coupled magnetic ions in dielectrics have been the subject of a great deal of experimental and theoretical work [1]. The properties of Jahn-Teller ions in particular are often very complicated and it is necessary to study them using several techniques.

The systems for which the most precise data exist and for which detailed theoretical models have been developed involve $(3d)^4$ ions in Al_2O_3 [2,3,4,5]. Very recently, uniaxial stress measurements [6,7] on these systems have confirmed that the multimode Jahn-Teller model satisfactorily described the simple magnetic properties of Cr^{2+}, Mn^{3+} and Ni^{3+} in Al_2O_3. It is found that the ion-lattice coupling is relatively strong so that Jahn-Teller effect dominates, the lattice vibrations are highly anharmonic, and that Cr^{2+} experiences a stronger trigonal field than Mn^{3+} in Al_2O_3.

Sheard [8] suggested that thermal expansion at low temperatures would provide useful information, especially when coupled with specific heat data, about the pressure dependence of the energy level splittings. In particular, the determination of the magnetic Grüneisen coefficients, γ_S, of the levels would provide a good test of the model.

2. Experimental Methods

Work has been carried out at Loughborough since 1977 to develop a high precision, low temperature, three-terminal capacitance dilatometer capable of detecting the contribution of small quantities ($\sim$ 10p.p.m.) of strongly coupled magnetic ions to the thermal expansion of their host lattices. This work was done in collaboration with the C.E.N. Grenoble (who measured the specific heat of the same samples) supported by the S.E.R.C. and the Scientific Affairs Division of N.A.T.O.

A dilatometer has been developed, and hand-built in the Physics Department, that is capable of reproducible measurements of specimen length as a function of temperature to $\pm$ 0.02Å. Most experiments are of long duration so the apparatus has been automated using a DAI computer and industrial control rack, all interface control gear and software having been developed 'in house'. The main object of constructing such a dilatometer was to obtain additional information on the properties of strongly coupled Jahn-Teller ions, particularly Cr^{2+} and Mn^{3+} ions in Al_2O_3.

---

* Present address - Oxford Instruments, Osney Mead, Oxford, U.K. OX2 ODX.

Initial experiments measured the thermal expansion of γ-irradiated Al_2O_3:Cr relative to H.C.O.F. copper and showed that a large positive peak in the thermal expansion was present at ∿ 4K, scaling with total chromium concentration |9|. This peak was attributed to the presence of Cr^{2+} (formed from Cr^{3+} by γ irradiation). Having established that the dilatometer was able to detect low temperature structures, the expansion cell was redesigned so that the thermal expansion of Al_2O_3:Cr was measured relative to 'pure' Al_2O_3, both after γ irradiation and after subsequent U-V bleaching |10|. These results confirmed our assignment of Cr^{2+} as the source of the positive peak and this technique has been applied to Al_2O_3 samples doped with different magnetic ions, such as Mn and V.

3. Results and Discussion

The measurements of the thermal expansion of Mn-doped and V-doped Al_2O_3 relative to 'pure' Al_2O_3 at low temperatures are shown in figures 1 and 2 respectively. The Mn-doped Al_2O_3 specimen was measured 'as received' and after further high temperature oxidising treatment. Earlier work |11| suggested that this would ensure the maximum concentration of Mn^{3+} in the sample and hopefully (since Mn^{3+} is isoelectronic with Cr^{2+}) would increase the size of the sharp positive peak observed at ∿ 3.8K in the thermal expansion data. To our surprise, the effect of the heat treatment, as shown in figure 1, was to move the peaked feature to ∿ 4.6K with no appreciable increase in the total area of the feature. We presume that this is caused by some annealing of the lattice.

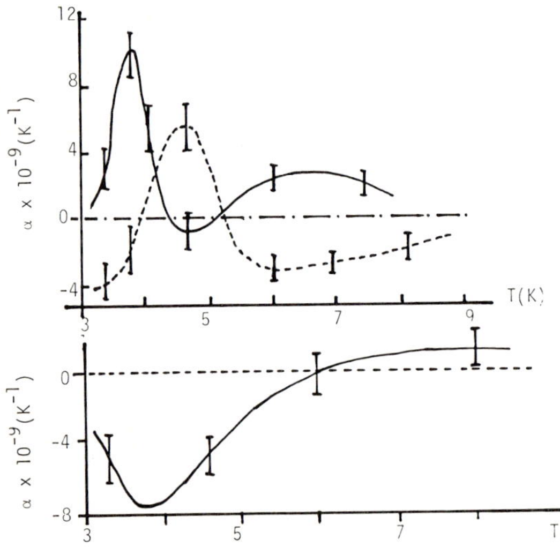

Fig. 1. Thermal expansion of Al_2O_3:Mn in the 'as received' state and after oxidising heat treatment

Fig. 2. Thermal expansion of Al_2O_3:V 'as received'

In summary, we have observed sharp positive peaks in the thermal expansion data (with negative wings and a width of ∿ 2K for both strongly coupled Jahn-Teller ions, Cr^{2+} and Mn^{3+} in Al_2O_3. Early work by Sheard and Bates |12| indicates that a negative wing on the low temperature side would be consistent with the current theoretical models. However, the peak would appear to be too narrow to be consistent with a simple Schottky peak, using different magnetic Grüneissen coefficients for each energy level.

In contrast, the thermal expansion data for 'as received' Al_2O_3:V^{3+} (essentially a simple three-level system) shows a small negative peak

∿ 3.5K (see figure 2). The peak maximum and width are consistent with a simple Schottky model.

4. Acknowledgements

We wish to acknowledge the financial support of the Science and Engineering Research Council and the Scientific Affairs Division of N.A.T.O. We thank the Royal Society for a travel grant and our colleagues in C.E.N. - Grenoble for useful discussions and the loan of specimens.

5. References

|1| Bates, C.A. (1978) Phys. Rep., 35, 187 - 304.

|2| Bates, C.A., Brauns, P., Fletcher, J.R. and Jaussaud, P.C. (1976) J. Physique, 37, 763 - 7.

|3| Flether, J.R., Grimshaw, J.M., Knowles, A.P. and Moore, W.S. (1980) J. Phys. C., 13, 6391 - 7.

|4| Bates, C.A. and Wardlaw, R.S. (1980), J. Phys. C., 13, 3609 - 24.

|5| Zoller, W., Dietsche, W., Kinder, H., de Goer, A-M. and Salce, B., (1980) J. Phys. C., 13, 3591 - 607.

|6| Abhvani, A.S. and Bates, C.A. (1981), J. Phys. C., 14, 2617 - 28.

|7| Salce, B., Abhvani, A.S. and Bates, C.A. (1983)m submitted to J. Phys. C.

|8| Sheard, F.W. (1971), Third Thermal Expansion Symp., Coming, N.Y. P.155.

|9| Brown, I.J. and Brown, M.A. (1981), Phys. Rev. Letts., 46, 835 - 8.

|10| Brown, I.J. and Brown, M.A. (1983), J. Phys. C., 16, 1031 - 7.

|11| Zoller, W., Dietsche, W., Kinder, K., de Goer, A-M. and Salce, B (1980), J. Phys. C., 13, 3591 - 607.

|12| Sheard, F.W. and Bates, C.A. (1983), Private communication.

Sound Velocity Measurements in Highly Oriented and Intercalated Graphite Specimen by Direct Electromagnetic Excitation of Ultrasound

K. de Groot, V. Müller, and D. Maurer

Institut für Atom- und Festkörperphysik B, Freie Universität Berlin, Arnimallee 14, D-1000 Berlin 33, W.-Germany

V. Geiser and H.-J. Güntherodt

Institut für Physik, Universität Basel, CH-4056 Basel, Switzerland

1. Introduction

Actually Graphite Intercalation Compounds (GIC) are of increasing practical and fundamental interest because of their highly anisotropic structure and high in-plane conductivity. Along the c axis they show a well-defined periodic sequence of graphite layers and ordered (or disordered) intercalant layers. Compared to the strongly coplanar bonded graphite layers, the interlayer interactions are relatively weak, thereby giving rise to highly anisotropic physical properties. In spite of numerous investigations, however, the origin of the long-range forces involved in the formation of the interplane intercalation structure, as well as the quasi 2D commensurate-to-glass phase transition /1/,/2/, are still not well understood. Ultrasonic measurements therefore are of particular interest since they provide direct information about the interaction forces among the intercalant and graphite layers. Moreover, the sound velocity is highly sensitive to cooperative phenomena and therefore to thermodynamical structural fluctuations appearing in the vicinity of a phase transition point.

In contrast to microscopic techniques - like ESR, NMR and Mössbauer effect - which, above all, are sensitive to short-range structural fluctuations, ultrasonic velocity measurements therefore should provide detailed information about critical phenomena in GIC's which otherwise are hard to obtain.

In order to excite the ultrasonic waves in the specimen, however, piezoelectric transducers cannot be applied for most of the GIC's because the great majority of graphite intercalation compounds are not stable in air and will react with the ultrasonic bonding agents. That's why successful ultrasonic velocity measurements on GIC's only have been reported for stage 1 and stage 2 $FeCl_3$ intercalated graphite at room temperature /3/. Nevertheless, these difficulties may be overcome by using a direct electromagnetic generation and detection technique of ultrasound /4/ where, by aid of an rf-electromagnetic field, an rf-electric current is excited (within the skin depth) at the surface of the specimen and, by applying a strong dc magnetic field oriented either parallel or perpendicular to the sample surface, longitudinal or transverse acoustic waves can be generated via the Lorentz force and the electron-lattice interaction. Accordingly the inverse process may be used to detect the ultrasonic waves by virtue of the ultrasonically induced electromagnetic field.

Here we report the first successful application of this technique to the determination of the temperature dependence of the velocity of longitudinal ultrasonic waves propagated along the c axis of HOPG, as well as of a stage 4 $SbCl_5$-graphite specimen. The measurements were performed in the temperature range from 30 to 273 K and to our knowledge this is the first time that both the velocity of longitudinal sound and its temperature dependence have been investigated for $SbCl_5$-graphite. Moreover, there is some evidence for a cricital phenomenon (phase transition) at T ≈ 135 K which has not been reported yet.

2. Experimental

Figure 1 shows a typical ultrasonic pulse echo train in HOPG and (see insert) the electromagnetic generation and detection scheme for longitudinal acoustic waves. The rf coil serves as "electromagnetic transducer" and is driven by a modified commercial ultrasonic pulse spectrometer (MATEC Inc., USA).

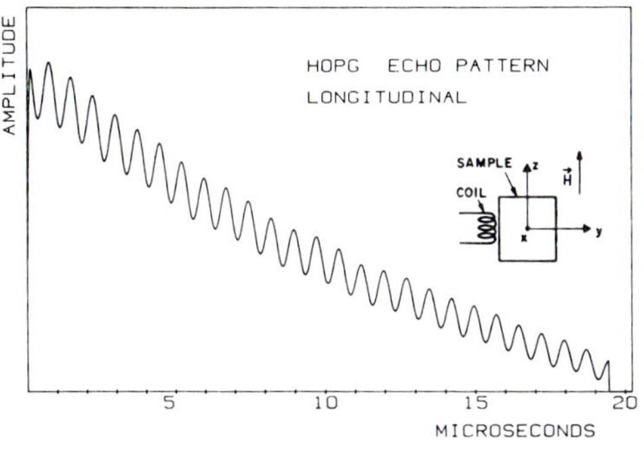

Fig. 1 Electromagnetically excited and detected ultrasonic pulse echo train in a HOPG specimen of 1.55 mm thickness and 5 MHz longitudinal acoustic waves propagated along the c axis at B_0 = 1.25 T

Before intercalation the quality of the HOPG samples (Union Carbide, grade ZYA) was checked in an ultrasonic velocity and attenuation experiment. The HOPG specimen were squares of about 10 x 10 mm with c-axis thickness ranging from 1.2 to 1.55 mm. The stage 4 $SbCl_5$-graphite was prepared by the two-zone vapour transport method /5/. The staging fidelity was established by X-ray diffraction measurements. After intercalation the as-grown GIC was sealed under vacuum into a rectangular quartz container.

3. Results and Discussions

Figure 2 shows for an external dc-magnetic field of 1.25 T the temperature dependence of the longitudinal sound velocity c_a in HOPG, and Fig. 3 - with reference to HOPG from which the GIC was prepared - the respective temperature dependence of the stage 4 $SbCl_5$-graphite sample. It should be realized that in both cases the velocity of longitudinal sound increases with decreasing temperature, but the variation with temperature in the $SbCl_5$-GIC (stage 4) is about one order of magnitude larger. The most striking feature however, is that for T $\gtrsim$ 150 K the sound velocity in $SbCl_5$-graphite is smaller and for T $\lesssim$ 120 K is larger than that one in HOPG, whereas the numerous studies of the elastic constants in various GIC's suggest /3/ that graphite becomes harder upon intercalation with donors and softer upon intercalation with acceptors. Accordingly, our results are consistent with this rule only if, on cooling the GIC, the acceptor-type intercalation compound $SbCl_5$-graphite (stage 4) undergoes an electronic phase transition from a hole-like to electron-like conductor with T ≈ 150 K and T ≈ 120 K as upper and lower limits of the transition temperature. However, from the preliminary data presented here no further conclusions can be drawn yet, although there is some evidence for a cricital phenomenon (critical slowing down?) between 120 K and 150 K where the sound velocity in HOPG and $SbCl_5$-graphite are expected to become identical.

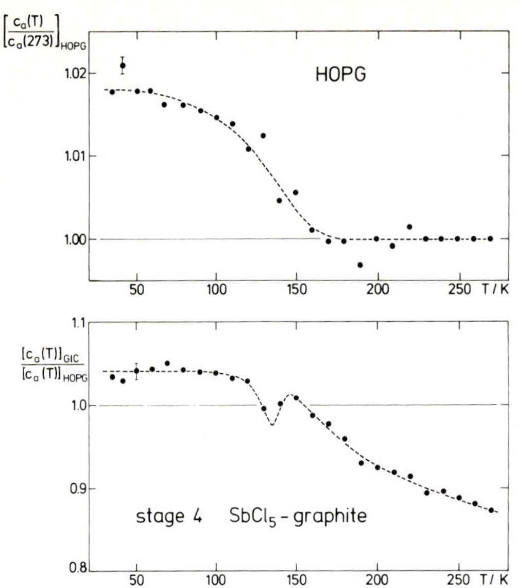

Fig. 2 Temperature dependence of the longitudinal sound velocity in HOPG

Fig. 3 Temperature dependence of the longitudinal sound velocity in stage 4 $SbCl_5$-graphite relative to HOPG

References

/1/ G. Timp, M.S. Dresselhaus, L. Salamanca-Riba, A. Erbil, L.W. Hobbs, G. Dresselhaus, P.C. Eklund and Y. Jye: Phys.Rev. B26, 2323 (1982)

/2/ P. Bak and T. Bohr: Phys.Rev. B27, 591 (1983)

/3/ D.M. Hwang, B.F. O'Donnell and A.Y. Wu: in Physics of Intercalation Compounds, ed. by L. Pietronero and E. Tosatti (Springer Series in Solid State Sciences 38, 1981) p. 193

/4/ K. Saermark and P.K. Larsen: Phys.Lett. 24A, 374 and 668 (1967); also see: J.R. Houck, H.V.Bohm, B.W. Maxfield and W.J. Wilkins; Phys. Rev.Lett. 19, 224 (1967)

/5/ K. Fredenhagen and G. Cadenbach: Z.allg.anorg.Chem. 158,249 (1926)

The Effect of Electron Relaxation on Damping of Long-Wavelength Phonons in Metals and Heavily Doped Semiconductors

I.P. Ipatova, A.V. Subashiev, and V.A. Shchukin

A.F. Ioffe Physical Technical Institute, Academy of Sciences of the USSR
Leningrad 194021, USSR

Electron-phonon interaction in metals and heavily doped semiconductors leads to the specific damping which could be the same order of magnitude as the anharmonic one or even larger. The effect is different in two opposite cases:

1. If $kl \gg max(\omega\tau, 1)$ then the large mean free path l of electrons enables some of them to propagate long enough in phase with the phonon($\omega = k v_F$, ω and k being the frequency and the wave vector of the phonon, v_F and τ being the Fermi velocity and the relaxation time of electrons). These electrons are subjected to the stationary phonon force. Therefore the dissipation of the phonon energy occurs /1/. The corresponding phonon damping is called Landau damping.

2. If $kl \ll max(\omega\tau, 1)$ then the short mean free path of electrons results in the random-phase regime. The phonon creates fluctuations of electron distribution function which relax due to the electronic collisions with impurities or phonons. The electron relaxation leads to the phonon damping in this case.

When the phonon frequency is low($\omega\tau \ll 1$) electrons relax to the local equilibrium defined by the instantaneous value of the phonon field. Corresponding kinetic equation linearized with respect to the electron-phonon interaction has the form

$$(-i\omega + i\underline{k}\underline{v} + \hat{I})\delta n_{\underline{p}} = i\omega(V_{\underline{p}} - e\varphi)\frac{\partial n^0_{\underline{p}}}{\partial \varepsilon} \quad (1)$$

where V_p is the phonon-induced energy shift for the electron with momentum $\underline{p}$, $\hat{I}$ is the collision integral, $n^0_{\underline{p}}$ is the equilibrium Fermi distribution function, φ is the electrostatic potential accounting for screening effects. Since the corresponding correction terms are of the order of $V_{\underline{p}}^2$ and do not appear in the linearized equation the same kinetic equation (1) proves to be valid at high phonon frequences too ($\omega\tau \gtrsim 1$).

The damping of the longwavelength phonon is proportional to the averaged energy transfer to the electronic subsystem:

$$\gamma = \frac{1}{2\hbar} \, Im \sum_{\underline{p}} V_{\underline{p}} \, \delta n_{\underline{p}} \, . \tag{2}$$

The isotropic fluctuations of the total electron density are screened out. But there are anisotropic fluctuations of the electron distribution function which give rise to non-zero phonon damping. They are classified according to irreducible representations Γ_i of the crystal point group. The phonon damping equals in this case /2/

$$\gamma = \sum_{\Gamma_i} \zeta_{\Gamma_i} \frac{\omega^2 \tau_{\Gamma_i}}{(1 + \omega \tau_{\Gamma_i})} \tag{3}$$

ζ_{Γ_i} being the constant of the electron-phonon interaction.

The scattering of electrons by thermal equilibrium phonons is the main process of the electron relaxation at some temperatures. The relaxation time τ decreases with temperature increase. Therefore when $T > \Theta_D$ (Θ_D is Debye temperature) the phonon damping could decrease with the temperature increase.

Nonmonotonous temperature dependence of the acoustic phonon damping has been observed by PINE in CdS /3/. The dependence has been explained by acoustoelectric effects /4/. The phonon damping decrease with the temperature increase has been recently observed for optical phonons by WIPT et al./5/ in V_3Si and by PONOSOV /6/ in Ru,Re,Os (see Fig.1).

The main reason of electron relaxation at low temperatures ($T \ll \Theta_D$) is the electron scattering by impurities. Since $\tau^{-1} \sim N_{imp}$ (N_{imp} is the concentration of impurities) the theory predicts that the phonon damping could decrease when N_{imp} increases.

There are out-of-phase by π fluctuations of the electron concentration in the neighbouring valleys of many-valley semiconductors. These fluctuations are not screened. The phonon damping is defined in this case by both the electron intervalley relaxation and the intravalley electron diffusion. The intervalley relaxation time in n-Ge is much larger than the intravalley one $\tau_{inter} \gg \tau_{intra}$. The damping of the phonon propagating along the fourth order axis in the collision-

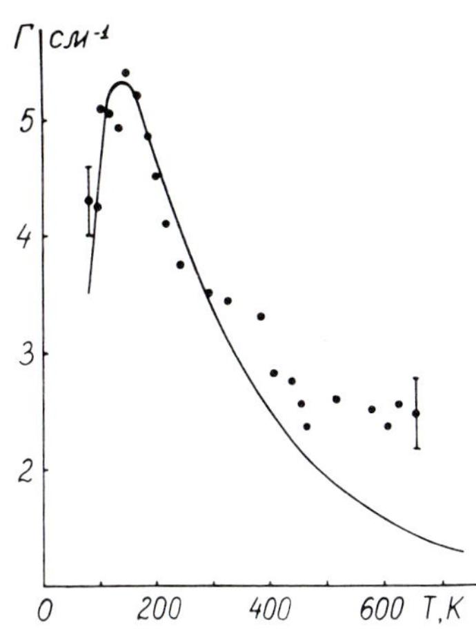

Fig.1 The temperature dependence of phonon damping in Os /6/

controlled regime ($\omega\tau_{intra} \ll 1$, $kl_{intra} \ll 1$) has the form

$$\gamma = \zeta \frac{\omega^2(\tau_{inter}^{-1} + \mathcal{D}k^2)}{\omega^2 + (\tau_{inter}^{-1} + \mathcal{D}k^2)^2} \tag{4}$$

where $\mathcal{D} = \frac{1}{3} v_F^2 \tau_{intra}$ is the diffusion coefficient.

REFERENCES

1. I.P.Ipatova, A.V.Subashiev: JETP 66 , 722 (1974)
2. I.P.Ipatova, A.V.Subashiev, V.A.Shchukin: FTT 24 , 3401 (1982)
3. A.S.Pine: Phys.Rev. B5 , 2997 , (1972)
4. A.R.Hutson, D.L.White: J.Appl. Phys. 33 , 40 (1962)
5. H.Wipt, M.V.Klein, B.S.Chandrasekhar, T.H.Geballe, J.H.Wernick: Phys. Rev. Lett. 41, 1752 (1978)
6. Yu.S.Ponosov: FTT 23, 1477 (1981)

Nuclear Acoustic Resonance Measurements of the Electron-Phonon Interaction in bcc Transition Metals

V. Müller, E.-J. Unterhorst, and W. Neumann

Institut für Atom- und Festkörperphysik (B), Freie Universität Berlin, Ar-Arminallee 14, D-1000 Berlin 33

1. Introduction

From pair-potential-model based analyses it has been deduced /1/ that in metals a substantial fraction of the outer electrons of atoms are highly incompressible but shape deformable. It is therefore expected that phonon-induced changes in the shape of the electron spatial charge distribution will play the dominant role in the volume conserving part of the electron-phonon interaction. Confining ourselves to cubic metals, it is easy to realize that changes in the shape of the electron cloud may only be accomplished by the shear part $\hat{\epsilon}$ (i.e. the volume conserving part) of the strain field ϵ, whereas changes in volume will originate from the isotropic so-called hydrostatic part $1 \cdot \mathrm{Tr}\,\epsilon$. If, on the other hand, the Wigner-Seitz cells of a cubic metal are (homogenously) strained by a long-wave-length ($\lambda \gg \lambda_{Debye}$) ultrasonic shear mode, then the lowest electric multipole field of each Wigner-Seitz cell is no longer the 16-pole field - as in the undistorted (or "hydrostatically"strained) cubic crystal - but the strain-induced dynamic electric field gradient (DEFG) tensor $\hat{\hat{V}}$ whose components can be determined in a nuclear acoustic resonance (NAR) experiment. NAR measurements of the DEFG in transition metals should therefore provide detailed informations about the electron-phonon interaction - in particular that of d electrons - which otherwise are hard to obtain.

Moreover, NAR experiments enable one to study the conduction-electron response to the two fundamental shear deformations - "tetragonal shear" (i.e. $\hat{\epsilon}_{zz} = -2\,\hat{\epsilon}_{xx} = -2\,\hat{\epsilon}_{yy} = 0,\ \hat{\epsilon}_{xy} = \hat{\epsilon}_{xz} = \hat{\epsilon}_{yz} = 0$) and "trigonal shear" (i.e. $\hat{\epsilon}_{xx} = \hat{\epsilon}_{yy} = \hat{\epsilon}_{zz} = 0,\ \hat{\epsilon}_{xy} = \hat{\epsilon}_{xz} = \hat{\epsilon}_{yz} = 0$) - separately. Regarding that for trigonal, as well as for tetragonal shear, a principal axes system (X,Y,Z) does exist in which the DEFG tensor becomes a digonal tensor, then the conduction-electron response to tetragonal and trigonal shear is most easily defined by /2/

$$r_{\substack{tetr\\(trig)}} = \left[(\hat{V}_{ZZ})_{ce} / ((1-\gamma_\infty)\,(\hat{V}_{ZZ})_{latt}\right]_{\substack{tetr\\trig}}$$

where we have made use of the usual ansatz: $\hat{V}_{ZZ} = (1-\gamma_\infty)\,(\hat{V}_{ZZ})_{latt} + (V_{ZZ})_{ce}$. Here $(\hat{V}_{ZZ})_{latt}$ is the lattice contribution of the screened metal ions, γ_∞ is the Sternheimer antishielding factor and $(V_{ZZ})_{ce}$ the conduction electron contribution within the Wigner-Seitz cell under regard which also comprises modifications by the distorted ion core.

2. Experimental

Successful DEFG-NAR experiments in bcc transition metals have been performed in Nb /3/, Mo /4/, Ta /5/, the dilute alloys $Nb_{1-x}Mo_x$ /6/ and recently also in V /2/.

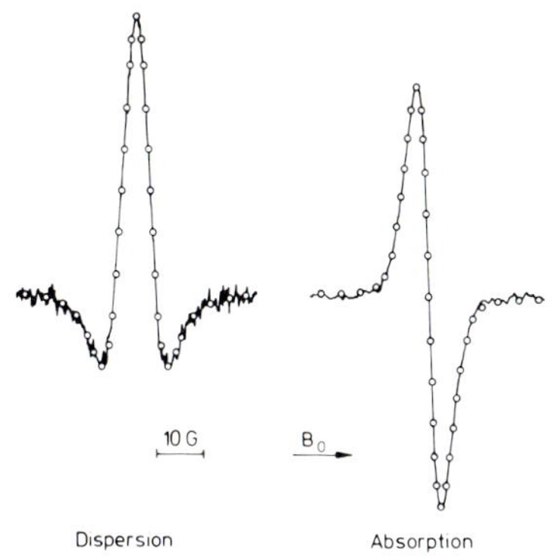

Fig. 1 Derivative of the $|\Delta m| = 2$ NAR absorption and NAR dispersion line of V^{51} averaged over 8 runs at T = 1.6 K. 19.9 MHz shear waves were propagated along [110]

Fig. 2 Angular dependences of the $|\Delta m| = 2$ NAR signal intensity in Ta for the dc magnetic field rotated in different crystal planes. Longitudinal acoustic waves were propagated along [110]

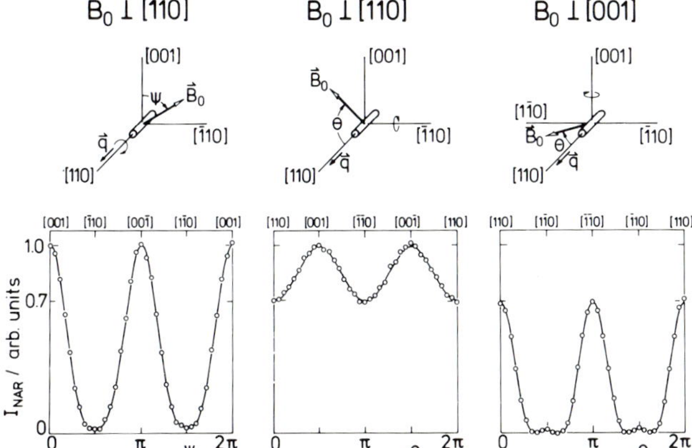

Fig. 1 shows a typical quadrupolar NAR absorption and NAR dispersion derivative signal $|\Delta m| = 2$ in a [110] single crystal of vanadium. Longitudinal acoustic waves were propagated along the [110] direction and the relevant DEFG components were determined from the angular dependences of the NAR signal intensity for the dc magnetic field rotated in different crystal planes. As an example Fig. 2 shows the relevant angular dependences of the $|\Delta m| = 2$ NAR signal intensity in Ta. The [110] single crystals were cylinders of about 1 cm in length and diameter. After grinding and polishing the end faces were flat and parallel to within 1 µm and were measured to be within $\pm$ 2° of the (110) crystal plane. In order to remove interstitial hydrogen and to reclude internal strains caused by crystal cutting and grinding, the specimens were carefully annealed in a vacuum of 10^{-10} Torr at high temperatures. After annealing a unilaterally gold-plated quartz piezoelectric transducer was bonded to the sample with polysulfide liquid polymers (Thiokol, LP 33).

3. Results and Discussion

The experimental results for the conduction-electron response quantities r_{trig} and r_{tetr} are shown in Fig. 3. The most striking feature is that, within experimental error, r_{trig} proves to be proportional to the density

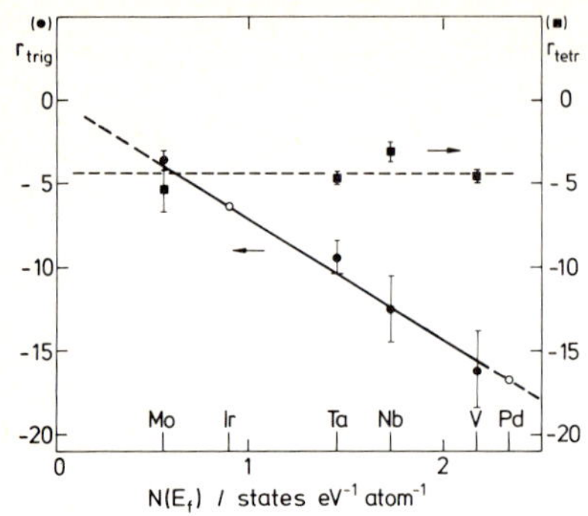

Fig. 3 Electron-phonon coupling in bcc transition metals for tetragonal and trigonal shear modes. The open circles are predicted data for fcc transition metals and were obtained on the supposition that fcc metals will follow the bcc systematics

of states $N(E_f)$ at the Fermi energy (note $N(E_f) \approx N_d(E_f)$) whereas r_{tetr} seems to be insensitive against $N(E_f)$. This disparity has recently been discussed by Bennemann /7/ who, by aid of energy arguments, concluded that due to the different coupling of s and d electrons to shear waves and volume deformations, this behaviour should be valid throughout the transition metal series. However, taking into account that changes in volume do not contribute to the electric quadrupole interaction and therefore cannot be observed in quadrupole NAR, it follows that the NAR results cannot be explained by differences in the coupling of s and d electrons to shear and volume deformations.

To throw more light upon the open questions, we have performed a multipole expansion of the electron-phonon interaction yielding /2/ that only the volume conserving part of radial distortions of the electron cloud should contribute to r_{tetr} and r_{trig}. With the assumption /8/ that the d-electron bonds will follow the relative displacements of the next nearest neighbours of the nucleus under regard and taking furthermore into account that tetragonal, as well as trigonal shear, will give rise to s-d transitions, we finally get /2/ for bcc transition metals

i) For trigonal shear r_{trig} is dominated by compressions and expansions of the d bonds ($\propto N_d(E_f) \approx N(E_f)$),

ii) For tetragonal distortions r_{tetr} is dominated by s-d transitions ($\propto N_s(E_f) \approx$ const.).

References

/1/ E.S. Machlin and S.H. Whang: Phys.Rev.Lett. 41, 1421 (1978)

/2/ V. Müller, E.J. Unterhorst, W. Neumann, G. Schanz and C. Schubert: Phys.Lett.A, accepted for publication (1983)

/3/ J. Pellison and J. Buttet: Phys.Rev.B 11, 48 (1975)

/4/ E. Fischer, V. Müller, D. Ploumbidis and G. Schanz: Phys.Rev.Lett. 40, 796 (1978)

/5/ B. Ströbel and V. Müller: Phys.Rev. B 24, 6292 (1982)

/6/ E.-J. Unterhorst, V. Müller, G. Schanz and D. Maurer: in Superconductivity in d- and f-Band Metals, ed. by W. Buckel and W. Weber, 389 (1982)

/7/ K.H. Bennemann: Phys.Rev.Lett. 42, 676 (1979)

/8/ M. Peter, W. Klose, G. Adam, P. Entel and E. Kudla: Helv.Phys.Acta 47, 807 (1974)

Revision of the Statistical Mechanics of Phonons to Include Phonon Linewidths*

W.C. Overton, Jr.

Los Alamos National Laboratory, Los Alamos, NM 87544, USA

1. Introduction

Zubarev[1] in 1960 obtained the "smeared" Bose-Einstein (B-E) function in order to take into account the fact that the eigenenergy associated with a fixed phonon wave vector q and fixed polarization index j is not precisely defined but, instead, is smeared by phonon-phonon and phonon-electron interactions. The ratio $\Gamma(qj)/\omega(qj)$ is often quite small, i.e., of the order of 0.01 or less, where Γ is the phonon linewidth and $\hbar\omega$ is the eigenenergy. However, in strongly anharmonic crystals Γ/ω may be as large as 0.3 at certain points of the Brillouin zone. In such dramatic cases, one would suspect that such phonon linewidths would have some observable effect on the thermodynamic properties. Zubarev represented the effect of "smearing" on the statistical properties by the infinite integral[1],

$$\bar{n} = \int_{-\infty}^{\infty} d\omega \, [\exp(\frac{\hbar\omega}{kT}) - 1]^{-1} L(\omega;\bar{\omega},\Gamma) \quad , \tag{1}$$

$$L(\omega;\bar{\omega},\Gamma) = (\Gamma/\pi)[(\omega - \bar{\omega})^2 + \Gamma^2]^{-1} \quad , \tag{2}$$

in which we have deleted the indices (q,j) for convenience. The term in square brackets in (1) is the usual B-E function, while in L, the usual Lorentzian function, $\bar{\omega}$ is the average or center frequency of the distribution.

Equation (1) is not usable as it stands. However, we have found a simple formula which is the exact equivalent of (1) by the use of contour methods. We obtain for the average B-E function,

$$\bar{n} = (e^{\bar{x}}\cos\gamma - 1)/(e^{2\bar{x}} - 2e^{\bar{x}}\cos\gamma + 1) - \sum_{\ell=1}^{\infty} 8\ell\pi\bar{x}\gamma/[(-4\ell^2\pi^2 + \bar{x}^2 + \gamma^2)^2 + 16\ell^2\pi^2\bar{x}^2] \tag{3}$$

where $\bar{x} = \beta\hbar\bar{\omega}$, $\gamma = \beta\hbar\Gamma$ and $\beta = 1/kT$. The summation part of (3) is due to poles on the imaginary axis of the ω plane. When we use (3) to derive the specific heat and entropy, we find that the entropy slope $\partial S/\partial T \to \infty$ as $T \to 0$ K, due to the summation part of (3). The first part of (3) is well behaved. Accordingly, the use of (1) is proven to be invalid.

2. A New Approach

Since we have shown that (3), which is equivalent to (1), is invalid, we have tried using the statistical mechanical expressions for the phonon partition function, entropy, and free energy, in the place of the B-E function in (1).

*Work performed under the auspices of the U.S.D.O.E.

Use of the two former functions leads to summations similar to that in (3) which are invalid. However, use of the phonon free energy per mode $\hbar\omega/2 + kT \ln[1 - \exp(-\beta\hbar\omega)]$ in (1) instead of the B-E function leads to a result that is well behaved at all temperatures. The purpose of this work is to derive the expression for the average free energy per mode $\bar{f}$ for a crystal having large phonon linewidths and to test the properties of the thermodynamic functions derivable from $\bar{f}$.

The procedure is to insert f in (1) and to assume ω is complex. The line integration is accomplished by contour integration over the ω plane. The counterclockwise contour for the upper half plane consists of a line ε above the real axis (with small semicircle at the origin) plus a large positive semicircle. Poles occur at $\omega = + 2n\pi i\ (kT/\hbar)$; n = 1,2,..., but the residues are zero for all n. The clockwise contour for the lower half plane consists of a line ε below the real axis (with small semicircle at origin) plus the large negative semicircle. Poles occur at $\omega = - 2n\pi i\ (kT/\hbar)$; n = 1,2,..., but, again, all of the residues are zero. The only residues of importance are at the poles $\omega = \bar{\omega} + i\Gamma$ (upper half plane) and $\omega = \bar{\omega} - i\Gamma$ (lower half plane). Results for the two half planes are averaged as discussed by Morse and Feshbach [2], a procedure which leads to,

$$\bar{f} = \hbar\bar{\omega}/2 + (kT/2) \ln [1 - 2 \exp(-\bar{x}) \cos\gamma + \exp(-2\bar{x})] \quad , \tag{4}$$

where $\bar{x} = \hbar\bar{\omega}/kT$ and $\gamma = \hbar\Gamma/kT$. This same procedure was used in deriving $\bar{n}$ in (3) from (1).

3. Thermodynamic Applications

The specific heat per mode is obtained by the formula $c_V = - T\, \partial^2 \bar{f}/\partial T^2$. However, the exact form of c_V depends on the assumptions made as to the temperature dependence of the anharmonic shift $\Delta(qj)$ and half width $\Gamma(qj)$. Maradudin and Fein [3] derived expressions for Δ and Γ and found approximations suitable for the very low temperature range and for the high temperature range with T > Debye Θ. At low temperatures, Δ and Γ are nearly independent of T while at high temperatures they vary linearly with T. They also calculated formulas for Δ and Γ for high temperatures based on an anharmonic potential function (Morse potential) which, in turn, was based on the heat of sublimation of lead. Using their data, we thus have a basis for testing the possible validity of (4), at least for the case of crystalline lead. Thus, for the high temperature approximation, we may write,

$$\Delta = d\,T;\quad \bar{x} = \hbar\omega_0/kT + \hbar d/k = x_0 + \delta;\ \Gamma = g\,T;\quad \gamma = \hbar g/k \ , \tag{5}$$

where ω_0 is the harmonic (or quasiharmonic) eigenfrequency. Using (5) in (4), we obtain the mode specific heat,

$$c_V = k\, x_0^2\, e^{\bar{x}}[(e^{2\bar{x}} + 1) \cos\gamma - 2e^{\bar{x}}]/(e^{2\bar{x}} - 2e^{\bar{x}}\cos\gamma + 1)^2. \tag{6}$$

Since Γ is small in the high temperature approximation, we may replace $\cos\gamma$ by $(1 - \gamma^2/2)$. The shift Δ is also small and we may replace $\exp(\bar{x})$ by $\exp(x_0)(1 + \delta + \delta^2/2)$. Thus, (6) can be rewritten in the form,

$$c_V/k = x^2\, e^x/(e^x - 1)^2 - \delta\, x\, e^x(e^x + 1)/(e^x - 1)^3 + (1/2)(\delta^2 - \gamma^2)\, x^2\, e^x(e^{2x} + 4\, e^x + 1)/(e^x = 1)^4 \quad , \tag{7}$$

in which x now denotes $x_0 = \hbar\omega_0/kT$. The leading term in (7) is the ordinary harmonic specific heat. When integrated over the spectrum, this leads to the Dulong-Petit value of 3R per mole for T > Debye Θ. Upon expanding further for the high temperature approximation, we observe that (7) can be expressed in the form,

$$c_v/k = c_{har}/k - (2\delta/x)(1 - x^4/40) + (3\delta^2/x^2 - 3\gamma^2/x^2)(1 + x^4/720). \quad (8)$$

Brockhouse et al. [4] have measured $\Gamma(qj)$ for lead by neutron scattering methods and find for the longitudinal mode at the [100] zone boundary and at 425 K the half width Γ = 1.35 tHz while the largest phonon frequency ω_L is 20.6 tHz, i.e., $\Gamma_{max}/\omega_L = 0.0655$. Thus, in view of (5), we may write $\Gamma(T) = (0.0655)(T/425)\omega_L$. For interior points we make the further approximation, $\Gamma(qj) = (0.0655)(T/425)\omega_L[\omega(qj)/\omega_L]$. Thus, $\gamma(qj) = (0.0655)(T/425)[\hbar\omega(qj)/kT] = (0.0655)(T/425)\,x$.

Since (4) and (7) correspond to the case in which volume V = V(0 K), we need only the cubic and quartic shifts while the shift due to thermal expansion does not enter. Following Ref. [3], we have for the quartic shift in lead, $\Delta_4 = (0.0151)(T/\Theta_\infty)\omega_L\,\lambda(qj)$, where λ = reduced frequency and $\lambda_{max} = 2$ and $\lambda/\lambda_{max} = \omega/\omega_L$. Thus $\Delta_4 = (0.0302)(T/\Theta_\infty)\omega(qj)$ so that $\delta = (0.0302)(T/\Theta_\infty)x$. Since the cubic contribution is about five times smaller, we make little error by assuming a similar formula. Based on [3], we approximate Δ_3 by $(-0.0034)(T/\Theta_\infty)\omega(qj)$. Thus, the net δ becomes $(+0.0134)(T/\Theta_\infty)\,x$, when one takes into account the fact that the shift in free energy is exactly one half that in neutron scattering theory[5]. Substituting in (8) leads to

$$c_v/k = c_{har}/k - (0.0268)(\Theta_0/\Theta_\infty)(1/x - x^3/240) + 3(0.0134)^2(\Theta_0/\Theta_\infty)^2(1/x^2) - (0.0129)(T/425)[1 + x^4/720], \quad (9)$$

where $\Theta_0 = h\omega_L/k = 157$ K. Integrating over a Debye spectrum, we obtain

$$C_v/3R - C_{har}/3R = -0.0392\,(\Theta_0/\Theta_\infty)(\Theta/T) - (0.0129)(T/425)^2 \quad . \quad (10)$$

Using the values 105 K and 143.4 K for Θ and Θ_∞, respectively, from [3] in (10), we obtain the results shown in Fig. 1. These indicate that inclusion of the phonon line width in the statistical mechanics of phonons improves the agreement with experiment.

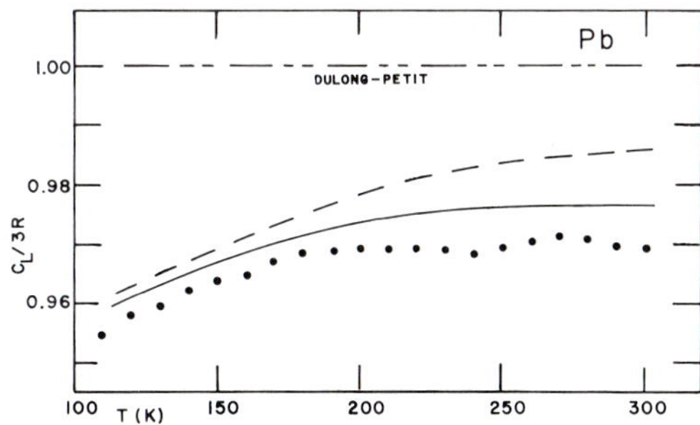

Fig. 1. Specific heat of lead. Dashed curve -- theory with shift correction only. Solid curve — theory with both shift and width corrections. Solid circles — experimental data of Meads, Forsyth, and Giauque: J. Am. Chem. Soc. 63, 1902 (1941), corrected to $C_{lattice}$ [T, V(0 K)]

References

1. D. N. Zubarev: [translation; Soviet Phys.--Usp. 3, 320 (1960)].
2. P. M. Morse and H. Feshbach: Theoretical Physics, McGraw Hill, New York, (1953).
3. A. A. Maradudin and E. A. Fein: Phys. Rev. 128, 2562 (1962).
4. B. N. Brockhouse, T. Arase, G. Caglioti, M. Sakamoto, R. N. Sinclair, and A. D. Woods: Inelastic Scattering of Neutrons, IAEA, Vienna (1961).
5. W. C. Overton, Jr.: J. Phys. Chem. Solids 29, 711 (1968).

Phonon Emission and Electron Heating in a Two-Dimensional Electron Gas

M.A. Chin, V. Narayanamurti, H.L. Stormer, and J.C.M. Hwang

Bell Laboratories, Solid State Electronics, Research Laboratory
Murray Hill, NJ 07974, USA

The transport properties of two-dimensional electrons formed at the heterojunction of GaAs and AℓGaAs has been the subject of intense experimental and theoretical interest over the last few years.[1] This interest has arisen because of the invention of modulation doping[2,3] which has led to spectacular improvements in the low-field, low-temperature mobility through the spatial separation of the current carrying electrons from their parent ionized donors. This mobility enhancement has been used to fabricate ultra-high speed field-effect transistors. The field dependence of the mobility[4] and the basic scattering mechanisms determining the mobility are, however, as yet imperfectly understood. Progress has recently[5] been made through the observation of photoluminescence spectra of modulation doped superlattices in the presence of an electric field. In this paper we report on complementary measurements made on a single interface by observing the energy relaxation of carriers by phonon emission. We correlate the phonon emission observed with changes in the mobility (μ) as a function of electric field (E). Earlier work[6] in this area involved phonon transmission through a super lattice which shows the ideality of MBE grown interfaces.

The modulation doped single heterostructures of GaAs/AℓGaAs were grown by molecular beam epitaxy on high purity, semi-insulating GaAs substrates of typical thickness 0.5 mm. Apart from low-field temperature dependence measurements, the electric field dependence of the mobility was measured under pulsed conditions at a lattice temperature of 1.5 K. Simultaneously the phonons generated during energy relaxation were detected on the opposite face of the GaAs sample by means of an aluminum bolometer. Even though measurements were made on several samples, with varying mobility, for the sake of simplicity we will present data on only one sample.

Figure 1 shows the low field mobility as a function of lattice temperature for a modulation doped GaAs/AℓGaAs heterojunction (areal density of 3.6×10^{11} cm^{-2}, three-dimensional density $\sim 2 \times 10^{17}$ cm^{-3}) and compares it with the mobility of high purity bulk GaAs (4×10^{13} cm^{-3}). At low temperatures the mobility advantage of modulation doping is apparent. At T ~1 K we reach a limiting mobility $\sim 1.6 \times 10^{6}$ cm^2/V sec. The solid lines in the figure represent theoretically expected phonon scattering due to polar longitudinal optic (LO) and piezoelectric acoustic phonons in bulk material. At the lowest temperature the mobility is affected by remnants of ionized impurity scattering but is within a factor of two of the limiting lattice mobility.

Fig. 2 shows the behavior of the field dependence of the mobility for a lattice temperature ~1.5 K. Even for extremely low fields ($\lesssim$1V/cm) we find that the mobility is field dependent and is to be contrasted with lower mobility samples.[5] Beyond about 3 V/cm, the mobility decreases quite rapidly and then slows down again for fields ($\lesssim$30 V/cm).

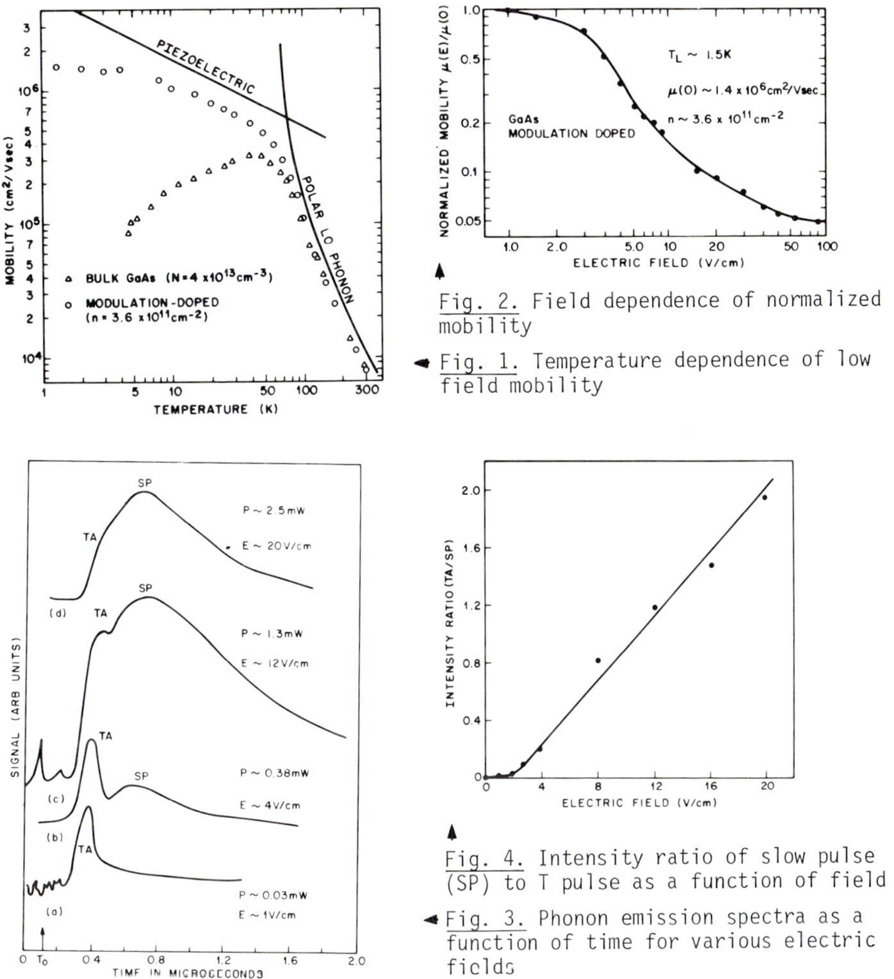

Fig. 2. Field dependence of normalized mobility

Fig. 1. Temperature dependence of low field mobility

Fig. 4. Intensity ratio of slow pulse (SP) to T pulse as a function of field

Fig. 3. Phonon emission spectra as a function of time for various electric fields

In Fig. 3 we show typical time of flight phonon spectra as a function of field. The propagation direction was (100). The power in each case is also indicated and was extremely low over the entire range. The 2-D heater size here was 0.1 mm^2. At low fields we observe a simple transverse (TA) mode while at higher fields the propagation changes quite rapidly and is dominated by a broad slow pulse (SP). The arrival time and the shape change markedly in the 3 to 12 V/cm region. At higher fields the arrival time of the slow pulse is essentially independent of field and reaches a limiting velocity with an apparent value of $\sim 0.8 \times 10^5$ cm/sec. The ratio of the peak heights of the SP to TA pulses are shown in Fig. 4.

It is important to emphasize here that the data are quite distinct from that observed previously[7] with metal heaters or low mobility 3-D, n layer heaters. With metal heaters, one observes both LA phonons and no SP pulses are observed unless the power density is at least two orders of magnitude higher than reported here.

We interpret the data as follows. The electron heating is determined by a balance between power supplied to the electrons by the electric field and the power lost by the electrons to the lattice. This is related by the relation $e\mu E^2 = P/N_e \propto (T_e - T_L)$ where T_e is the electron temperature. From the field dependence of μ we estimate that T_e changes from ~40 K to 80 K as the field changes from ~3 V/cm to ~20 V/cm. This is the regime where the phonon scattering changes from piezoelectric to polar LO phonons. We interpret the TA phonons observed at low fields as arising from piezoelectric scattering. This is also consistent with the absence of LA phonons which are not generated piezoelectrically for this direction. At higher fields the LO phonons are generated which rapidly decay[9] into longer lived slow transverse phonons which have the high density of states.

Quantitative comparisons between theory and experiment similar to that done previously[8] for 3-D n layer heaters requires measurements as a function of orientation, carrier density and energy selective detectors. Such measurements are currently underway.

References

1. For a review, see H. L. Stormer, Proceedings of the 13th Int'l Conference on the Physics of Semiconductors, Kyoto, Japan, 1980.
2. R. Dingle, H. L. Stormer, A. C. Gossard, and W. Wiegmann, Appl. Phys. Lett. 33, 665 (1978).
3. H. L. Stormer, R. Dingle, A. C. Gossard, W. Wiegmann, and M. D. Sturge, Sol. St. Comm. 29, 705 (1979).
4. T. J. Drummond, M. Keever, W. Kopp, H. Morkoc, K. Hess, B. G. Streetman, and A. Y. Cho, Elect. Lett. 17, 545 (1981).
5. J. Shah, A. Pinczuk, H. L. Stormer, A. C. Gossard, and W. Wiegmann, App. Phys. Lett. 42, 55 (1983) and to be published.
6. V. Narayanamurti, H. L. Stormer, M. A. Chin, A. C. Gossard, and W. Wiegmann, Phys. Rev. Lett. 43, 2012 (1979).
7. For a review see V. Narayanamurti Science 213, 717 (1981).
8. M. Lax and V. Narayanamurti, Phys. Rev. B24, 4692 (1981) and to be published.
9. R. G. Ulbrich, V. Narayanamurti and M. A. Chin, Phys. Rev. Lett. 45, 1432 (1980).

Density and Field Dependence of the Phonon-Limited Hot-Electron Temperature in n-Si Inversion Layers

R.A. Höpfel, E. Vass, E. Gornik

Institut für Experimentalphysik, Universität Innsbruck, Schöpfstraße 41
A-6020 Innsbruck, Austria

Abstract

Electron temperatures in n-inversion layers of (100) Si MOSFET's are measured as a function of the electric field by analysing the absolute intensities of the broadband thermal far infrared emission from the two-dimensional electron system. The electron heating ΔT is found to be proportional to the square root of the input power in the experimentally accessible range of $2K \leq \Delta T \leq 40K$ at 4.2 K lattice temperature. Data in the whole electron concentration range (1 to 5 x 10^{12} cm^{-2}) are presented resulting in a concentration-independent heating of $\Delta T/(e\mu E^2)^{1/2} = (1.0 \pm 0.2) \times 10^{-2}$ $Ks^{1/2}(eV)^{-1/2}$. An increase of the electron heating (especially at high electron concentrations) is observed when a negative substrate bias is applied. The experimental results can be explained by including screening of the electron-phonon interaction in the two-dimensional carrier system.

1. Determination of Electron Temperatures from FIR Emission

In order to evaluate the thermal far infrared emission from a two-dimensional (2D) system of temperature T and carrier concentration n_s the absorptivity is calculated as a function of the dynamical conductivity $\sigma(\omega)$ giving

$$A(\omega) = \frac{4.\mathrm{Re}F}{|\sqrt{\varepsilon} + 1 + F|^2} , \qquad (1)$$

where $F = \sigma(\omega)/\varepsilon_0 c$ and ε being the dielectric constant of the substrate. A Drude form of the dynamical conductivity $\sigma(\omega)$ in n-Si inversion layers has been confirmed by microwave and FIR transmission experiments performed by Allen et al. |1| up to frequencies of 40 cm^{-1}. As a consequence of Kirchhoff's law the broadband emission intensity in thermal equilibrium is given by

$$I(\omega) = I_{BB}(\omega).A(\omega) \qquad (2)$$

(I_{BB} ... emission intensity of the black body). I_{BB} is a strong function of the electron temperature T_e, therefore it is possible to measure T_e from the absolute intensity $I(\omega)$.

2. Experimental

The far infrared emission from the Si MOSFET is detected with a high purity n-type GaAs detector at its photoconductivity peak at 35.5 cm^{-1} (4.4 meV). The responsivity is measured by using a carbon-glass bolometer as a reference source and, in addition, by comparison with InSb cyclotron emission at saturation electron temperature |2|. Equations (1) and (2) - valid for normal incidence - can be used, since FIR emission perpendicular to the surface is mainly detected as a consequence of waveguide losses and detector orientation.

The samples investigated are large (100) Si MOSFET's with gate areas of 2.5 x 2.5 mm^2. The gate is a FIR-transparent 50 Å thick Ti layer on top of the silicon oxide (d_{ox} = 1400 Å). The 2D electron system is heated by applying electric pulses of 0.5 V up to 10 V along the source-drain contacts.

3. Results

We investigated MOS samples with mobilities from 2000 to 10000 cm²/Vs. Figure 1 shows a plot of ΔT as experimentally determined versus the input power for two different sorts of samples at several electron concentrations. In the figure also the slope for $\Delta T \propto (e\mu E^2)^{1/2}$ is given (dashed line). It can be seen that the electron heating ΔT is - in the whole range of temperature (2K to 40K) - exactly proportional to the square root of the input power $e\mu E^2$. Rather surprisingly, the electron heating is - within a possible error of 20 % - independent on the electron concentration n_s in the range of 1 to 5 x 10^{12}cm^{-2}. From the experimental data the following relation can be deduced:

$$\Delta T / (e\mu E^2)^{1/2} = (1.0 \pm 0.2) \times 10^{-2} \mathrm{K.s}^{1/2}.(\mathrm{eV})^{-1/2} \ . \qquad (3)$$

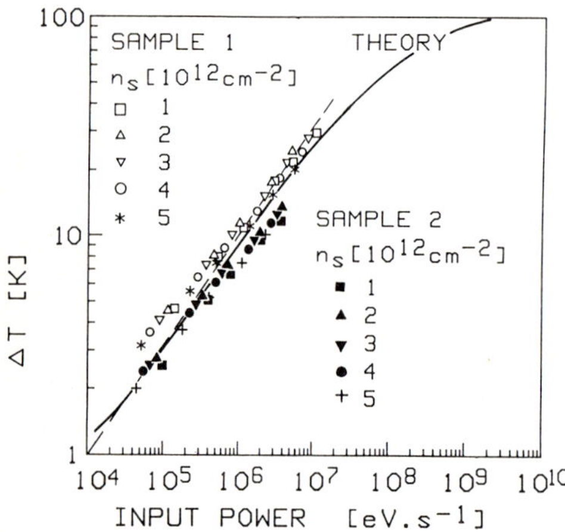

Fig. 1. Electron temperature determined according to (2) from the absolute intensities of the FIR emission for two samples (1: peak mobility 10.000, 2: 6.500 cm²/Vs). The full curve shows the theoretical behaviour according to (6)

Our results are somewhat lower than those obtained by different methods at low electron temperatures by FANG and FOWLER |3|, KAWAJI and KAWAGUCHI |4|, and agree with those of HÖNLEIN and LANDWEHR |5|. In the high-temperature range the observed values are somewhat higher than those obtained from subband emission experiments |6|.

In addition, we have performed emission experiments applying reverse substrate bias voltages of -10 V and -20 V. Figure 2 (below) shows the original emission signals, the increase of the electron heating is plotted above for three values of n_s at constant $e\mu E^2$. Especially at a high electron density, the substrate effect strongly increases the electron heating.

4. Theory

In the following we show that the electric field (E-) as well as n_s dependence of the electron temperature T_e can be attributed to the interaction of the inversion layer electrons with longitudinal acoustic phonons. In order to simplify the calculation we assume that all electrons are in the lowest electric sub-

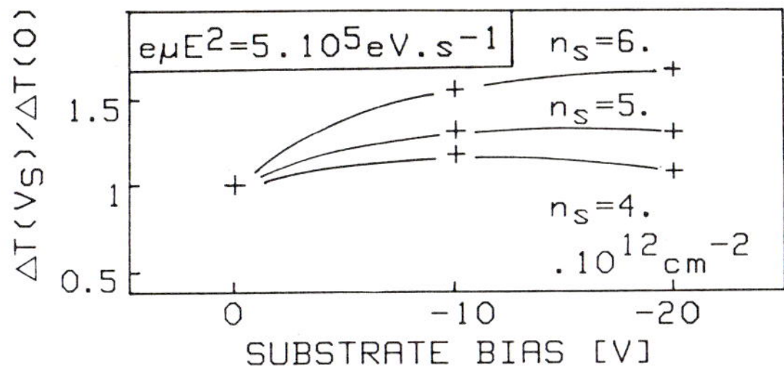

Fig. 2. Substrate bias dependence of the FIR emission at 35.5 cm^{-1} (lower curves) as a function of the electron concentration. The upper diagram shows the relative increase of the electron heating

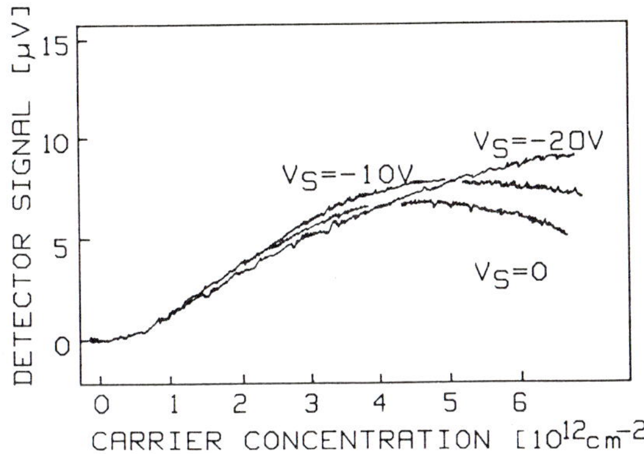

band. Furthermore we assume that the phonon wavevectors $\vec{q}$ are strictly two-dimensional and have no component perpendicular to the inversion layer. Then the 2D average energy loss rate of an electron to the phonon system, which is needed for the calculation of T_e from the power balance equation, is given by

$$\langle -\frac{d\varepsilon}{dt}\rangle = \frac{1}{n_s}\int \frac{d^2q}{(2\pi)^2}\left[n_q^0(T_e)-n_q^0(T)\right]\cdot\frac{2\pi}{\hbar}\sum_k \left|\frac{M_{kq}}{\varepsilon(q,\langle z\rangle)}\right|^2 \delta(\varepsilon_{k-q}-\varepsilon_k-\hbar\omega_q) \qquad (4)$$

where $n_q^0(T)$, M_{kq}, $\varepsilon(q,\langle z\rangle)$ and ε_k denote the phonon occupation number at the lattice temperature T, the bare bulk electron-acoustic-phonon matrix element, the dielectric function of the 2D electron gas and the energy of an electron with the wavevector k. $\langle z\rangle$ is the width of the inversion layer. If the influence of screening is suppressed ($\varepsilon(q,\langle z\rangle)\to 1$), we obtain in the limiting cases of weak and intermediate electron heating:

$$\langle -\frac{d\varepsilon}{dt}\rangle = \frac{2D^2 m k_B^3 T^3}{3\pi\rho\hbar^4 c_s^3}\left[A\left(\frac{\Delta T}{T}\right)^2 + B\left(\frac{\Delta T}{T}\right)\right] \qquad (5a)$$

where D, m, c_s and ρ are the acoustic deformation potential, the effective electron mass, the sound velocity and the volume mass density. The coefficients A and B are in the order of 1. For strong electron heating ($T_e \gg T$) we obtain

$$\langle -\frac{d\varepsilon}{dt}\rangle = \frac{D^2 m k_B^3 T_e^3}{3\pi\rho\hbar^4 c_s^3} \quad . \qquad (5b)$$

From (5) it follows that the electron temperature should increase proportional to E^2 and E for weak and intermediate electron heating, respectively. In fig. 1 the theoretical electron temperatures are plotted versus the input power $e\mu E^2$ (full curve) and are compared with the experimental data. The deformation potential used in the calculation was chosen as D=8.5 eV to obtain the best fit to the experimental data points.

From (5) it can be seen furthermore that the 2D energy loss rate does not depend on the electron concentration n_s if screening is not taken into account. The influence of this effect can be demonstrated clearly in the substrate bias experiment (fig. 2). With increasing n_s und bias voltage the dielectric function $\varepsilon(q,\langle z\rangle)$ increases due to the decrease of the channel width $\langle z\rangle$, which leads to a stronger screening of the electron-phonon interaction.

Acknowledgements

The MOS samples were made at Bell Laboratories, Murray Hill, USA, and kindly provided by Prof. D.C. Tsui, Princeton University. This work was supported by the Fonds zur Förderung der Wissenschaftlichen Forschung, AUSTRIA (S 22/05).

References

|1| S.J. Allen, D.C. Tsui, F. DeRosa: Phys. Rev. Lett. 35, 1359 (1975)
|2| E. Gornik: J. of Magnetism and Magn. Mat. 11, 39 (1979)
|3| F.F. Fang, A.B. Fowler: J. Appl. Phys. 41, 1825 (1970)
|4| S. Kawaji, Y. Kawaguchi: Lecture Notes in Physics Vol. 177, 53 (1983)
|5| W. Hönlein, G. Landwehr: Surface Science 113, 260 (1982)
|6| E. Gornik, D.C. Tsui: Solid State Electronics 21, 139 (1978)

Amplification of Total-Reflection-Mode Surface Phonons in n-Type InSb Films*

C.-C. Wu

Department of Applied Mathematics, National Chiao Tung University
Hsinchu, Taiwan, Republic of China

Acoustic waves can be propagated along the boundary of an elastic half-space [1]. The surface phonons are the quanta of elastic waves that satisfy the proper boundary condition on solid surfaces. In an elastic medium with a stress-free plane boundary, the waves are reflected on the boundary and the electrons localized in the surface region can be affected by the reflection of waves. The total-reflection (TR) mode due to the localized nature of wave functions in the vicinity of the solid surface becomes important when the sound velocity of propagation c lies between the transverse sound velocity c_t and the longitudinal sound velocity c_ℓ [1]. It has been shown that the effect of nonparabolicity on amplification of surface phonons in n-type InSb becomes considerably important for the Rayleigh waves [2]. In this paper, we investigate the amplification characteristics of TR-mode surface phonons in n-type InSb films using the quantum mechanical treatment in the GHz region such that $q\ell > 1$, where q is the wave number of surface phonons and ℓ is the mean free path of electrons. The energy band structure of semiconductors is assumed to be nonparabolic. Both deformation-potential and piezoelectric couplings are considered in our numerical calculations.

The configuration of the amplifier is shown in Fig. 1(a). A thin layer with the thickness d of a piezoelectric semiconductor is grown epitaxially on an insulating substrate with the same elastic properties as the semiconductor layer [3]. The Cartesian coordinates are fixed so that the material occupies the half-space $z > 0$ and has the stress-free surface parallel to the xy plane. If a shear wave polarized in the vertical plane (SV wave) of a medium is incident on the stress-free boundary surface, the SV wave and a pressure wave (P wave) come out as the reflected waves from the surface. The longitudinal P wave is totally reflected by the surface as shown in Fig. 1(b). It is assumed that the potential along the z axis is a square well which has infinitely high potential barriers at $z = 0$ and $z = d$. The field operator $\Psi(\vec{r})$ of conduction electrons in the second quantized form can be written as [4]

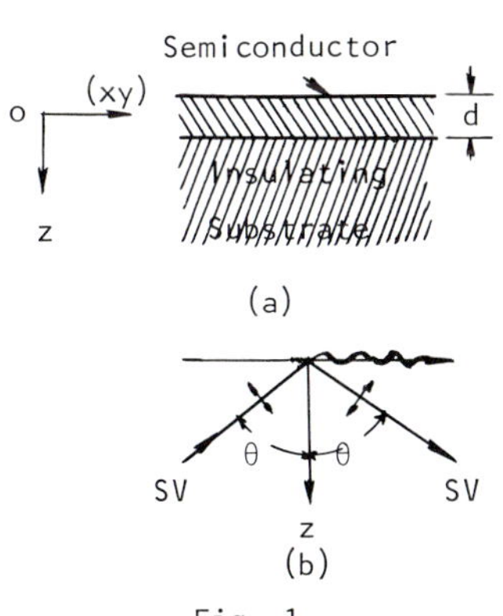

Fig. 1

* Partially supported by National Science Council of China in Taiwan

$$\Psi(\vec{r}) = \left(\frac{2}{V}\right)^{\frac{1}{2}} \sum_{n=1}^{\infty} \sum_{\vec{k}} b_{\vec{k}n} \exp(i\vec{k}\cdot\vec{x}) \sin\left(\frac{n\pi z}{d}\right) \tag{1}$$

where $\vec{r} = (\vec{x},z) = (x,y,z)$, $\vec{k} = (k_x,k_y)$, and V is the volume of the film. $b_{\vec{k}n}$ is the operator of conduction electrons satisfying the commutation relation of Fermi type. The energies of the conduction electrons $E_{\vec{k}n}$ for the nonparabolic band structure are given by the relation [2]

$$E_{\vec{k}n}\left(1 + \frac{E_{\vec{k}n}}{E_g}\right) = \frac{\hbar^2 \vec{k}^2}{2m^*} + \frac{\pi^2 \hbar^2 n^2}{2m^* d^2}, \qquad n = 1, 2, 3, \ldots \tag{2}$$

where E_g is the energy gap between the conduction and valence bands, m* is the effective mass of conduction electron. The quantization of the elastic-wave field $\vec{u}(\vec{r},t)$ can be expanded in terms of the coefficient a_J as [1,5]

$$\vec{u}(\vec{r},t) = \sum_J \left(\frac{\hbar}{2\rho\omega_J S}\right)^{\frac{1}{2}} [a_J \vec{u}_J(\vec{r}) \exp(-i\omega_J t) + (\text{Hermitian conjugate})] \tag{3}$$

where ρ is the mass density of the medium, $J = (\vec{q},c,m)$ is a suitable set of quantum numbers, $\vec{q} = (q_x,q_y)$ is the wave vector parallel to the surface, c is the phase velocity defined by $\omega_J = c|\vec{q}| = cq$, m specifies the TR-mode surface phonons, S is the surface area of the film, a_J is the operator of surface phonons obeying the commutation relation of Bose type, and $\vec{u}_J(\vec{r})$ is the wave function for the TR mode.

In the piezoelectric semiconductors, the conduction electrons interact with the TR-mode surface phonons through the deformation-potential and piezoelectric couplings. Using the Green's function method with the Born approximation [2,4], the amplification coefficient α_i for the i-type electronic screening effect induced by the acoustic vibration of longitudinal waves or transverse waves in the piezoelectric coupling can be obtained. We present numerical results as shown in Figs. 2 and 3. The relevant values of physical parameters for an n-type InSb thin film grown epitaxially on a semi-insulating InSb substrate are taken to be: d = 10 μm, $\beta_p = 1.8 \times 10^4$ esu/cm^2 for $\vec{q}$ // [110], $m^* = 0.013m_0$ (m_0 is the mass of free electron), $\rho = 5.8$ gm/cm^3, $\varepsilon_0 = 18$, $n_0 = 1.75 \times 10^{14}$ cm^{-3}, $c = 2.5 \times 10^5$ cm/sec, $c_\ell = 3.76 \times 10^5$ cm/sec, $c_t = 1.61 \times 10^5$ cm/sec, C = 4.5 eV, and $E_g = 0.2$ eV.

In Fig. 2, we show the amplification coefficient as a function of frequency with the parameter x = (v/c) - 1 = 10 or the applied electric field

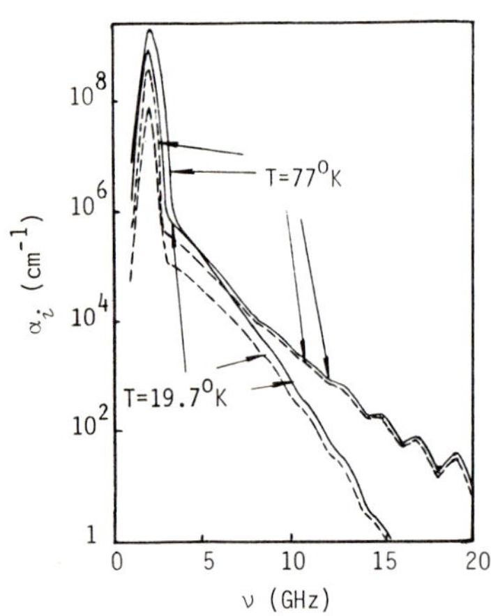

Fig. 2 Amplification coefficient of TR-mode surface phonons versus frequency in n-type InSb with x = 10 (E = 4.62 V/cm). Solid curve : amplification coefficient for the screening effect of the transverse field, dashed curve : amplification coefficient for the screening effect of the longitudinal field

E = 4.62 V/cm, where v is the drift velocity of conduction electrons. It can be seen that the amplification coefficient decreases with the temperature. This means that the effect of amplification of surface phonons will be diminished considerably at very low temperatures. Moreover, the amplification coefficient with the transverse dielectric function is larger than that with the longitudinal dielectric function owing to the larger electronic screening effect for the longitudinal waves. We can also see that the amplification coefficient oscillates with the frequency and rapidly decreases with increasing the frequency in the high-frequency region. Figure 3 shows the amplification coefficient versus the drift parameter x or applied electric field E with ν = 3 GHz. It is found that the amplification coefficient increases with the applied electric field and then decreases with the field. This is different from the result for the Rayleigh wave [2] in which the amplification coefficient increases monotonically with the applied electric field.

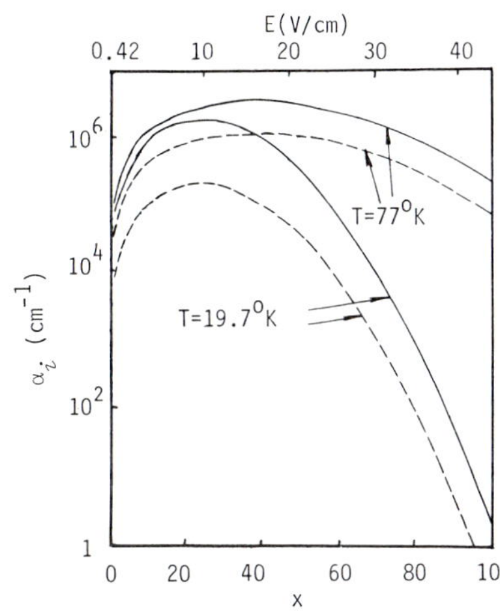

Fig. 3 Amplification coefficient of TR-mode surface phonons versus drift parameter or applied electric field in n-type InSb with ν = 3 GHz. Solid curve : amplification coefficient for the screening effect of the transverse field, dashed curve : amplification coefficient for the screening effect of the longitudinal field

References

1. H. Ezawa, Ann. Phys. (N.Y.) 67, 438 (1971)
2. C. C. Wu and J. Tsai, J. Phys. C : Solid State Phys. 15, 4939 (1982)
3. S. M. Sze, Physics of Semiconductor Devices, Wiley Interscience, New York (1981)
4. S. Tamura and T. Sakuma, Phys. Rev. B 16, 3638 (1977)
5. T. Nakayama and T. Sakuma, J. Appl. Phys. 47, 2263 (1976)

Time-Resolved Photoluminescence and Phonon Transport in Amorphous Si:H Films

U. Strom, J.C. Culbertson, P.B. Klein, and S.A. Wolf
Naval Research Laboratory, Washington, DC 20375, USA

We have measured, under as nearly identical conditions as possible, both time-resolved photoluminescence and phonon emission in a-Si:H excited by a pulsed dye laser (8 ns, 4000-7500 Å). PL was detected with a cooled S1 photomultiplier; phonon response was measured with a superconducting NbN bolometer. Most measurements were made at 2 K. Two films of a-Si:H were studied with thickness of ~1 and 4 μm, respectively. The films, deposited by plasma decomposition of silane onto 1 mm thick sapphire substrates, were comparable to the "NRL" samples listed in Ref. 1.

The CW spectrum of the PL has a peak at ~1.4 eV. A typical time-resolved PL spectrum at T = 2.1 K with excitation wavelength of 5100 Å and PL energy of 1.43 eV is shown in Fig. 1. The characteristic fast (~8 nsec) and slow (10^{-7} to 10^{-3} sec) components of the PL are very similar to that observed previously [2-5].

The phonon signal associated with the PL measured with a NbN bolometer under identical experimental conditions as for Fig. 1 is shown in Fig. 2 (curve b). Normalized phonon responses for higher and lower excitation intensities are also shown. For 5100 Å exciting light we estimate a penetration depth ~2000 Å in a-Si:H at 2 K. For the lowest laser pulse energy (Fig. 2a) the phonons are ballistic. We have also observed a comparable ballistic phonon signal in the 1 μm thick a-Si:H film for 4000 Å light, which has a penetration depth of less than 1000 Å. The observed ballistic

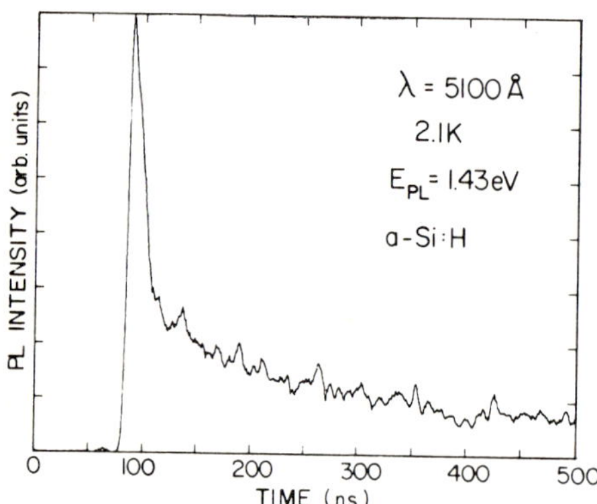

Fig. 1. Decay of PL. Incident laser pulse energy 0.02 μJ. Laser spot diameter 0.23 mm

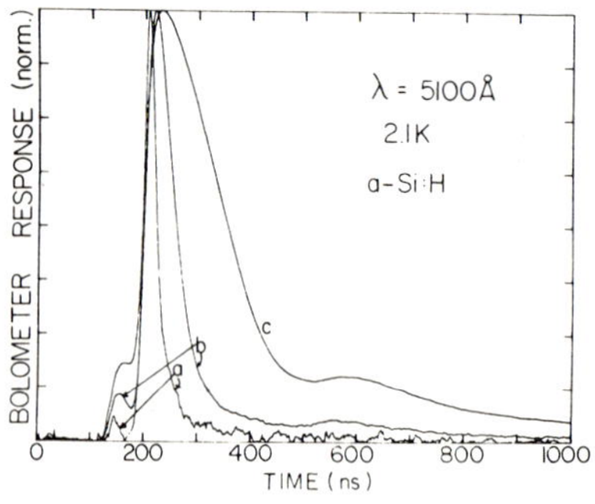

Fig. 2. Normalized bolometer response for identical conditions to Fig. 1. Laser pulse energies: (a) 0.003 μJ, (b) 0.02 μJ, (c) 0.08 μJ

phonon signal is primarily due to the initial rapid thermalization (2.4 eV excitation, ~1.9 eV band gap) of the excess energy supplied to the carriers by the exciting laser pulse. The bulk of this phonon signal is generated within the optical penetration depth at times much less than 10 nsec, traverses the remainder of the a-Si:H film and is subsequently transmitted through the 1 mm thick sapphire with the characteristic sound velocities corresponding to LA and TA phonons in sapphire. The weak tail observed at longer times in trace (a) may contain small diffusive contributions or possibly a competitive nonradiative process. We conclude that the thermalization process represented by trace (a) in Fig. 2 leads to acoustic phonons which have a mean free path (m.f.p.) greater than ~1 μm at 2 K. We find that a 4 μm thick film shows considerable broadening of the ballistic pulses at 2 K for comparable power levels, indicating that the m.f.p. is not significantly greater than 1 μm. If it is assumed that a-Si:H exhibits acoustic phonon scattering as in a-SiO_2 (at 1 K) [6] our results imply that the photo-generated phonons in a-Si:H are down-converted to frequencies below ~250 GHz. Alternatively, higher phonon frequencies may be generated, but this would require that the phonon m.f.p. in a-Si:H is substantially longer than in glassy SiO_2.

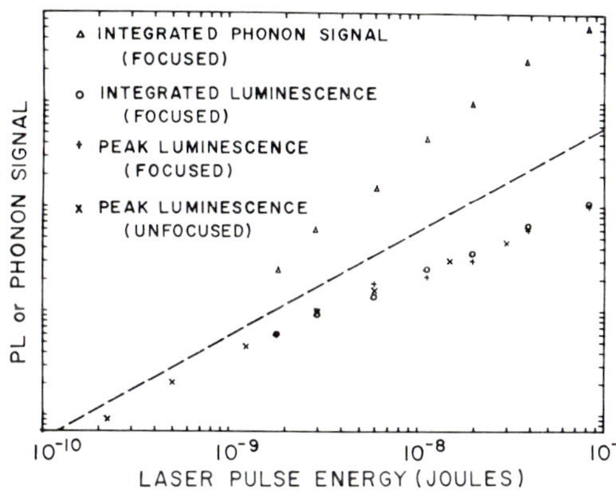

Fig. 3. PL and phonon signal as function of incident laser pulse energy. Dashed line represents response which is linear with laser pulse energy

We will discuss now the effects of high laser excitation. As shown in Fig. 2, we observe the following changes in the low power phonon spectrum with increasing excitation level: (a) an increasing width of both LA and TA peaks; (b) an increasing amplitude of LA relative to TA phonons; (c) a shift to longer times of both LA and TA peaks; (d) a more pronounced signal (near 600 ns) due to phonons which are generated in the a-Si:H film, subsequently reflected at the bolometer/sapphire and then again at the a-Si:H/sapphire interfaces; finally, (e) an integrated phonon signal (integration up to ~2 μsec) which increases faster than linear with excitation intensity. In contrast to the dramatic changes observed in the phonon spectra we find that the PL decay is completely insensitive, within experimental error, to the laser intensity. However, we find that the magnitude of the peak or integrated PL increases less than linear in excitation energy, in agreement with other PL studies [2-5]. The dependence of PL and integrated phonon intensity on exciting laser pulse energy is shown in Fig. 3. Qualitatively, we observe that the loss in PL at high laser intensities is reflected by an integrated phonon signal which increases faster than linear in laser intensity.

In principle, the results in Fig. 3 can lead to a direct determination of the radiative quantum efficiency, provided a linear low power regime can be determined for both PL and phonon signals. The above experimental results are qualitatively similar to those for other laser excitations ranging from 4000 to 5800 Å which are estimated to correspond to nearly an order of magnitude different excitation volume. In addition, the observation of the ballistically reflected phonon corresponding to three sapphire thicknesses precludes the possibility that the observed increased width of the phonon pulse is due to diffusive scattering in the sapphire substrate.

The power-dependent phonon width is very similar to observation of the so-called "hot spot" in crystalline Ge by Greenstein et al. [7]. In fact, from additional measurements from the data used to generate Fig. 2 we find an approximate relation for the excess width $\Delta\tau$ induced by the excitation intensity E: $\Delta\tau \cong 2E$ sec/Joule. Greenstein et al. use a 100 ns laser pulse and smaller spot size (60 μm vs our 230 μm) and consequently higher energy density for crystalline Ge. Scaling our results to their power regime suggests that the power-induced width in a-Si:H is of comparable magnitude to the entire phonon width in crystalline Ge which includes both delayed ballistic and diffusive contributions. Our parallel PL and phonon study strongly suggests that the observed phonon power-dependent width is thermal in nature and essentially uncoupled from the PL channel. This conclusion was also reached in context of the "hot spot" in crystalline Ge [7]. The fundamental difference between c-Ge and a-Si:H is the scale of phonon m.f.p. over which the heat can be described as ballistic, diffusive and stored.

In summary, after excitation of a-Si:H with above band gap light the following occurs: (1) Phonons due to initial thermalization of carriers have mean free paths $\gtrsim 1$ μm, provided the injected e-h pair density is less than $10^{18}cm^{-3}$. Simultaneously, the PL intensity increases linearly with laser power. (2) For carrier densities $>10^{18}cm^{-3}$, a "hot spot" develops in a-Si:H, the PL intensity increases sublinearly with power and the associated phonon signal exhibits a power-dependent width and increasing integrated intensity. (3) The dramatic changes in the time evolution of the phonon emission take place in the a-Si:H film at comparable times during which e-h pairs recombine radiatively (i.e. 10 ns-1μs). The lack of correlation between the shape of the PL decay and the phonon emission provide strong evidence that the PL and nonradiative channels are essentially uncoupled in this time regime.

This work was sponsored in part by the Office of Naval Research.

References

1. W.E. Carlos and P.C. Taylor, Phys. Rev. B 26, 3605 (1982).
2. C. Tsang and R.A. Street, Phys. Rev. B 19, 3027 (1979).
3. W. Rehm and R. Fischer, phys. stat. sol. (b) 94, 595 (1979).
4. B.A. Wilson, P. Hu, J.P. Harbison and T.M. Jedju, Phys. Rev. Lett. 50, 1490 (1983).
5. J. Shah, B.G. Bagley and F.B. Alexander, Jr., Sol. State Commun. 36, 199 (1980).
6. W. Dietsche and H. Kinder, Phys. Rev. Lett. 43, 1413 (1979).
7. M. Greenstein, M.A. Tamor and J.P. Wolfe, Phys. Rev. 26, 5604 (1982).

Surface Acoustic Waves in Metals

J. Heil, I. Kouroudis, C. Lingner, and B. Lüthi

Physikalisches Institut, Joh.-Wolfgang-Goethe-Universität
Robert-Mayer-Straße 2-4, D-6000 Frankfurt am Main, Fed. Rep. of Germany

Using ZnO sputtering and photolithographic techniques we were able to produce surface acoustic wave (SAW)- structures directly on to high quality surfaces of single crystals of aluminium and gallium. Using single-phase array transducers we were able to excite SAW in the frequency range between 10 MHz and 60 MHz. Using low temperatures and high magnetic fields we were able to observe a number of magnetoacoustic effects, some of which have their counterpart in volume waves and some of which are unique for SAW. We give a brief survey of these effects.

In Fig. 1 we show experimental results for aluminium, where one can clearly discern three distinct effects: geometric resonances, a nonreciprocal effect, and de Haas-Shubnikov oscillations in the velocity of SAW. In Fig. 1 we give relative velocity changes as a function of applied magnetic field at 4K. The surface is a (011) plane. The propagation direction is in [100] and the magnetic field is applied in the plane of the surface along the [0$\bar{1}$1] direction.

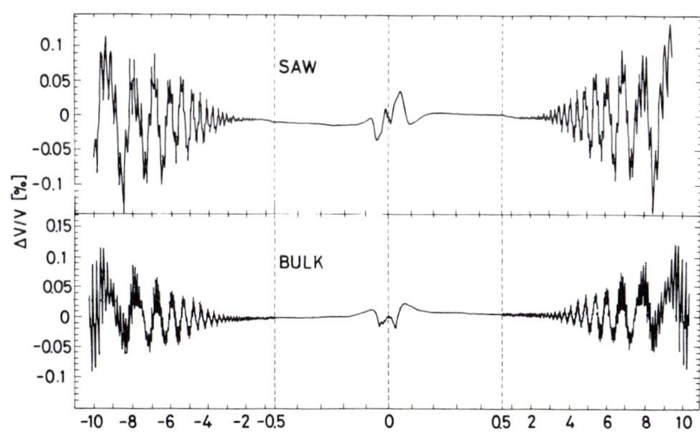

Fig. 1 Relative velocity changes for SAW and bulk waves (c_{44}) for Al at 4K as a function of magnetic field B

The oscillations at small fields (below 0.15 T) are geometric oscillations where $kR \geq 1$ with R the cyclotron radius of the conduction electrons. The magnetic field values for these resonances determine the Fermi surface k vectors in certain directions. We find that the geometric resonances observed with SAW occur at exactly the same field positions as those for bulk waves for the same frequencies. They also agree with previous bulk measurements done on Al [1].

The second effect one notices for fields up to 0.5 T is a nonreciprocal effect, i.e. the SAW velocity $v(k,B) \neq v(-k,B)$ and $v(k,B) \neq v(k,-B)$. This effect we discussed before in detail [2,3]. Because of the lack of mirror symmetry at the surface the SAW velocity is nonreciprocal with respect to propagation direction reversal or magnetic field direction reversal. Physically it arises because the sense of rotation for electrons on cyclotron orbits changes with field reversal and the sense of rotation for the elliptical particle motion in SAW is changed with reversal of the propagation direction. A quantitative theory was developed before [2] and the observed effect is of the order of 0.1% in the relative velocity. We get an observable effect because of the strong deformation potential coupling in Al. A previous search for nonreciprocal effects in the paramagnetic phase of $CeAl_2$ was unsuccessful [4], because of the much smaller magnetoelastic coupling of the SAW strain field to the 4f - Ce^{3+} ion. Clearly bulk waves show only reciprocal behaviour. The frequency and the temperature dependences of this nonreciprocal effect are in agreement with theory [3].

A third effect visible in Fig. 1 are the de Haas-Shubnikov oscillations which appear in the SAW velocity for fields larger than 1 T. They are again a manifestation of the strong electron-phonon coupling and of the high quality of our sample. These quantum effects for the electrons show up in the SAW velocity via the conductivity tensor which determines attenuation and dispersion of the SAW in metals [3]. The de Haas - van Alphen periods determined from these oscillations agree well with those attributed to extremal orbits of the monster surface for electrons in the third Brillouin zone in Al [5]. In the region of the de Haas - van Alphen effect ($\omega_c\tau > 1$) the nonreciprocal effect has completely disappeared [2].

Similar results have been observed in gallium. In Fig. 2 we show an analogous result for SAW propagation on a b plane of single crystal Ga. Again the three effects mentioned above for Al appear also in Ga. Some geometric oscillations are clearly distinguished. The nonreciprocal effect is present too although the additional reciprocal effect is larger than in the case of Al. The de Haas - van Alphen effect is also present and starts at fields of 0.4 T already. A comparison with bulk waves gives again analogous results as for Al: geometric oscillations and de Haas - Shubnikov oscillations give identical information for the two types of waves. A more quantitative interpretation will be given elsewhere [3].

Nonreciprocal propagation effects are not confined to the case of surface acoustic waves in pure metals. Nonreciprocal effects have been observed or predicted for dipolar surface spin-waves in ferromagnets and antiferromagnets and for magnetoplasmon - surface polaritons in metals and semiconductors. They follow from a general symmetry argument involving the absence of mirror symmetry on the surface which does not change k but changes B into -B [6].

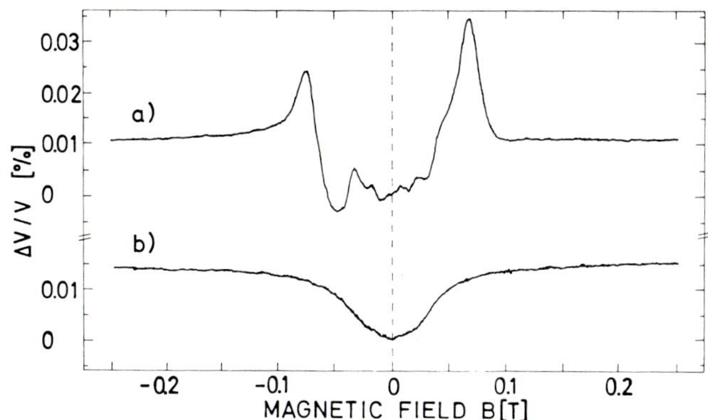

Fig. 2 Relative velocity changes as a function of magnetic field B for T = 4K for single crystal Ga. Surface is b plane, propagation direction of SAW is c direction and B is applied in ±a direction for case a) and in ±b direction for case b)

This research was supported by the SFB 65 Frankfurt-Darmstadt.

References

1 G.N. Kamm, H.V. Bohm: Phys. Rev. 131, 111 (1963)
2 J. Heil, B. Lüthi, P. Thalmeier: Phys. Rev. B25, 6515 (1982)
3 J. Heil, I. Kouroudis, B. Lüthi, P. Thalmeier: to be publ.
4 C. Lingner, B. Lüthi: Phys. Rev. B23, 256 (1981)
5 N.W. Ashcroft: Phil. Mag. 8, 2055 (1963)
6 R.Q. Scott, D.L. Mills: Phys. Rev. B15, 3545 (1977)

Discussions

PHONON ATTENUATION IN HEAVILY DOPED p-TYPE SEMICONDUCTORS page 304
T. Sota, K. Suzuki, D. Fortier

J.P. Maneval: Due to the range of temperature you consider it is likely that the dominant coupling goes from piezoelectric to deformation potential coupling depending upon the material. Did you verify this?

T. Sota: I think that the deformation potential coupling is more effective in III-V materials with zincblende structure rather than the piezoelectric one, provided that the carrier concentration is very high.

BALLISTIC PHONON TRANSPORT IN Ge:P UNDER MAGNETIC FIELD page 307
T. Miyasato, M. Tokumura, K. Suzuki

L.J. Challis: I did not understand your figure showing that you get a change in attenuation even if a^* is constant. Does this not make 4Δ constant as well?

T. Miyasato: To explain the experimental results, the transmitted heat pulse should be more increased with magnetic field, so we assumed that the a^* is constant because the shrinkage of a^* reduces the transmission of heat pulses in the theory of heat-pulse scattering by neutral-donor electrons.
Of course, the (100) magnetic field dependence of 4Δ was taken into account using Aggarwal´s data.

HOLE-PHONON INTERACTION IN WURTZITE-TYPE SEMICONDUCTORS page 310
M. Singh, J. Leotin

M. Wagner: You have used the 2nd-order Born approximation, but have claimed also that you are able to evaluate resonance scattering processes. However, this cannot be done within a 2nd-order Born approximation; you would need a kind of phenomenological damping constant.

M. Singh: In the present paper we have calculated only the elastic scattering relaxation rate using a 2nd-order Born approximation. I did not claim but I just mentioned that one can also calculate other scattering processes such as the resonance scattering without mentioning the method. I agree with you that if one wants to calculate the resonance scattering relaxation rate one needs a kind of phenomenological damping constant.

PHONON EMISSION AND ELECTRON HEATING IN A TWO-DIMENSIONAL ELECTRON GAS page 328
M.A. Chin, V. Narayanamurti, H.L. Stormer, J.C.M. Hwang

J.C. Hensel: It is a bit surprising that you observe phonon emission normal to the plane of the 2 DEG, a process forbidden by the conservation of quasimomentum. Perhaps what you are seeing are finite angle phonons scattered normal to the plane and if you were to look at finite angles the intensities might increase appreciably.

W. Dietsche: Do you expect the frequency spectrum of the emitted phonons to look similar to the one observed by Carlsson and Segmüller, PRL 28, 175 (1972)?

V. Narayanamurti: For low fields one should observe a strongly greater distribution for $q \simeq 2 K_F$. The distribution should, however, be highly anisotropic in a pure sample.

DENSITY AND FIELD DEPENDENCE OF THE PHONON-LIMITED HOT-ELECTRON TEMPERATURE IN n-Si INVERSION LAYERS page 331
R.A. Höpfel, E. Vass E. Gornik

W. Dietsche: How does the deformation potential of 8.5 eV which you obtained compare with the one deduced from high temperature mobility data by other people?

R.A. Höpfel: The temperature dependence of the mobility could not be used to determine the deformation potential as it was shown by Stern-Wheeler (PRL 1982).

J.C. Hensel: A reply to Dietsche´s comment: As I pointed out earlier, it has not been possible to determine the electron-phonon interaction for the 2 DEG in Si from electrical transport measurements. It has been shown, mainly by F. Stern and R.G. Wheeler in separate papers several years ago, that the temperature dependence of the mobility is almost completely dominated by scattering processes other than acoustic phonon.

J.P. Maneval: By what method did you measure the electron concentration in order to derive the energy input per electron? Can the corresponding uncertainty limit the validity of your conclusions as to the mechanisms of electron heating?

R.A. Höpfel: The electron concentration is determined by the oxide capacity and the gate voltage, and was additionally measured by Shubnikov- de Haas oscillations. The uncertainty is negligibly small.

TIME-RESOLVED PHOTOLUMINESCENCE AND PHONON TRANSPORT IN AMORPHOUS Si:H FILMS page 338
U. Strom, J.C. Culbertson, P.B. Klein, S.A. Wolf

J.P. Maneval: Can you comment on the way you use NbN bolometers from 2 to 12 K?

U. Strom: The bolometer resistance is current dependent and essentially constant below 10 K . However, the bolometer can be nonlinear as a function of bias current and incident phonon flux. So care has to be taken to operate in a linear regime. Typically, a dynamic range of two orders of magnitudes in laser pulse power can be realized.

R.P. Huebener: Can you comment on the linearity of your bolometer?

U. Strom: The bolometer does exhibit nonlinear behaviour at sufficiently high incident phonon fluxes. At sufficiently low power, a regime can be realized where the phonon response is linear. (See Maneval's comments).

K. Dransfeld: Does the phonon generation by light depend on the hydrogen content?

U. Strom: This has not been examined at present.

B. Golding: It is well known that the microstructure of a-Si:H is very inhomogeneous with a great deal of void volume. Is it possible that the heat generated may not propagate out of the a-Si:H and perhaps persist for long times as highly excited local modes as in, for example, H?

U. Strom: We do not observe a very large long-time heat component. But, it is certainly possible that the small tail observed at longer times after the ballistic pulse may contain some locally retained heat.

A.F.G. Wyatt: Do you know whether the recombination that gives rise to a photon occurs at the same region in the crystal as the recombination that creates a phonon?

U. Strom: For 4000 Å excitation light the light penetrates ≃ 10% of the 1 μm thick film. However, at 5800 ≃ the penetration depth light and hence the region of recombination are comparable.

SURFACE ACOUSTIC WAVES IN METALS page 341
J. Heil, I. Kouroudis, C. Lingner, B. Lüthi

K. Weiss: Do you think that it would be possible to have an acoustic analog to the Ghantmaker size effect so as to couple acoustic waves from one side of the sample to the other by a cyclotron resonance?

J. Heil: We have thought about the effect and we are in the process of doing it.

Part VIII

Defects

Chairmen: **T. Ishiguro C. Laermans J. P. Maneval K. Suzuki**

Phonon Scattering by Dislocations

A.C. Anderson

Department of Physics and Materials Research Laboratory
University of Illinois, 104 S. Goodwin
Urbana, IL 61801, USA

1. Introduction

This discussion of the phonon-dislocation interaction will focus on the frequency range of thermal phonons, namely $\approx 10^9 - 10^{12}$ Hz. The paper is not comprehensive, as a detailed review is available [1,2]. Rather, the emphasis is on the limitations which exist in our understanding of the phonon-dislocation interaction, an understanding that has accrued over the past thirty years. In brief, it is not as yet possible to predict theoretically the thermal conductivity of any deformed sample.

Most investigations of the phonon-dislocation interaction have utilized thermal conductivity measurements as it is possible, with little effort or expense, to scan a broad range in phonon frequencies. For convenience of discussion we will use the Debye model of phonon thermal transport since most data have been obtained at temperatures T < 10K, for which there is little phonon dispersion. Also, since the phonon-dislocation interaction is only weakly frequency dependent, it is possible to use the dominant-phonon approximation in which the important phonons at temperature T are assumed to have a unique frequency of $\nu \approx 10^{11}$ T [Hz]. The resulting form for the thermal conductivity, κ, is

$$\kappa = 4.08 \times 10^{10}\, T^3\, \ell/\bar{v}^2 \text{ [W/mK]} \tag{1}$$

where ℓ is the frequency-dependent phonon mean free path and $\bar{v}$ is an average phonon velocity.

The phonon mean free path is determined by the strongest phonon scattering mechanism. For dislocation interactions, the many proposed scattering mechanisms can be divided into two general categories, namely static scattering and dynamic scattering. By static scattering we mean that the dislocation remains sessile and the phonons are scattered or "refracted" by the static strain field surrounding the dislocation. In the dynamic interaction the dislocation can "flutter" in some manner at a fixed resonant frequency ν_o. The amplitude of this motion is much smaller than a crystal-lattice spacing [3], as the resonance is in thermal equilibrium at the low temperature of the measurement. A phonon having the frequency ν_o can be absorbed by the dislocation, then reradiated in a different direction. The dynamic scattering mechanism is much stronger than the static mechanism for phonons having frequencies near ν_o.

Metals frequently have been used to study the phonon-dislocation interaction since large densities of dislocations may be inserted easily. We therefore begin with a discussion about metallic alloys.

2. Normal Metals

Both conduction electrons and phonons contribute to the thermal conductivity of metals, but only the phonon contribution is of interest. Hence the metals are generally alloyed to reduce the electronic contribution, then the Wiedemann-Franz law and the measured electrical conductivity are utilized in subtracting off the electronic contribution to κ. Of course electrons also scatter phonons and decrease ℓ, but this effect can be corrected for or ignored if the dislocation density is sufficiently large that dislocation scattering dominates ℓ.

A qualitative composite of measurements on copper alloys [4-9] is shown by the solid lines in Fig. 1. The conductivity κ has been divided by T^3 so that the effective frequency dependence of ℓ may be shown on the same plot. Since the data have been divided by N, the density of dislocations, only those measurements in which N was measured or estimated could be represented in Fig. 1.

The static scattering mechanism for isolated, sessile dislocations is found by all authors to have a theoretical frequency dependence of $1/\nu$. Two recent computations [10,11], which presumably have benefitted from earlier work, are represented by dotted lines 1 and 2 in Fig. 1. The computations may explain the data of curve A, but not of curves B and C.

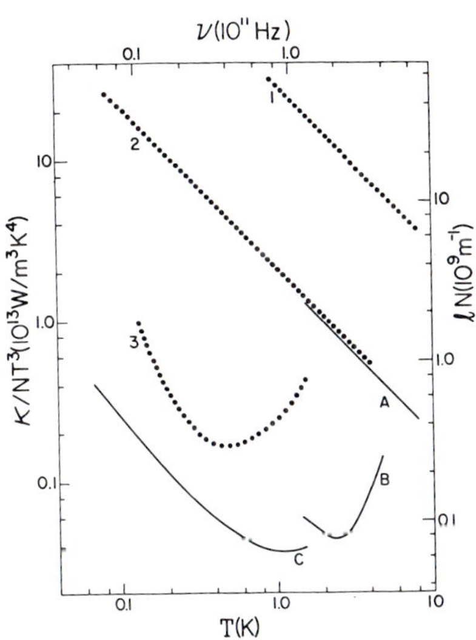

Fig. 1. Phonon thermal conductivity κ of copper alloys, divided by T^3 and the dislocation density N. The right plus top scales provide the frequency dependence for the effective phonon mean free path ℓ. The experimental data are shown by the solid lines. A, composite from Refs. 4-7 (within a factor of roughly 2); B, Ref. 8; C, Ref 9. The dislocation density for curve C was smaller than for curves A and B. Theory is represented by the dotted lines. 1, Ref. 10; 2, Ref. 11; 3, rough estimate based on Ref. 13 as discussed in text

A "fluttering" dislocation should contribute an Einstein term to the specific heat. If the mobile, isolated dislocation is fixed at a statistical distribution of pinning points [12], the specific heat contribution should be nearly linear in T at temperatures above $T \approx 10^{-11}\ \nu_o$ [K] where ν_o is the average resonant frequency of a dislocation segment of length L,

$$\nu_o \approx \overline{v}/2L. \qquad (2)$$

Such a contribution to the specific heat has been observed [13] in cold-worked (pure) copper, with $\nu_o \approx 3 \times 10^{10}$ Hz. Using this value for ν_o, a very rough estimate of ℓ may be obtained [1,9] and is represented by curve 3 in Fig. 1.

The frequency (or temperature) dependence is similar to that for the experimental data of curves B and C, and the magnitude is in better agreement than that of the static mechanism, curves 1 or 2.

The collected data of Fig. 1 do not provide strong support for either the static or dynamic scattering processes. It should be recognized that the experimental results rely on an accurate subtraction of the electronic contribution from the measured κ, and this requires that the laboratory temperature scale closely approximates the absolute thermodynamic temperature scale embedded in the Wiedemann-Franz law [9]. In brief, definitive measurements are difficult to obtain for a normal metal. The problems associated with conduction electrons may be avoided in superconductors at $T < T_c$, the superconducting transition temperature.

3. Superconducting Metals

The phonon thermal conductivity for superconducting tantalum is shown in Fig. 2. Curve A is for a sample not intentionally deformed, while curves B and C are for different regions in the same sample after deformation [14].

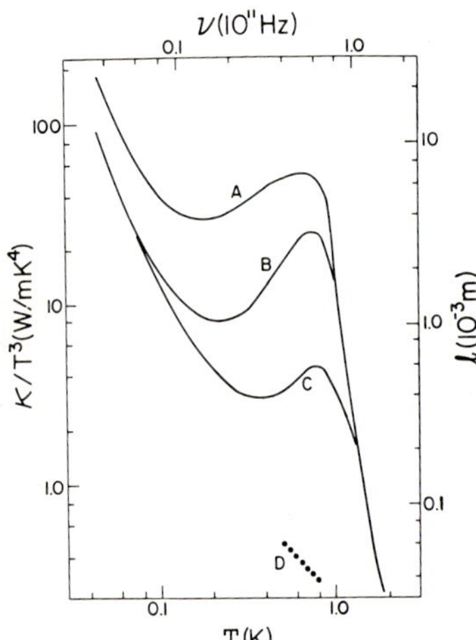

Fig. 2. Phonon thermal conductivity of superconducting tantalum, divided by T^3. The right plus top scales provide the frequency dependence for the effective phonon mean free path ℓ. Curves A,B,C from Ref. 14, deformed by bending; curve D from Ref. 19, deformed by cold-rolling. The strong phonon scattering which occurs above $\approx$ 1 K is due to conduction electrons (T_c = 4.5K)

The sample was then sectioned and etched to reveal an etch-pit density of $\approx 3 \times 10^{11}/m^2$ for curve C, which should approximate the density of isolated dislocations. The static scattering mechanism cannot account for either the magnitude or the frequency dependence found in Fig. 2. On the other hand, the dynamic scattering mechanism can account for both the frequency dependence and magnitude if an average resonant frequency of $\nu_o \approx 2 \times 10^{10}$ Hz is assumed. Since the minima in Fig. 2 increase in temperature with increasing dislocation density, it would appear that ν_o also increases with dislocation density.

Measurements on superconducting niobium [15] and aluminum [16] display similar evidence for the dynamic scattering mechanism. Unfortunately, the possible contribution of fluttering dislocations to the specific heat of superconducting tantalum or niobium can be masked by the presence of even a trace impurity of hydrogen [17], and so the specific heat has not confirmed the presence of fluttering dislocations.

The above remarks apply to samples of low dislocation densities, $N < 10^{12}/m^2$. At larger densities in niobium [18] the ℓ is similar in behavior to curve D of Fig. 2, which was obtained from a cold-rolled sample of tantalum [19]. The phonon scattering at these large dislocation densities is readily explained by the static scattering mechanism [18]. Clearly there is a need to observe phonon transport in tantalum or niobium over a broad temperature range as the dislocation density is increased, systematically, over a broad range. A complicating detail is that the defect structure, introduced into the sample by deformation, is sensitive to the presence of impurities [20].

Lead, having a high T_c, should be an ideal material for the investigation of dislocations. However, lead can deform under its own weight, and can also anneal at room temperature. Thus it is difficult to control the density of dislocations intentionally introduced. Measurements that have been made [14,21] find better agreement with the dynamic scattering mechanism than with the static scattering process.

In summary, the work on metals, both normal and superconducting, presents conflicting evidence on the relative importance of the static and dynamic phonon scattering mechanisms. Most data obtained at low dislocation densities ($< 10^{12}/m^2$) appear to be consistent with the dynamic mechanism, while most data obtained at high dislocation densities seem to indicate a static mechanism. It should be noted that such a division by dislocation density is not consistent with all available data [1].

4. Ionic Solids

Lithium fluoride was used in early investigations of dislocations [22]. It was natural, therefore, that LiF should later be used in studies of the phonon-dislocation interaction [23]. There can be little doubt that phonon scattering in LiF is caused by the dynamic mechanism.

To discuss the data for LiF, it is necessary to rewite Eq. 1 as

$$\kappa = 1.36 \times 10^{10}\ T^3\ [(\ell/v^2)_{ST} + (\ell/v^2)_{FT} + (\ell/v^2)_L] \qquad (3)$$

where the three terms represent the slow transverse, fast transverse and longitudinal phonon contributions to κ. It is also convenient to discuss separately the temperature ranges above and below $\simeq$ 1K. Below 1K only the thermal conductivity technique has been applied while above 1K, both thermal conductivity and ballistic phonon techniques have been used.

Below $\simeq$ 1 K, deformation of LiF reduces κ by a small factor of 2-5 as shown in Fig. 3, independent of dislocation density N. Yet it also has been shown that κ remains proportional to the thickness of the sample, just as in the undeformed state, [24-26]. These results may be understood with the help of Eq. 3. The ST mode has the most appropriate strain field to interact with dislocations in LiF. As a result, the ST phonons are so strongly scattered by dislocations that they make no contribution to κ. The ℓ_{FT} and ℓ_L are still determined by scattering from sample surfaces, and hence κ continues to scale with the size of the sample. Adding more dislocations does not alter this situation. The scattering of the ST mode is so strong that only the dynamic mechanism can account for it.

Exposure of deformed Lif to γ irradiation adds pinning points to the existing dislocations, thus reducing L of Eq. 2 and increasing the average

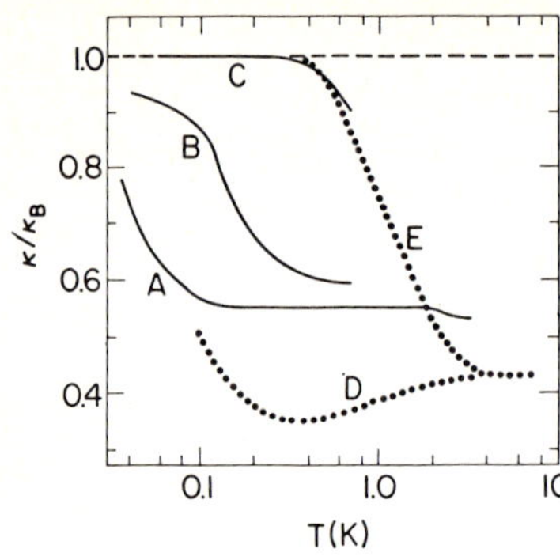

Fig. 3. Thermal conductivity of deformed LiF crystals divided by the value κ_B for the undeformed sample as limited by boundary scattering. Curves A, B and C are for a sample deformed by shear [Ref. 24], D and E are for a sample deformed by bending [Ref. 25]. Curve B followed exposure to 1000 R of γ irradiation, curve C to 136000 R, and curve E to 180000 R

resonant frequency ν_o. This effect is seen in Fig. 3. The thermal conductivity is restored to the predeformation value of $\kappa/\kappa_B = 1$, first at low tempertures (corresponding to low ν_o) and then at higher temperatures as irradiation progresses. Following extensive irradiation, the dislocations remain but do not scatter phonons within the resolution of the measurement. This weak scattering by highly pinned dislocations is consistent with the static scattering mechanism.

Above $\approx$ 1 K the picture is more complex. The amount that κ is depressed by deformation depends on the purity and condition of the sample [26]. Nevertheless the qualitative behavior, as deduced from thermal conductivity measurements, is similar to that observed below 1 K [27]. Some phonons are strongly scattered by dislocations, others have a mean free path as large as $\simeq$ 1 cm and are scattered by sample surfaces even in heavily deformed samples. The details are exposed by ballistic phonon imaging measurements [23].

For convenience, we may consider the imaging technique to employ a small, pulsed, thin-film heater on one side of a sample and an array of bolometer detectors on the opposite face of the sample. The arrival time of a ballistic phonon at a detector determines the phonon mode, and the position of the detector establishes the direction of propagation of that phonon. The output of the detector array for transverse phonons is shown in Fig. 4a for an undeformed sample of LiF. The intensity pattern is caused by phonon focusing. Deformation of the LiF results in the pattern of Fig. 4b, which demonstrates that even in a sample deformed by 10%, a fraction of the phonons propagate through the sample without scattering. The increased phonon scattering (decreased intensity) over a portion of Fig. 4b can be explained in detail by the dynamic scattering mechanism, and depends on phonon mode, direction of propagation, and dislocation orientation.

At small dislocation densities, the measured scattering strength deduced from Fig. 4 is in good agreement with the dynamic scattering theory. However, at large dislocation densities ($N > 10^{12}/m^2$) the scattering strength, as deduced from κ or from ballistic measurements, is much too strong to be explained by the dynamic scattering of phonons from isolated dislocations. It is speculated [12] that the additional scattering is caused by a large density of dislocation dipoles. The presence of dislocation dipoles is not revealed in the etch-pit counts used to measure N. The dislocation dipoles should contribute to the specific heat. However, with the dislocation densities available in these samples of LiF, the speculated increase in specific heat is too small to measure [28].

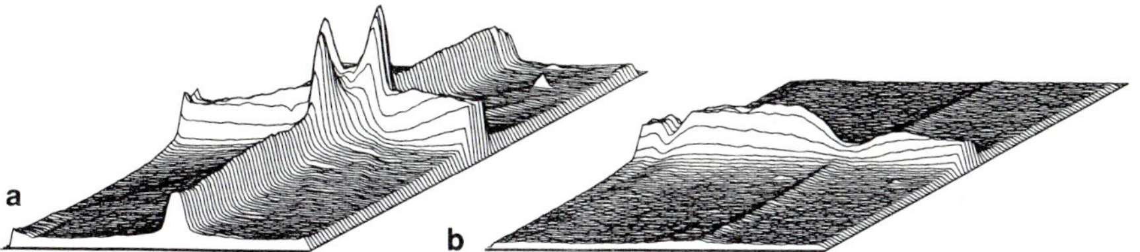

Fig. 4. Intensity of the ballistic-phonon signal arriving on one face of a LiF sample, from Ref. 23. a) No deformation, expected singularities resulting from phonon focusing. b) Same as (a), but following deformation by ≈ 10%

In summary, the work on LiF has revealed that the dominant phonon scattering mechanism is dynamic unless the dislocations are heavily pinned. The static mechanism is too weak to observe for the experimental resolution that has been available. It is likely that other ionic solids will exhibit similar behavior.

5. Covalent Solids

In a covalent material such as germanium, a dislocation is constrained in a deep potential-energy minimum. As a result of this restraint on the dislocation, deformation is introduced only at elevated temperatures (≈ 600°C), and any dislocation resonance is expected to lie at a high frequency corresponding to T > 3K. Indeed, the scattering of thermal phonons by dislocations in deformed germanium is weak [29,30] and is consistent in temperature dependence and in magnitude with the static scattering mechanism [11]. Less definitive results have led to the same conclusion for silicon [31]. It would be helpful to have additional measurements on highly deformed samples, obtained over a broader temperature range.

6. Conclusions

Evidence presently available suggests that the static mechanism of phonon scattering by dislocations is dominant in covalently bonded Ge, but the dynamic scattering mechanism dominates in ionic LiF. With metals, normal or superconducting, both mechanisms seem to be present, with perhaps the dynamic mechanism being more important at low dislocation densities. It is also noted that the κ produced by deformation often depends on sample purity and history, and on the process or type of deformation imposed.

Clearly, much work remains to be done in this field. First, the experimentalists should delineate more clearly where and why the dynamic and static scattering mechanisms are important. Second, the theorists should converge on the magnitude to be expected for phonon scattering from dislocations. After these two steps are completed, it should be possible, third, to investigate the details of the phonon scattering mechanisms. An example is the modification in phonon scattering to be expected when dislocations are not isolated, but lie in close proximity with other dislocations [10,12].

Until the work outlined above is completed, the thermal-conductivity community should be dissuaded from the common and indiscriminate use of a phonon scattering term proportional to $1/\nu$ to account for a thermal resistance arising from a speculated presence of dislocations. Even if there is some evidence that a static scattering mechanism may be present in the sample, the theoretical magnitude of the $1/\nu$ dependence is still uncertain.

Acknowledgement

The author's work on dislocations has been supported by the Materials Sciences Division of the United States Department of Energy under contract DE-AC02-76ER01198.

References

1. A.C. Anderson: in Dislocations in Solids, ed. by F.R.N. Nabarro (North Holland, New York) in press
2. A brief review is provided by R. Berman: Thermal Conduction in Solids (Clarendon, Oxford 1976)
3. K. Ohashi and Y.H. Ohashi: Phil. Mag. A38, 187 (1978)
4. J.N. Lomer and H.M. Rosenberg: Phil. Mag. 4, 467 (1959)
5. W.R.G. Kemp, P.G. Klemens and R.J. Tanish: Phil. Mag. 4, 845 (1959)
6. R. Zeyfang: Phys. Stat. Sol. 24, 221 (1967)
7. P. Charsley, J.A.M. Salter and A.D.W. Leaver: Phys. Stat. Sol. 25, 531 (1968)
8. M. Kusunoki and H. Suzuki: J. Phys. Soc. Japan 26, 932 (1969)
9. J.L. Vorhaus and A.C. Anderson: Phys. Rev. B 14, 3256 (1976)
10. F.L. Madarasz and F. Szmulowicz: Phys. Rev. B (to be published)
11. R.A. Brown: J. Phys. (Paris) 42, C6-271 (1981)
12. G.A. Kneezel and A.V. Granato: Phys. Rev. B 25, 2851 (1982), and papers cited therein
13. J. Bevk: Phil. Mag. 28, 1379 (1973) 14. S.G. O'Hara and A.C. Anderson: Phys. Rev. B 10, 574 (1974)
14. S.G. O'Hara and A.C. Anderson: Phys. Rev. B 10, 574 (1974).
15. A.C. Anderson and S.C. Smith: J. Phys. Chem. Solids 34, 111 (1973)
16. S.G. O'Hara and A.C. Anderson: Phys. Rev. B 9, 3730 (1974)
17. G.J. Sellers, A.C. Anderson and H.K. Birnbaum: Phys. Rev. B 10, 2771 (1974)
18. W. Wasserbach: Phil. Mag. A 38, 401 (1978)
19. M. Ikebe, N. Kobayoshi and Y. Muto: J. Phys. Soc. Japan 37, 278 (1974).
20. W. Wasserbach: Phys. Stat. Sol. B 84, 205 (1977)
21. L.P. Mezhov-Deglin: Soc. Phys. JETP 50, 369 (1979)
22. W.G. Johnston and J.J. Gilman: J. Appl. Phys. 30, 129 (1959)
23. G.A. Northrop, E.J. Cotts, A.C. Anderson and J.P. Wolfe: Phys. Rev. B 27, 6395 (1983), and papers cited therein
24. A.C. Anderson and M.E. Malinawski: Phys. Rev. B 5, 3199 (1972)
25. E.P. Roth and A.C. Anderson: Phys. Rev. B 20, 768 (1979)
26. E.J. Cotts, D.M. Miliotis and A.C. Anderson: Phys. Rev. B 24, 7336 (1981)
27. E.J. Cotts, S.E. Shore and A.C. Anderson: J. Phys. Chem. Solids (to be published)
28. E.J. Cotts and A.C. Anderson: Phys. Rev. B 24, 7329 (1981)
29. M. Sato and K. Sumino: J. Phys. Soc. Japan 36, 1075 (1974)
30. M.P. Zaitlin and A.C. Anderson: Phys. Rev. B 10, 580 (1974)
31. E.P. Roth and A.C. Anderson: Phys. Stat. Sol. B 93, 261 (1979)

The Acoustic Paramagnetic Resonance of Cr^{2+} in n-Type GaAs

A.S. Abhvani, C.A. Bates, P.J. King, D.R. Pooler, V.W. Rampton, and P.C. Wiscombe

Department of Physics, University of Nottingham, University Park Nottingham, NG7 2RD, England

P. Bury

Department of Physics, Technical University of Advanced Transport Engineering CS-010 88 Zilina, Czechoslovakia

ABSTRACT

We have compared the acoustic paramagnetic resonance spectra found from unilluminated n-type chromium-doped gallium arsenide with the predictions of a dynamic Jahn-Teller model for Cr^{2+}.

Gallium arsenide which is doped with both chromium and silicon in proportions which produces semi-insulating material exhibits very intense and complex acoustic paramagnetic resonance (APR) spectra which are weakly modified by illumination [1]. If the proportions are such as to produce n-type material a weak APR spectrum results if the sample is cooled in the dark to helium temperatures. The intense spectrum seen in semi-insulating material can however be induced by suitable illumination.

We attribute the weak spectrum to Cr^{2+} at gallium sites of T_d symmetry. Such a spectrum is not found in chromium-free samples and it correlates well with the presence of the acoustic relaxation peak due to Cr^{2+} in doped samples [2].

A typical "weak spectrum" is shown in figure 1. The acoustic frequency used was 9.46 GHz, the acoustic propagation was in the [001] direction and the magnetic field was at 45^0 to that direction in the (110) plane. The temperature was 1.85K. The absorption spectrum consists of a broad and asymmetric background (A), a number of sharp lines (B) and a smaller number of broader but extremely field-angle-dependent lines. Only one of these can be seen in figure 1 and is marked C. The peak value of A occurs at a field

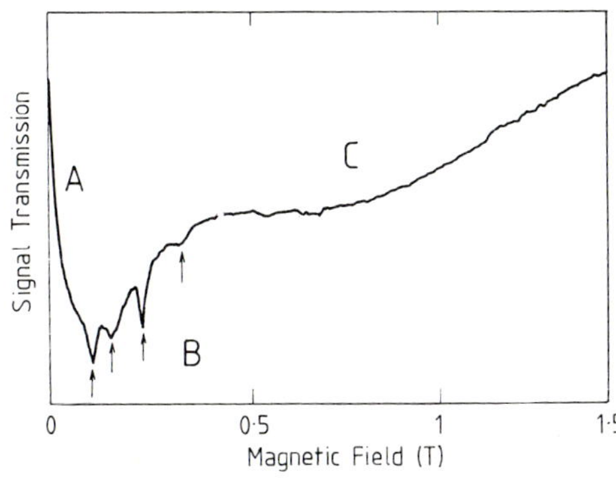

Figure 1. A typical APR spectrum

value which is almost independent of field angle. The shape of the upper tail varies considerably with field angle however. The bulk of the absorption is of type A. The sharp lines B occur in pairs having axial symmetry about the cube axes and their position may be described by $B_0\sec\theta$ and $2B_0\sec\theta$ where $B_0 = 0.0865 \pm 0.0008T$ and where θ is the angle between the magnetic field and the <001> directions. The weak and broad features C can be described by an angular dependence $\sec(\theta + \pi/4)$ in the (010) plane and corresponding features occur at angles $\sec(\theta + 35.3^o)$ in the (110) plane. θ is in each case the angle between the magnetic field and the [001] direction. The entire spectrum is found to have an amplitude dependence varying as 1/T in the range 1.6K to 4.2K, the line widths being independent of temperature.

We have attempted to interpret these results using a dynamic Jahn-Teller model for Cr^{2+} previously developed to explain the electron paramagnetic resonance (EPR) results [3][4][5]. Early static models in which the Cr^{2+} centre remains in one of three tetragonally distorted configurations are clearly inappropriate since transition probabilities for acoustic perturbations are zero and APR absorption is not possible. The dynamic model allows tunnelling between wells corresponding to the three distortions and finite APR absorption is possible. When random strains are introduced into the model it is found that the EPR comes from sites where one well lies considerably below the other two and most sites contribute. The APR on the other hand comes from the special sites described below and is thus rather weak.

An analysis of the dynamic model shows that the sharp features B and the broader features C are due to sites with random strains where the two lowest wells are closely equal in energy and lie significantly below the third. An approximate analytical method which retains only large terms in the Hamiltonian yields the following results. With the magnetic field in the (010) plane resonances at $B_0\sec\theta$ and $2B_0\sec\theta$ and at $\sqrt{2}B_0\sec(\theta + \pi/4)$ are expected. It is possible to obtain a good fit to the experimental value of B_0 by adjusting γ, the first-order Ham reduction factor. Since the EPR is not appreciably dependent on this factor it is still possible to simultaneously produce a good fit to the EPR. A detailed computer calculation shows that C should be broader than the lines B as is found experimentally.

With the field in the (110) plane, lines at $B_0\sec\theta$ and $B_0(8/3)^{\frac{1}{2}}\sec(\theta + 35.3)$ are predicted in good agreement with experiment. However lines at $2B_0\sec\theta$ are not predicted by transitions from within the ground state levels. Such lines are predicted to arise from the first excited states but as these are at 167 GHz above the ground states these lines would not be expected to have the observed dependence on temperature.

An attempt has been made to calculate the contribution to the APR spectrum from all regions of the strain plane, using a Monte Carlo method. The method failed to converge after 1000 points and 40 hours of computing time. Points in the strain plane were therefore selected from regions which were found to be more important to the APR spectrum. This includes the region of low strain where the three wells are of almost equal energy, expected to contribute much of the feature A, and the regions close to those where two wells are lower and of almost equal energy, expected to provide the lines B. As a result we were able to obtain broad agreement with the shape of the background peak A and to show that the lineshape predicted by the model for A varies with field angle in a similar way to that found experimentally. The statistics were still sufficiently poor however that we could not predict the lineshapes of B and C by this method.

ACKNOWLEDGEMENTS

ASA, PCW and DRP are grateful to the SERC for financial support. PB would like to thank the British Council and the Czechoslovakian Academy of Science for a scholarship. We are also most grateful to the Plessey Company for their gift of samples.

REFERENCES

1. Bury, P., King, P.J., Rampton, V.W. and Wiscombe, P.C. 1982 Acta Phys. Slovaka 32 17-23.

2. Abhvani, A.S., Austen, S.P., Bates, C.A., Parker, L.W. and Pooler, D.R. 1982 J. Phys. C., Solid St. Phys., 15 2217-2231.

3. Tokomoto, H. and Ishiguro, T. 1979 J. Phys. Soc. Japan 46, 84-91.

4. Krebs, J.J. and Stauss, G.H., 1977 Phys. Rev. B16 971-973.

5. Kauffmann, U. and Schneider, J., 1982 Advances in Electronics and Electron Physics 58 81-141.

Magnetic-Field-Dependent Phonon Scattering by Li Ions in Si

L.J. Challis, A.P. Heraud, V.W. Rampton, and M.K. Saker

Department of Physics, University of Nottingham, University Park
Nottingham, NG7 2RD, England

M.N. Wybourne

GEC Hirst Research Centre, Wembley, England

The Li ion acts as a shallow donor in Si and its ground state has been studied by far infra-red, [1], EPR, [2], APR, [3], ultrasonics, [4] and thermal conductivity [5,6]. It appears to be in an interstitial site and its levels are inverted with respect to those of substitutional group V impurities. The lowest levels are the approximately degenerate ^2E and $^2T_2$ states and the $^2A_1$ level lies at an energy $\Delta \sim 440$GHz. This appears to be the smallest chemical shift for shallow donors in silicon and makes it of potential value for frequency crossing experiments. The sizes of the spin-orbit coupling constants λ, λ' within $^2E + {}^2T_2$ are not known but expected to be small. If $\lambda, \lambda' \to 0$, APR attenuation should be Kramers forbiiden and so very weak. The rather strong APR signals observed by WIGMORE [3] at $g \sim 2$ therefore suggest significant spin-orbit mixing. This should lead to fine structure around $g \sim 2$ [7] from which λ, λ' and the E-T_2 splitting should be obtainable if not smeared out by strain. No such structure was seen by WIGMORE although he did observe additional lines corresponding to $g \sim 1.38$ and 0.97 which are not explicable by this model. No anisotropy measurements were made. Evidence that about 20% of the Li ions experienced large strain splittings ($\sim$ 50 GHz) in the samples used for thermal conductivity experiments [5,6] comes from the measurements below IK. Above IK, the scattering at Δ = 440 GHz is dominant but at lower temperatures, where the effect of this becomes weak, the strong scattering observed is attributable to processes within the $^2E + {}^2T_2$ states.

In the present work we have made an investigation of the phonon scattering in a <110> axis silicon sample cut from Wacker Chemitronic material and containing less than 5×10^{15} cm^{-3} of C and O impurities. Li was introduced by diffusion and from the room temperature resistivity we estimate its concentration to be 9×10^{15} cm^{-3}. Its APR spectrum is shown in fig. 1 for longitudinal ultrasonic waves at 10.3GHz propagating along the <110> axis. In this plot the magnetic field made an angle of 65° with the direction of propagation and the temperature was 1.62K. The spectrum consists of two sharp lines at $g \sim 2.03$ and separated by about 2×10^{-3}T with a background of weaker resonances spread over more than 0.1T. At some magnetic field orientations the background can be seen to contain at least six component lines. No lines at higher field - up to 2T - were found and there was an increase in attenuation at resonance rather than a decrease as found by WIGMORE [3]. It should also be noted that the complex EPR spectrum observed by WATKINS and HAM [2] lies entirely within the span of the sharp APR lines seen here.

Within the range of measurement (1 - 10K), the thermal conductivity was similar to those of the lightly doped samples measured by FORTIER and SUZUKI [5] appropriately scaled for differences in Li concentration. The conductivity is shown in fig. 2 as a function of magnetic field applied normal to the <110> direction of heat current; there was no detectable

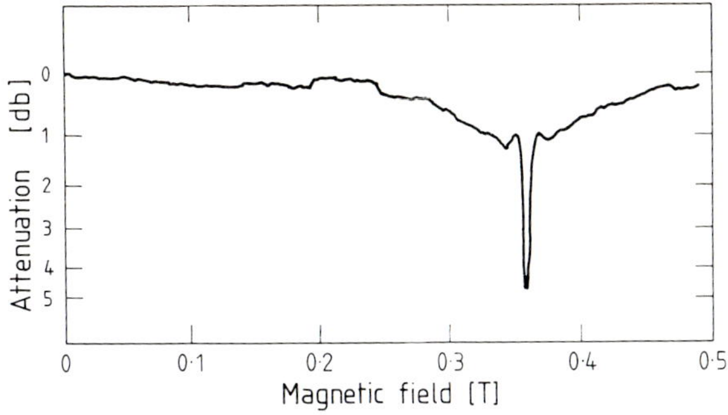

Figure 1. Attenuation of 10.3GHz longitudinal waves through Si(Li)

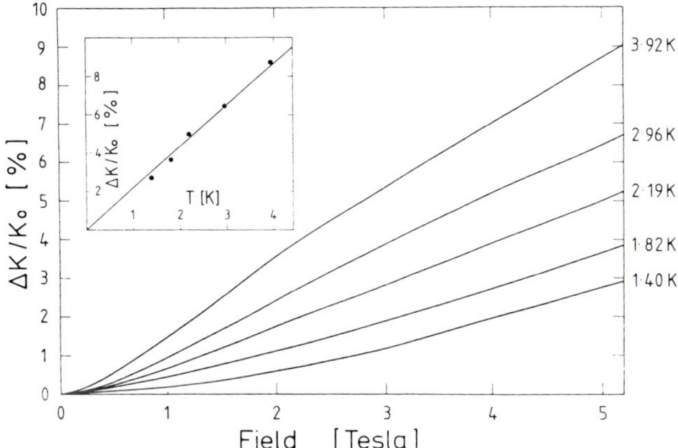

Figure 2. Magneto-thermal conductivity of Si(Li). The detail shows the change in 5T

dependence on field direction in the (110) plane. The change in conductivity produced by a field of 5T is shown as a function of temperature in the detail to the figure.

Increases in thermal conductivity and heat pulse transmission have also been seen in Ge(As) and partially explained by wave-function shrinkage [8]. The magnetic field compresses the donor wave function and so increases the central cell correction and hence the chemical shift which in this case is the ^2E - $^2A_1$ splitting ($\Delta \tilde{\propto} a^{*-3}$). For $T \lesssim 5K$, Δ lies beyond the peak in the heat current so an increase in its frequency reduces its effect on the thermal conductivity as observed. However, at higher temperatures, the reverse effect should occur and this is presently being investigated.

Finally we note that we should expect that splitting the $^2E + {}^2T_2$ states by a magnetic field would produce an increase in the scattering and so a reduction in conductivity. Since an increase in conductivity is observed this must be a weaker process than caused by changes in Δ. This is not surprising since the additional ground-state scattering depends on the rather weak spin-orbit coupling. Even so we might expect to see weak frequency crossing at low fields. No such crossings were observed, suggesting that they are either too weak or perhaps too broad because of random strain.

We are very grateful to Dr E. C. Lightowlers, and Mr W.B. Roys for their advice and assistance in preparing the samples, to Mr B.J. Kiff for his help in making some of the measurements and to S.E.R.C. for their support.

References

[1] R.L. Aggarwal, P.Fisher, V. Mourzine and A.K. Ramdas, Phys. Rev. 138, A882, 1965.

[2] G.D. Watkins and F.S. Ham, Phys. Rev. B1, 4071 (1970).

[3] J.K. Wigmore, Phys. Letts. 26A, 15 (1967).

[4] M. Pomerantz, Phys. Rev. B1, 4029 (1970).

[5] D. Fortier and K. Suzuki, Phys. Rev. B9, 2530 (1974).

[6] A. Adolf, D. Fortier, J.H. Albany and K. Suzuki, Phys. Rev. B21, 5651 (1980).

[7] K. Suzuki and N. Mikoshiba, J. Phys. Soc. Jap. 27, 1207 (1969).

[8] L. Halbo and R.J. Sladek, Proc 10th Int. Conf. on Physics of Semiconductors, Cambridge, Mass 1970, eds. S.P. Keller, J.C. Hensel and F. Stern (U.S.A.E.C. Division of Technical Information, Springfield, Va.), p826; L. Halbo, Proc. 2nd Int Conf. on Phonon Scattering in Solids, Nottingham, 1975, eds. L.J. Challis, V.W. Rampton, A.F.G. Wyatt, (Plenum, New York) p. 346; A. Kobayashi and K. Suzuki, Phys. Rev. B20, 3278 (1979); F. Miyasato, M. Tokumura, M. Toguchi and F. Akao, J. Phys. Soc. Japan 50, 1986 (1981); T. Miyasato, M. Tokumura and F. Akao, J. de Physique C6, 658 (1981).

Influence of Defects on the Splitting of the Acceptor Ground State in Silicon

A. Ambrosy and K. Laßmann
Physikalisches Institut der Universität Stuttgart, Pfaffenwaldring 57
D-7000 Stuttgart 80, Fed. Rep. of Germany

A.M. de Goër and B. Salce
Centre d'Etudes Nucléaires de Grenoble, F-38000 Grenoble, France

H. Zeile
Valvo Röhren-und Halbleiterwerke der Philips GmbH, Stresemannallee 101
D-2000 Hamburg 54, Fed Rep. of Germany

The distribution of strain fields from specified defects in otherwise pure silicon crystals is reflected in the resulting distribution of splittings E of the partially orbitally degenerate Γ_8 ground state of effective mass acceptors. The spectral density N(E) can be probed by resonant scattering of $h\nu$ = E ultrasonic [1] or 3.8 kT ≅ E thermal phonons [2]. The results of both methods (in the following: $\alpha(\nu)$ and $\kappa(T)$) are compared for Si(B) and Si(In) crystals containing definite amounts of C and O [Tab.1]. Monte Carlo calculations (M.C.) for point defects and for 60^0 dislocations in Si (isotropic approximation) have been made to obtain N(E) as well as $D_1(E)$, the latter being the mean coupling constant of a specified phonon (L,T1,T2,[100],[110],[111]) to splittings at E. This average coupling for all 9 types of phonons practically does not depend on E and therefore presumably also for the thermal phonons. Thus, though in both experiments $D_1^2(E)\cdot N(E)$ is measured the form of N(E) is preserved and D_1 can be estimated by integrating over the whole distribution normalizing with $n_a=\int N(E)dE$. The calculated distribution for point defects derives from Lorentzians and from Gaussians for dislocations. Fig.1 shows $D_1^2\cdot N(E)/n_a$ for Si(B) crystals with various concentrations of point defects from $\alpha(\nu)$. Analysis shows that only for the crystal with the highest concentration of point defects (S 80) $D_1^2\cdot N(E)/n_a$ can be fitted by the calculated "Lorentzian". The smaller the concentration the more "Gaussian" is mixed into N(E) merging into the quasi Gaussian residual distribution of unknown origin of the pure crystal S 87. S125 and S 54 of Fig.2

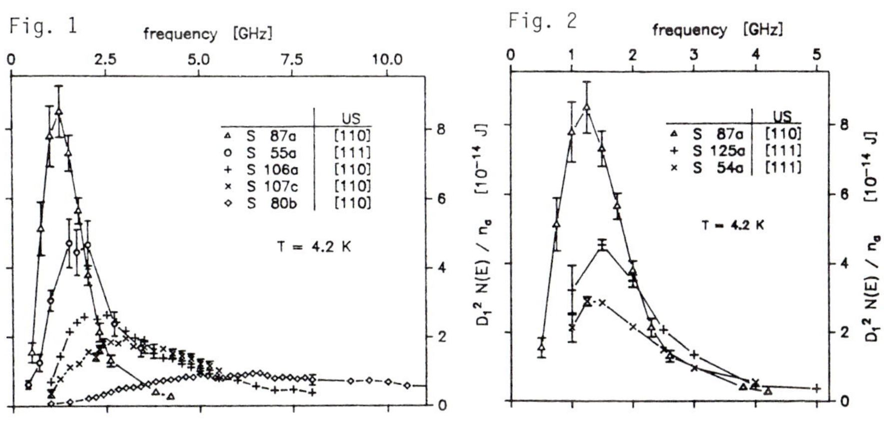

Fig.1. Splitting distribution of Si(B) with point defects

Fig.2. Splitting distribution of pure Si for different concentration of boron

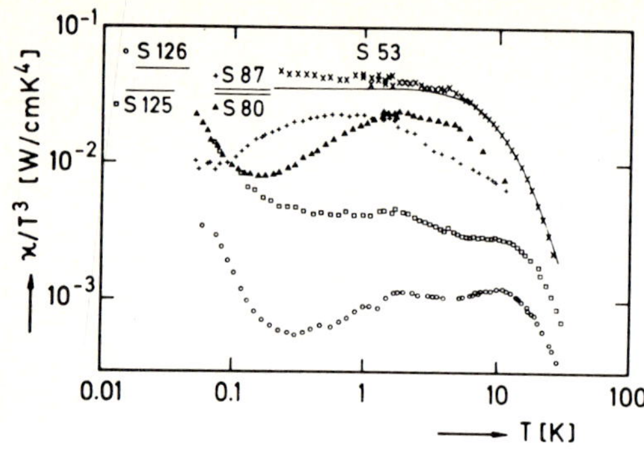

Fig.3. Thermal scattering in Si(B)

are pure crystals with high [B], where a dynamic B-B interaction had been found by hole burning [3]. Possibly an interaction shows up also in these curves, not so much in the form of N(E) but in a reduction of D_1. Analogous results in the form $\kappa(T)/T^3$ vs. T are shown for crystals with as sawn side faces in Fig.3. The extra scattering below 1K is due to the split levels, whereas the depression around 5K - going monotonously with [B] - is interpreted as due to Jahn-Teller resonance scattering [4]. Boundary scattering for the defect-free and undoped S 53c is found to be temperature dependent and about 20 % larger than the theoretical Casimir term. A check with chemically polished side faces and also with an extra dummy collar showed that there is only negligible change for the S 80 but that thermometer mounting and strains from side face damage may have a significant influence for S 87 with narrow distribution. For the S 53 with polished side faces $\kappa(T)/T^3$ increases by about 1.6 at 2 K to 3.4 at 60 mK. These facts make it difficult to analyse narrow distributions of small acceptor concentrations where the extra scattering is at low temperatures and weak. In crystals with high [B] (S 125, S 126) the scattering regime is much larger than expected from $\alpha(\nu)$. Again the S 80 can be fitted by the calculated distribution but with a maximum at 8.3 GHz instead of about 6.5 GHz as obtained from $\alpha(\nu)$. The experimental/theoretical average coupling constants are 1.22/1.2 eV and 1.94/2.22 eV for $\kappa(T)$ and $\alpha(\nu)$ respectively.

Table 1: Si(B) samples

sample	ρ[Ωcm]	concentrations [10^{21}m^{-3}]			ν_{max} in GHz from	
		[B]	[O]	[C]	$\alpha(\nu)$	$\kappa(T)$
S 53*	>2500.0	0.005	< 0.5	–		
S 54*	0.3	85.0			1.2	
S 55	1.6	9.0	< 2.0	75.0	1.7	
S 80	10.85	1.2	570.0	41.0	6.5	8.3
S 87*	2.5	5.4	< 2.0	< 5.0	1.2	4.6
S 106		5.0	<20.0	200.0	2.2	
S 107		1.8	<20.0	280.0	2.8	
S 125*	0.74	24.0			1.2	
S 126	0.3	85.0	200.0	<50.0		10.4

* Wacker chemitronic, waso quality

For Si(In) the deformation potential constants being about 0.6 of those of Si(B), one would expect the distributions and the phonon scattering to be correspondingly smaller. A possible In-In interaction should set in at a concentration about 8 times larger than for B because of the smaller Bohr radius. In Tab. 2 reliable figures for the concentration can be given only for samples S 52 and S 123 where temperature-dependent Hall measurements have been made [5]. The results of $\kappa(T)$ are normalized to the pure case to show the additional scattering [Fig.4]. The Jahn-Teller resonance at about 20 K increases monotonously with [In] whereas the scattering at low temperatures is due to the splitting distributions. For S 123 from $\alpha(\nu)$ we get E_{max} = 3.2 GHz as compared to ~5 GHz from $\kappa(T)$. Estimating the elastic strength A from the covalent radii gives A_{In}/A_C= 0.68 and taking into account the deformation potential constants one would obtain E_{max}=0.5 GHz from M.C. For the S 52 with high concentration of point defects a fit somewhat broader than "Lorentzian" gives E_{max} = 14.5 GHz, whereas from M.C. one would expect $E_{max} \leqslant$ 9 GHz if the indicated figures for [In] and [O] are correct. The deformation potential constant from this analysis is 0.24eV as compared to the theoretical value 0.72 eV. Since [In] is rather high the reduction could be due to an In-In interaction in analogy to the case of boron.

Table 2: Si(In) samples

sample	ρ[Ωcm]	concentrations [$10^{21}m^{-3}$] [In]	[O]	[C]	ν_{max} in GHz from $\alpha(\nu)$	$\kappa(T)$
S 33	2.0		700.0			
S 52	0.1	500.0	800.0			14.5
S 123	1.29	100.0	<10.0	< 1.0	3.2	≃ 5
S 196	0.55					

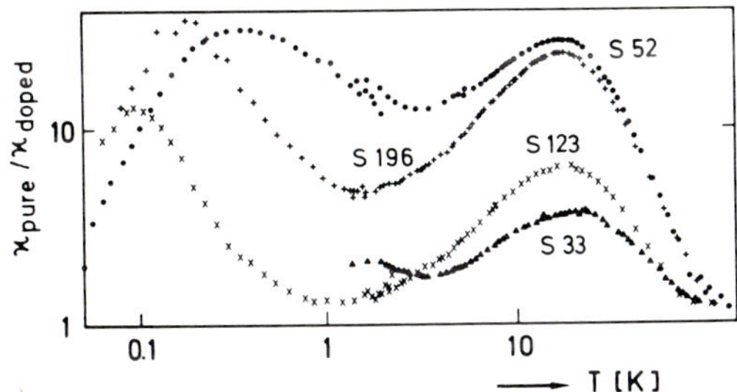

Fig.4. Thermal scattering in Si(In)

1 H.Zeile, K.Laßmann, Phys. Stat. Sol. (b) 111,555(82)
2 A.M.de Goer, M.Locatelli and K.Laßmann, J. de phys. 42 C 6-235(81)
3 H.Zeile, U.Harten and K.Laßmann, Phys. Stat. Sol. (b) 111,213(82)
4 J.Maier and E.Sigmund, these Proceedings
5 S 52 obtained by J.S.Blakemore (sample 260 in PR B4 1873(71))
S 123 obtained by R.Helbig, Univ. Erlangen (sample Ru 237/1-1b)

Phonon Scattering at Electronically Degenerative Systems: An Application to the Defect Systems Si(In), Si(B), and GaAs(Mn)

J. Maier and E. Sigmund

Institut für Theoretische Physik, Universität Stuttgart, Pfaffenwaldring 57
D-7000 Stuttgart 80, Fed. Rep. of Germany

1. *Introduction*

Phonon scattering experiments at the Γ_8 ground state of acceptors in cubic semiconductors as Si(In), Si(B), and GaAs(Mn) show a specific resonance structure in the meV range [1], [2], which can be explained by a dynamical Jahn-Teller effect [3] .However, due to the large extension of the defect wave function (5 - 15 Å) the influence of random internal fields is reinforced, which causes a static splitting of the Γ_8 state. We present a first step to a unified description of the phonon scattering processes including the influences of static fields together with the dynamical electron-phonon-interaction.

2. *Phonon Scattering at the Jahn-Teller Defect*

Neglecting random internal fields the acceptor systems are described by the Γ_8 multimode Hamiltonian:

$$H = \sum_{q,\lambda} \left(D_\varepsilon \sum_{i=1}^{2} \hat{\rho}_i \, r^i_{q\lambda} + D_\tau \, \hat{\rho}_3 \sum_{j=1}^{3} \hat{\sigma}_j \, s^j_{q\lambda} \right) Q^{q\lambda} \tag{1}$$

where D_ε and D_τ are the deformation potentials of the coupling to e and t phonons. The description of the defects is by the effective-mass approximation, where the Bohr radius a^* accounts for the defect extension. Within this approach the cut-off function f(q), which is the main part of the projection coefficients $r^i_{q\lambda}$ and $s^i_{q\lambda}$, diminishes the influence of large-λ phonons [4].

The relaxation rate of a phonon q, λ,is calculated by a high-order Green's function formalism. The accuracy of the calculation is checked by the validity of the sum rules for the first 2 moments. The final result for the mean scattering rate can be brought to the Lorentzian-like form [4]:

$$\frac{P(\omega,T)}{(\omega^2-\Delta^2(\omega,T))^2 + \Gamma^4(\omega,T)} \quad . \tag{2}$$

In fig. 1 the temperature dependence of the maximum scattering rate is shown. The increase in the relaxation rate with higher temperatures is due to the stronger fluctuations of the coupled phonon system. The influence of the Bohr radius is shown in fig. 2, where the maximum scattering rate and the resonance frequency are plotted against the defect extension. In the typical range of Bohr radii we have $\omega_{res} \sim 1/a^*$ for fixed parameters otherwise. The thermal conductivity of GaAs(Mn) is shown in fig. 3. Beside the calculated relaxation due to the Jahn-Teller effect the relaxation rates of boundary,

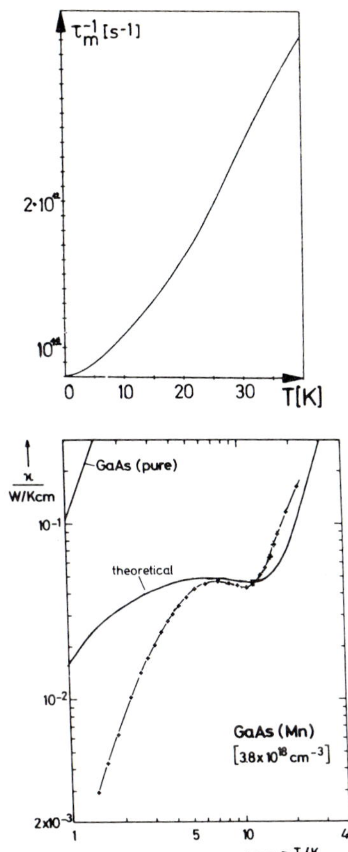

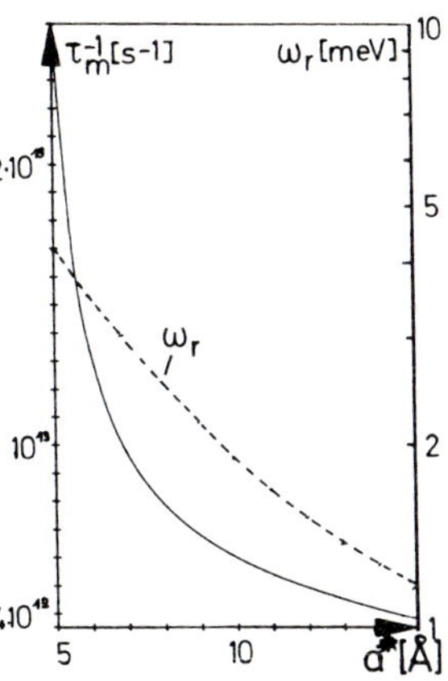

Fig. 1: Temperature dependence of the maximum scattering rate for GaAs(Mn)

Fig. 2: The influence of the maximum scattering rate (full line) and the resonance frequency (dashed line) dependent on the Bohr radius

Fig. 3: Comparison of the theoretical and experimental thermal conductivity curves of GaAs(Mn) [2]

isotope, and Umklapp scattering are introduced. The discrepancy of the experimental and theoretical results at lower temperatures is due to the splittings caused by random internal fields.

3. *Influences of Static Fields*

To include the influence of elastic fields onto the phonon scattering mechanism we have to start with the Hamiltonian of the form [5]:

$$H = H_{e-p} + \varepsilon_1 \hat{\rho}_1 + \varepsilon_2 \hat{\rho}_2 + \hat{\rho}_3 (\varepsilon_3 \hat{\sigma}_1 + \varepsilon_4 \hat{\sigma}_2 + \varepsilon_5 \hat{\sigma}_3) \tag{3}$$

where H_{e-p} denotes the pure electron-phonon part as discussed above. ε_i are the symmetry-adapted strain coefficients. Applying the scattering formalism in the same way as in the pure Jahn-Teller case, we arrive at the following formula for the mean scattering rate:

$$\frac{P(\omega,T,\varepsilon) + A(\varepsilon,T)}{(\omega^2-\Delta_s^2-\Delta^2(\omega,T,\varepsilon))^2 + \Gamma^4(\omega,T,\varepsilon)} \quad . \tag{4}$$

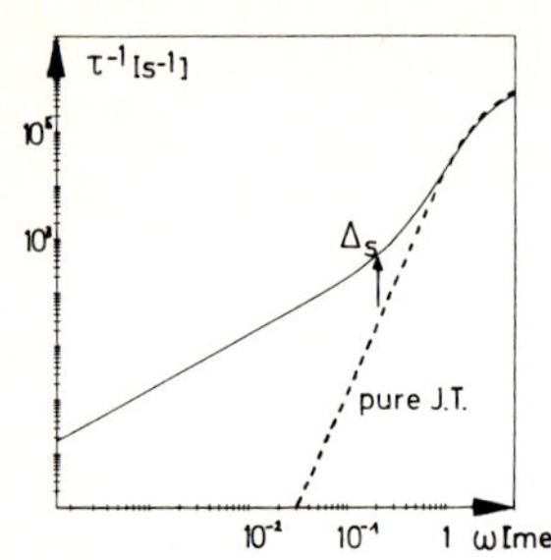

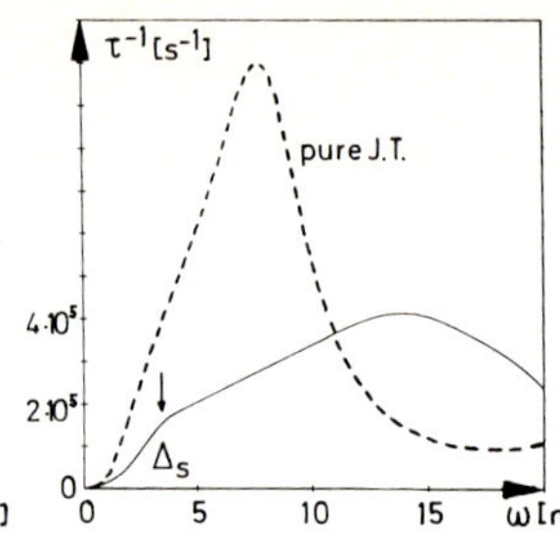

Fig. 4, 5: The influence of static fields on the relaxation rate in Si(In) (Δ_s denotes the static splitting)

For vanishing strain fields expression (2) is regained. All inelastic parts as received in a perturbational approach [6] are included in the modified functions $\Delta(\omega,T,\varepsilon)$, $\Gamma(\omega,T,\varepsilon)$, and $P(\omega,T,\varepsilon)$. The additional terms are due to the direct resonance processes. τ^{-1} influenced by a weak elastic field is drawn in fig. 4. The Jahn-Teller resonance scattering is nearly unaffected, the increase in the scattering rate at low frequencies is due to direct processes. Fig. 5 shows the result where the static energy splitting approaches the Jahn-Teller resonance energy. Compared with the pure Jahn-Teller case the relaxation rate is now broadened and lowered and the maximum is shifted to higher energies.

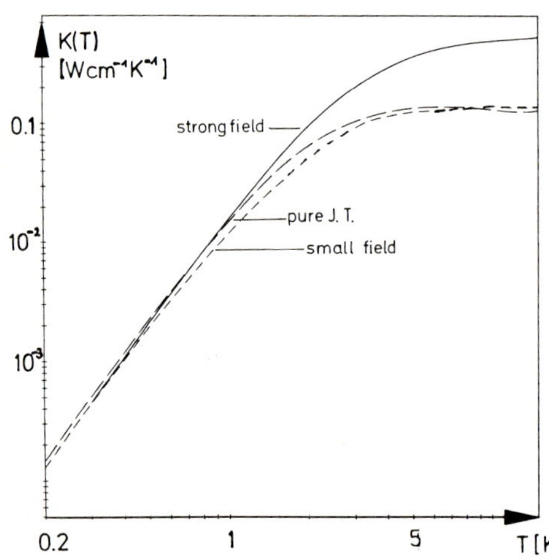

Fig. 6: The thermal conductivity curve of Si(In) under the influence of elastic fields compared with the pure Jahn-Teller case. In concentration assumed as $5\cdot10^{19}$ cm^{-3}. Field strengths are the same as in Figs. 4 and 5, respectively

In fig. 6 the thermal conductivity including the pure Jahn-Teller interaction as well as a static strain field is shown. For large external fields the thermal conductivity is increased in the region of the Jahn-Teller resonance due to the partial destruction of the Jahn-Teller effect, whereas in the weak-field case the thermal conductivity is lowered in the region of low temperatures. This tendency can be seen in experiments with static strain fields [7]. However, to describe the real experimental situation a whole distribution of internal elastic fields must be introduced in the theory. This will be done in the near future.

References

1. Holland, M.G., Proc. 7th Int. Conf. Semicond., Paris (1964)
   Schad, Hp., Laßmann, K., Phys. Lett. 56A, 409 (1976)

2. de Combarieu, A., Laßmann, K., in: Phonon Scattering in Solids, New York (1976)
3. Sigmund, E., Laßmann, K., in: Phonon Scattering in Cond. Matter New York (1980)
4. Sigmund, E., Laßmann, K., phys. stat. sol.(b) 111, 631 (1982)
Maier, J., Sigmund, E., to be published (1983)
5. Bir, G.L., Soviet Physics JETP, 24, 372 (1967)
6. Suzuki, K., Mikoshiba, N., J. Phys. Soc. Japan, 31, 44 (1971)
7. Salce, B., private communication (1982)

Magnetothermal Conductivity of Boron-Doped-Silicon

L.J. Challis and A.P. Heraud

Department of Physics, University of Nottingham, University Park
Nottingham, NG7 2RD, England

A recent comparison by SINGH and VERMA [1] of the detailed SUZUKI-MIKOSHIBA theory (SM) [2] of the magnetothermal conductivity of semiconductors containing acceptors with Γ_8 ground states indicated that it was in poor agreement with the experiments on Si(B) by CHALLIS and HALBO [3]. It seems likely [4] that the splitting of the Γ_8 by neighbouring C and O impurities could have been significant in samples of this period and we report here measurements on a purer sample S87d whose splittings are small, 1.3GHz, and have been determined by ultrasonic and low-temperature thermal conductivity measurements [4]. The sample contained 5.4 x $10^{15}cm^{-3}$B. The resonant scattering from the Γ_8 splittings appears to be negligible above 0.5K where K/T^3 becomes constant (de Goër, private communication). The fall in K/T^3 above 1K is attributable to scattering by boron acceptors and theoretically $\tau_i^{-1} = C_i\nu^2$ [5] at frequencies well below $\sim$ 400GHz where we can neglect the cut-off in τ_i^{-1} and resonant scattering due to vibronic levels [6]. The present measurements in the range 1-4K were made after the surface of the sample had been roughened and the conductivities are somewhat lower ($\sim$ 30%). A reasonable fit to these data could be obtained using the values for $D = D_u^a / D_{u'}^a = 0.84$ and $D_{u'}^a = 3.8$eV used by SUZUKI and MIKOSHIBA [5] to fit earlier data by HOLLAND and NEURINGER [7] (these are close to values found since by applying uniaxial stress in far infra-red, Raman and EPR experiments [8] which have weighted means of $D \sim 0.62$ and $D_{u'}^a \sim 3.7$eV). However, comparable fits could be obtained with a range of D if $D_{u'}^a$ is suitably adjusted (D_u^a and $D_{u'}^a$ are deformation potentials describing coupling to E and T_2 modes respectively).

The solid lines in fig. 1 show the magnetothermal conductivity for B||<001> (normal to the heat current along <110>). The broken curves show values computed from SM theory using D = 0.78 and $D_{u'}^a = 3.6$eV which gave slightly better agreement with the data (including the zero field values, K_0) than the values given above. The other parameters used were the g values measured by NEUBRAND [9], a* = 17Å (this gave slightly better agree- to K_0 at higher temperatures than a* = 15Å used in ref.[5]) and L_B = 1.91mm ($\sim$0.6 L (Casimir)).

The overall size of the change is in reasonable agreement with experiment but the minima occur at higher fields and the behaviour at low fields is quite different. A likely source of error here is our neglect of the one-phonon absorption process which normally dominates at low resonance frequencies. Disagreement would be expected at the higher fields shown due to neglect of vibronic levels, uncertainty in the form of the cut-off and second-order Zeeman splittings. (Measurements up to 13T will be given elsewhere).

To test our programme we have computed the magnetothermal conductivity with parameters used by SUZUKI [10] and SINGH and VERMA [1] in their

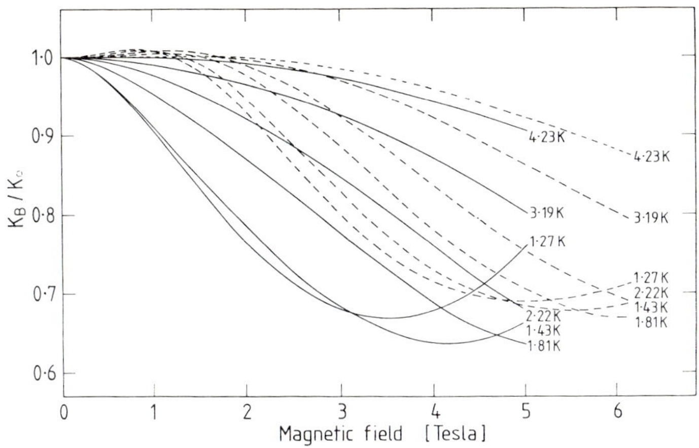

Fig. 1 The magnetothermal conductivity of Si(B)

computations. We get agreement with SUZUKI but not with SINGH and VERMA (our curves are appreciably closer to the earlier experimental data [3] though still not in agreement) and attribute this partly to their neglect of inelastic scattering since the expressions in SM indicate this provides a fraction $(2 + 5D^2 + D^4)/(3 + 6D^2 + D^4) \sim 70\%$ of the scattering above the resonances. We also disagree with the form of τ^{-1} shown in their figures [11].

The expressions in the SM theory refer to the case B||<001>. The resonant frequencies are isotropic to within 3% [10] but state mixing occurs as the field moves away from the <001> axis. The matrix elements of the strain Hamiltonian between these mixed states have been calculated by ZEILE and LASSMANN [12] and we have used these to calculate the dependence of τ^{-1} on the direction of field in the (110) plane. Calculations of $K(B,\theta)$ are in progress and should provide a further test of the SM theory. These have not been completed but we have compared the data with a convenient approximation which in principle gives directly a value for D.

At low concentrations, the principal effect of splitting the Γ_8 into 4 is to provide resonant scattering at 2 frequencies $\nu_{12} = 16.4B$ and $\nu_{13} = 31.5$ B GHz/T. At low fields, the effect of resonant elastic scattering at ν_{13} is predominant and the reduction in conductivity it produces is proportional to the width Δ of the hole it causes in the heat current. So the angular dependence of K is proportional to that of Δ. It can be shown from this that $[K_0 - K(B,\theta)]/K_0 \propto 1 + F \sin^2\theta(3\cos^2\theta + 1)/4$ where $F = (1 - D^2)/(1 + D^2)$ and θ is the angle of field in the (110) plane relative to the <001> direction (we note that $\frac{1}{4}\sin^2\theta(3\cos^2\theta + 1) = \ell^2m^2 + m^2n^2 + n^2\ell^2$ where ℓ,m,n are the direction cosines of B relative to the cubic axes). From this

$$\frac{K(B,o) - K(B,\theta)}{K_0 - K(B,o)} = F \sin^2\theta(3\cos^2\theta + 1)/4 \qquad (1)$$

which is shown plotted in fig.2(b) for 3 field values; $K(B,\theta)$ is shown as $f(\theta)$ for B = 2T in fig.2(a). Equation (1) is clearly not obeyed. For 2 field values the points do not fall on straight lines and also the slopes of F of the best lines change with field. However, the field plane appear-

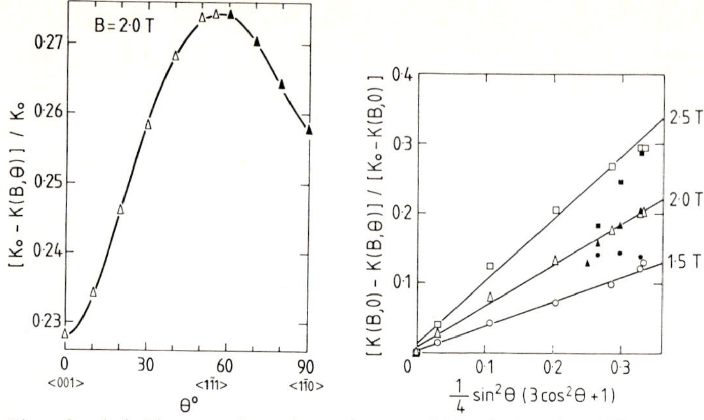

Fig. 2 (a) The angular dependence $K(B,\theta)$ in the (110) plane at 1.3K and 2T, (b) $K(B,\theta)$ plotted at 1.3K as described in text; the open and closed symbols are for $\theta <$ and $> 55^0$ respectively. The slopes F of the 'best lines' correspond to $D \sim 0.7$, 0.5 and 0.25 for 1.5, 2.0 and 2.5T respectively

ed to be slightly misoriented from the (110) crystal plane since the conductivity differed somewhat between the $\langle 1\bar{1}1\rangle$ and $\langle\bar{1}11\rangle$ directions; the values used in fig.2 are mean values. Further measurements on other samples are in progress.

We are very grateful to Dr K. Lassmann (Stuttgart) and Dr A.M. de Goër (CEN, Grenoble) for samples and discussions, Dr M.N. Wybourne (GEC) for discussions and S.E.R.C. for financial support.

References

1 S.Singh and G.S. Verma : Phys. Rev. B22, 6350 (1980).
2 K. Suzuki and N. Mikoshiba : J. Phys. Soc. Japan 31, 44 (1971) and 32, 586 (1972). The paper contains 1 of the 6 expressions for inelastic scattering and the others were kindly sent to us by Dr. Suzuki.
3 L.J. Challis and L. Halbo : Phys. Rev. Letts. 28, 816 (1972).
4 A.M. de Goër, M. Locatelli and K. Lassmann : J. de Physique C6, 235 (1981).
5 K. Suzuki and N. Mikoshiba : Phys. Rev. B3, 2550 (1972).
6 A. Ambrosy, A.M. de Goër, K. Lassmann and B. Salce : these proceedings.
7 M.G. Holland and L.J. Neuringer : Proc. Inf. Conf. on Semiconductor Physics, Exeter, 1962, ed. A.C. Stickland (Institute of Physics, London) p.474.
8 H.R. Chandrasekhar, P. Fisher, A.K. Ramdas and S. Rodriguez : Phys. Rev. B8, 3836 (1973) ; J.M. Cherlow, R.L. Aggarwal and B. Lax: Phys. Rev. B7, 4547 (1973) ; H. Neubrand Phys. Stat. Sol. (b) 90, 301 (1978).
9 H. Neubrand : Phys. Stat. Sol. (b) 86, 269 (1978).
10 K. Suzuki : Memoirs of School of Science and Engineering, Waseda University, 40, 29 (1976).
11 Figs. 1 and 2 show 3 resonance frequencies - SM predict 2 although there is a misprint in equation B5 where $1/\bar{t}^2_{23}$ should be $1/\bar{t}^2_{12}$ - and show slope changes in place of resonances.
12 H. Zeile and K. Lassmann : Phys. Stat. Sol. (b) 111, 555 (1982).

An Additive Conservation Law for Phonon Collision Operator in Molecular Crystals

B. Perrin

D.R.P., LA. 71, Université Paris VI, Tour 22, 4 Place Jussieu
F-75230 Paris Cedex 05, France

1 INTRODUCTION

Ultrasonic measurements made on some molecular crystals have shown that these compounds exhibit abnormally large sound absorption and have also revealed in some cases an ultrasonic relaxation {1}. If we try to account for this large absorption only by phonon viscosity, we have to assume the existence of slow relaxation times ($\sim\tau$) among the eigenvalues of the Phonon Collision Operator (PCO). In a previous work {2} we have given an expression of the ultrasonic attenuation coefficient in terms of small phenomenological eigenvalues of the PCO. This expression accounts well for the relaxation strength and for the dependence of the relaxation on the anisotropy of the crystal. Now the problem is to find the source of these small eigenvalues.

Up to now the idea which has prevailed was that these slow relaxation times expressed a slow transfer of energy between the internal (index I) and the external (index E) vibrational modes of the crystal. In fact this idea, first suggested by LIEBERMANN {3} was directly issued from what happens in the gaseous and liquid phases but has never been supported by a careful investigation of the properties of the PCO in molecular crystals. Recent advances in the experimental determination of lifetimes of zero-wave-vector internal phonons by non-linear or time-resolved spectroscopy {4} - {6} and also in computations of these lifetimes {7} in molecular crystals throw a new light on the problem and call in question the assumption usually admitted. The lifetimes determined up to now lie in the range 1-10 ps at high temperature and in what follows we show that these values are inconsistent with a slow transfer of energy between internal and external modes. The different processes in which an internal phonon may take part in three-phonon interactions are described in Fig.1. The processes (a) which only concern internal phonons imply the existence of Fermi resonances (which are fortuitous) among the vibrational states of the molecule so that in general external phonons always take part in the building up of internal phonons lifetimes. Thus, these lifetimes being rather short, it may be concluded that there is an efficient (or fast) energy exchange between the two groups of modes and the systematic occurrence of slow relaxation times in the PCO of molecular crystals must be explained in another way.

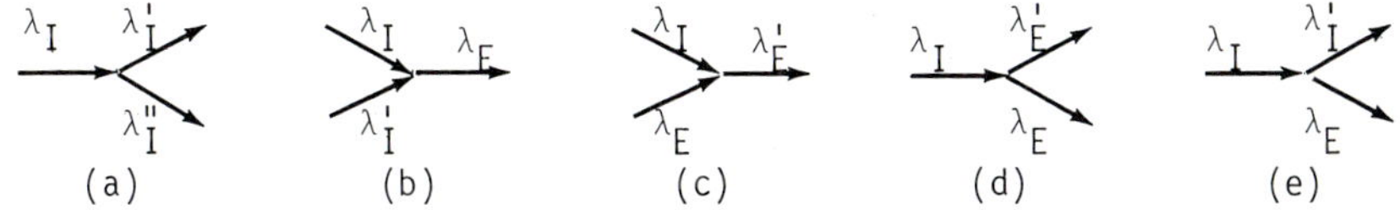

Fig. 1 - Different three-phonon processes in which an internal phonon λ_I may take part

2 INTERNAL PHONONS NUMBER CONSERVATION LAW

We suggest that the typical structure of the vibrational spectrum of molecular crystals shown on Fig. 2 may be the source of a new conservation law for three-phonon interactions. We have seen that the absence of Fermi resonances forbid the (a) processes. In the same way, the internal modes usually having higher frequencies than those of external modes, processes (b) and (c), are also forbidden. Now we see that if the gap between internal and external modes is such that $\omega_o > 2\omega_m$ energy conservation for (d) processes cannot be satisfied. Thus the only available processes are (e) processes for which the number of internal phonons is conserved. When the above conditions are fulfilled[1] the PCO Mo restricted to three-phonon interactions obeys a conservation law of the number of phonons belonging to the whole set of internal modes.

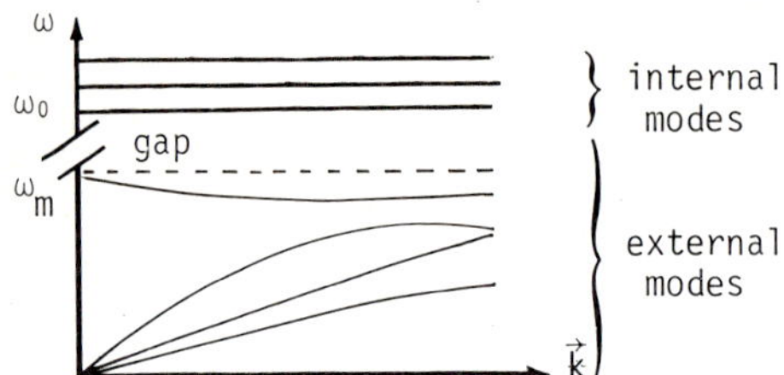

Fig. 2 - Typical dispersion curves for the vibrations of a molecular crystal

Then we have to introduce non-zero chemical potential in the internal phonon populations and these are only four- and higher-order processes (less efficient than three-phonon processes) which relax the chemical potential to zero within a slow time τ which hides the conservation law. These higher-order processes add to Mo a PCO M_1 we can treat as a perturbative term in the effective PCO $M = Mo + M_1$.

3 EIGENFUNCTION ASSOCIATED with the SMALL EIGENVALUE

In the following we sketch a pertubation method applied to M which leads to the eigenfunction X_τ associated with τ. The eigenvalue spectrum of Mo is a quasi-continuum spectrum which extends down to zero (without gap) {8},{9}; moreover zero is a two-degenerated eigenvalue with two eigenfunctions -[2]-

Xo and X'_1 which can be obtained respectively from the two following conservation laws that Mo obeys :

- the energy conservation of the whole phonon system (E+I);
- the conservation of the number of internal phonons.

M_1 conserving the energy of the whole phonon system Xo is also a zero eigenvalue eigenfunction of M_1. Thus, X_τ must be given to the first order in an elementary pertubation method by the function

$$X_1 = X'_1 - (X'_1 . Xo)\, Xo \qquad (1)$$

where $(X'_1 . Xo)$ is a scalar product; it is clear from (1) that X_1 and Xo are orthogonal. Moreover

$$-1/\tau = (X_1 .\ M_1(X_1)). \qquad (2)$$

However some problems arise in applying the perturbation method due to the gapless spectrum of Mo. Let S be the subspace associated with the low-lying eigenvalues μ_α of Mo ($|\mu_\alpha| \leq 1/\tau$). It can be shown that

$$X_\alpha \in S \qquad (X_\alpha .\ M_1(X_\alpha))) \ll (X_\alpha .\ Mo\ (X_\alpha)). \qquad (3)$$

-1- It can be shown that a conservation law may be obtained under less drastic conditions than $\omega_o > 2\,\omega_m$ by defining new sets of modes E' and I'.
-2- In fact Xo and X'^o_1 are eigenfunctions of the symmetric operators $\widetilde{M}o$ and $\widetilde{M}_1$ (defined by Eq. (15) of Ref. {2})which have a set of orthogonal eigenfunctions.

From (3) we may deduce two consequences :
- (2) actually gives a first-order approximation for τ;
- (1) gives the expression of X_τ to the first order except for the projection of X_τ on S ; the dimension of S being small compared with that of the whole space, one might expect only little trouble from this part of X_τ.
Now it is straightforward to use (1) to obtain a new expression of the ultrasonic relaxation strength which can be compared with experimental results as a test of the above theory.

4 CONCLUSION

In this work, we have shown that the systematic occurrence of slow relaxation times in the PCO of molecular crystals might be related to a hidden conservation law obeyed by three-phonon interactions in these crystals : under some assumptions concerning the vibrational spectrum, the number of internal phonons is conserved by three-phonon scattering.

Two consequences of this result may be mentioned :
- At low temperature up-conversion processes become negligible ; internal phonons belonging to the lowest internal mode must be affected by the phonon bottleneck linked with this hidden conservation law and their lifetimes must increase up to the range of time τ.
- Non-equilibrium internal phonons excited in time-resolved spectroscopy experiments must exhibit a non-zero chemical potential within the scale of τ times.

References

1 A. E. Victor and R. T. Beyer, J. Acoust. Soc. Am. 54, 1639(1973)
2 B. Perrin, Phys. Rev. B 24, 6104(1981)
3 L.N. Liebermann, Phys. Rev. 113, 1052(1959)
4 A. Laubereau, G. Wochner and W. Kaiser, Opt. Commun. 14, 75(1975)
5 P.L. Decola, R.M. Hochstrasser and H.P. Trommsdorff, Chem. Phys. Letters 72, 1(1980)
6 Ben H. Hesp and Dowe A. Wiersma, Chem. Phys. Letters 75, 423(1980)
7 R. Righini, P.F. Fracassi, R.G. Della Valle, to be publised
8 J. Jäckle, Phys. Cond. Matter, 11, 139(1970)
9 F.A. Buot, J. Physics, C5, 5(1972)

Discussions

PHONON SCATTERING BY DISLOCATIONS page 348
A. C. Anderson

H. Maris: Why is there such a large uncertainty in the theoretical predictions of the scattering cross-section?

A. C. Anderson: You should ask the author of the theoretical papers. It may be due to an accumulation of several approximations.

THE ACOUSTIC PARAMAGNETIC RESONANCE OF Cr^{2+} IN n-TYPE GaAs page 355
A. S. Abhvani, C. A. Bates, P. Bury, P. J. King,
D. R. Pooler, V. W. Rampton, P. C. Wiscombe

K. Lassmann: What are your assumptions about the distributions of the strain types?

P. J. King: Very simple assumptions were made about the strain distribution. We assumed no angular variation in the Q_Θ, Q_ε plane and the distribution fell off from the origin as a Gaussian with width 90 GHz/V_E. Although we have some idea of typical strain magnitudes in these crystals we have no idea of the angular variation. One must note that the strain distribution may not even have the symmetry of the crystal but may be determined by, for example, growth conditions.

T. Ishiguro: 1) I wish to know the Cr concentration of your sample. Actually how about the concentration dependence?
2) Have you checked the effect of light illumination?

P. J. King: 1) The measurements were made on samples cut from a boule which is n-type at one end and semi-insulating in its other extreme. The region used had a silicon (donor) concentration of 2×10^{17} cm^{-3} and a chromium concentration of 7×10^{16} cm^{-3}. We have not had access to samples with a very wide range of Cr concentration. 2) As I have mentioned, a very strong and complex APR spectrum is found in semi-insulating material. This has been reported by ourselves and other workers and is so far unexplained although it may be due to trigonally distorted Cr^{2+} sites. The same spectrum may be induced in n-type material by illumination.

MAGNETIC-FIELD-DEPENDENT PHONON SCATTERING BY page 358
Li IONS IN Si
L. J. Challis, A. P. Heraud, V. W. Rampton, M. J. Saker,
M. N. Wybourne

K. Lassmann: What was the relation of the concentrations of Li and O in your samples?

L. J. Challis: The Li concentration was 9 x 10^{15} cm^3 and the C and O concentrations are below the detection limit of 5 x 10 ^{15} cm^3. Experiments are planned on samples made from the same Si but with different Li concentrations as well as with older Si likely to contain higher concentrations of C and O.

J. K. Wigmore: The only point not covered by Prof. Challis in his excellent critique was that the APR lines I observed were very power dependent indeed, and I attributed the high field line, not observed in his recent experiments, to double quantum transitions I want to ask whether he has observed only power dependent effects in his good samples of Si:Li?

V. W. Rampton: We have not yet checked the dependence of the APR on ultrasonic power or Li concentration. We intend to do this soon.

M. Singh: The experiments are done till 5 T. You said that there is a compression of electronic orbit due to magnetic field. In my opinion in this range of magnetic field there should be a negligible effect on compression of electronic orbit due to magnetic field.

L. J. Challis: The increase of conductivity with field seems most easily interpreted as due to an increase in the zero-field splitting. The association of this with wave function compression is by analogy with the experiments on germanium although of course this effect may be appreciably less in this system since a^* is smaller.

INFLUENCE OF DEFECTS ON THE SPLITTING OF THE ACCEPTOR GROUND STATE IN SILICON page 361
A. Ambrosy, A.M. de Goër, K. Lassmann, B. Salce, H. Zeile

S. Hunklinger: Does the coupling constant depend on concentration and is there any explanation for this behaviour?

K. Lassmann: As yet S 54 and S 125 of Fig. 2 are the only examples measured by ultrasonics giving a reduced coupling at high boron concentration provided the distribution does not extend significantly beyond the measuring range. From additional APR and relaxation attenuation measurements at magnetic fields up to 0.7 T there is also no indication of a broad distribution in contrast to the results from $\kappa(T)$. We have no estimate whether B-B neighbouring could affect the effective coupling so strongly.

D. Golding: Recent infrared dichroism measurements by Stavola et al.(Appl. Phys. L. 42, 73 (83)) have shown that heat treatment of Si has a striking effect on the distribution of oxygen (O). Relatively low temperature anneals can lead to clustering due to an enhanced diffusivity. Is it possible that oxygen could play a significant role in explaining the apparently "hidden strain variable" in your experiments?

K. Lassmann: A clustering of O at these low levels is improbable. Annealing a sample of S 87 (see (1), p. 563) did not change the distribution.

PHONON SCATTERING AT ELECTRONICALLY DEGENERATE SYSTEMS: AN APPLICATION TO THE DEFECT SYSTEMS Si(In), Si(B), AND GaAs(Mn) page 364
J. Maier, E. Sigmund

L.J. Challis: 1) Is it possible to explain physically why both the strength and the linewidth of the scattering rate are temperature dependent? 2) Does the resonant increase in scattering due to the Jahn-Teller level always coincide closely with $qa^* \sim 1$ or can there ever be two maxima in the scattering rate?

J. Maier: 1) The temperature dependence of the linewidth is obvious by the explanation of the splitting by the fluctuations of the phonon bath. These fluctuations increase with increasing temperature (nevertheless the zero-point fluctuations give a large amount to the splitting). The strength of the scattering is directly connected with these fluctuations. But I know no direct physical explanation besides the more formal fact that if the real part of a Green's function shows a temperature dependence, so does the imaginary part. 2) There is no direct connection between the resonance energy and the cut-off. But the cut-off function forces the resonance to lower values. The larger the Bohr radius is the more the resonance is driven to the cut-off (and, obviously) to smaller values. The question of 2 maxima we did not discuss so far, but for some parameter choices we got a small second maximum above the first. But at the moment we are not quite sure if this is due to an incomplete inclusion of higher order terms (2-phonon processes).

MAGNETOTHERMAL CONDUCTIVITY OF BORON-DOPED SILICON page 368
L.J. Challis, A.P. Heraud

S. Singh: 1) What is the impurity concentration of your sample? 2) Have you taken into account both the elastic and inelastic scattering or simply elastic scattering?

A.P. Heraud: In our theoretical calculation of the magnetothermal conductivity both elastic and inelastic scattering processes are considered.
The boron impurity concentration is 5.4×10^{15} cm^{-3}.

M. Singh: What is the physical significance of κ vs Θ?

A.P. Heraud: As the magnetic field moves off the $^{2}001^{3}$ axis the transitional probabilities change as states mix. Thus the amount of scattering will change. We have shown that, in principle, a direct determination of D is possible from this anisotropy.

Part IX

Two-Level Systems

Chairmen: **W. Arnold K. Dransfeld B. Golding**

Low-Energy Excitations in Disordered Solids: New Aspects

S. Hunklinger

Institut für Angewandte Physik II der Universität Heidelberg
Albert-Überle Straße 3-5, D-6900 Heidelberg, Fed. Rep. of Germany

1. Introduction

About ten years ago ZELLER and POHL [1] published their pioneering work which stimulated the investigation of the low-temperature properties of amorphous materials. Their study and that of STEPHENS [2] demonstrated that amorphous solids exhibit an anomalously high specific heat and a rather low thermal conductivity. Furthermore the magnitude of these thermal quantities was hardly dependent on the chemical nature of the amorphous solids. One year later, in 1972, PHILLIPS [3] and ANDERSON et al. [4] independently developed the tunneling model. It turned out that this phenomenological model describes the experimental results most successfully. It is based on the assumption that in an amorphous network small groups of atoms can tunnel between two positions of equilibrium (see Fig. 1a). Thus two-level tunneling states are formed which give rise to the enhanced specific heat and which scatter phonons strongly. Shortly afterwards an important consequence of this model could be verified. In ultrasonic experiments carried out by HUNKLINGER et al. [5] and GOLDING et al. [6] it was found that the low-temperature acoustic absorption can be saturated. This observation shows that the low-energy excitations do indeed act like two-level tunneling states as postulated by the tunneling model.

Since that time many experiments and theoretical studies have been performed and a wealth of exciting and surprising phenomena has been discovered. Because of the brevity of this review only a small fraction of the activities in this field can be discussed. Therefore many important and interesting new developments are not even mentioned. The most important question, however, is still unanswered: what is the atomic nature of the tunneling particle? The answer cannot be found by experiments alone. Theoretical studies of local structural rearrangements are needed in which the free energy is hardly changed and thus can give rise to the formation of tunneling states.

2. "Universality" of the Low-Temperature Anomalies

Before starting with my discussion I want to call to mind some of the basic facts. For more details see for example [7]. My description of the low-temperature anomalies is based on the Tunneling Model (TM). One of its central assumptions is the uniform distribution of the asymmetry Δ and the tunneling parameter λ:

$$P(\Delta,\lambda)\, d\Delta d\lambda = \bar{P}\, d\Delta d\lambda \qquad (1)$$

where $\bar{P}$ is constant. This distribution leads to an energy density of states $n(E) \simeq n_0$, that is nearly constant. The tunneling states (TS) interact strongly with phonons. Because of their two-level nature, the resonant absorption co-

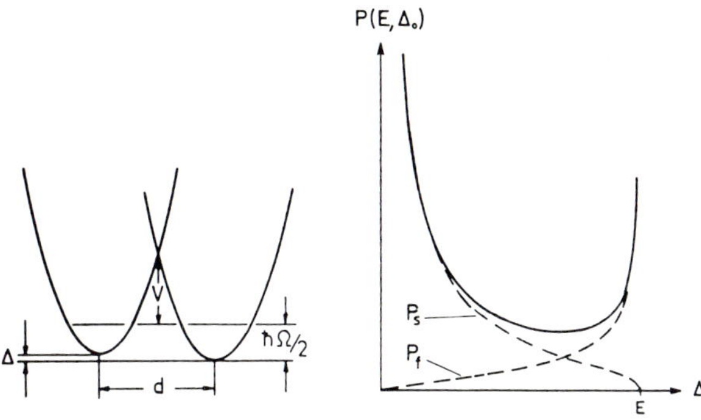

Fig. 1a) Double well potential with barrier height V, asymmetry Δ, and distance d. $\hbar\Omega/2$ is the ground-state energy of the tunneling particle with mass m. The tunnel splitting is given by $\Delta_0 = \hbar\Omega\,\exp(-\lambda)$, where the tunneling parameter $\lambda = d\sqrt{2mV}/\hbar$. The total energy splitting E of the tunneling system is given by $E^2 = \Delta^2 + \Delta_0^2$.
b) Schematic diagram of the distribution function $P(E,\Delta_0)$ (full line) and of the two branches P_f and P_s (dashed lines)

efficient decreases at higher intensities, i.e. the absorption can be saturated. Besides this one-phonon process also a relaxation absorption is observed [8] due to the modulation of the energy splitting of the TS by the sound wave. Qualitatively the same behaviour is also observed for the dielectric properties [9].

As already mentioned it was found rather early that the constant density of states n_0 or the "magnitude" of the anomalies look like a "universal" quantity. Recent measurements of the acoustic properties of thin films of a-Si and a-Ge [10], however, have demonstrated that n_0 is considerably reduced in amorphous solids with fourfold coordinated covalent bonds. This conclusion originally drawn from acoustic measurements has been confirmed by specific heat measurements [11]. Thus an upper limit for n_0 can be given that is an order or magnitude smaller than in vitreous silica. Similarly in a-As [12], which is threefold coordinated, n_0 is reduced by roughly a factor of three. Clearly, the "magnitude" of the low-temperature anomalies does depend on the chemical nature. In random structures with a high local rigidity, i.e. covalently bonded amorphous solids with higher coordination numbers, the number of low-energy excitations is smaller than for example in silicate glasses or polymers. Obviously in materials with high coordination numbers the local rearrangement of clusters of atoms is energetically less favourable than in "ordinary" amorphous solids. This leads to the question to which physical parameter the magnitude of n_0 is related. Very recently it has been reported that $n_0 \propto T_g^{-1}$ [13]. The qualitative explanation of this relation is rather simple. It is assumed that the total number of TS $N = \int n(E)dE$ is independent of T_g, but the energy scale over which these states are distributed is determined by T_g. Thus $n_0 \propto T_g^{-1}$ in agreement with theoretical considerations based on the description of the TS by the free volume theory of glasses [14].

A series of experiments has demonstrated that also crystals can exhibit low-temperature anomalies analogous to those of amorphous solids. This phenomenon originates from the fact that to some extent disorder is also possible in crystals. A typical representative is the fast ionic conductor β-alumina [15]. This crystal contains "two-dimensional amorphous areas", which are regularly arranged between spinel blocks. In these layers alkali ions are randomly distributed and their tunneling gives rise to low-energy excitations.

In contrast, disorder in one-dimensional structures does not result in a "glassy" behaviour. Specific heat measurements of one-dimensional fast ionic conductors do not show a linear temperature dependence but rather a Schottky anomaly [16]. In addition thermal conductivity experiments indicate that these excitations are only weakly coupled to phonons.

There are many other examples of "glassy" low-temperature properties of crystals. For instance the dielectric constant of cyclohexanol [17] shows the same low-temperature variation as glasses. In this "glassy crystal" the molecules occupy regular lattice sites. Disorder exists in orientation and/or molecular conformations. Similar anomalies are observed for crystals of highly doped alkali halides like KCl:OH [18] or hydrogen-doped Li_3N [19]. In both cases the tunneling entities are identical with the impurities. The randomly varying distance between them results in statistically varying strength of interaction and thus gives rise to a uniform distribution of the asymmetry [20]. In contrast the distribution of the tunneling parameter λ is not well understood.

In a series of experiments it has been shown that hydrogen in crystalline niobium also gives rise to TS, provided that the metal also contains interstitial or substitutional impurities like O, N or Ti. These centers trap hydrogen and form tunneling centers which have been detected by measurements of the specific heat and in neutron scattering experiments [21]. In Nb with 5% Ti the density of states of these TS is more or less uniform [22]. It seems, however, that these TS couple only weakly to phonons and do not lead to those anomalies in thermal conductivity and acoustic properties which are characteristic for amorphous solids. Accidentally, present hydrogen might also influence the specific heat of metallic glasses, but so far this possibility has not been studied systematically.

An elegant way to vary the concentration of specific defects without changing the crystalline structure noticeably is by irradiation. Irradiation by γ rays, electrons and neutrons enhances the specific heat of natural quartz [23]. A "typical glassy" behaviour is found after neutron irradiation. γ and electron irradiation seems to give rise only to weakly coupling TS whose density of states is not as uniform as in glasses. Probably neutrons destroy the lattice locally and generate "amorphous islands" which behave like small inclusions of vitreous silica. This interpretation is supported by x-ray studies which show a broadening of the Bragg peaks in neutron-irradiated samples but no effect after electron irradiation [24].

3. Uniformity of the Distribution Function

In amorphous dielectrics the relaxation time of TS is determined by the one-phonon process [7,8]:

$$\tau^{-1} = u^2\tau_m^{-1} = A\, u^2E^3 \coth E/2\,kT \qquad (2)$$

where $u = \Delta_o/E$, the ratio between tunneling and total energy splitting (see Fig. 1a). For fixed values of E and T there exist a minimal relaxation time τ_m if u = 1. The constant A contains material parameters that will not be discussed here. If we replace the parameters Δ and λ in (1) by u and E we obtain

$$P(u,E)dudE = \bar{P}/Eu\sqrt{1-u^2}\, dudE. \qquad (3)$$

This function is schematically shown in Fig. 1b. Since $\tau \propto u^{-2}$ (see (2)), the relaxation times exhibit a similar distribution for fixed values of E. An interesting consequence of the distribution of τ is that the observed specific heat should depend on the time of measurement. One expects [25]

$$C_V = \frac{\pi^2 k^2 T}{12} \bar{P} \ln(4t/\tau_m) . \quad (4)$$

After several attempts which failed it has been observed recently [26]. Qualitative agreement with (4) has been obtained for times of measurement between 10 μs and 10^4s. Thus an important prediction of the TM has been confirmed experimentally.

Let us discuss in more detail the distribution function given by (3). For this purpose we split the function into two parts (see Fig. 1):

$$P(u,E) = P_f + P_s = \frac{\bar{P}}{E} \left(\frac{u}{\sqrt{1-u^2}} + \frac{W\sqrt{1-u^2}}{u}\right) . \quad (5)$$

The first term P_f represents mainly TS with large values of u which relax fast. The second term P_s describes the slowly relaxing systems for which $u \simeq 0$. To allow for a variation of the relative weight of the branches a factor W has been introduced. For W = 1 we come back to the original TM.

The contribution of the two branches to the thermal and acoustic properties can be calculated separately. Most striking is the result that both branches lead to the same temperature and frequency dependence of the resonant and the low-temperature relaxation absorption and the corresponding velocity change. But only the slow branch gives rise to the time dependence of the specific heat, briefly discussed above. For a quantitative determination of the weighting parameter W acoustic experiments are more suitable. For the increase of the sound velocity at low temperature one finds:

$$\left.\frac{\Delta v}{v}\right|_{res} = \frac{\bar{P}\gamma^2}{\rho v^2} \left(\frac{2}{3} + \frac{W}{3}\right) \ln \frac{T}{T_o} . \quad (6)$$

γ is the acoustic coupling constant, ρ the mass density, v the sound velocity and T_0 an arbitrary reference temperature. At higher temperatures the relaxation process dominates the contribution of the resonant interaction and the velocity decreases [27]. If $\omega\tau_m << 1$ this leads to

$$\frac{\Delta v}{v} = \left.\frac{\Delta v}{v}\right|_{res} - \left.\frac{\Delta v}{v}\right|_{rel} = \frac{\bar{P}\gamma^2}{\rho v^2} \left(\frac{2}{3} - \frac{7}{6}W\right) \ln \frac{T}{T'_o} . \quad (7)$$

T'_o is a reference temperature related to A, ω and T_o. In this temperature range only the slow branch contributes to the relaxation absorption:

$$\alpha_{rel} = \frac{\bar{P}\gamma^2}{\rho v^2} \frac{\pi\omega}{2v} W . \quad (8)$$

Fig. 2 shows the attenuation and sound velocity measured at very low frequencies [28]. Clearly all the effects described by (6) - (8) are observed. Notice that the logarithmic decrease is only observed up to temperatures of 1.5 K. Since the slope of the decrease is sensitive to the relaxing process, this deviation indicates that more-phonon processes become dominant above this temperature. Since the maximum of the velocity of sound shifts to higher temperatures with increasing frequency, only measurements in the kHz range are suitable for the observation of the logarithmic decrease.

From Fig. 2 we deduce the value W = 1 in agreement with the original TM. Very recently attenuation measurements on vitreous silica have been carried

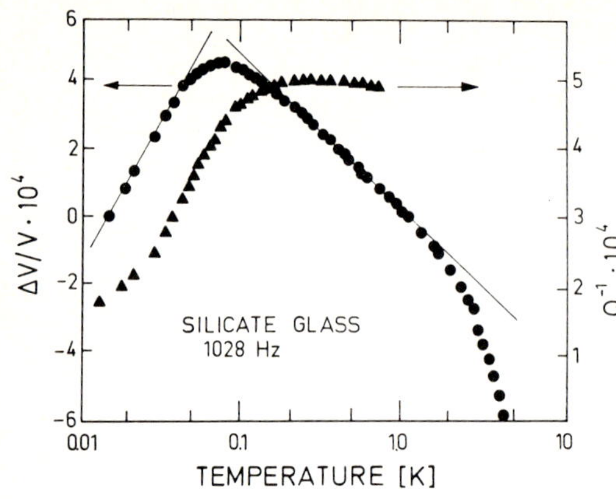

Fig. 2 Sound velocity and internal friction of a silica-based insulating glass

out down to 5 mHz [29] and still (8) gives the correct description. This means that the distribution shown in Fig. 1b is valid down to values of $u \simeq 10^{-6}$! Nevertheless a cut-off at still smaller values of u must exist since otherwise the number of TS would go to infinity. Is this agreement fortuitous for silicate glasses? Since for other materials measurements only exist at higher frequencies, we have to rely on (6) and (8). For polymers [30] and β-Al_2O_3 [27] this analysis leads to $W \simeq 3$. Obviously in these materials much more slowly relaxing TS exist than suggested by the orginal TM!

4. Theoretical Developments

Besides the experimental work also theoretical efforts have been made during the last years to understand the puzzling low-temperature anomalies of amorphous solids. The aim of most authors was to improve, extend or justify the existing TM. For example the effect of environmental vibrations on the tunneling process has been taken into account [31]. In other studies acoustic and dielectric low-temperature properties have been calculated for very high frequencies ($\hbar\omega > kT$) [32]. In both cases temperature and frequency dependence of the relevant quantities are changed, but so far no critical comparison with existing experimental results has been carried out.

A central, but open question is that of the origin of the TS. For example it has been proposed that in amorphous semiconductors, atoms with crosslink relatively large clusters are rather mobile and give rise to TS [33]. In another attempt the free volume theory of glasses has been extended to make the existence of TS plausible [14]. The relation $P \propto 1/T_g$ mentioned above follows naturally from this theory. Originally such a relation has been extracted from measurements of the thermal conductivity [34] and was also found by specific heat and acoustic measurements [13]. However, there are many other experiments that are in contradiction to such a relation. Very attractive is the idea that the existence of TS is intimately connected with the topology of the amorphous structure. It has been shown that in all random structures with odd numbered rings there exist disclination lines with two degenerated states of the same elastic energy. The motion of these odd lines is frozen in but tunneling is possible between each pair of metastable states [35]. I want to point out here that all the theoretical models developed so far only give qualitative arguments. They are not yet able to provide the experimentalist with numbers, which can be experimentally verified.

5. Glassy Metals

Acoustic and thermal studies on superconducting glasses have given a wealth of information on the TS-electron coupling [36]. Its theoretical treatment [37] in analogy to the Korringa relaxation has been confirmed. In particular the puzzling intensity dependence of the ultrasonic attenuation in PdSi at MHz frequencies has been explained [38]. It seems that in normal conducting glasses the relaxation absorption becomes intensity dependent because of the fast relaxation of the TS via their interaction with free electrons. One aspect is still under discussion: it is not clear whether this interaction can also explain the low-temperature anomalies of the electrical resistivity. Although it seems that it has only very weak influence on electrical properties, the possibility of its enhancement by many-body effects cannot be ruled out [39].

Surprising and not yet understood are most results obtained in annealing experiments. In general the low-temperature thermal conductivity of metallic glasses increases after annealing [40,41,42]. A simple explanation would be that the density of states of the TS is reduced [43]. However, measurements of the specific heat exhibited controversial results. Annealing seems to reduce the specific heat only of sputtered samples [41], whereas rapidly quenched samples do not show a variation [42]. On the other hand annealing has also significant influence on the electrical and superconducting properties of these materials [44]. The discussion of these new aspects is, however, beyond the scope of this brief review.

6. Optical Properties

Finally I want to mention briefly optical studies of the TS. In these experiments optically active molecules are embedded in amorphous matrices. Because of the strain sensitivity of the electronic levels an inhomogeneous broadening of the optical absorption line is caused by the varying local environment. Measurements by fluorescence line-narrowing and hole-burning have shown that in addition also the homogeneous linewidth is considerably broadened. According to very recent measurements [45] this width decreases steadily with decreasing temperature down to 0.4 K as shown in Fig. 3. It is likely that the homogeneous broadening is due to the interaction between TS and the impurity molecules. Pressure fluctuations at the site of the impurity molecule are generated by neighbouring TS due to the absorption and reemission of thermal phonons. These fluctuations modulate the electronic level splitting of the

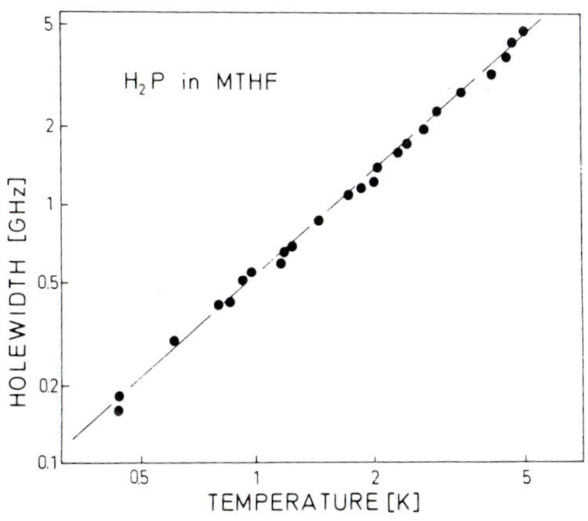

Fig. 3 Hole width as a function of temperature for H_2P-MTHF (Free-base Porphin in 2-Methyltetrahydrofuran) glass. Full line represents the theoretical fit

impurity molecule and lead to a broadening of the measured line width. The theoretical treatment [46] is very similar to that of the time-dependent specific heat. The calculated curve based on this theory is shown in Fig. 3 as a full line. If the explanation given above is correct, such optical techniques would be extremely useful because they can also be applied at higher temperature. In this way it might be possible to obtain information on the density of states of the TS at higher energy splittings where almost nothing is known so far.

7. Summary

During the last years many new and exciting experimental and theoretical studies have been carried out. In this review important contributions like studies of the coherent effects in semiconductors [47] or thermal expansion [48] could not be treated because of space restrictions. The puzzling low-temperature anomalies of amorphous solids are best described by the tunneling model although the tunneling particles are still unknown.

References

[1] R.C. Zeller, R.O. Pohl: Phys.Rev. B4, 2029 (1971)
[2] R.B. Stephens: Phys.Rev. B8, 2896 (1973)
[3] W.A. Phillips: J.Low Temp.Phys. 7, 351 (1972)
[4] P.W. Anderson, B.I. Halperin, C.M. Varma: Phil.Mag. 25, 1 (1972)
[5] S. Hunklinger, W. Arnold, S. Stein, R. Nava, K. Dransfeld: Phys.Lett. A42, 253 (1972)
[6] B. Golding, J.E. Graebner, B.I. Halperin, R.J. Schutz: Phys.Rev.Lett. 30, 223 (1973)
[7] S. Hunklinger, W. Arnold: In Physical Acoustics, ed. by W.P. Mason, R.N. Thurston (Academic, New York 1976) Vol. XII, p. 155
[8] J. Jäckle: Z.Phys.257, 212 (1972)
[9] see for example: S. Hunklinger, M.v.Schickfus: In Amorphous Solids, ed. by W.A. Phillips (Springer, Berlin 1981)
[10] M.v.Haumeder, U. Strom, S. Hunklinger: Phys.Rev.Lett. 44, 84 (1980); J.Y. Duquesne, G. Bellessa: J.Phys.C 16, L 65 (1983); K.L. Bhatia, S. Hunklinger: Solid State Commun. 47, 489 (1983)
[11] H.v. Löhneysen, H.J. Schink: Phys.Rev.Lett. 48, 1121 (1982)
[12] M.T. Loponen, R.C. Dynes, V. Narayanamurti, J.P. Garno: Phys.Rev. B 25, 4310 (1980)
[13] A.V. Raychandhuri, R.O. Pohl: Solid State Commun. 37, 105 (1980); P. Doussineau, M. Matecki, W. Schön: J.Physique 44, 101 (1982)
[14] M.H. Cohen, G.S. Grest: Phys.Rev.Lett. 15, 1271 (1980)
[15] see for example: U. Strom: Solid State Ionics (to be published)
[16] H.v. Löhneysen, H.J. Schink, W. Arnold, H.U. Beyeler, L. Pietronero, S. Strässler: Phys.Rev.Lett. 46, 1213 (1982); H.J. Schink, H.v. Löhneysen: to be published
[17] G.P. Singh, R. Vacher, R. Calemczuk: J.Physique 43, C9-525 (1982)
[18] M. Saint-Paul, M. Mesa, R. Nava: Phys.Lett. 92A, 466 (1982)
[19] T.Baumann, M.v.Schickfus, S.Hunklinger: Physica 108B, 1267 (1981)
[20] M.W.Klein, B.Fischer, A.C.Anderson, P.J.Anthony: Phys.Rev. B 18,5887 (1978)
[21] H.Wipf:Proc.Int.Conf.Diff. in Metals and Alloys,Tihany 1982:to be publ.
[22] K. Neumeier, H. Wipf, G. Cannelli: Phys.Rev.Lett. 49, 1423 (1982)
[23] see for example: A.M. de Goer, M. Locatelli, C. Laermanns: J. Physique 42, C6-78 (1981); M. Hofacker, H.v.Löhneysen: Z.Phys. B 42, 291 (1981; M. Saint-Paul, J.C. Lasjausnias, M. Locatelli: J.Phys.C 15, 2375 (1982)
[24] D. Grasse, D. Müller, H. Peisl, C. Laermans: J.Physique 43, C9-119 (1982)
[25] J.L. Black, B.I. Halperin: Phys.Rev.B 16, 2879 (1977)

[26] M.T. Loponen, R.C. Dynes, V. Narayanamurti, J.P. Garno: Phys.Rev. Lett.45, 457 (1980); M. Meissner, K. Spitzmann: Phys.Rev.Lett. 46, 265 (1981); J.Zimmermann, G. Weber: Phys.Rev.Lett. 46, 661 (1981)
[27] see for example: P. Doussineau, C. Frénois, R.G. Leisure, A. Levelut, J.-Y. Prieur: J. Physique 41, 1193 (1980)
[28] A.K. Raychandhuri, S. Hunklinger: J. Physique 43, C9-485 (1982)
[29] H. Tietje, M.v.Schickfus, E. Gmelin, H.-J. Güntherrodt: to be published
[30] G. Federle, S. Hunklinger: J.Physique 43, C9-505 (1982)
[31] V.N. Flurov, L.I. Trakhtenberg: Solid State Comm. 44, 187 (1982)
[32] V.L. Gurevich, D.A. Parshin: Solid State Comm. 43, 271 (1982)
[33] J.C. Philipps, Phys.Rev. B 24. 1744 (1981)
[34] C.L. Reynolds Jr.: J.Non-Cryst. Solids 30, 371 (1978)
[35] D.M. Dufy, N.Rivier: Physics 108B, 1261 (1981)
[36] G.Weiß, S.Hunklinger, H.v.Löhneysen: Physica 109&110B, 1946 (1982); W.Arnold, A.Billmann, P.Doussineau, A.Levelut: Physica 109&110B, 2036 (1982)
[37] J.L.Black, P. Fulde: Phys.Rev.Lett. 43, 453 (1979)
[38] W.Arnold, P.Doussineau, A.Levelut: J.Physique 43, C9-553 (1982)
[39] K. Vladar, A.Zawadowski: Solid State Comm. 41, 649 (1982)
[40] P.Esquinazi, M.-E.de la Cruz, A.Ridner, F. de la Cruz:Solid State Comm. 44, 941 (1982); E.J.Cotts, A.C.Anderson, S.J. Poon: to be published
[41] A.Ravex, J.C. Lasjannias, O. Bethoux; J.Phys. F (to be published)
[42] S. Grondey, H.v. Löhneysen, J.J. Schink: Z.Phys. B (to be published)
[43] M. Banville, R. Harris: Phys.Rev.Lett. 44, 1136 (1980)
[44] P. Esquinazi, M.E. de la Cruz, F. de la Cruz: Physics 108B, 1215 (1981)
[45] H.P.H.Thiyssen,S.Völker,M.Schmidt,H.Port: Chem.Phys.Lett. 94, 537 (1983)
[46] S. Hunklinger, M. Schmidt: to be published
[47] D.L. Fox, B.Golding, W.H. Haemmerle: Phys.Rev.Lett. 49, 1356 (1982)
[48] D.A. Ackermann, A.C. Anderson: Phys.Rev.Lett. 49, 1176 (1982); W. Kaspers, R. Pott, M.D. Herlach, H.v. Löhneysen:Phys.Rev.Lett. 50, 433 (1983)

Anomalous Low-Temperature Ultrasonic Behaviour in a Fluoride Glass Containing Mn

P. Doussineau, A. Levelut, and W.D. Wallace[1]

Laboratoire d'Ultrasons[2], Université Pierre et Marie Curie, Tour 13
4 place Jussieu, F-75230 Paris Cedex 05, France

M. Matecki

Laboratoire de Chimie Minérale D[2], Université de Rennes Beaulieu
F-35042 Rennes Cedex, France

1 INTRODUCTION

It is well established that amorphous systems have a characteristic ultrasonic behaviour at temperatures of a few K which is satisfactorily explained by the two-level system (TLS) model [1]. Since for the most part these have been non-magnetic systems, it is of interest to know if the addition of a magnetic ion such as Mn would change the ultrasonic behaviour. Fluoride glasses provide a means to study this point. Recent measurements of the magnetic susceptibility of fluoride glasses containing Mn have shown evidence of a spin-glass state at low temperatures [2]. By way of comparison, ultrasonic studies of a fluorozirconate glass (without Mn) are in substantial agreement with the TLS theory [3].

We have measured the ultrasonic velocity and attenuation at low temperatures in a fluoride glass, $(ThF_4)_{0.2}(BaF_2)_{0.1}(MnF_2)_{0.5}(AlF_3)_{0.2}$, containing 14 atomic per cent Mn. We find a typical glass-like behaviour for the ultrasonic attenuation but the temperature variation of the velocity differs from the usual behaviour in a relatively narrow temperature range near 1 K.

2 RESULTS

2.1 The ultrasonic measurements were made using standard pulse-echo techniques with longitudinal waves between 200 MHz and 600 MHz. The variation of the attenuation with temperature on a log-log scale is shown in Fig. 1. Below 1 K the attenuation varies as T^n with n approximately 3. Our results are not inconsistent with a frequency independent attenuation below 0.4 K. The value of the attenuation at the temperature-independent plateau around 2 K varies as the ultrasonic frequency to within experimental errors. As with other glasses, the attenuation rises again when the temperature is increased still further.

2.2 The variation of velocity with temperature is given on a semi-log scale in Fig. 2. There is a small, approximately linear, increase on this graph up to about 0.3 K. At high temperatures ($T > 3$ K) the velocity decreases as a linear function of T : this behaviour at low and high temperatures is typical of amorphous materials, although the total variation at low temperatures is about five times smaller than that found in the fluorozirconate glass [3]. Perhaps more important, however, is the flattening of the velocity *vs* temperature data between high and low temperatures ($0.3 < T < 3$ K). This is not observed in other glasses, which instead have a velocity-temperature graph that rounds off smoothly between the high- and low-temperature regions.

2.3 We have found that applying a magnetic field of 1.3 T perpendicular to the ultrasonic wave vector decreases the longitudinal ultrasonic velocity by

---

[1] Permanent address: Department of Physics, Oakland University Rochester, MI 48063, USA

[2] Associated with the Centre National de la Recherche Scientifique

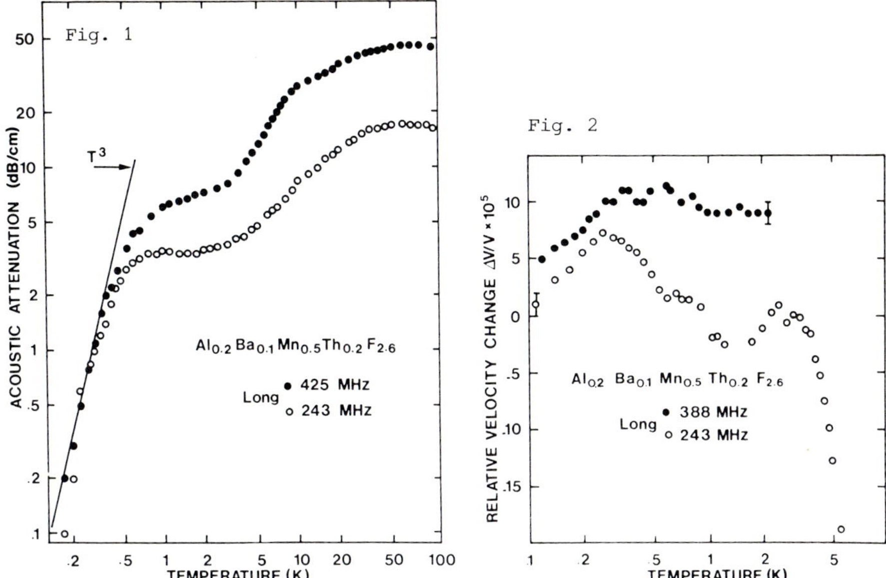

Fig. 1. Ultrasonic attenuation as a function of the temperature in a log-log plot for two frequencies of a longitudinal wave propagating in the glass $Al_{0.2}Ba_{0.1}Mn_{0.5}Th_{0.2}F_{2.6}$

Fig. 2. Phase velocity change of a longitudinal ultrasonic wave propagating in the glass $Al_{0.2}Ba_{0.1}Mn_{0.5}Th_{0.2}F_{2.6}$ in a semi-log plot. The two sets of points are arbitrarity shifted one from the other. The zero of the velocity scale is arbitrary

about three parts in 10^5 at 1.5 K. No change in velocity is seen if the field is applied parallel to the wave vector and the velocity change with magnetic field has not saturated at 1.3 T.

3 ANALYSIS of DATA

Resonant absorption of phonons by the TLS results in an attenuation which is saturable and is not observed here. The corresponding fractional velocity change, $\Delta v/v$, is found to be [1] :

$$\frac{\Delta v}{v} = C\,\ell n(T/T_0) \qquad (1)$$

where T_0 is an arbitrary reference temperature and C is a constant that depends on the polarization. We observe this behaviour in our sample.

The modulation of the TLS splitting by the ultrasonic wave is a second process which results in a relaxational attenuation. This attenuation α is given by the following expressions [4]. (T_1^m is the minimum value of the distribution of relaxation times which characterize the TLS of energy E).

For $\omega T_1^m \gg 1$ (low temperatures) :

$$\alpha = \frac{\pi^4}{96} C \frac{K_3}{v} T^3 \qquad (2)$$

and for $\omega T_1^m \ll 1$ (high temperatures) :

$$\alpha = \frac{\pi\,\omega}{2\,v} C\,. \qquad (3)$$

Here, K_3 is a constant involving the coupling of the phonons to the TLS, and C is the same constant which appears in (1).

Thus, the attenuation measured at the plateau can be used in (3) to calculate C and this result compared to that determined from (1). From (3) we obtain the value : $C \simeq 1.6 \cdot 10^{-4}$. From (1) we find $C \simeq 0.8 \cdot 10^{-4}$. While it is not unusual to have a discrepancy of a factor of 2 using these two equations, both values are small compared to that measured in the non-magnetic fluorozirconate glass, where a value is found from (1) of $4.2 \cdot 10^{-4}$ [3]. i.e., the low-temperature velocity change in our manganese fluoride glass is 5 times smaller. Indeed, typical values in glasses seem to be $C = 3$ or $4 \cdot 10^{-4}$ [1,3].

4 INTERPRETATION and CONCLUSION

The results presented here can be said to be anomalous compared to the behaviour found in other, non-magnetic amorphous systems. The behaviour is glass-like except in the vicinity of 1 K. In addition, there is a weak dependence of the velocity on magnetic field. Further, it is known that a similar fluoride glass with about the same concentration of manganese undergoes a spin-glass transition at 1.25 K [2]. Our results are therefore not inconsistent with what might be expected if there was a velocity decrease caused by a magnetic transition near 1.3 K. Ultrasonic measurements in CuMn [5,6] have shown that near the spin-glass transition temperature the ultrasonic velocity also exhibits a decrease from a reasonable extrapolation of the higher temperature variation to the vicinity of the transition temperature. Confirmation of our interpretation, of course, can only come from a measurement of the magnetic susceptibility.

REFERENCES

1. S. Hunklinger and W. Arnold: in Physical Acoustics, edited by W.P. Mason & R.N. Thurston (Academic Press, N.Y. 1976), vol. XII, p. 155
2. C. Dupas, J.P. Renard, G. Fonteneau, J. Lucas: J. Magn. Mat. **27**, 152 (1982)
3. P. Doussineau and M. Matecki: J. Physique Lett. **42**, L - 267 (1981)
4. J. Jäckle: Z. Phys. **257**, 212 (1972)
5. G.F. Hawkins and R.L. Thomas: J. Appl. Phys. **49**, 1627 (1978)
6. G.F. Hawkins, R.L. Thomas and A.M. de Graaf: J. Appl. Phys. **50**, 1709 (1979)

Vibrational Dynamics of Lithium Ions in β-Alumina Crystals

R. Di Valerio, A. Fontana, G. Mariotto, and M. Montagna
Dipartimento di Fisica, Università di Trento and Unità GNSM-CNR di Trento
I-38050 Povo, Italy

In the β-aluminas the monovalent conducting cations occupy low density mirror planes between spinel blocks made by aluminum and oxygen ions. The mobile ion may occupy different sites in the conduction plane, denoted as Beever-Ross (BR), anti-Beever-Ross (aBR) and mid-oxygen (mO) [1]. The relative occupation probability depends on the cation properties and changes with both the temperature and the degree of non-stoichiometry. Vibrational frequencies of the cation motions fall in the region below 100 cm^{-1} [2,3], while the spinel modes have frequencies up to ~1000 cm^{-1}. In fact the cations feel quite smooth potentials because of the weak bonds with the surrounding ions. On the contrary the Li^+ ion vibrational frequency was found to be anomalously higher (~380 cm^{-1}) than that expected on the basis of vibrational frequency of Na^+ and the mass ratio [4]. This suggests that the lithium containing β-alumina has peculiar structural properties. Moreover heat capacity [5], dielectric constant [6], neutron [7] and X-ray [8] diffraction measurements indicate that out of the mirror plane sites are occupied by most of the Li^+ ions. KANEDA et al. [4] have calculated the potential energy experienced by the Li^+ ion as a function of the displacement from the mirror plane along the crystallographic c direction above and below the BR sites. The results of their calculation are resumed in Fig. 1: the Li^+ ion, due to its small radius, tends to escape from the conduction plane approaching the spinel blocks, and feels a potential having two minima at ± 0.8 Å away from the mirror plane. The two charged layers of Li^+ ions give rise to the asymmetry in the potential energy. By means of this two-wells potential energy, KANEDA et al. were able to fit the anomalously high vibrational frequency observed in the a(c,c)a' Raman spectrum of Li β-alumina crystals.

In the present work we extend the model in order to study the vibrational dynamics of this system as a function of the temperature. The Schrödinger

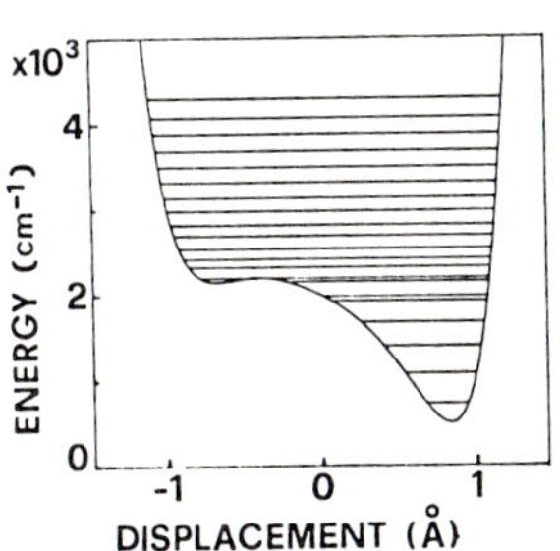

Fig.1. Potential energy for $^6Li^+$ ion as a function of the displacement from the mirror plane along c crystallographic direction. Vibrational energy levels are also shown

equation has been resolved by means of Hamiltonian diagonalization in a harmonic oscillator basis [9], using standard techniques. The results for energy levels are reported in Fig. 1. Once the eigenvalues and eigenfunctions are known, it is straightforward to calculate the shape of the Raman spectra at every temperature, simply by taking into account the thermal population distribution.

Raman spectra in different polarization are recorded in both ^{6}Li and ^{7}Li enriched β-alumina crystals in the temperature range from 30 to 700K. The a(c,c)a' low-temperature spectra agree with those of KANEDA et al.: a band is observed at 370 cm^{-1} and 390 cm^{-1} in $^{7}Li^{+}$ and $^{6}Li^{+}$ respectively with the expected scale of mass (Fig. 2). This vibrational mode is also observed in the a(a',a')c polarization. No other feature, whose frequency scales with the lithium mass, is observed in all the scanned temperature range. The intensity of the Li^{+} ion vibrational mode strongly decreases with temperature, until it disappears in the underlying background. The a(c,c)a' $^{6}Li^{+}$ spectra results are more suitable for a quantitative analysis because of the relatively higher intensity and better defined shape. Two of these spectra are reported in Fig. 3. The integrated intensities of the band are reported in Fig. 4 as a function of temperature, together with the result of the two-wells potential model (lower line). The overall behavior of the data is well reproduced by the model, while it cannot be reproduced by the harmonic oscillator calculated intensity which is also reported as a comparison (upper line). This is still true even if we consider the data indetermination.

No evidence of Raman transitions between excited states was observed even at the highest temperatures. The lack in the Raman spectra of the features corresponding to such transitions may be due to the disorder present in our non-stoichiometric crystals. In fact Li^{+} ions are expected to experience site-dependent potentials. Our calculations [10] show that the fundamental transition is not very much affected by local strains. On the contrary the transitions between excited levels up to the saddle point are very sensitive to the details of the potential shape. They are expected to give a very broad-

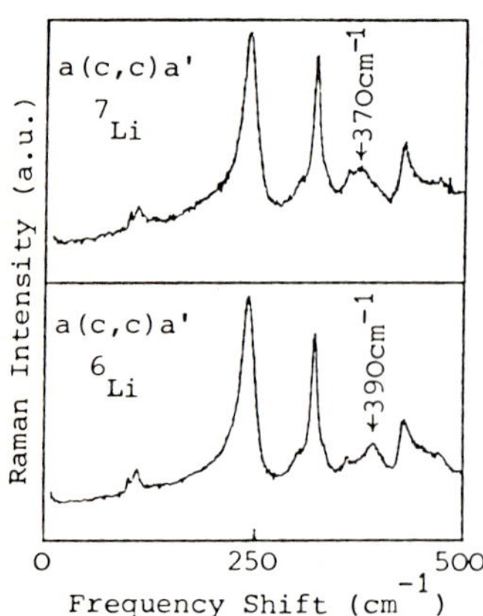

Fig.2. Polarized Raman spectra at 30 K on Li β-alumina crystals

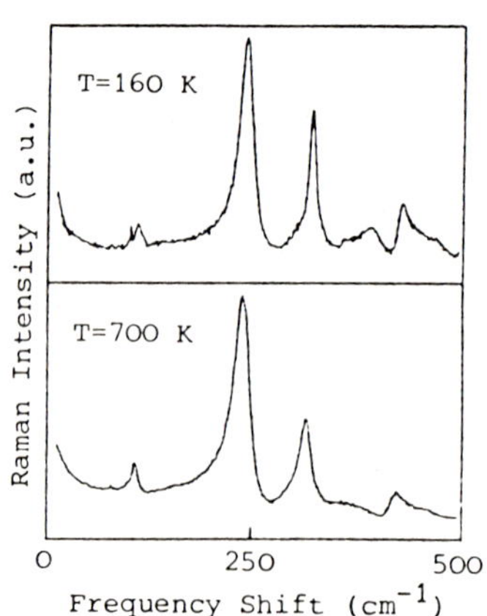

Fig.3. a(c,c)a' Raman spectra on ^{6}Li β-alumina crystals

ened band in the region below 300 cm^{-1}, probably too low to be observed. On the other hand transitions between higher excited states, above the saddle point, should give a more sharp band in the 120 cm^{-1} region, shifting toward higher frequencies as the temperature increases. Even at the highest temperatures of our experiment these higher excited levels were not enough populated to give observable effects. A band at ~ 120 cm^{-1} was observed in the a(c,c)a' Raman spectra of mixed Na-Li-β-Alumina crystals [11]. In this system the asymmetry of the potential is reduced and the activation energy of the highly excited levels is lower. The band intensity and peak frequency behavior, as a function of temperature, is very well reproduced by our model.

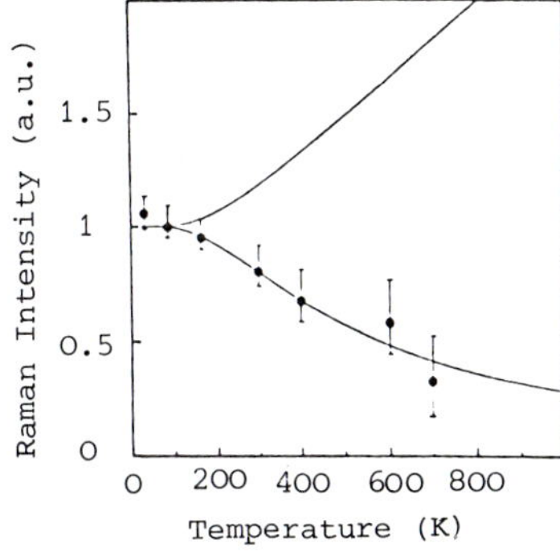

Fig.4. Intensity of the ^{6}Li Raman band as a function of temperature (dots), together with the results of our model (lower line). Expected intensity for the linear harmonic oscillator is also reported (upper line)

[1] J.H. Kennedy: "Solid Electrolytes, Topic in Applied Physics, Vol. 21", ed. by S. Geller (Springer, Berlin 1977) pp. 105-141

[2] C.H. Hao, L.L. Chase and G.D. Mahan: Phys. Rev. B 13, 4306 (1976)

[3] S.J. Allen, Jr., A.S. Cooper, F. Rosa, J.P. Remeika and S.K. Ulasi: Phys. Rev. B 17, 4031 (1978)

[4] T. Kaneda, J.B. Bates and J.C. Wang: Solid State Commun. 28, 469 (1978)

[5] D.B. McWhan, C.M. Varma, F.L.S. Hsu and J.P. Remeika: Phys. Rev. B 15, 553 (1977)

[6] P.J. Antony and A.C. Anderson: Phys. Rev. B 19, 5310 (1979)

[7] B.C. Tofield and G.C. Farrington: Nature (London) 278, 438 (1979)

[8] A. Marini, G. Floor, V. Massarotti, A. McGhie and G.C. Farrington: J. Electrochem. Soc. (in press)

[9] A.L. Somorjai and D.F. Hornig: J. Chem. Phys. 36, 1980 (1962)

[10] R. Di Valerio, A. Fontana, G. Mariotto and M. Montagna: to be published

[11] M. Villa, J.B. Bjorkstam, G. Mariotto, A. Fontana and E. Cazzanelli: J. Chem. Phys. 76, 2804 (1982)

Low-Energy Excitations in $(KBr)_{1-x}(KCN)_x$ in the Orientational Glass State [1]

M. Meissner [2], J.J. de Yoreo, and R.O. Pohl

Cornell University, Ithaca, NY 14853, USA

S. Susman

Argonne National Laboratory, Argonne, IL 60439, USA

CN^- ions substituting in small concentrations for Br^- ions in KBr are known to possess low energy, quasirotational tunneling and librational states. For molar concentrations $x>0.01$, $(KBr)_{1-x}(KCN)_x$ is believed to enter into a quadrupolar glass phase at low temperature [1,2]. We have measured the low-temperature thermal conductivity and specific heat of this mixed crystal system for x=0.25 and 0.5, and have found it to possess the same low-energy excitations as are seen in amorphous solids.

Figure 1 shows the thermal conductivity of $(KBr)_{1-x}(KCN)_x$ for both low and high CN^- concentrations. For $x<0.01$, the thermal conductivity is strongly reduced below that of pure KBr, and varies proportionally to x^{-1}. However, as x increases beyond 0.01 the conductivity increases, indicating a disappearance of individual CN^- tunneling states. Comparison with the conductivity of PMMA shows the similarity of $(KBr)_{1-x}(KCN)_x$ to real glasses for $x>0.25$. The specific heat, measured on both long- and short-time scales, also shows the disappearance of the individual CN^- tunneling states. In Fig. 2 the long-time specific heat C_l (~ 10 sec) is shown for a pure KBr crystal and four $(KBr)_{1-x}(KCN)_x$ samples with $x=8.5 \times 10^{-4}$, 1.5×10^{-4}, 0.25 and 0.5. The specific heats of the samples with x=0.25 and 0.50 are nearly identical, and are smaller than that of the sample with $x=1.5 \times 10^{-4}$ (150 PPM) below 0.5K. The specific heat is very similar to that of an amorphous solid, like PMMA [3]. Between 55 and 1500 mK it can be described by

$$C_V = 38.4T^{1.08} + 100T^3 + C_D, \text{ in erg g}^{-1}\text{ K}^{-1} \tag{1}$$

where $C_D=170T^3$ is the Debye contribution for the x=0.25 sample, as calculated from the measured speeds of sound [2]. For the pure KBr crystal calculations on the basis of elastic constants give $C_D=62T^3$ erg g^{-1} K^{-4} [4], while for our long-time specific heat data (5 sec) $C_l=80T^3$ erg g^{-1} K^{-4} for $T<1K$. It is conceivable that this increase over C_D is due to impurities; however, it is difficult to see why this should lead to a T^3 specific heat. The specific heats of the pure KBr crystal and of the sample with x=0.25 were also measured on short-time scales (50 µsec and >200 µsec respectively) (see Fig. 2). For the pure KBr crystal some overshoot in the temperature profiles was seen, which leads to a short-time specific heat, C_s, smaller than C_D. It is likely that this overshoot is actually the result of a slow phonon thermalization process, i.e., the time scale of the experiment may be less than the time required to produce the equilibrium phonon distribution in the crystal. The short-time specific heat for the x=0.25 sample shows a much stronger time dependence similar to that

[1] Work supported by the National Science Foundation, Grant No. DMR-82-07079

[2] Permanent address: Institut für Festkörperphysik, Technische Universität D-1000 Berlin, Fed. Rep. of Germany

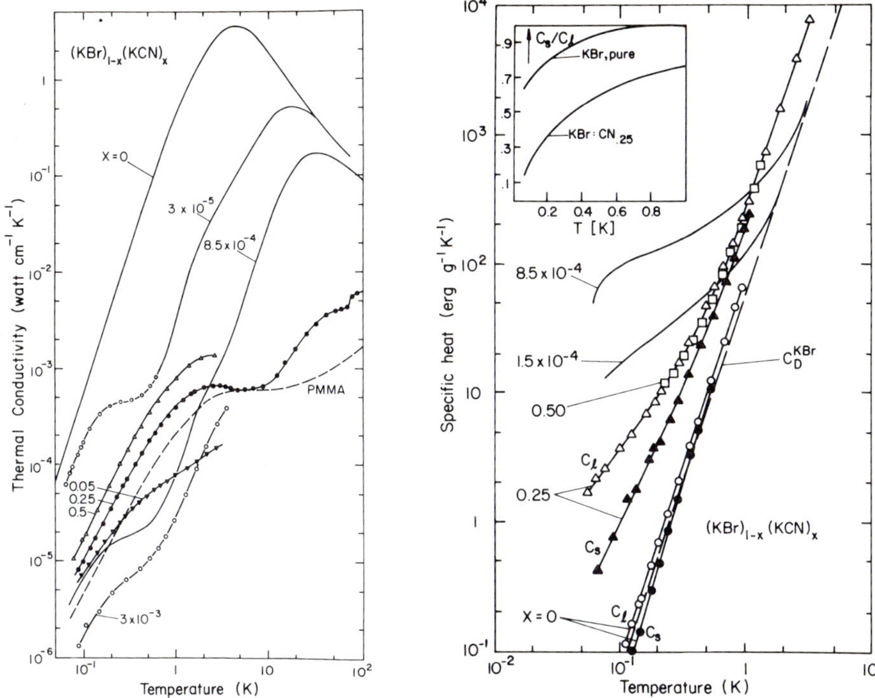

Fig. 1. Thermal conductivity of $(KBr)_{1-x}(KCN)_x$. Solid curves for x=0; $3x10^{-5}$; $8.5x10^{-4}$, after Ref. [8]. Extension of data for $x=3x10^{-5}$ to low temperatures and data for $x=3x10^{-3}$, this study (open circles). For x=0.25 and 0.5, the conductivity below 0.5K varies at $T^{1.8}$ and $T^{2.0}$, respectively; a plateau occurs at ≈ 10K. These features are characteristic for amorphous solids. Thermal conductivity of polymethylmethacrylate (PMMA) for comparison (dashed line), after Ref. [3]

Fig. 2. Specific heat of $(KBr)_{1-x}(KCN)_x$. Solid curves for $x=8.5\cdot10^{-4}$ and $1.5x10^{-4}$ after Ref. [9]. For x=0.25 and pure KBr the specific heat was measured on short- and long-time scales: $C_l^{0.25}$: >50 msec; $C_s^{0.25}$<1 msec; C_l^{KBr}: > 1sec; C_s^{KBr}: < 50 μsec. The theoretical Debye specific heat for pure KBr (Θ_D=172K) is shown for comparison (dashed line). The inset shows the ratio C_s/C_l for pure KBr and $(KBr)_{.75}(KCN)_{.25}$ calculated from short- and long-time specific heat data, C_s and C_l

seen in glasses [5]. The data were taken on time scales which varied from 200 μsec at 55 mK to 1 msec at 1K because of the variation of the thermal diffusivity of the sample. On time scales greater than 50 msec, the specific heat was observed to be time independent and agreed with that taken on 50 sec time scales. The inset to Fig. 2 shows how the measured specific heat is greatly reduced at short times.

One might try to understand these results by assuming that the motions of all but a small fraction of the CN^- ions are frozen at low temperatures. In this picture, the thermal conductivity is the result of resonant scattering of phonons by the tunneling states of the remaining individual CN^- ions. Since the phonon scattering arises from the most rapidly relaxing fraction of these states, the short-time specific heat puts an upper limit on the number of such states involved in the scattering of phonons. A comparison of the specific heat data for the x=0.25 sample with that of the dilute samples puts this upper limit at $x \sim 10^{-5}$. Fig. 1 shows that the effect of such a small number of CN^- tunneling ions on the thermal conductivity is more than an order of magnitude less than is observed for the x=0.25 sample. For the x=0.50 sample, this discrepancy is twice as large. Thus, while according to the picture used here, less than 0.001% of the CN^- ions can tunnel on short-time scales at high concentrations, their coupling to the phonons must be more than an order of magnitude larger than for the CN^- ions at low concentration. Furthermore, there is, of course, no reason to believe that C_S represents a lower limit of the short-time specific heat. (The time scale, 200 μsec, was dictated by the thermal diffusion time through our sample which was 0.7 mm thick.) We thus conclude that tunneling by residual isolated individual CN^- ions is not responsible for the phonon scattering observed in the thermal conductivity for x>0.25.

We propose that the excitations of this system are similar to those proposed for spin glasses or for interacting elastic dipoles [6,7]. When the quadrupolar motion freezes out, the system does not freeze into an absolute ground-state configuration, but into one of many nearly degenerate configurations. As long as the barrier to reorientation is small, the system can make transitions from one local minimum to another. This motion gives rise to collective excitations which produce the enhanced specific heat.

References

1. J. M. Rowe, J. J. Rush, D. G. Hinks, and S. Susman, Phys. Rev. Lett. 43, 1158 (1979).
2. R. Feile, A. Loidl, and K. Knorr, Phys. Rev. B26, 6875 (1982).
3. R. B. Stephens, Phys. Rev. B8, 2896 (1973).
4. J. T. Lewis, A. Lehocsky, and C. V. Briscoe, Phys. Rev. 161, 877 (1967).
5. M. Meissner and K. Spitzmann, Phys. Rev. Lett. 46, 265 (1981). M. T. Loponen, R. C. Dynes, V. Narayanamurti, and J. P. Garno, Phys. Rev. B25, 1161 (1982).
6. P. W. Anderson, B. I. Halperin, and C. W. Varma, Phil. Mag. 25, 1 (1972).
7. B. Fischer and M. W. Klein, Phys. Rev. Letters 43, 289 (1979).
8. W. D. Seward and V. Narayanamurti, Phys. Rev. 148, 463 (1966).
9. P. P. Peressini, Ph.D. Thesis, Cornell Univ. 1973, unpublished.

Spectral Hole Burning of Dyes, Probing Phonon Processes at Surfaces and in Amorphous Systems

U. Bogner and G. Röska

Institut für Physik III, Universität Regensburg, Universitätsstraße 31
D-8400 Regensburg, Fed. Rep. of Germany

Selectively laser-excited aromatic dye molecules, embedded in low concentrations in monomolecular layer systems of Langmuir-Blodgett films at the surface of crystals, provided a phonon memory [1], i.e., a new method of detecting high-frequency acoustic phonons up to the terahertz range. The experimental results of this phonon memory and unusual spectroscopic observations in polymers have been explained by a model [1] of matrix-shift variations by phonon-induced transitions in the asymmetric double-well potentials [2] which were originally introduced in order to explain anomalous acoustical and thermal properties of amorphous solids by tunneling processes in two-level systems. The model has been confirmed by direct experimental evidence of the matrix-shift variations [3]. The model also explains persistent spectral hole burning [4] of optical centers in solids and its temperature dependence and large thermal line broadening [5].

In the present paper we report on the investigation of the phonon-induced transitions in which the barrier of the double-well potentials is crossed and which correspond to an inelastic type of phonon-scattering. The investigations were performed in particular by the heat-pulse technique with high time resolution or by increasing the temperature. The sample consists of a Langmuir-Blodgett film of Cd-arachidate doped with perylene molecules. This monomolecular layer system was prepared at the surface of a 50 Ω constantan heater film evaporated on a sapphire single crystal.

The experimental procedure is demonstrated schematically in Fig. 1 in an energy level diagram and in the diagram - see inset (above right) - of the density of occupation N(E) of the statistically distributed energy le-

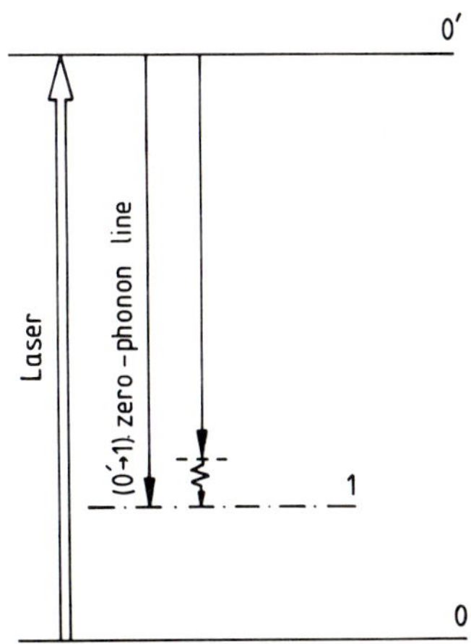

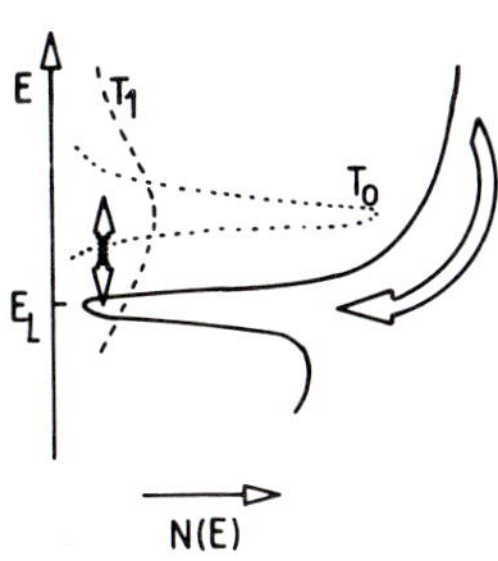

Fig. 1. Energy level diagram and density N(E) of occupation of levels

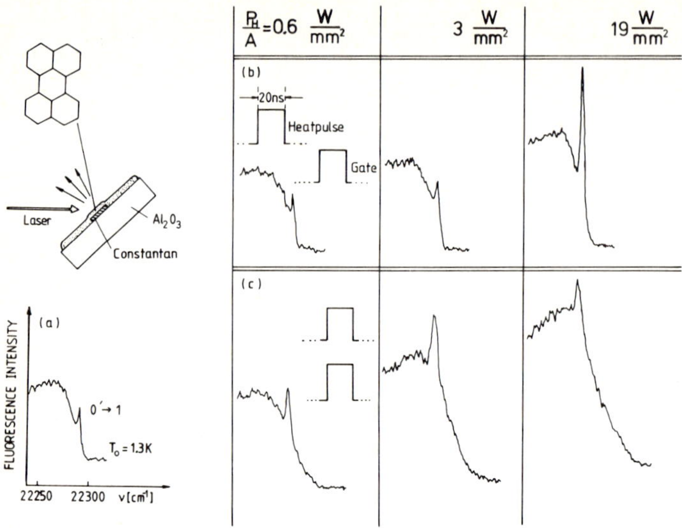

Fig. 2. Effects of heat pulses on the fluorescence of perylene molecules

vels of the dye molecules. By selective excitation with laser photons of the energy $E_L = h\nu_L$, in the inhomogeneously broadened absorption range of the pure electronic transition (0' ← 0), a narrow spectral hole is burned in N(E). The occupation of the levels in the center of the hole is probed by measuring the fluorescence in the spectral range of the first vibronic zero-phonon line (0' → 1) shown in Fig. 2 a, which is shifted 355 cm^{-1} towards longer wavelengths compared with λ_L = 441.56 nm of the exciting He-Cd laser. The fluorescence is measured by a single-photon counting system with a gate (width: ≅ 20 ns) which could be delayed with respect to the voltage pulses (pulse width: ≅ 20 ns, repetition rate: 200 kHz) applied to the constantan heater. In Fig. 2 b the fluorescence was recorded while the gate was shifted beside the heat pulse, i.e., we measured the phonon memory which is caused by irreversible refilling of the spectral hole (see bent arrow in Fig. 1). In Fig. 2c the gate and the heat pulse are coincident, i.e., we measured the real-time effects of the heat pulse on the fluorescence which are due in particular to a reversible filling in the center of the hole because of phonon-induced line broadening of the absorption profile (see dotted profile for the bath temperature T_0 and dashed profile for $T_1 > T_0$ in the inset of Fig. 1) of the molecules beside the spectral hole. In both cases the power density P_H/A of the heat pulses was varied. In Fig. 2 b and 2 c the spectra are shown for P_H/A = 0.6 W/mm²; 3 W/mm² and 19 W/mm², which correspond to calculated heater temperatures T_H = 8; 12; and 20 K respectively. The fluorescence in the anti-Stokes region of the zero-phonon lines in the spectra of Fig. 2 is mainly caused by phonon-assisted fluorescence excitation of dye molecules in higher lying levels $E_h > E_L$, which cannot absorb a laser photon without simultaneous absorption of a phonon. If the Langmuir-Blodgett film is prepared on the sapphire surface opposite that containing the heater it is necessary to shift the gate by the time of flight of the longitudinal or transversal acoustic phonons in order to obtain real-time phonon effects. It can also be demonstrated that the system can be used for phonon detection and phonon spectroscopy.

From the observation that the increase of the intensity of the (0' → 1) zero-phonon line is linear with increasing heater temperature, we conclude that the density of states of the double-well potentials, concerning their barrier height, is constant in the energy range corresponding to phonon frequencies of 0.5 THz up to 2 THz.

We have performed also measurements with dye molecules adsorbed at the surface, e.g., of a sapphire single crystal or at the surface of the microcrystals in polycrystalline paraffins. Concerning the physical nature of the double-well potentials the results suggest a microscopic model in which double-well potentials are formed by adsorbed molecules which were bound by weak intermolecular forces to surfaces or interfaces, e.g., at the boundary of the small free volume regions present in all disordered materials.

With perylene molecules in their two discrete sites in the bulk of n-heptane crystals we studied also selective laser excitation in the phonon sideband. If the wavelength of the exciting dye laser lies in one of the maxima of the phonon-sidebands the generation of monochromatic phonons is indicated by a strong fluorescence peak in the anti-Stokes region of the zero-phonon lines; this peak depends on the concentration of the perylene. The results are explained by linear electron-phonon interaction involving pseudolocalized phonons.

Financial support by the Deutsche Forschungsgemeinschaft is gratefully acknowledged.

1. U. Bogner: Phys. Rev. Lett. **37**, 909 (1976).
2. P. W. Anderson, B. I. Halperin, and C. M. Varma: Philos. Mag. **25**, 1 (1972); W. A. Philips, J. Low. Temp. Phys. **7**, 351 (1972).
3. U. Bogner and R. Schwarz: Phys. Rev. B **24**, 2846 (1981).
4. B. M. Kharlamov, R. I. Personov, and L. A. Bykovskaya: Opt. Spectrosc. **39**, 137 (1975).
5. U. Bogner and G. Röska: J. of Luminescence **24**, 683 (1981).

Evidence of Two-Level Systems in Electrolyte Glass by Brillouin Scattering

J. Pelous, R. Vacher, and A. Essabouri
Laboratoire de Spectrométrie Rayleigh-Brillouin, E.R.A. C.N.R.S.
U.S.T.L., Pl. E. Bataillon, F-34060 Montpellier Cedex, France

U. Reichert and M. Schmidt
Institut für Angewandte Physik II, Universität Heidelberg
Albert-Überle-Straße 3-5, D-6900 Heidelberg, Fed. Rep. of Germany

At low temperature, glasses exhibit anomalous thermal and acoustic properties due to localized excitations coupled with phonons. The possible relation between these excitations characteristic of disorder and the glass transition temperature T_g was suggested by REYNOLDS (1) from a plot of the phonon mean free path deduced from the thermal conductivity *vs.* T_g.

Another study by RAYCHAUDHURI and POHL (2) suggests that the excess specific heat observed at low temperature in nitrate glasses varies linearly with the reciprocal glass transition temperature T_g^{-1}. Finally, DOUSSINEAU *et al.* (3) have performed acoustic measurements at ultrasonic frequencies in a series of fluorozirconate glasses in which T_g varies with composition. They deduced the density of states $\bar{P}$ of defects in the tunneling state picture from simultaneous measurements of the velocity and attenuation of longitudinal and transverse acoustic waves. Their measurements, plotted together with other results for silicate glasses, indicate a variation of $\bar{P} \sim T_g^{-1}$ for T_g ranging from about 500 K to 1500 K, nearly independent of composition. A theoretical description of the correlation between T_g and the configurational states associated with tunneling systems has been proposed by COHEN and GREST (4) on the basis of the free volume theory, and predicts $\bar{P} \sim T_g^{-1}$ weakly dependent on chemical composition.

The purpose of the present study is to extend the experimental investigations to glasses with very low T_g, namely electrolytic solutions of alkali halides in water. By Brillouin scattering we have measured the hypersonic velocity and attenuation of 21 GHz longitudinal waves in the range 1-20 K. Glasses are easily obtained by cooling down $(LiC\ell)_x(H_2O)_{1-x}$ mixtures. The results obtained for a solution with $x = 0.14$ (T_g = 140 K) are given in Figs. 1 and 2.

The velocity *vs.* temperature curve (Fig. 1) shows a maximum near 7 K, and a logarithmic variation is observed between 1 and 6 K. On the other hand, the attenuation in Fig. 2 varies as T^{-1} below 3 K. This behaviour, which is observed for the first time in an electrolyte glass, can be described by assuming a resonant interaction of phonons with two-level systems (TLS) present in the glass. In the framework of the tunneling model (5) the resonant attenuation α_{res} of phonons with frequency ω is given by

$$\alpha_{res} = (\pi C\omega/v)\ \tanh\ (\hbar\omega/2kT)\ .$$

Here v is the sound velocity and $C = \bar{P}\gamma^2/\rho v^2$, where γ is a deformation potential and ρ the mass density.

The relative velocity change is $\Delta v_{res}/v = C\ \ell n(T/T_0)$ where T_0 is some fiducial temperature.

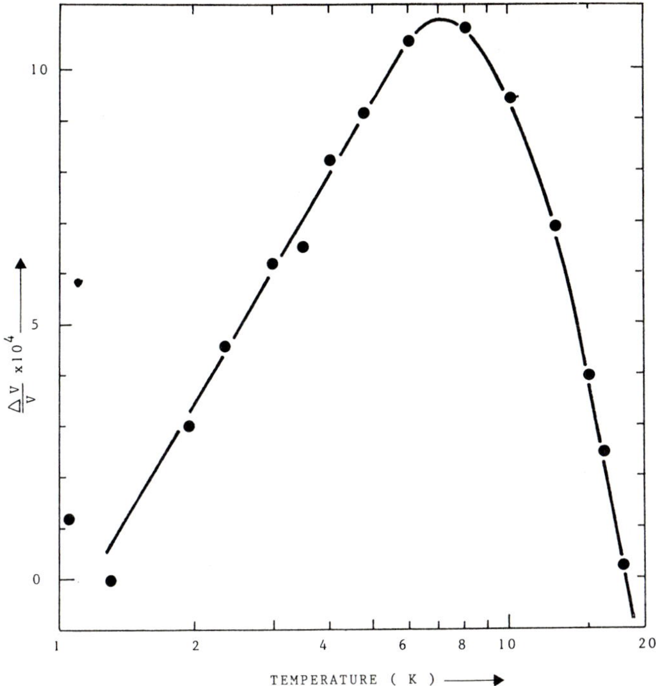

Fig. 1 - Relative velocity change of longitudinal acoustic waves in the glass formed from aqueous solution of LiCℓ with a molar content of 14 % . The straight line in the low-temperature part represents the logarithmic variation while the curved solid line is only a guide to the eye

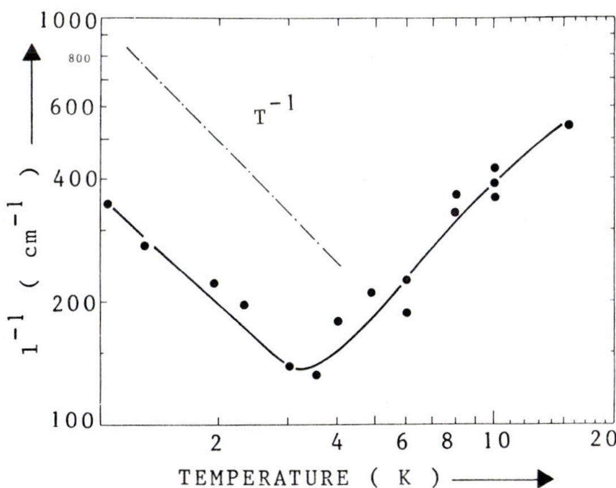

Fig. 2 - Inverse mean free path of 22 GHz longitudinal acoustic phonons in the glass formed from aqueous solution of LiCℓ with a molar content of 14 %. The solid line is a guide to the eye

The experimental results give two independent values for C, which are in good agreement. From the velocity measurements, we obtain $C = (6.7 \pm 0.3) \times 10^{-4}$, while $C = (6.5 \pm 0.7) \times 10^{-4}$ is found from the attenuation values. This result can be compared to $C = 2.8 \times 10^{-4}$ in vitreous silica ($T_g = 1500$ K). As we do not have a separate determination of $\bar{P}$ and γ from our experiments, let us assume that $\bar{P} \sim T_g^{-1}$, leading to $\bar{P}_{LiC\ell-H_2O} \simeq 10\ \bar{P}_{SiO_2}$. The above value for C in the two glasses leads to $\gamma_{LiC\ell-H_2O} \simeq (1/5)\ \gamma_{SiO_2}$. Therefore, the assumption $\bar{P} \sim T_g^{-1}$ implies that γ varies strongly with the glass transition temperature, which seems difficult to understand from a physical point of view. The analysis of ultrasonic results in fluorozirconate glasses indicates a linear variation of γ with T_g (3), even if one must keep in mind that there is a large uncertainty in the experimental determinations of γ.

In order to test the influence of concentration on the TLS parameters, we have performed a second series of experiments on a glass formed from a $(LiC\ell)_x(H_2O)_{1-x}$ mixture with a molar content $x = 0.2$ ($T_g = 148$ K). The results are qualitatively similar to those shown in Figs. 1 and 2 for $X = 0.14$. However, the parameter C is seen to be very sensitive to concentration as we find $C_{0.2} \simeq (4.3 \pm 0.3) \times 10^{-4}$. Taking into account the composition dependence of ρ and v, we get $(\bar{P}\gamma^2)_{0.2} / (\bar{P}\gamma^2)_{0.14} \simeq 0.8$. Therefore, while T_g increases only by 5 % when x changes from 0.14 to 0.2, $\bar{P}\gamma^2$ decreases by about 25 %. If we still assume that $\bar{P}$ is proportional to T_g^{-1}, γ is now found to decrease with increasing T_g, in disagreement with the results discussed above.

We are therefore led to the conclusion that T_g is not the only parameter determining the characteristics of two-level systems in glasses. The electrolyte concentration seems also to have an influence on the TLS parameters.

REFERENCES

1 C.L. Reynolds : J. Non-Crust. Solids 30, 371 (1979).

2 A.K. Raychaudhuri and R.O. Pohl : Solid State Commun. 37, 105 (1981) ; Phys. Rev. B 25, 1310 (1982).

3 P. Doussineau, M. Matecki and W. Schön : J. Physique 43, C9-493 (1982).

4 M.H. Cohen and G.S. Grest : Phys. Rev. Lett. 45, 1271 (1980) ; Solid State Commun. 39, 143 (1981).

5 P.H. Anderson, B.I. Halperin and C.M. Varma : Phil. Mag 25, 1 (1972) ; W.A. Phillips : J. Low Temp. Phys. 7, 351 (1972).

Acoustic Absorption Due to Hydrogen Tunneling in $NbN_{0.0015}H_{0.0025}$

J.L. Wang: Pohl Institute, Tongji University, Shanghai, China

G. Weiss: Institut für Angewandte Physik II der Universität Heidelberg
Albert-Überle-Straße 3-5, D-6900 Heidelberg, Fed. Rep. of Germany

H. Wipf: Physik-Department E19, Technische Universität München
D-8000 München, Fed. Rep. of Germany

A. Magerl: Institut Laue-Langevin, F-38042 Grenoble, France

Some years ago it was found that doping of niobium with hydrogen gives rise to excess specific heat around and below 1 K [1]. At corresponding temperatures the thermal phonon mean free path was reduced by an order of magnitude. This has been interpreted as resonant scattering of thermal phonons by the discrete set of energy levels associated with interstitial motion of H or D atoms [2]. Meanwhile it could be demonstrated that these thermal anomalies induced by hydrogen are strongly intensified by the additional presence of N interstitials [3,4]. N (or O) impurities act as trapping centers, and at low temperatures the hydrogen is still able to tunnel between adjacent interstitial sites around the trapping atoms. This tunneling has directly been observed in inelastic neutron scattering experiments on Nb doped with O and H [5].

It has been demonstrated already that acoustic measurements are well suited to study the tunneling centers formed by OH or OD pairs in Nb [6,7]. In a recent paper the change of sound velocity due to the resonant interaction between the sound wave and the OH tunneling systems has been reported [8]. However, until now nobody has taken into account the influence of the conduction electrons on the dynamical behaviour of these hydrogen tunneling systems. By the ultrasonic study of amorphous metals and especially amorphous superconductors it has already been proved that the relaxation rates of the tunneling systems present in these materials are mainly governed by conduction electrons [9].

In this contribution we report on ultrasonic absorption measurements of Nb doped with 0.15 at % N and 0.25 at % H. The doping is achieved in an analogous procedure as described in ref. [5]. The experiments are done between 15 MHz and 195 MHz. Some of our results are presented in fig. 1. On cooling, the ab-

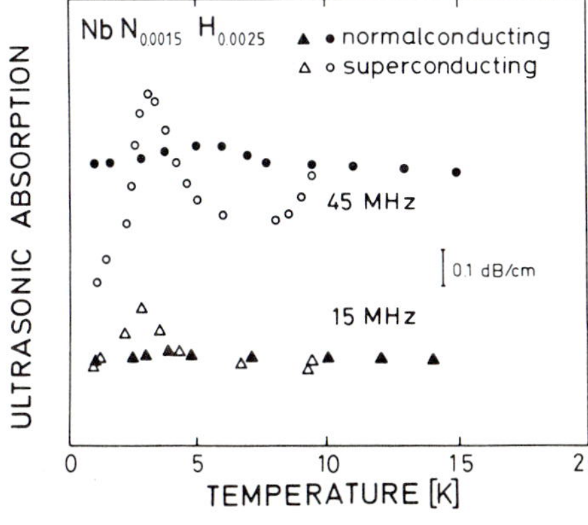

Fig. 1: Ultrasonic absorption of Nb doped with 0.15 at % N and 0.25 at % H in the superconducting (open symbols) and in the normal conducting state (full symbols). A magnetic field of 0.8 Tesla has been applied to maintain the normal state down at 0.5 K. The data of the two frequencies are separated by an arbitrary value

sorption drops directly at $T_c = 9.2$ K and, in the superconducting state (open symbols in fig. 1), shows a maximum around 3 K. This relaxation peak is in agreement with previous measurements and has been attributed to the tunneling motion of trapped hydrogen [6]. The peak observed in recent internal friction measurements [7] probably originates from a different microscopic tunneling system although it has to be ascribed to OH and NH pairs, too. This peak also appears around 3 K but at frequencies 1000 times lower than ours. The decrease at T_c is well understood. It is due to the interaction of the sound wave with the electron gas and its modification when the metal becomes superconducting [10]. When a magnetic field is applied to suppress the superconductivity, this part of the absorption remains high. Most striking, however, is the complete disappearance of the 3 K peak in the normal state (full symbol of fig. 1). This behaviour clearly demonstrates the importance of conduction electrons for the dynamics of the hydrogen tunneling.

The 3 K relaxation peak in the superconducting state can be analysed after subtracting the attenuation caused by the free electrons, which in fact varies quadratically with the sound frequency for $T > T_c$, and the residual absorption. The remaining absorption is proportional to the frequency, and its maximum shifts from 2.7 K at 15 MHz to 4 K at 195 MHz. We can describe this behaviour quite well assuming tunneling systems with a single energy splitting E. The relaxation absorption due to such a two-level system is given by

$$\alpha = A \frac{\exp(E/kT)}{kT(\exp(E/kT) + 1)^2} \frac{\omega^2 \tau}{1 + \omega^2 \tau^2} \qquad (1)$$

where A contains the density of tunneling states and their deformation potential. The second factor describes the change of occupation with energy at a given temperature T. The last factor is the well-known relaxation term which has a maximum if the sound frequency $\omega/2\pi$ is equal to the relaxation rate τ^{-1} of the tunneling states [11]. In our case, taking into account only the interaction with conduction electrons, the relaxation rate can be written as [12]

$$\tau^{-1} = KT/1 + \exp(\Delta/kT). \qquad (2)$$

Here K is a constant containing the electron density of states at the Fermi level and the strength of the interaction between tunneling systems and electrons. Δ is the energy gap according to BCS theory. Equation (2) holds for $E \to 0$ and describes the rapidly varying relaxation rate in the superconducting state. Since $E \neq 0$, a more complicated expression than (2) has to be used taking into account the real density of states at the gap edges [12].

Using (1) and (2) we can easily explain the frequency dependence of the relaxation peak around 3 K. The experimental peaks are just a little broader than the calculated ones. The agreement could certainly be improved by introducing tunneling systems with asymmetric potentials. This would lead to a variation of the relaxation rates without change of the energy splitting E. Unfortunately (1) is rather insensitive with respect to E. Nevertheless, the deduced values are in agreement with other experiments. From inelastic neutron scattering, for example, an energy splitting of 0.19 eV could be found for the OH center [5].

If the sample is forced into the normal conducting state, the relaxation rate of the tunneling systems remains high and the condition $\omega\tau = 1$ is fulfilled at much lower temperatures. This explains the "disappearance" of the 3 K peak when a magnetic field is applied.

In summary, our experiment clearly demonstrates the influence of the conduction electrons on hydrogen tunneling. Evidently, interpretations of the dynamical experiments on hydrogen tunneling motion in niobium at low temperatures have to take into account the rapidly varying relaxation rates below T_c due to the freezing of the conduction electrons into the BCS ground states.

References

1 G.J.Sellers, A.C.Anderson, H.K.Birnbaum: Phys.Rev.B 10, 2771 (1974)
2 S.G.O'Hara, G.J.Sellers, A.C.Anderson: Phys.REv. B 10, 2777 (1974)
3 C.Morkel, H.Wipf, K.Neumaier: Phys.Rev.Lett. 40, 947 (1978)
4 M. Locatelli, K.Neumaier, H.Wipf: J.de Physique 39, C6-995 (1978)
5 H.Wipf, A.Magerl,S.M.Shapiro, S.K.Satija, W.Thomlinson: Phys.Rev.Lett. 46, 947 (1981)
6 D.B.Poker, G.G.Setser, A.V.Granato, H.K.Birnbaum: Z.Phys.Chem. 116, 39 (1979)
7 G.Cannelli, R.Cantelli: Solid State Comm. 43, 567 (1982)
8 G.Bellessa: J.d.Physique 44, L-387 (1983)
9 G.Weiss, S.Hunklinger, H.v.Löhneysen: Physica 109&110 B+C, 1946 (1982)
10 W.P. Mason in Physical Acoustics, Vol. IVA, p. 299, ed. W.P.Mason, Academic Press, N.Y. (1966)
11 S. Hunklinger, W.Arnold in Physical Acoustics, Vol. XII, eds. W.P.Mason and R.N.Thurston, Academic Press, N.Y. (1976)
12 J.L.Black, P.Fulde: Phys.Rev.Lett. 43, 453 (1979)

Phonon Scattering in Phosphorous-Implanted Silicon

M. Grimshaw and G. Feuillet

Centre d'Etudes Nucléaires de Grenoble, Service des Basses Températures
Laboratoire de Cryophysique, 85 X
F-38041 Grenoble Cedex, France

Introduction

In this paper we present measurements of the scattering of phonons of frequency 100 to 300 GHz in amorphous silicon created by ion implantation. The thermal conductivity of a wide range of amorphous solids has remarkably similar form [1], being proportional to T^2 below 1 K and almost independent of temperature in the region 1 to 10 K, forming the so-called plateau. Although not well understood, this plateau may well be connected with the Two-Level Systems (T.L.S.) that are commonly thought to account for the T^2 dependence below 1 K. It is not yet clear whether these T.L.S. actually occur in 4-coordinated amorphous networks, although recent work [2] indicates their presence in amorphous germanium.

The experimental technique that we employ, devised by Dietsche and Kinder [3], for a study in a-SiO_2, uses phonons generated and detected by means of superconducting tunnel junctions and allows us to measure the frequency dependence of the phonon scattering in the range of frequencies relevant to the plateau region in thermal conductivity. In our study we have produced the amorphous layer in the interior of the crystal which has the advantage of reducing possible problems due to scattering at the sample surfaces and, moreover, both generators can be evaporated onto crystalline substrates.

Experimental

Samples of zone-refined n-type Si (resistivity $\sim$ 4000 Ω.cm) were cut and polished on (100) faces. Phosphorous implantation was carried out on half of one of the (100) faces. We used a double implantation (P^+ and P^{2+}) at 200 KV which creates a heavily damaged region about 0.5 μm thick in the interior of the crystal. Two samples were implanted at doses :

$$D_{p+} = 1.5\ 10^{15}\ cm^{-2}\ ;\ D_{p2+} = 3\ 10^{15} cm^{-2}\ ;\ I = 0.5\ \mu A/cm^2\ \text{(amorphous)}$$

$$D_{p+} = 7.5\ 10^{13}\ cm^{-2}\ ;\ D_{p2+} = 1.5\ 10^{14} cm^{-2}\ ;\ I = 1\ \mu A/cm^2\ \text{(non-amorphous)}.$$

Two generator junctions, for which we used either Sn or a PbIn alloy were then evaporated simultaneously on the implanted and non-implanted parts of the crystal, together with a single heterojunction of Al-PbBi or Al-PbIn on the opposite face of the crystal. The alternative arrangement of a single generator and two detectors is unsatisfactory because of the strong dependence of the detector sensitivity to slight differences in junction characteristics.

The phonon detection onset of the heterojunction depends on its bias and can thus be varied [4]. In order to avoid the usual problems due to the large uncertainties in the position of the phonon spectra baselines, the phonon intensity after passage through the implanted layer is compared with that obtained through the pure crystal by successive measurements of the height of the step at the detection onset. The frequency dependence of the attenuation due to the implanted layer is then obtained by repeating the measurement with different detector biases.

Results

The results to date concern mainly the heavily implanted sample and are shown in Fig.1(a) in the form of the ratio A/B of the signal obtained at the implanted side to that obtained at the non-implanted side as a function of frequency. The different sets of points were obtained for different sets of junctions on the same sample. All show the same frequency dependence but are displaced vertically from each other. This is shown more clearly in Fig.1(b) where the results for each experiment have been multiplied by a factor which makes the high frequency part of the curve pass through 1.0. It can be seen that the attenuation of the phonons is strongest at the low frequency end of our range where it reaches about 40 %. The multiplying factor, apparently uncorrelated with differences in the generator I-V characteristics can change overnight when the sample warms up to about 77 K, which eliminates any explanation in terms of misalignment of the junctions. A possible explanation is that there is strong scattering at the junction-sample interface [5], which differs for the two generators and which can be modified by strains set up by differential thermal expansion.

One set of measurements has been performed on the sample implanted at lower doses. The results indicate scattering of a similar form, although about three times less strong. Measurements on a non-implanted sample have shown that the ratio remains constant and approximately equal to 1.0 over the whole range of frequencies, as shown by the dashed line in Fig. 1(b).

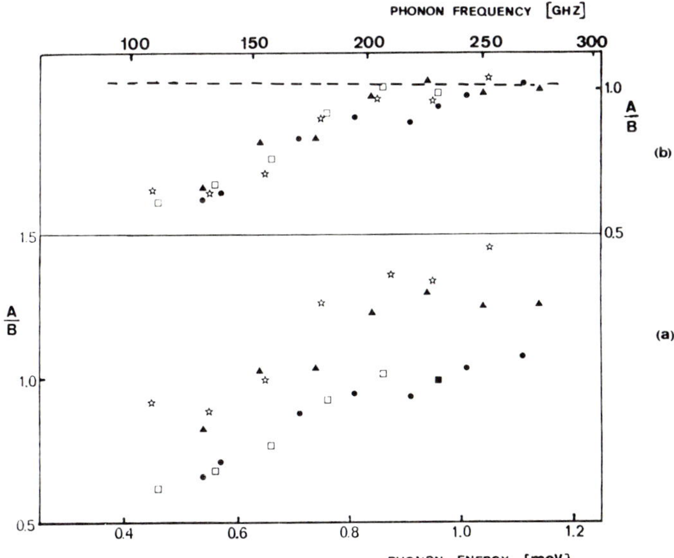

Fig.1 Frequency dependence of the phonon attenuation through the amorphous layer

Discussion

The scattering expected from an amorphous solid increases rapidly with frequency [3], which is the opposite to what we observe. From the results of surface acoustic wave measurements [6], one would expect the scattering due to the T.L.S. to be at least 12 times less strong in a-Si than in a-SiO_2. If, then, the scattering in our frequency range is due to the T.L.S., we can use the results of [3] to estimate that, at our maximum frequency of 300 GHz, the attenuation would be less than 8 %, which would be difficult to detect, especially as the exact form of the low frequency scattering is unknown. We have not as yet determined the origin of this scattering, but it would appear, from the fact that a similar scattering was detected in the non-amorphous sample, that it is not directly connected with the amorphous state. It is however possible that small amorphous regions are formed at lower doses than that required for the continuous amorphous layer. Alternatively we could be seeing resonantlike scattering from clusters of defects, which would be required to have dimensions of the order of 10 nm, in the damaged regions on each side of the amorphous layer. A further possibility is that the scattering is due to electronic levels of a defect either inside or outside the amorphous layer. We propose to study samples implanted with different ions and/or annealed to shed light on this question.

We thank J.P. Gailliard (CENG/LETI) for implanting the samples and it is a pleasure to acknowledge helpful discussions with A.M. de Goër and experimental assistance from P. Geslin.

References

1 A.C. ANDERSON, Amorphous Solids, Low Temperature Properties, Ed. W.A. Philips, 1981
2 J.Y. DUQUESNES, G. BELLESSA, J. Phys. C 16 L 65, 1983
3 W. DIETSCHE, H. KINDER, Phys. Rev. Let. 43, 1413, 1979
4 W. DIETSCHE, Phys. Rev. Let. 40, n° 12, 786, 1978
5 H.J. TRUMPP, W. EISENMENGER, Z. Physik B28, 159, 1977
6 M. VON HAUMEDER, U. STROM, S. HUNKLINGER, Phys. Rev. Let. 44 n° 2 84, 1980

Relaxation Ultrasonic Attenuation Measurements in Quartz Slightly Disordered by Neutron Irradiation

C. Laermans and V. Esteves

K.U. Leuven, Dept. Natuurkunde, V.S.H.D., Celestijnenlaan 200D
B-3030 Leuven, Belgium

1. INTRODUCTION

Acoustic saturation effects [1] and coherent resonant phenomena [2] observed in slightly neutron irradiated crystalline quartz bring out, unambigously, the existence of excitations similar to those found in glasses. The tunneling model [3,4] originally put forward for the explanation of the nearly universal behaviour of thermal, acoustic and dielectric properties in glasses, and its simplified version the two-level systems (TLS) model [5], successfully account for the unusual low-temperature ultrasonic behaviour of the neutron-damaged quartz. Complete knowledge of the microscopic origin of these excitations is however not achieved yet. It is believed that at low irradiation dose ($< 3 \times 10^{19}$ neutrons/cm^2) point defects and defective clusters of mean radius $r \simeq 12$ Å or localized amorphous regions form [6]. However, it is not clear yet whether the TLS have to be necessarily in glasslike regions [7,8]. It thus appears that the further study of crytals damaged by neutron bombardment might be quite informative for a better understanding of the unusual low-temperature behaviour of the dynamic properties of glasses.

2. SAMPLE CHARACTERIZATION

Three synthetic single quartz crystals were irradiated up to a fast neutron dose ($E \geq 0.3$ MeV) of 6.6, 12 and 24 x 10^{18} neutrons/cm^2. They are labeled NIRQ1, NIRQ2 and NIRQ3 and were irradiated in the same nuclear cycle, in order to avoid the influence of a difference in reactor spectrum and flux calibration. In addition NIRQ1 was irradiated at the same time as a sample (N_3) used in previous low temperature thermal conductivity measurements [9].

3. EXPERIMENTAL RESULTS

The attenuation for longitudinal acoustic waves of frequency $\omega/2\pi$ = 490 MHz was measured as a function of temperature in the range 1.4 - 4 K. For each sample, a temperature-independent residual attenuation, attributed to geometric factors, was subtracted from the measured values so that only the temperature-dependent part of the attenuation was plotted in Fig. 1. Data on vitreous silica [10] and irradiated vitreous silica [11] are shown for comparison. It is expected that vitreous silica as the most widely studied glass might be contrasted to quartz, its crystalline counterpart. The dashed lines represent T^3.

4. DISCUSSION

Since by now it is generally accepted that the low temperature anomalies in neutron irradiated quartz are due to TLS similar to these in glasses

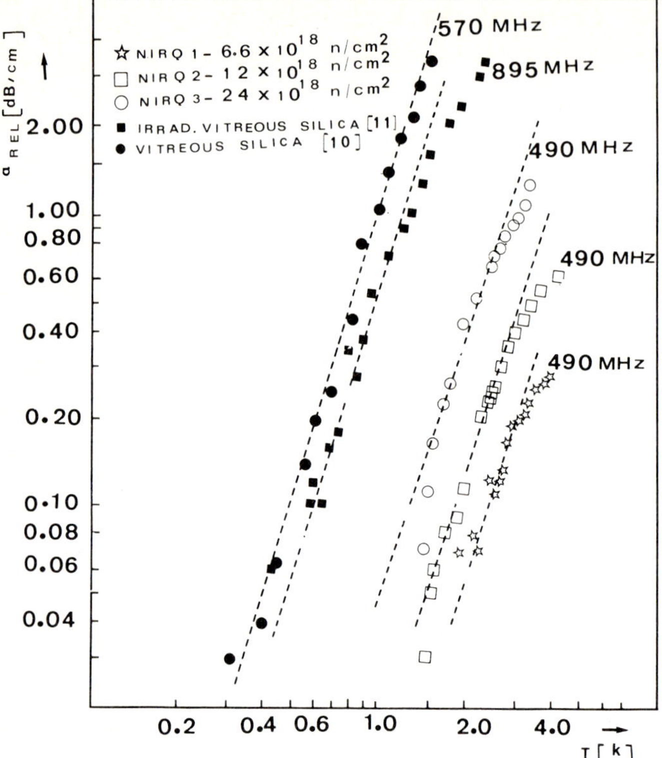

Fig. 1 : Ultrasonic attenuation as a function of temperature

[1,2,9,12,13] we will discuss the data in the TLS framework [5]. In the high power limit (TLS saturated) the attenuation is attributed to a one-phonon relaxation process. In the low temperature regime ($\omega\tau_m \gg 1$) the relaxation attenuation is given by $\alpha_{rel} \propto n_0 \overline{M} D^2 \omega^{\circ} T^3$ where n_0 is the density of states of the TLS, $\overline{M}$ and D are coupling constants. As it can be seen in Fig. 1 the T^3 dependence fits the data in the temperature range 1.4 - 3.2 K. The bending of the curves for T > 3.2 K is attributed to other processes. In order to see the dose dependence the attenuation in neutron irradiated quartz, normalized to that in vitreous silica, was plotted in Fig. 2 (since $\alpha_{rel} \propto \omega^{\circ}$ we used the data at 570 MHz from Fig. 1 which are from ref. 10). A linear dose dependence is found for the used dose. For NIRQ1 we can make an estimate of D_L. From earlier echo experiments it was found that for a quartz sample irradiated to a similar dose (and in the same reactor) the coupling constant M_L is similar to that in vitreous silica. $\overline{M}$, which is a kind of mean value, can then also be taken the same as in vitreous silica. n_0 is found from thermal conductivity measurements on a sample (N_3) which was irradiated at the same time as NIRQ1 to the same dose and was located in the same holder : n_0 (NIRQ1) / n_0 (VIT.SIL.) = 0.048 [9]. It then follows that D_L(NIRQ1) / D_L (VIT.SIL.) ≃ 0.4. (For this estimate the sound velocities have been taken to be the same in neutron irradiated quartz and vitreous silica, which is only approximately true).

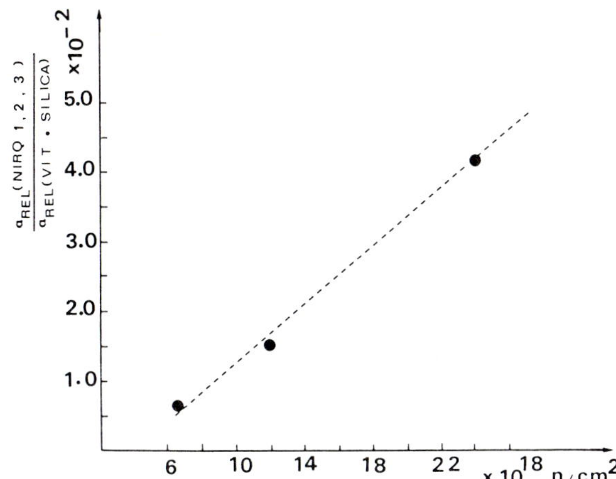

Fig. 2: Relative relaxation attenuation versus neutron irradiation dose

ACKNOWLEDGEMENTS

One of us (V.E.) would like to thank INIC - Instituto Nacional de Investigaçao Cientifica (Portugal) for financial support.
The authors also would thank the Belgian IIKW for financial support, and SCK/CEN Mol for the irradiation of the samples.

REFERENCES

1. C. Laermans, Phys. Rev. Lett. 42, 250 (1979)
2. B. Golding, J.E. Graebner, W.H. Haemmerle and C. Laermans, Bull. Am. Phys. Soc. 24 495 (1979); B. Golding and J.E. Graebner, in "Phonon Scattering in Condensed Matter", H.J. Maris, ed. (Plenum N.Y., 1980), p.11
3. P.W. Anderson, B.I. Halperin and C.M. Varma, Philos. Mag. 25, 1 (1972)
4. W.A. Phillips, J. Low Temp. Phys. 7, 351 (1972)
5. S. Hunklinger and W. Arnold : in Physical Acoustics, Vol. XII, ed. by W.P Mason, R.N. Thurston (Academic, New York 1976) p. 155
6. D. Grasse, O. Kocar, J. Peisl, S.C. Moss and B. Golding, Phys. Rev. Lett., 46, 261 (1981)
7. A.C. Anderson, J.A. McMillan and F.J. Walker, Phys. Rev. B24, 1124 (1981)
8. From thermal conductivity experiments in electron-irradiated quartz : C. Laermans, A.M. de Goër and M. Locatelli, Phys. Lett. 80A 331 (1980), which does not show these amorphous regions : D. Grasse, M. Müller, H. Peisl and C. Laermans : J. de Physique, 43, C9 - 119 (1982)
9. A.M. de Goër, M. Locatelli and C. Laermans, J. Physique 42,C6-78 (1981)
10. J. Jäckle, L. Piché, W. Arnold and S. Hunklinger, J. Non-cryst. Sol. 20, 365 (1976)
11. C. Laermans, L. Piché, W. Arnold and S. Hunklinger in : "Non-crystalline Solids, Ed. G.H. Frischat (Trans. Tech., 1977) p. 562
12. J.W. Gardner and A.C. Anderson, Phys. Rev. B 23, 474 (1981)
13. M. Saint-Paul and J.C. Lasjaunias, J. Phys. C14, L365, (1981)

Low-Energy Excitations in Zr-Based Amorphous Alloys Studied by Thermal Conductivity and Specific Heat

J.C. Lasjaunias, A. Ravex, and O. Béthoux

Centre de Recherches sur les Très Basses Températures, CNRS, BP 166 X
F-38042 Grenoble Cedex, France

For temperatures well below their superconducting transition T_c, the thermal properties of amorphous superconducting alloys become very similar to amorphous insulators, that means governed by the low-energy excitations (Two-Level Systems) successfully proposed by the tunneling model [1] : $T^{1.9}$-T^2 regime for thermal conductivity below about 1 K and a residual specific heat anomaly roughly linear with T and almost universal in magnitude, of the order of a few erg/g.K at 100 mK.

We present specific heat (C) and thermal conductivity (K) data related to T.L.S. for a few Zr-Cu sputtered amorphous alloys, with Zr concentration of 76-80 at %, of different thermal history : $Zr_{76}Cu_{24}$-I in its "as-prepared" state and after annealing 1 hr at 200°C, below the crystallization temperature of $\simeq$ 320°C ; $Zr_{76}Cu_{24}$-II aged at room-temperature ; $Zr_{80}Cu_{20}$ "as-prepared". Also data for a $Zr_{76}Ni_{24}$ "as-prepared" [2] are reported for comparison. All alloys have T_c included between 3 and 3.5 K [2]. Details about the preparation and characterization of these samples are given elsewhere [2,3].

In Fig.(1) are reported specific heat data below 0.8 K only. In order to obtain a precise analysis, measurements were performed to about 7 K, which includes the normal state range where C follows the usual $\gamma T + BT^3$ relation. Below 0.5 K the electronic term C_{es} vanishes exponentially and the remaining contributions are phonons (indicated by solid lines) and T.L.S. : $C = \beta T^3 + C_{TLS}$. Below 0.1 K the specific heat ceases to decrease, which indicates the progressive influence of a nuclear hyperfine contribution ($\sim T^{-2}$) assigned to ^{91}Zr nuclei (for a discussion of this term and its relation to T.L.S., see ref. [3]).

The first striking feature is the large spreading of the amplitude of the anomaly mainly due to T.L.S. between 0.1 and 0.3 K, for this series of ZrCu samples of close chemical composition. Including ZrNi, this spread reaches a factor of five at 0.1 K. This is at variance with the insulating glasses (e.g. oxide glasses, fluoride glasses) where continuous variations of chemical composition (or annealing near T_G) has much less effect on the TLS density of states ; however universality of the magnitude of the anomaly (about 1 erg/g.K at 0.1 K) remains. This effect is corroborated by the thermal conductivity data (Fig.(2)). Well below T_c (in the present case below 1 K), due to condensation of normal electrons the dominant phonon scattering process is the resonant scattering by TLS, characterized by a $T^{1.9}$-T^2 regime obeyed in this series of alloys. There appears a good correlation between the C and K raw data for the TLS regimes : a larger value for K corresponding to a lower for C. This will be confirmed in the precise analysis.

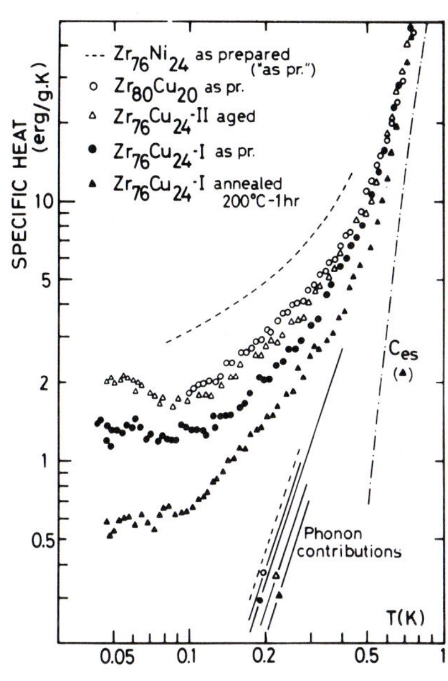

Fig. 1 : Specific heat data

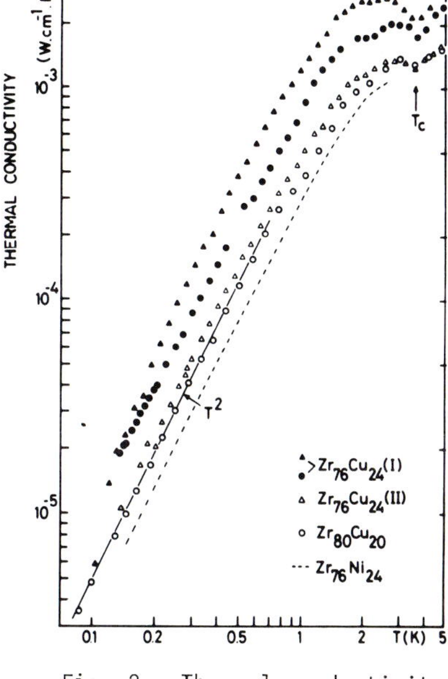

Fig. 2 : Thermal conductivity data

Secondly, a large effect of decrease of the TLS density of states occurs upon annealing (sample $Zr_{76}Cu_{24}$-I). Similar increase of K was detected in different melt-spun Zr_xCu_{1-x} alloys (x = 0.6, 0.7, 0.74 [4]) but *not* correlated to the specific heat, which remained unchanged.

In order to give precise numerical TLS parameters, we have to separate the phonon contribution βT^3 in the specific heat data. In the series of ZrCu, contrary to ZrNi or ZrPt [2], the $C_{TLS}(T)$ variation is very sensitive to this phonon term above 0.3 K : if one uses for β the value B defined in the normal state above T_c, it results in an abnormal behaviour for C_{TLS} which *decreases* with T above 0.3 K [5]; it means that the actual phonon term to be used in this T range is probably smaller than B. Indeed a complete analysis of the thermodynamics of superconductivity based on the B.C.S. model (jump at T_c, parameters of C_{es}) leads to a β value smaller than B, which restores a monotonous increasing variation for C_{TLS} up to $\sim$ 0.5 K, in agreement with all other results of amorphous materials. Since a significant variation of the sound velocity has never been detected in the T_c range for other Zr-based alloys, this suggests the presence in BT^3 of an extra-phononic contribution which rapidly vanishes in the superconducting state, probably of electron-phonon interaction origin .

Generally, the $C_{TLS}(T)$ variation for the "as-prepared" samples is slower than aT (i.e. $\sim T^{0.5}$ to $T^{0.7}$ [2]) but a linear dependence is almost verified after annealing which indicates, in the frame of the tunneling model, the tendency for the TLS density of states n(E) to a more energy-independent distribution, simultaneously to its decrease for $\frac{E}{k_B} \lesssim 1$ K.

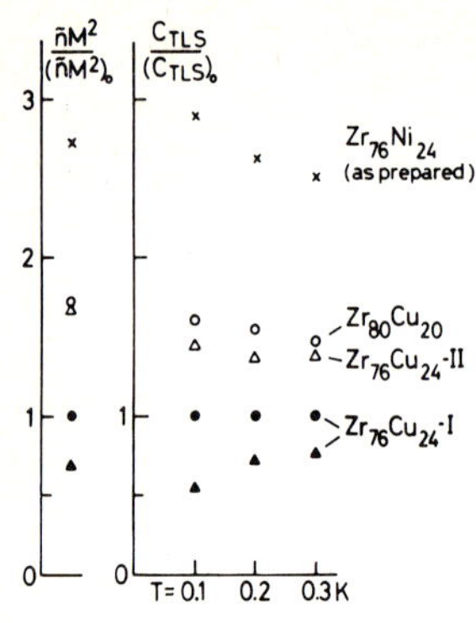

Fig. 3 : Correspondence between the reduced TLS specific heat (at three temperatures) and thermal conductivity coupling parameter $\overline{n}M^2$ for the ZrCu samples and ZrNi [ref. 2]. The reference values $(C_{TLS})_0$ and $(\overline{n}M^2)_0 = 4.1 \times 10^7$ erg.cm^{-3} are those of "as-prepared" $Zr_{76}Cu_{24}$-I

From the quasi T^2 regime of K(T) we calculate the coupling parameter $\overline{n}M^2$ which mainly determines the variations of K in this series of samples ($K \propto \frac{k_B^3}{\hbar^2} \cdot \frac{\rho v_D}{\overline{n}M^2} T^2$, with ρ = density, v_D = mean sound velocity calculated from $\beta = (2\pi^2 k_B^4)/(5\hbar^3 \rho v_D)$), where $\overline{n}$ represents a constant density of states of TLS, the most strongly coupled to phonons, which is only a fraction of n(E) active in specific heat, and M a mean phonon-TLS coupling constant. Fig.(3) demonstrates the excellent numerical correlation between both thermal properties of this series of alloys, which strongly supports their common origin in the tunneling states.

In conclusion, we have pointed out the sensitivity of the TLS density of states to the conditions of preparation and thermal history, contrary to insulating glasses : this is probably a direct consequence of metallic bonding. Moreover this property is perhaps related to the high concentration of Zr atoms which could determine the TLS characteristics [5].

REFERENCES

[1] : "Amorphous Solids : Low Temperature Properties", ed. W.A. Phillips (Springer - Verlag, Berlin, 1981).

[2] : A. Ravex, J.C. Lasjaunias, O. Béthoux, Physica 107 B, 395-397 (1981) and Solid State Comm., 40, 853 (1981).

[3] : J.C. Lasjaunias and A. Ravex, J. Phys. F : Met. Phys. 13, L 101(1983).

[4] : S. Grondey, H.v. Löhneysen, H.J. Schink and K. Samwer, Z. Physik (in press).

[5] : A. Ravex, J.C. Lasjaunias, O. Béthoux, to be published in J. Phys. F.

Low-Temperature Thermal Properties of Amorphous Zr_xCu_{1-x} After Structural Relaxation

H.J. Schink, S. Grondey, and H.v. Löhneysen

2. Physikalisches Institut der RWTH Aachen, Templergraben 55
D-5100 Aachen, Fed. Rep. of Germany

K. Samwer

1. Physikalisches Institut der Universität
D-3400 Göttingen, Fed. Rep. of Germany

1 Introduction

Localized low-energy excitations dominate many low-temperature properties of amorphous solids. Their existence in amorphous metals is well established /1/. In the tunneling model /2/, these excitations are viewed as two level systems (TLS) arising from tunneling of single atoms or small groups of atoms between the two wells of a double well potential. This model describes many experiments successfully, in particular the coupling of TLS to phonons as is observed,e.g.,in the thermal conductivity. Unfortunately, however, the microscopic nature of the TLS is not known.

As there is some evidence which links the low-temperature properties, e.g., the magnitude of the quasi-linear specific heat attributed to TLS, to the phenomena occurring at the glass transition temperature T_g /3/, it might be of interest to induce a structural relaxation of a metallic glass close to T_g and look for changes in the low-temperature properties, in particular in the specific heat C and the thermal conductivity κ. In this paper, we report on such a study performed on glassy Zr_xCu_{1-x} which was annealed at temperatures between 0.7 and 0.8 T_g /4/. Such a heat treatment is known to induce mainly a topological structural relaxation without much changing the chemical short-range order. While there have been some reports of measurements of κ in annealed metallic glasses /5,6/, a comparative study of annealing effects on C and κ has previously only been reported for sputtered amorphous metals /7/, but not for liquid-quenched metallic glasses.

2 Results

The specific heat C and the thermal conductivity κ of a $Zr_{0.70}Cu_{0.30}$ sample are shown as functions of temperature in Figs. 1a and 1b, respectively. First C and then κ of the as prepared sample were measured. Annealing was done with the Cu clamps for the κ measurements remaining on the sample in order to avoid any change in the geometry factor affecting the accuracy of κ. Then κ and finally C were remeasured. Further experimental details can be found elsewhere /4/ and will not be repeated here.

The linear specific heat observed (Fig. 1a) well below the superconducting transition temperature T_c = 2.55 K can be attributed to TLS /8/. Upon annealing at 150°C, T_c has decreased to 2.39 K, but the specific heat for T << T_c remains unchanged within the error of ∿ 20 %. The increase of the thermal conductivity just below T_c (vertical arrows in Fig. 1b) indicates that the phonon contribution κ^{ph} dominates κ^e due to conduction electrons. Below about 1 K, $\kappa \equiv \kappa^{ph}$ varies as T^2 which is typical of phonon scattering by TLS. In this temperature range κ has increased by 70 %,i.e.,well beyond any measuring error. For other Zr_xCu_{1-x} samples annealed at higher temperature, an increase of up to 100 % has been found /4/.

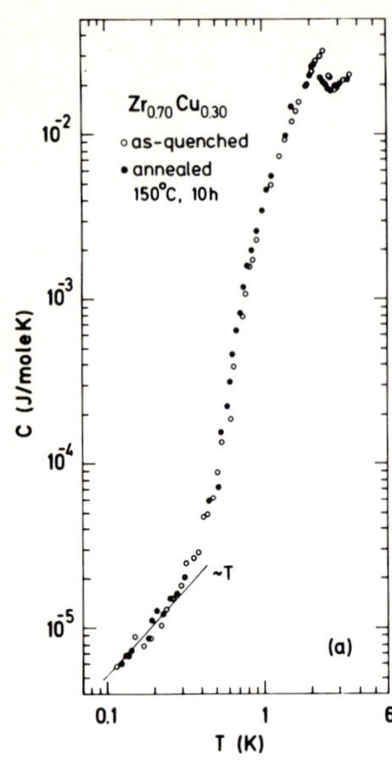

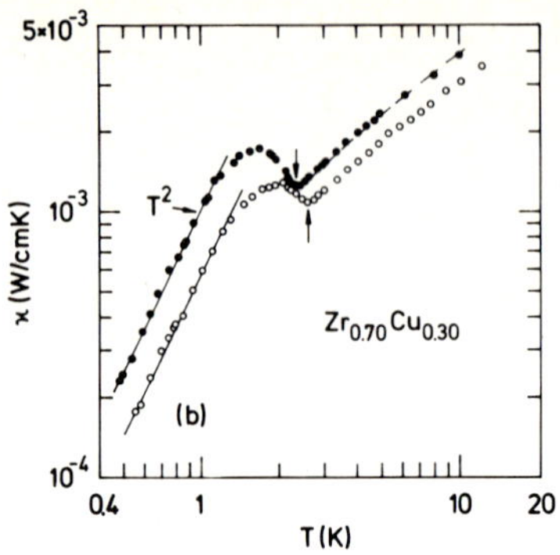

Fig. 1 (a) Specific heat C of amorphous $Zr_{0.70}Cu_{0.30}$ as a function of temperature T before and after annealing (b) Thermal conductivity κ as a function of temperature T. Symbols are the same as in Fig. 1a

3 Discussion

The enhancement of the thermal conductivity appears to be in contradiction to the specific heat results, if all excitations are coupled with the same strength to phonons. In this case the density of states of excitations (i.e., specific heat) and the scattering coefficient (i.e., inverse thermal conductivity) should be directly proportional to each other. However, our results can be understood in the frame of the TLS tunneling model. An essential feature of this model is a wide distribution of relaxation times τ of TLS which results from a broad distribution of asymmetries Δ and tunneling parameters λ of the double-well potentials /9/. While a phonon-scattering experiment (like thermal conductivity or ultrasonic attenuation) probes the strongly coupled (i.e., symmetric) TLS, the specific heat is determined by all TLS which can be excited within the measuring time t, i.e., for which $\tau < t$. Hence the different behavior of C and κ can be interpreted assuming a change in the relaxation time spectrum $P(\tau)$ upon annealing.

As annealing of metallic glasses leads to a release of local strains with a small overall densification, it is conceivable that the atomic tunneling motion is hindered by an enhancement of either the potential barriers or the asymmetries. On the other hand, it could be argued that the release of local strains leads to an increase of the number of symmetric TLS by reducing the asymmetries. Our results, however, appear to rule out the latter possibilities.

A final point is concerned with the thermal conductivity above T_c where κ^{ph} is limited - besides "intrinsic" scattering - by scattering from conduction electrons. Very likely the latter is not changed upon annealing since the electron density of states at the Fermi level and the Debye temperature remain unchanged /4/. (Also, κ^e is essentially unchanged upon annealing because the resistivity changes only by ~ 0.4 %.) While the "intrinsic" scattering at low temperatures can be attributed to TLS as discussed above, it is not clear what limits the thermal conductivity above a few K where κ^{ph}

generally exhibits a temperature-independent plateau. Among other explanations, phonon scattering by TLS with a quadratic density of states has been invoked /10/.

From our results a correlation between phonon scattering at low temperatures and at higher temperatures $T > T_c$ can be inferred. If the same decrease by 70 % of intrinsic scattering as found at low T is present also above T_c, the thermal conductivity of the annealed sample can be calculated from the data for the untreated metallic glass, cf. dashed line in Fig. 1b. The good agreement between experiment and calculation does indeed suggest a correlation between TLS scattering and the plateau in κ^{ph}.

References

1 J.L. Black: In: Glassy Metals I, H.J. Güntherodt, H. Beck (eds.), p. 167, (Springer Berlin 1981);
H.v.Löhneysen: Phys. Rep. 79, 161 (1981);
G.Weiss, S. Hunklinger, H.v.Löhneysen: Physica 109 u. 110 B, 1946 (1982)

2 P.W. Anderson, B.I. Halperin, C.M. Varma: Philos. Mag. 25, 1 (1972);
W.A. Phillips, J. Low Temp. Phys. 7, 351 (1972)

3 A.K. Raychaudhuri, R.O. Pohl: Phys. Rev. B 25, 1310 (1982)

4 A more complete account of the work presented here has recently been published: S. Grondey, H.v.Löhneysen, H.J. Schink, K. Samwer, Z. Phys. B 51, 287 (1983)

5 J.R. Matey, A.C. Anderson: Phys. Rev. B 17, 5029 (1978)

6 P. Esquinazi, M.E. de la Cruz, A. Ridner, F. de la Cruz: Solid State Commun. 44, 941 (1982)

7 A. Ravex, J.C. Lasjaunias, O. Béthoux: Solid State Commun. 40, 853 (1981); and to be published

8 K. Samwer, H.v.Löhneysen, Phys. Rev. B 26, 107 (1982)

9 J. Jäckle: Z. Phys. 257, 212 (1972)

10 M.P. Zaitlin, A.C. Anderson: phys. stat. sol. (b) 71, 323 (1975)

Time-Dependent Specific Heat of Vitreous Silica Between 0.1 and 1 K

W. Knaak and M. Meissner

Institut für Festkörperphysik, Technische Universität Berlin
D-1000 Berlin 12, Fed. Rep. of Germany

The anomalous thermal and acoustic properties of amorphous materials below 1K have been rather successfully explained by assuming a system of tunneling states with a wide distribution of energy and relaxation times [1]. A consequence of this standard tunneling model is that the linear term of the specific heat should depend logarithmically on time [2]. This time-dependence of the specific heat arises because the low energy excitations have relaxation times $\tau \geq \tau_{min}$ (for vitreous SiO_2 $\tau_{min}=4.6 \cdot 10^{-10} \cdot T^{-3}$ sec K^3 [3]). If the sample is heated uniformly throughout on short time scales $t \lesssim \tau_{min}$ the coupling of the low energy excitations to the phonons is observable, which results in an increase of the specific heat with increasing time.

Only two experiments have been reported on the observation of time-dependent specific heat in glasses on short time scales [4,5]. Using thin samples and spatially uniform heat pulses, thermal equilibrium was achieved on µsec time scales. The analysis of the subsequent temperature profiles showed a logarithmic time dependence of the specific heat for the timedecade ~10 µsec to ~100 µsec only. For $t \gtrsim 100$ µsec a much stronger time dependence, i.e. larger coupling of the low energy excitations was observed [4]. These results were in disagreement with the standard tunneling model. Furthermore, the arrival time of the heat pulses, on the time scale of µsec, showed to be inconsistent with the results obtained from temperature profiles after the diffusion process. In several attempts the temperature vs. time profiles have been analyzed only with the negative result that a time dependence of the specific heat is not observable within the diffusive increase of the temperature [4-7].

In order to resolve these questions we have extended the time scale of our experimental technique [5]. Time resolved specific heat measurements (Fig.1) were performed on thin samples ($0.5 \times 1.4 \times 1.4$ mm^3, $m \cong 200$ mg) of vitreous silica (Suprasil and Suprasil W). As a test of our experimental technique single-crystalline quartz ($25 \times 25 \times 25$ mm^3, $m \cong 44$ g) was remeasured [8]. Through the proper choice of copper wires as weak links, sample-to-bath time constants as large as $\tau_R \cong 0.5$ sec at 100 mK were obtained. From the extrapolation of the long-time exponential temperature decay to t=0, ΔT_l was determined and used to calculate the long-time specific heat C_l. From the observed maximum overshoot temperature ΔT_s the short-time specific heat C_s was computed. For the quartz crystal the short-time temperature profiles are complicated by ballistic phonons. As the distance heater-thermometer is 25 mm in z-axis direction, arrival times $t_B=4.0$ µsec and 5.8 µsec were detected, which agree with literature data for the energy velocities of longitudinal and transverse phonons [9]. The ballistic signal is followed by a rapid decay to a plateau-like profile around 20 µsec, which is defined as the short-time temperature rise ΔT_s.

In Fig.2 the results for C_l and C_s are shown for vitreous SiO_2 (Suprasil) and the quartz crystal. The long-time specific heat of Suprasil agrees rea-

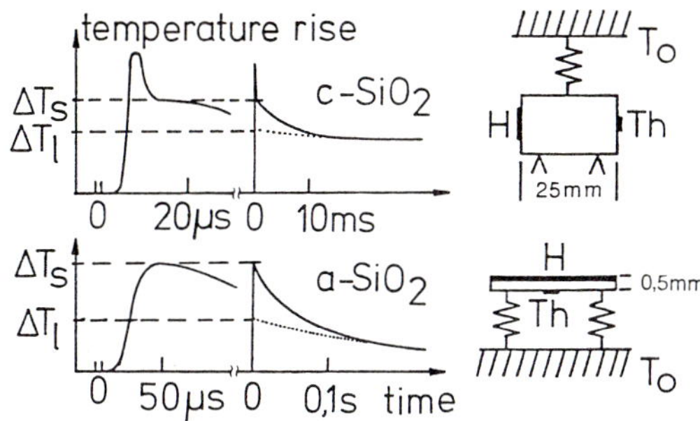

Fig.1. Temperature profiles and experimental setup for both the quartz and vitreous silica sample. ΔT_l is the temperature rise extrapolated to t=0 from long-time exponential decay, ΔT_s is the maximum overshoot temperature on short-time scales

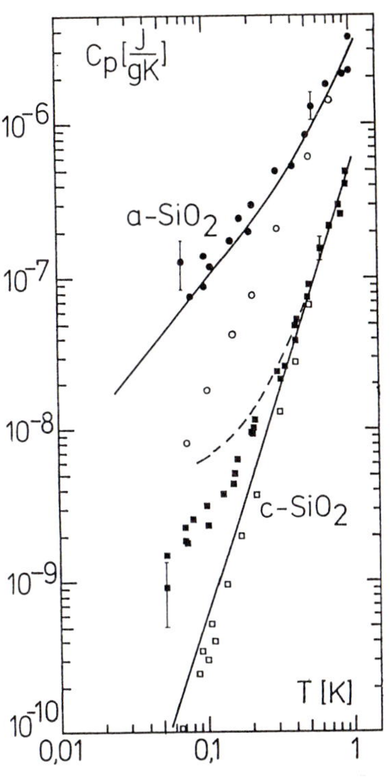

Fig.2. Long-time specific heat C_l (closed symbols) and short-time specific heat C_s (open symbols) for vitr. silica (Suprasil) and the quartz crystal. Upper curve: long-time data for Suprasil by Ref. [10]; lower straight line: calculated Debye specific heat for quartz, $C_D=5.7\cdot10^{-7}\cdot T^3$ J/gK^{-4}; dashed line: long-time data on the same quartz crystral, after [8]

sonably well with earlier measurements [10]. The short-time specific heat data show an increasing decoupling effect with decreasing temperature (at 70 mK C_s is ~9 times smaller than C_l). These C_s values are still considerably larger than the Debye value $C_D=8\cdot10^{-7}\cdot T^3$ J/g K^4, but one should expect a much larger decoupling for sample arrangements with shorter diffusion times. For the quartz sample the long-time specific heat deviates from the T^3 behaviour below 200 mK, as reported in the original work by ZELLER and POHL [8]. There it had been suggested that impurities might be the origin of these anomalies. As a test of our experimental technique it is important to note that the short-time specific heat is in reasonable agreement with the elastic limit. Thus we conclude that the additional excitations, which contribute to C_l of this quartz sample, couple to the phonons on msec time scales. Similar results have been observed on high-purity crystals of Ge [4,11] and KBr [11].

In order to study the time dependence C(t) we have plotted $(\Delta T_l(t)/\Delta T_s(t))\cdot C_l$ versus log t for the 3 materials investigated at T_o=100 mK (Fig.3). A straight line in this plot corresponds to a logarithmic time dependence with an appropriate value for the density of states $\bar{P}$ of the low energy excitations. For comparision curve (a) was calculated for a single-exponential coupling process. If one compares the time dependence for Suprasil and quartz in this way, the difference in the relaxation process is evident for time $t\lesssim10$ msec: the glass shows a logarithmic time dependence with $P=(3.5\pm0.5)\cdot10^{38}$ J^{-1} cm^{-3}, whereas the quartz crystal relaxes in an almost single-exponential process ($\tau\cong1.5$ msec) from the C_D value to a long-time specific heat $C_l\cong4.4\cdot C_D$.

Between ~10 msec and ~200 msec the coupling in the glass (Suprasil and Suprasil W) appears to be much larger than predicted by the standard tunnel-

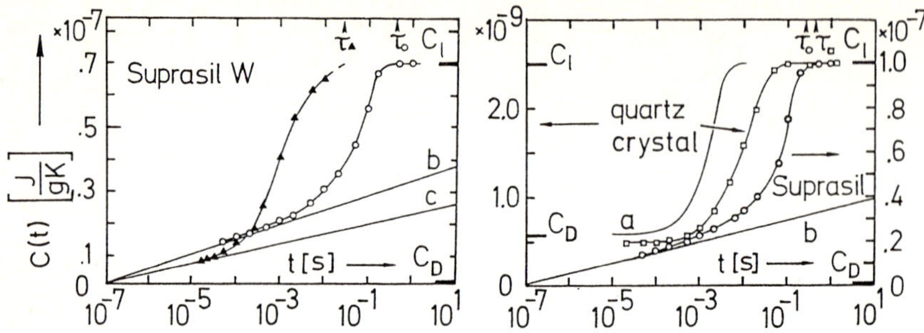

Fig.3. Time-dependent specific heat C(t) vs. log t for Suprasil W (left graph), Suprasil and quartz (right graph) defined as $C(t)=(\Delta T_1(t)/\Delta T_S(t))\cdot C_1$. Open symbols: our data; ▲ : LOPONEN et al. [4]. Arrows drawn at $\tau_{\square}$=0.6 sec, τ_0=0.5 sec, $\tau_{\blacktriangle}$=30 msec indicate the different sample-to-bath time constants. Curve (a) is calculated for a single exponential relaxation process with τ=1 msec; straight lines (b) and (c) show logarithmic time dependence according to the tunneling model [3] with $\overline{P}$=3 and $2\cdot10^{38}$ J^{-1} cm^{-3}, respectively

ing model. The present work shows similar behaviour for both types of vitr. silica: a logarithmic time dependence up to ~10 msec. From LOPONEN's data [4] a logarithmic time dependence is observed only up to 100 µsec. A larger coupling apparently sets in for t≳1 msec and any time dependence vanishes above 10 msec. We conclude that the experiments are in agreement with the logarithmic time dependence according to the standard tunneling model only as long as the sample-to-bath time constant τ_R is much larger than the actual experimental time of observation, i.e. the more the measurement approaches an adiabatic one. It appears that the diffusion process of the phonons within the sample influences the relaxation of the low energy excitations. The physical origin of this effect is still not understood.

References

1. For a review see: Amorphous Solids ed. by W.A.Phillips, Springer 1981.
2. P.W. Anderson, B.I. Halperin, C.M. Varma, Phil. Mag. 25, 1 (1972); W.A. Phillips, J.Low Temp. Phys. 7, 351 (1972).
3. J.L. Black, Phys. Rev. B 17, 2741 (1978).
4. M. Loponen, R. Dynes, V. Narayanamurti, J. Garno, Phys.Rev.B 25,1161(1982).
5. M. Meissner and K. Spitzmann, Phys.Rev.Lett. 46, 265 (1981).
6. W.M. Goubau and R.H. Tait, Phys. Rev. Lett. 34, 1220 (1975).
7. R.B.Kummer, R.C. Dynes and V.Narayanamurti, Phys.Rev.Lett.40, 1187 (1978).
8. R.C. Zeller and R.O. Pohl, Phys. Rev. B 4, 2029 (1971).
9. R.J. von Gutfeld and A.H. Nethercot Jr., Phys.Rev.Lett. 12, 641 (1964).
10. J.C.Lasjaunias, A.Ravex, M.Vandorpe and S.Hunklinger, Solid State Comm. 17, 1045 (1975); W.Block, thesis, Technische Universität Berlin (1982).
11. W. Knaak, M. Meissner and R.O. Pohl, to be published.

Phonon-Dispersion Measurements in Glasses

M. Rothenfusser, W. Dietsche, and H. Kinder

Physik Department, Technische Universität München
D-8046 Garching, Fed. Rep. of Germany

The phonon-dispersion relation is one of the most fundamental properties of the solids. In particular, its knowledge is required for the explanation of the low-temperature thermal properties. Unfortunately, this relation can not be measured in glasses with neutron-scattering techniques because of the absence of a lattice periodicity. In addition, if the glass is only available as thin film, not even the sound velocity, the slope of the dispersion curve at the low-frequency end, can be determined by the standard ultrasonic techniques.

We measured the phonon transmission through thin films and observed phonon interferences [1] similar as with a Fabry-Perot filter. From these interferences information about the phonon dispersion was obtained.

The experimental setup is shown as the inset in Fig. 1. A crystal of 4 mm thickness was used as the base material for a glass film of about 500 Å thickness. Monochromatic phonons [2] were generated with a PbBi tunnel junction [2] atop of the film. After transmission through the film and the base material, the phonons were detected by an Al tunnel junction. The phonon transmission through the glass film is maximum whenever the phonon frequency coincides with that of an eigenstate of the glass film. The wave vectors of the eigenstates are $k_n = (\pi/d^*)n + k_o$, where d^* is the phonon-path length in the film[1], n is an integer, and k_o is a phase shift which depends on the boundary conditions.

We studied several types of amorphous materials:

Vitreous silica: The films were prepared by thermally oxidizing a Si base crystal. This process yields a high-purity glass which is equivalent to fused silica. Transverse phonons only were observed due to phonon focusing if the phonon beam was directed along the [100] direction. This direction was normal to the crystal faces . Figure 1 shows the experimental results. The intensity of the transmitted phonons is plotted vs frequency. Above the detection threshold nine maxima and minima were observed which demonstrate the existence of the phonon interferences. A similar result was obtained for the longitudinal phonons. In this case generator and detector were placed along the [111] direction of the same sample. In this direction focusing favors the longitudinal mode. Again phonon interferences are visible (Fig. 2).

In Fig. 3 the frequencies of the maxima (o) and minima (•) are plotted vs their respective wave vectors. The zero of the abscissa is set arbitrary at the moment. From the slopes of the two experimental data sets we obtain

[1] Generally d^* is not equal to the thickness because of nonnormal incidence of the phonons and consequent refraction.

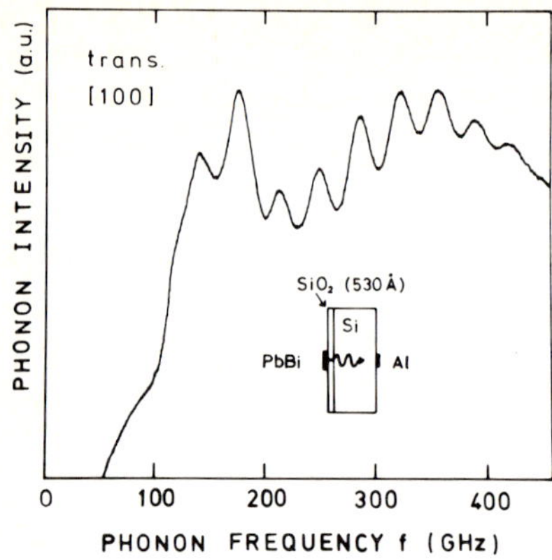

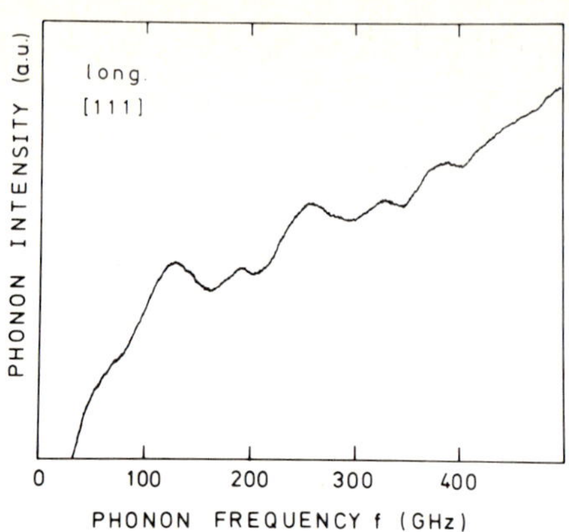

Fig 1. Intensity of transverse phonons transmitted through a film of vitreous silica. Inset: sample setup

Fig.2 Same as Fig.1 but for longitudinal phonons

the sound velocities $V_L = (5840 \pm 300)$ m/s and $V_T = (3700 \pm 120)$ m/s for the longitudinal and transverse phonons, respectively. These values agree very well with the ones obtained with ultrasonics, $V_L^{US} = 5850$ m/s and $V_T^{US} = 3750$ m/s. Thus we assume that there are no anomalies in the frequency range below 100 GHz where we have no data. Thus we can set the zero of the k scale by extrapolating to the origin of coordinates.

From the linearity of the phonon-dispersion curves in Fig. 3, we can draw conclusions about the thermal properties of glasses: the excessive T^3 term [3] which is ubiquitous in amorphous solids cannot be explained by softened plane-wave phonon modes [4] but must be due to localized states.

Besides this type of vitreous silica we also prepared SiO_2 films by electron-beam evaporation under high-vacuum (HV) and UHV conditions. The dispersion of the transverse phonons was again found to be linear but the sound velocities were much smaller. We found $V^{UHV}_T = 3330$ m/s and $V^{HV}_T = 2530$ m/s. This is a remarkable reduction of the sound velocity which was caused just by different preparation methods.

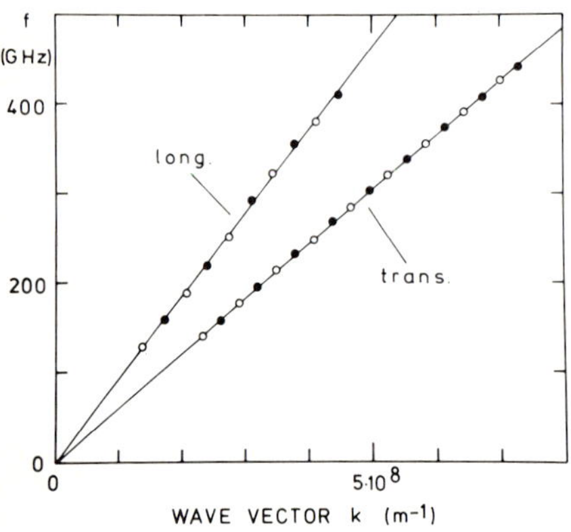

Fig. 3 Phonon-dispersion relation in vitreous silica for both phonon modes obtained from plotting the maxima (o) and minima (•) from Figs. 1 and 2 vs their respective k vectors

A reduction of the sound velocity was already found by KRAUSE [5] if the SiO_2 glass contained OH impurity ions . Extrapolating his results over two orders of magnitude we need to have 20 % OH in our HV films. From infrared absorption we estimate that we had an OH content of 1 to 5 %. The agreement is not unreasonable keeping the crudeness of the estimates in mind.

Amorphous Ge: We evaporated Ge films under UHV conditions with an electron beam gun. The base crystal was Al_2O_3 in this case. In this experiment phonon focusing allowed the observation of both longitudinal and transverse phonons at normal incidence in one run. The dispersion was for both modes linear up to 500 GHz and the sound velocities were $V_L=(4000\pm300)$m/s and $V_T=(2360\pm180)$m/s. These values are 19% (L) and 25 % (T) lower than the ones expected for polycrystalline Ge. For comparison KING et al. [6] measured the specific heat and found a 16 % reduction of the Debye temperature in their amorphous (sputtered) Ge films.

After taking these data we recrystallized the Ge film at 550° C and repeated the phonon measurements. The value of the longitudinal phonons was now $V_L=(5100\pm500)$ m/s. The interference patterns of the transverse phonons contained a superposition of two modes. Fourier transformation revealed $V_{T1}=(3450\pm500)$ m/s and $V_{T2}=(2550\pm200)$ m/s. We assume that the films recrystallized with the [110] direction normal to the surface. This kind of preferential crystallization had been observed before by other workers [7]. The sound-velocity values of bulk crystalline Ge in [110] direction are generally about 5 % larger, i.e. within our experimental errors.

In conclusion, we have demonstrated that the method of phonon interference is capable of determining the phonon dispersion relation in the frequency range of several times 100 GHz. The upper frequency limit of the method is set by: (i) the roughness of the films which in our experiments must have been better than 20 Å and (ii) the mean free path which must have been longer than three times the film thickness, i.e. longer than 1500 Å at 400 GHz.

References

1. M. Rothenfusser, W. Dietsche, and H. Kinder: Phys. Rev. B 27, 5196 (1983)
2. H. Kinder: Phys. Rev. Lett. 28, 1564 (1972)
3. R. C. Zeller and R. O Pohl: Phys. Rev. B4, 2029 (1971)
4. D. P. Jones, J. Jäckle, and W. A. Phillips: Phonon Scattering in Condensed Matter, ed by H. J. Maris (Plenum, N. Y., 1980), p. 49
5. J. T. Krause: J. Appl. Phys. 42, 3035 (1971)
6. C. N. King, W. A. Phillips, and J. P. deNeufville: Phys. Rev. Lett. 32, 538 (1974)
7. J. C. C. Fan, H. J. Zeiger, R. P. Gale, and R. L. Chapman: Appl. Phys. Lett. 36, 158 (1980)

The Influence of Two-Level States on the Thermal Conductivity of Amorphous Materials

D.E. Farrell*, J.E. de Oliveira, and H.M. Rosenberg

The Clarendon Laboratory, University of Qxford, Parks Road
Oxford OX1 3PU, England

1 Introduction

The thermal properties of almost all amorphous materials at low temperatures exhibit certain very general features. Below about 1 K the specific heat has a contribution linear in the temperature T in addition to the ordinary phonon (Debye)T^3 term, and the thermal conductivity in this region varies as T^n where n is usually about 1.8. At slightly higher temperatures, usually between 5 and 10 K, the thermal conductivity passes through a plateau-like region in which its value is almost constant, before it starts to increase again at higher temperatures.

The behaviour below 1 K has been explained [1,2] as being due to the presence of localised two-level states (TLS), which because the material is amorphous, cover a wide band of energies. As the temperature is raised these higher energy states are populated and if the assumption is made that the density of these states as a function of energy is constant, then it can be shown that this leads to the linear specific heat which is observed experimentally. The actual physical effect which gives rise to these TLS has not been established, but it is usually suggested that they might be caused by slight rearrangements of the atoms or molecules with respect to one another and which would have very slightly different energies. The thermal conductivity behaviour below 1 K is thought to be due to the scattering of phonons by these localised TLS and if this is so, then it can be shown that this would lead to a T^2 temperature dependence, which is roughly what is observed.

The present work has been done to try and shed more light on the nature of the TLS and in particular to see whether changes in the thermal conductivity are accompanied by corresponding variations in the specific heat. At present measurements have not been taken below 2 K and so the influence of the TLS is not completely dominant, as it is below 1 K, but nevertheless the effects due to them can still be observed at these slightly higher temperatures.

2 The Specimens

Measurements have been taken on a series of epoxy-resin samples made from Araldite MY 750 resin which has been hardened with a simple primary diamine hardener, ethylene diamine $H_2N(CH_2)_2NH_2$. During the curing process each of the hydrogen atoms in the NH_2 groups binds itself to the epoxy group which is present at either end of the resin molecule and so a very tight cross-linked array is formed. Specimens were made with the correct quantity of hardener to ensure exact stoichiometry, so that all the bonds and cross-links were made and then further samples were prepared with excess hardener - up to 4 times stoichiometry - and also with too little hardener - down to $\frac{1}{4}$ the

*Permanent address: Department of Physics, Case Western Reserve University Cleveland, OH 44106, USA

stoichiometric quantity. The samples were cast in the form of rods about 25 mm long and 8 mm in diameter.

3 Results

The thermal conductivity versus temperature curves are shown in Fig. 1. The most outstanding feature is that as the quantity of hardener is increased beyond the stoichiometric amount the thermal conductivity is considerably enhanced by about a factor of almost three. In addition a small maximum appears in what was the plateau region. If too little hardener is used the thermal conductivity is very much reduced.

The specific heat results are shown as a plot of C/T^3 versus T in Fig. 2. They show that as the amount of hardener is increased above the stoichiometric amount the specific heat decreases.

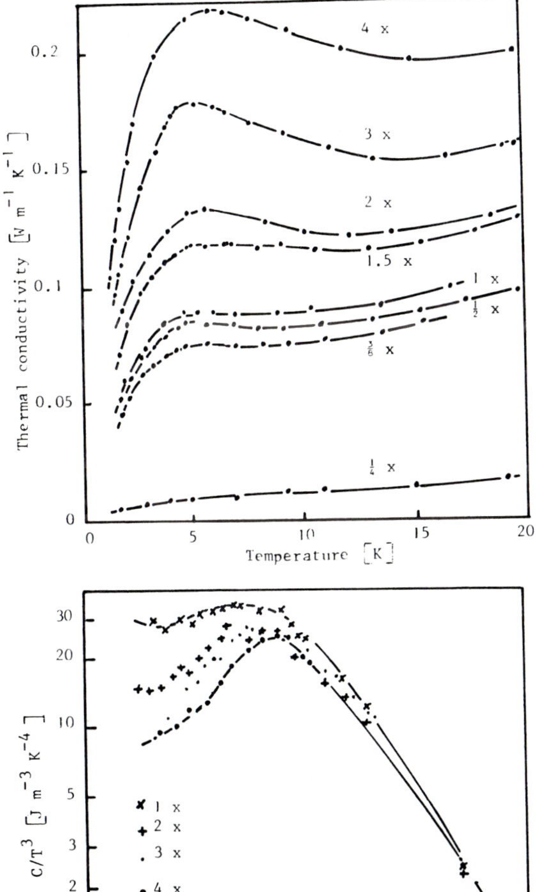

Fig. 1. The thermal conductivity of the epoxy samples as a function of hardener concentration

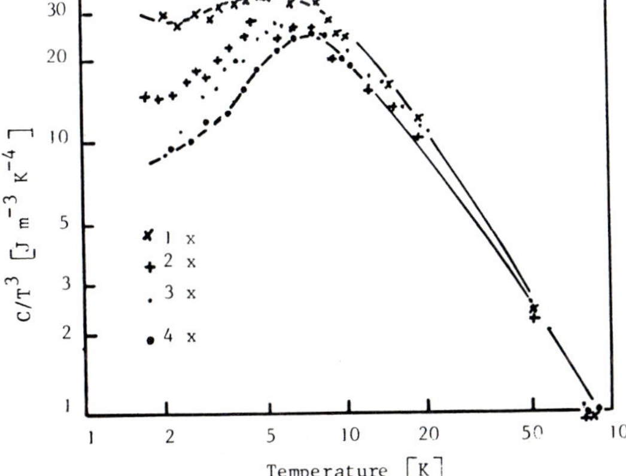

Fig. 2. The (specific heat)/T^3 of the epoxy samples as a function of hardener concentration

4 Discussion

The thermal conductivity κ can be discussed in terms of the kinetic theory formula $\kappa = Cvl/3$, where C is the specific heat per unit volume, v is the phonon velocity and l is the phonon mean free path. The substantial increase in the thermal conductivity which is observed for those specimens with more than the stoichiometric amount of hardener would imply an increase in either v or l, since C, as Fig. 2 shows, actually decreases. Whilst we have not yet measured the sound velocity of these samples we have measured it for a sample with double the stoichiometric quantity of another diamine hardener (1-12 diamino-dodecane - which also has a higher conductivity) and it is virtually unchanged. It is therefore unlikely that in the present specimens v changes very much as the amount of hardener is increased. The only other parameter which is left is the phonon mean free path, l. In order for this to increase the phonon scattering must be correspondingly reduced.

If we assume that the phonon scattering is dominated by interactions with the TLS then an increase in l would imply a decrease in the density of TLS and hence a decrease in the specific heat. This is just what we observe. It would appear that the presence of extra hardener molecules above the stoichiometric amount inhibits the motion of the cross-linked network and hence the density of TLS will decrease.

5 Conclusions

Although this work is still in progress it does seem that the correlation between an increase in the thermal conductivity and a corresponding decrease in the specific heat does suggest that the TLS are responsible for phonon scattering in these amorphous materials. For confirmation further experiments must still be done particularly in the region below 1 K where the influence of the TLS should be dominant.

6 Acknowledgements

J.E. de O. is in receipt of an award from the Conselho Nacional de Desenvolvimento Científico e Tecnólogico, Brasil.

7 References

1. P.W. Anderson, B.I. Halperin and C.M. Varma: Phil. Mag 25, 1 (1972)
2. W.A. Phillips: J. Low Temp. Physics 7, 351 (1972)
3. S. Kelham and H.M. Rosenberg: J. Phys. C. 14, 1737 (1981)

Low-Frequency Elastic Loss in Dielectric and Metallic Glasses at Low Temperature

H. Tietje[1], M.v. Schickfus[2], E. Gmelin[1], and H.-J. Güntherodt[3]

[1]Max-Planck-Institut für Festkörperforschung, Heisenbergstr. 1
D-7000 Stuttgart 80, Fed. Rep. of Germany

[2]Institut für Angewandte Physik II, Universität Heidelberg
D-6900 Heidelberg, Fed. Rep. of Germany

[3]Institut für Physik, Universität Basel
Basel, Schweiz

Since a few years it is a well established fact that the low-temperature properties of amorphous dielectrics and metals are largely determined by low-energy tunneling systems [1]. To agree with the observed anomalies in the thermal and acoustic properties, a particular distribution of the parameters of the tunneling centers has been assumed. The tunneling systems are characterized by the asymmetry Δ of the potential wells and by the tunnel splitting Δ_0 due to the overlap of the wavefunctions in the two wells. Their coupling to phonons or conduction electrons leads to a relaxation time $\tau = \tau(E,\Delta_0)$. The distribution of the parameters Δ and Δ_0 can be transformed into a distribution of the relaxation times $p(E,\tau^{-1}) = 1/2\bar{p}\tau^{-1}(1-\tau/\tau_{min})^{-1/2}$ [2], where $E=(\Delta^2+\Delta_0^2)^{1/2}$ is the energy of the tunneling center and τ_{min} the shortest relaxation time for a given energy reached at $\Delta_0=E$. This distribution diverges for $\tau \to \tau_{min}$ and $\tau \to \infty$, so that for a finite density of states $n(E) = \int\bar{p}(E,\tau^{-1})d\tau^{-1}$ some kind of a cutoff has to be introduced for large values of τ or, equivalently, for small values of Δ_0. Until now, however, this cutoff value of τ was not known. The time dependence of the specific heat [3] has shown only qualitatively that large values of τ must exist. Based on measurements of the specific heat it has been argued that $\Delta_{o,min} \approx 15$ mK [4].

We have quantitatively investigated the relaxation behaviour of glasses in the range of very long relaxation times by measuring the elastic absorption at **very low** frequencies (5×10^{-3} to 30 Hz). In this frequency range and at our measuring temperatures (.7 K < T < 6 K) the absorption is determined by the relaxation of the tunneling systems. For both dielectric and metallic glasses, the condition $\omega\tau_{min} \ll 1$ is valid. If the predictions of the tunneling model concerning the distribution $p(E,\tau^{-1})$ remain valid for such long relaxation times, one finds for the absorption

$$Q^{-1} = (\pi\bar{p}\gamma^2)/(2\rho v^2) \qquad (1)$$

where γ is the elastic deformation potential of the tunneling centers, v the velocity of the sound and ρ the mass density. It should be emphasized here that the absorption is dominated by tunneling systems with $0 < E < kT$ and $(\Delta_0/E)^2 < \omega\tau_{min}$ and that eq.1 is valid for both dielectric and metallic glasses.

In our experiment the absorption was measured through the elastocaloric effect. A sinusoidal stress $\sigma = \sigma_0\sin(\omega_S t)$ causes a temperature rise of the sample, either a rod of vitreous silica Suprasil I or a ribbon of the metallic glass $Pd_{0.775}Si_{0.165}Cu_{0.06}$. The samples were mounted in the vacuum chamber of a cryostat and clamped at both ends to copper blocks at constant temperature. Stress was applied from outside the cryostat through a bellows-sealed stainless steel wire. The temperature variation of the sample was measured with an exposed-element germanium resistor using a Wheatstone bridge and recorded with a desktop computer.

As a typical experimental result, Fig.1 shows the temperature variation of a-SiO_2 at 1,9 K and 0,007 Hz with σ_0 = 3 MPa. Because of our low experimental frequency, a periodic contribution of the relaxation process at 2 ω_S is still visible. An additional amount of heat at ω_S is caused by the anharmonicity of the sample [5]. The absolute value of $\dot{q}$, the heat generated by the absorption process, was determined by comparing the equilibrium temperature rise of the sample with that caused by an electrical heater.

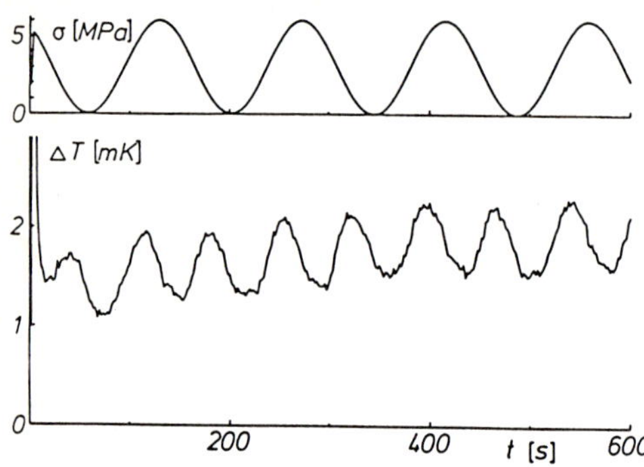

Fig.1: Temperature variation of a vitreous silica sample at 1,9 K. Stress is applied at t = 0 with a frequency of 7 mHz and an amplitude σ_0 = 3 MPa

In Fig.2 we show the frequency dependence of $\dot{q}$, the heat produced by the relaxation process, for a-SiO_2 and for PdSiCu at T = 3.6 K. From the linear frequency dependence of $\dot{q}$ it is obvious that the prediction $Q^{-1} \propto \omega^0$ of the tunneling model is very well obeyed since $\dot{q} = \omega\sigma_0^2 Q^{-1}/4Y$. Similar results have been obtained at all other temperatures. From the slope in Fig.2 and from the experimental parameters Q^{-1} can be determined. The temperature dependence of Q^{-1} is shown in Fig.3. For both dielectric and metallic glass, at most a very weak temperature dependence can be detected within the

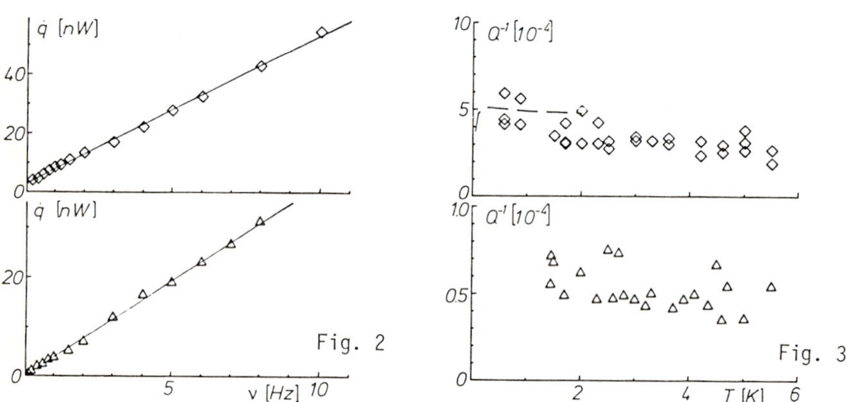

Fig.2: Frequency dependence of the heat produced at 3,6 K for a-SiO_2 (top) and for PdSiCu (bottom). The stress amplitudes were 0,45 and 23,3 MPa, respectively

Fig.3: Temperature dependence of the absorption Q^{-1} for a-SiO_2 (top) and for PdSiCu (bottom). The data are computed from q(ν) in the frequency range from 5 mHz to 30 Hz. The dashed line indicates the data of [6] taken at 1 kHz

accuracy of the data. Almost the same Q^{-1} has also been found in a vibrating reed experiment around 1 kHz in a silica-based glass [6]. According to eq.1 the coupling parameter $\bar{p}\gamma^2$ can be derived from the absorption data. We find $\bar{p}\gamma^2 = 1.9\times10^8$ erg/cm^3 for a-SiO_2 in fair agreement with measurements of the velocity of sound [7]. For PdSiCu we find $\bar{p}\gamma^2 = 7\times10^7$ erg/cm^3, again in good agreement with ultrasonic data [8].

Our results confirm the predictions of the tunneling model in a range of the distribution $p(E,\tau^{-1})$ which has not been investigated until now. If we assume an energy of a tunneling center of, say, 0.1 K, the tunnel splitting must extend below $\Delta_0 = 3\times10^{-5}$ K for SiO_2 and below 5×10^{-6} K for PdSiCu. Therefore a cutoff of the distribution $\bar{p}(E,\tau^{-1})$ must be well beyond the limits shown by our experiment.

References

1. For recent review articles see: "Amorphous Solids: Low-Temperature Properties"; W.A. Phillips, ed.; Springer Verlag, Heidelberg 1981.
2. J. Jäckle; Z. Physik 257, 121 (1972)
3. M.T. Loponen, R.C. Dynes, V. Narayanamurti, J.P. Garno; Phys. Rev.B25, 1161 (1982)
   M. Meissner, K. Spitzmann; Phys.Rev.Lett. 46, 265 (1981)
   J. Zimmermann, G. Weber; Phys.Lett. 86A, 32 (1981)
4. J.E. Lewis, J.C. Lasjaunias; J. de Physique C6 39, C6-965 (1978)
5. H. Tietje, M.v.Schickfus, E. Gmelin;J.de Physique C9, 43, C9-529 (1982)
6. A.K. Raychaudhuri, S. Hunklinger; J. de Physique C9, 43, C9-485 (1982)
7. S. Hunklinger, W. Arnold; in "Physical Acoustics", R.N. Thurston, W.P. Mason eds., Vol.12, p.155 (Academic Press, New York 1976)
8. G. Bellessa, O. Bethoux; Phys.Lett.62A, 125 (1977); the value given in this paper has to be increased by a factor of two, see G. Weiss, S. Hunklinger, H. v. Löhneysen; Physica 109 & 110B, 1946 (1982)

Phonon Absorption Due to Two-Level Systems in Metallic Glasses

N. Thomas

Physics Department, Birmingham University, P.O. Box 363
Birmingham B15 2TT, England

Phonons in metallic glasses at low temperatures are scattered by two-level systems (TLS) arising from atomic tunnelling states [1, 2]. The TLS themselves are heavily damped by inelastic scattering of conduction electrons [1,2], and phonon absorption by these TLS cannot realistically be treated using elementary perturbation theory. A more sophisticated theory has been developed [3] to describe resonant phonon scattering by heavily damped TLS and this is extended here to include relaxation absorption.

We treat this as a many-body problem and take as the unperturbed system a symmetrical TLS of energy splitting $2\Delta_0$ embedded in a sea of electrons and phonons; both the TLS-electron and TLS-phonon interactions are then approximately off-diagonal. The TLS asymmetry necessary for relaxation absorption is generated by an additional static off-diagonal perturbation, and the full interaction Hamiltonian for a TLS in a sample of volume Ω becomes

$$H_{int} = \sum_q M_q \eta_q \sigma_x + \frac{i}{\Omega} \sum_{kk'} W_{kk'} a^{\dagger}_{k'} a_k \sigma_x + \Delta\sigma_x \ . \qquad (1)$$

The first term here represents coupling to phonons of wave vector q with strain η_q and coupling constant M_q. The second term describes scattering of electrons from k to k', and the third term produces the TLS asymmetry. All these interactions are proportional to σ_x, so the imaginary-time spin-flip propagator, defined by

$$G(\tau) = -\theta_\tau \langle\sigma_x(\tau)\sigma_x(0)\rangle - \theta_{-\tau}\langle\sigma_x(0)\sigma_x(\tau)\rangle \ , \qquad (2)$$

plays a crucial role in the theory [3]. The final result for the phonon scattering rate (taking $\hbar = 1$) is

$$\frac{1}{\tau_{ph}} = 2\pi \left(\frac{M_q^2\omega}{2\rho_o \Omega v_s^2}\right) \rho(\omega)\tanh(\beta\omega/2) \ , \qquad (3)$$

where ρ_o is the density of the glass, v_s is the speed of sound, $\rho(\omega)$ is the TLS spectral density, and $\beta = 1/kT$. The spectral density is defined using the Lehmann representation for G as

$$\rho(\omega) = \frac{1}{Z}\sum_{ij} e^{-\beta E_i}|\langle\psi_i|\sigma_x|\psi_j\rangle|^2(1 + e^{-\beta\omega})\delta(\omega - \overline{E_j - E_i}) \ , \qquad (4)$$

where Z is the partition function of the interacting system, which has exact eigenstates $|\psi_i\rangle$ and $|\psi_j\rangle$ with energies E_i and E_j respectively.

Equation (3) has the same form as Fermi's Golden Rule, but with $\rho(\omega)$ playing the role of the density of final states. Note that $\rho(\omega)$ is defined in (4) using exact eigenstates, so the lifetime broadening of the TLS, however large, does not upset energy conservation. Furthermore, since it applies to asymmetrical TLS, (3) must describe both resonant and relaxation absorption of phonons.

For an isolated asymmetrical TLS, $\rho(\omega)$ is temperature independent and consists of a delta function at $\omega=0$ of weight $8\Delta^2/E^2$ and delta functions at $\omega=\pm E$ each of weight $4\Delta_o{}^2/E^2$, where $E=2\sqrt{(\Delta_o{}^2+\Delta^2)}$; the delta function at $\omega=0$ arises from the fact that $\langle\psi_i|\sigma_x|\psi_i\rangle$ is non-zero. A plausible form for $\rho(\omega)$ in the presence of interactions is shown in Fig. (1).

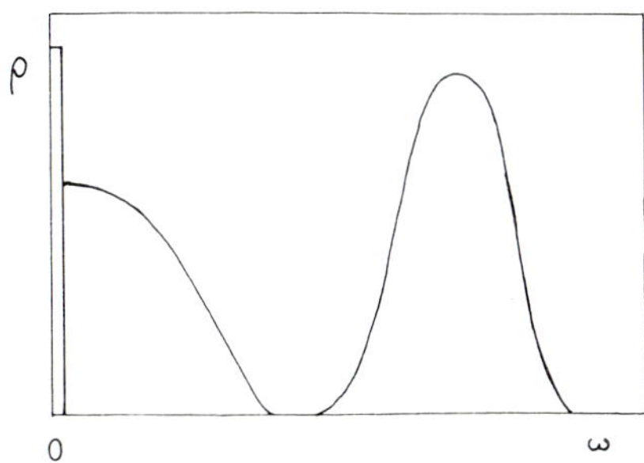

Figure (1): Spectral density for an asymmetrical TLS

The high-frequency delta function is broadened to form the TLS resonant-absorption peak. There is still a delta function at $\omega=0$, but this develops wings, which produce the relaxation absorption. However, at T=0 we find (for $\omega>0$)

$$\rho(\omega) = \sum_i |\langle\psi_o|\sigma_x|\psi_i\rangle|^2\delta(\omega-E_i) = N(\omega)|\langle\psi_o|\sigma_x|\psi_i\rangle|^2 , \qquad (5)$$

where $|\psi_o\rangle$ is the interacting ground state and $N(\omega)$ is the total density of states. In an insulating glass, $N(\omega)$ vanishes as $\omega\to 0$, and therefore the relaxation absorption vanishes at T=0; relaxation absorption at higher temperatures in an insulating glass arises from $\langle\psi_i|\sigma_x|\psi_j\rangle$ between excited states.

In a metallic glass, $N(\omega)$ does not vanish as $\omega\to 0$. If, as seems likely, $\langle\psi_o|\sigma_x|\psi_i\rangle$ is non-zero for the low-lying excited states, then the low-frequency absorption will remain even at T=0. This occurs because the strain changes the electric dipole moment of the asymmetrical TLS, but the conduction electrons cannot respond adiabatically to this change even at T=0. If we assume that the delta function in $\rho(\omega)$ carries negligible weight, then the relaxation absorption should be given by

$$\frac{1}{\tau_{rel}} = \frac{8\Delta^2 M_q{}^2}{E^2\rho_o\Omega v_s{}^2}\,\frac{\omega T_1}{(1+\omega^2 T_1{}^2)}\tanh(\beta\omega/2) , \qquad (6)$$

where T_1 is the TLS lifetime. This differs from the usual formula [1] in that there is no $\mathrm{sech}^2 \beta E/2$ factor, so the relaxation absorption does not vanish at T=0. This is clearly important for ultrasonics in metallic glasses, where $\omega T_1 < 1$ at T=0. Indeed, if the TLS is very heavily damped at T=0, the peaks in $\rho(\omega)$ may merge, and resonant and relaxation absorption can then no longer be separated.

1. B. Golding, J.E. Graebner, A.B. Kane and J.L. Black: Phys. Rev. Lett. 41, 1487 (1978)

2. J.L. Black: in *Glassy Metals I* (eds. H.J.Güntherodt and H. Beck, Springer 1981)

3. N. Thomas: Phil. Mag. B (in press)

Resonant Interaction of Acoustic Waves with Two-Level Systems in a Fluorozirconate Glass

R. Vacher[1], J. Pelous[1], M. Schmidt[2], P. Doussineau[3], A. Levelut[3]

1 INTRODUCTION

Amorphous materials have very peculiar thermal properties at low temperatures. For instance the specific heat contains, in addition to the Debye contribution ($C_D \sim T^3$), a supplementary part which varies linearly with the temperature T (1). The acoustical properties are also specific. For instance the low temperature sound velocity first increases as the logarithm of T, then passes through a maximum which is weakly frequency dependent and finally decreases rapidly with T (1).

These properties (and others) are explained by the two-level system (TLS) theory which postulates the existence of low energy excitations with a quasi constant density of states n extending up to kT_g (T_g is the glassy temperature). The constancy of n is fundamental for the predictions of a specific heat varying linearly with T and a sound velocity increasing as C $\ell n T$ through a resonant interaction between the TLS and the elastic wave. This interaction is also responsible for a resonant attenuation which varies at low temperatures as $C\,\omega^2 T^{-1}$. C is the same coefficient as above ; it is proportional to the constant density of states.

Recently, the specific heat **C** of a fluorozirconate glass (V-52 of composition $(ZrF_4)_{57.5}(BaF_2)_{33.75}(ThF_4)_{8.75}$) has been measured (2). The result is intriguing : **C** varies almost linearly with T from the lowest measured temperature up to 0.5 K where the linear term rapidly disappears, leaving for $T \gtrsim 1.5$ K only a T^3 contribution ; moreover this linear part is 20 times larger than in a classical glass such as silica. An hypothesis may explain this result : the TLS are concentrated in the low energy range ; the density of states is constant and 20 times larger than in silica but it has a cut-off at the energy equivalent temperature $T_{c-o} \simeq 1$ K. This hypothesis can be tested by ultrasonic experiments. In fact the resonant contribution to the sound velocity measured at a frequency ν corresponding to a temperature $T_\nu = h\nu/k$ small compared to the cut-off T_{c-o} must give a variation $\Delta v/v = C\,\ell n(T/T_0)$ for $T_\nu < T < T_{c-o}$. On the contrary, for a sound frequency in the cut-off region or higher, a different behaviour is expected. This velocity change $\Delta v/v$ would no longer vary as $\ell n T$ because of the rapid variation of n in this energy range. This test is precisely the aim of the present article. It extends to higher frequencies previous ultrasonic measurements on the same material (3).

---

[1]Laboratoire de Spectrométrie Rayleigh-Brillouin, E.R.A. C.N.R.S., U.S.T.L. place E. Bataillon, F-34060 Montpellier Cedex, France

[2]Institut F. Angew Physik II, Universität Heidelberg, Albert-Überle Str. 3-5 D-6900 Heidelberg, Fed. Rep. of Germany

[3]Laboratoire d'Ultrasons, E.R.A. C.N.R.S., Université Pierre et Marie Curie Tour 13, 4 place Jussieu, F-75230 Paris Cedex 05, France

2 EXPERIMENTAL RESULTS

The experiments were done with ultrasonic longitudinal waves in the range 100-500 MHz and by Brillouin scattering corresponding to 25 GHz longitudinal phonons. The samples used were cut from the same ingot (4). The frequencies measured in temperature units ($T_\nu = h\nu/k$) are typically 25 mK for 500 MHz and 1.2 K for 25 GHz. The results of the velocity measurements for three frequencies are given in Fig. 1 on a semi-log plot. In this plot the increase of the velocity is well represented by a straight line. Moreover, the slopes are the same within experimental errors, for the three frequencies. The common slope gives for coefficient C the value $C = (4.0 \pm 0.15) \cdot 10^{-4}$. As usual the temperature $T_M(\nu)$ at which the maximum occurs increases weakly with the frequency ν. With the Brillouin scattering technique the resonant attenuation is not saturated and can be measured. It corresponds in Fig. 2 to the increase of the mean free path when the temperature is lowered below 2 K. From this measurement, it is found that $C = (3.5 \times 0.7) \cdot 10^{-4}$. This value is in reasonable agreement with that deduced from the velocity measurements.

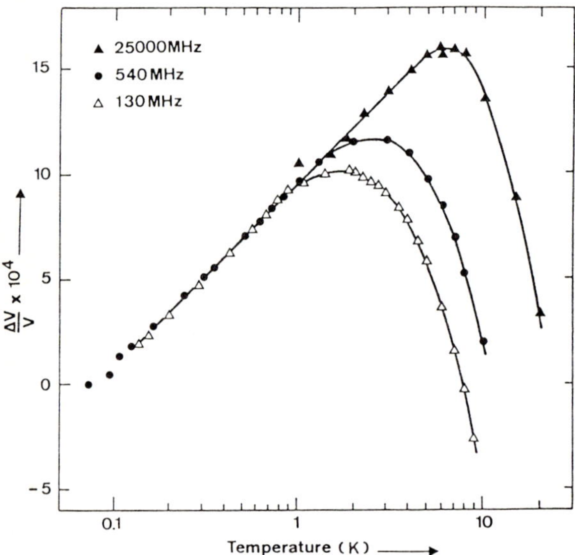

Fig. 1 - Relative velocity change of longitudinal acoustic waves in the glass $(ZrF_4)_{57.5}(BaF_2)_{33.75}(ThF_4)_{8.75}$ as a function of the temperature in a semi-log plot. The data at 130 MHz and 540 MHz have been obtained by an ultrasonic method and the data at 25000 MHz by Brillouin scattering. The straight line in the low temperature part represents the logarithmic variation while the curved solid lines are only guides to the eye

3 DISCUSSION

The velocity measurements presented here show that the density of states of the systems to which the elastic waves are coupled is constant at least up to the energy corresponding to the highest temperature where the variation is logarithmic. This energy equivalent temperature is about 6 K (or 75 GHz). The resonant attenuation measurement done at 25 GHz can be considered as a confirmation of the above results for the particular frequency $T_\nu = 1.2$ K.

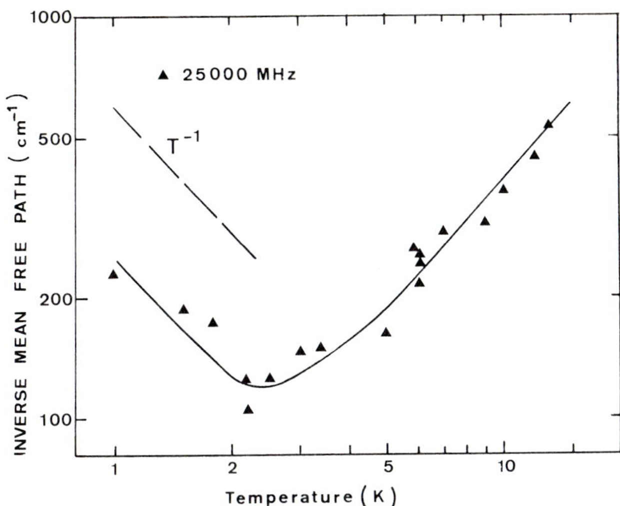

Fig. 2 - Inverse mean free path of 25000 MHz longitudinal acoustic phonons in the glass $(ZrF_4)_{57.5}(BaF_2)_{33.75}(ThF_4)_{8.75}$ as a function of the temperature. The solid line is a guide to the eye

It can therefore be inferred from our experiments that the density of states of the TLS of the glass V-52 is constant at least up to an energy T_ν = 6 K. Moreover, since the coefficient C is of the usual order of magnitude, it is likely that the density of states is of the same order of magnitude as in standard glasses. Therefore the strong anomaly observed at low temperatures in the specific heat is not due to the TLS. Its origin probably lies in extraneous degrees of freedom (impurities ?) not coupled to elastic waves.

REFERENCES

1 See review of papers in Amorphous Solids, Topics in Current Physics, ed. by W.A. Phillips (Springer, 1981), Vol. 24

2 J.C. Lasjaunias and M.A. Grosdemouge : J. Non Cryst. Solids 54, 183 (1983)

3 P. Doussineau and M. Matecki : J. Physique Lett. 42, L-267 (1981)

4 The V-52 glass was prepared and kindly provided by M. Matecki (Laboratoire de Chimie Minérale D, Rennes)

Heat Treatment Effects on the Phonon-Electron Contribution to Thermal Conduction in Amorphous $Zr_{70}Cu_{30}$

P. Esquinazi and F. de la Cruz

Centro Atómico Bariloche[1], Instituto Balseiro[2]
8400 Bariloche, Rio Negro, Argentina

The phonon conduction in amorphous metals is limited by disorder and electron scattering. The first attempt to separate both contributions has been made by H.v. Löhneysen et al.[1]. Assuming the validity of Matthiessen's rule

$$\kappa_g^{-1} = (\kappa_g^e)^{-1} + (\kappa_g^d)^{-1} \tag{1}$$

where κ_g^e and κ_g^d are the phonon thermal conduction limited by electron and disorder scattering respectively. By means of a strong magnetic field the superconductivity was suppressed and the thermal conductivity in the normal state was measured [1]. With these measurements and assuming (1) to be valid they obtained k_g^e below $T_c = 2.7$ K. The temperature dependence of κ_g^e was found to be linear, in agreement with Pippard model [2]. Using a similar procedure [3], we were able to determine κ_g^e in the normal and in the superconducting state. It was shown [3] that its temperature dependence below T_c was in excellent agreement with the BRT predictions [4]. This result establishes an independent check of the validity of expression (1). In Fig.1 we have plotted the results for the phonon conductivity of one sample of $Zr_{70}Cu_{30}$, as quenched. In the same figure it can be seen κ_g^d and κ_g^e obtained as was indicated previously.

Annealing [5] induces a systematic increase of the thermal conductivity. In Fig.1 we show κ_g for the same sample after 28 days of annealing at room temperature, curve B, and after 20 hours of annealing at 250°C, curve D. The thermal conduction represented by curve D increases in the whole temperature range and its thermal conductivity above T_c is of the order of κ_g^e from the as quenched sample. The increase in thermal conduction below 1K, where $\kappa_g \simeq \kappa_g^d$, was assumed to be due to a reduction in the density of the two-level systems (TLS). Any reasonable extrapolation of κ_g^d above T_c shows that to obtain the experimental κ_g it is necessary to increase the κ_g^e, obtained from the as quenched sample. Annealing does not introduce fundamental changes in the temperature dependence of the phonon conductivity. This shows that although there is a factor of 3 in the change of the conductivity, curve D, the change hardly affects the relative weight of the different scattering processes. We believe this is the fundamental point to be taken into account to understand the effect of annealing on the thermal conduction properties. As a consequence, this

[1] Comisión Nacional de Energía Atómica

[2] Comisión Nacional de Energía Atómica and Universidad Nacional de Cuyo

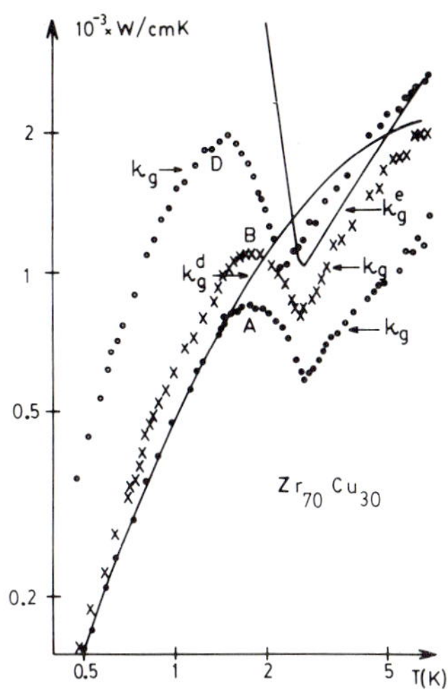

Figure 1: Phonon thermal conductivity as a function of temperature for an as quenched sample, curve A; after 28 days at room temperature, curve B, and with 20 hours at 250°C, curve D. Full lines represent κ_g^d and κ_g^e for curve A

result, that seems to be typical [6] of amorphous metals, can hardly be explained by changing only the density of states of TLS.

We propose a qualitative and simplified explanation based on a description of the thermal conduction process in amorphous insulators, made by Zaitlin and Anderson [7]. The thermal conductivity can be expressed by

$$\kappa_g = \frac{1}{3} \int_0^{\omega_D} C(\omega)\ v\ \ell(\omega)\ d\omega \tag{2}$$

where $C(\omega)$, v and $\ell(\omega)$ are the phonon specific heat, sound velocity and mean free path respectively and ω_D is the Debye frequency. If the mean free path is reduced to zero at a frequency ω_0, the effect would be equivalent to an introduction of a frequency cut-off in the integral of expression (2). In Fig. 2 we plotted the κ_g^d obatined from expression (2) with

$$\ell_g^d = \left(\frac{h\ \omega}{k\ A}\ \mathrm{tgh}\ \left(\frac{h\ \omega}{2kT}\right)\right)^{-1} \qquad \omega \leq \omega_0\ ; \qquad \ell_g^d = 0 \qquad \omega > \omega_0 \tag{3}$$

where the notation is that given in [7]. The non-resonant term has been suppressed for simplicity. To simulate the phonon-electron scattering we introduce a phonon mean free path [2] given by

$$\ell_g^e = 10^{20}\ \omega^{-2}\ \mathrm{cm} \tag{4}$$

where the absolute value is chosen to fit the order of magnitude of the experimental data.

Assuming the validity of Matthiessen's rule

$$\ell(\omega)^{-1} = \left(\ell_g^d(\omega)\right)^{-1} + \left(\ell_g^e(\omega)\right)^{-1} \tag{5}$$

with this $\ell(\omega)$, using expression (2) to (4) with A=1.2 10^{-3} cm K and $\omega_0=6.10^{11}$ seg^{-1}, we obtain curve A in Fig.2. Curve B has the same parameters but a decrease in the density of TLS, A=2,9 10^{-3} cm K. Curve C is obtained with the same parameter as B but with an increase in $\omega_0=9 \cdot 10^{11}$

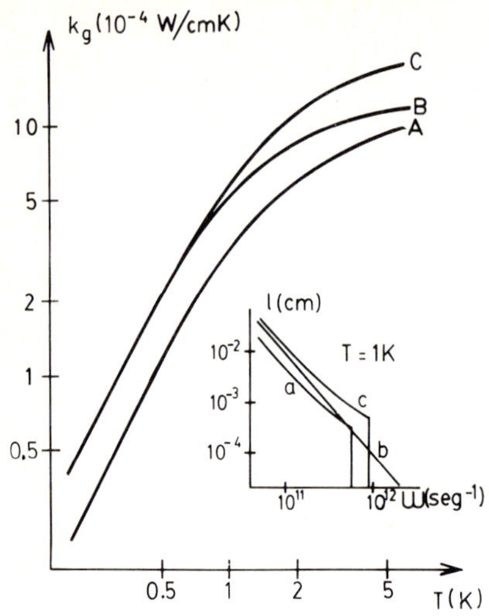

Figure 2: Phonon thermal conductivity using expression 2) to (5). Curve A and B differ in the density of TLS. Curve C is the same as B with a higher cut-off frequency. Insert shows the phonon m.f.p. as a function of frequency. Curves A and C are obtained from expression (3) using $\omega_0 = 6\times10^{11}$ seg^{-1} and $A = 1.2\times10^{-3}$ cm K for A and $\omega_0 = 9\times10^{11}$ seg^{-1} and $A = 2.9\times10^{-3}$ cm K for C, evaluated at T=1K. Curve B represents expression (4)

seg^{-1}. Notice that the introduction of ℓ_g^{c} makes curve B not parallel to curve A. This is not the tendency shown by the experimental curve D when compared to A in Fig.1, as discussed previously. We see that an increase in ω_0, curve C in Fig.2, simulates the experimental results increasing the thermal conductivity by approximately a constant factor in the whole temperature range. As a consequence we see that the increase at low temperature is mainly due to the decrease in the density of TLS, while the increase at high temperatures is due to the contribution of higher frequency phonons incorporated by the rise of ω_0. Within the assumptions of the model we can understand the behaviour of the experimental data shown in Fig.1. Curve B can be understood if it is assumed that at the beginning of the thermal heat treatment there is an increase of ω_0 at nearly constant density of TLS.

It was shown [3] that κ_g^d in the non-annealed sample was very well described by a theoretical dependence proposed in [7], there is no doubt that the theory provides enough parameters to fit, as well, the results of the relaxed samples. For this reason we make no quantitative fitting of the experimental data.

[1] H.v. Löhneysen, D.M. Herlach, E.F. Wassermann and K. Samwer, Solid State Commun. 39, 591 (1981).

[2] See, for example, A.B. Pippard, **The dynamics of Conduction Electrons** (Gordon and Breach, New York, 1965).

[3] P. Esquinazi and F. de la Cruz, Phys.Rev.B 27, 3069 (1983).

[4] J. Bardeen, G. Rickayzen and L. Tewordt, Phys.Rev. 113, 982 (1959).

[5] P. Esquinazi, M.E. de la Cruz, A. Ridner and F. de la Cruz, Solid State Commun. 44, 941 (1982).

[6] A. Ravex, J.C. Lasjaunias and O. Béthoux, submitted to J. Phys.F.
S. Grondey, H.v. Löneysen, H.J. Schink and K. Samwer, to be published in Z. Physik.
A. Ravex, J.C. Lasjaunias and O. Béthoux, Solid State Commun. 40, 853 (1981).
J.R. Matey and A.C. Anderson, J. Non-Cryst. Solids 23, 129 (1977).

[7] M.P. Zaitlin and A.C. Anderson, Phys.Rev.B 12, 4475 (1975).
M.P. Zaitlin and A.C. Anderson, Phys. Status Solidi B 71, 323 (1975).

Disorder-Induced Light Scattering in α-AgI

E. Cazzanelli, A. Fontana, and G. Mariotto

Dipartimento di Fisica, Università di Trento and Unità GNSM-CNR di Trento
I-38050 Povo, Italy

V. Mazzacurati, G. Ruocco, and G. Signorelli

Dipartimento di Fisica, Università la Sapienza and Unità GNSM-CNR di Roma
I-00185 Roma, Italy

Raman results [1,2] on α-AgI, in the frequency range 7÷165 cm^{-1}, show that both the integrated Raman intensity and the depolarization ratio $R=I_{ZX}/I_{ZZ}$ (which was also found to be constant versus frequency) are strongly temperature dependent, decreasing up to 450 °C. In order to explain this unusual behaviour we proposed [3] a model, in which all of the observed light scattering is explained by a simplified charge-induced dipole mechanism (C.I.D.). In such a way it is possible to relate the two measured quantities with the arrangements of the mobile cations around each iodine ion. The observed temperature dependences were theoretically accounted for by introducing a correlation in the silver ion distribution, which is destroyed at increasing temperature. The spectral intensity was tentatively attributed to the vibrational dynamics of iodine atoms, while the motion of the silver ions has been neglected as source of light scattering in the frequency range measured. In the aim to clarify the problem of the scattering contributions at lower frequency we have calculated the integrated intensity as well as the depolarization ratio both from iodine vibrational dynamics and from silver motions [4]. The polarizability term due to the silver dynamics shows calculated values of ${}^{Ag}R$ quite different from the ones related to the iodine term: ${}^{Ag}R(180\,°C)=0$ and ${}^{Ag}R(450\,°C)=1.5$ [4], while the iodine contribution leads to ${}^{I}R(180°C)=1.05$ and ${}^{I}R(450\,°C)=0.69$ [3]. Moreover, the measured depolarization ratio down to 2 cm^{-1} presents no frequency dependence [4]. This fact further supports the hypothesis that the light scattering contribution from the silver dynamics is confined below the frequency range we have explored, so all of the intensity above 2 cm^{-1} is primarily due to one-phonon density of states, Raman allowed by the disordered anisotropic configurations of the silver ions around the iodine.

In principle the Raman scattering intensity can be written as:

$$I_{\gamma\gamma'}(\omega) \propto \Sigma_{il,i'l'} \exp\{-i\vec{q}\cdot[\vec{x}_{il} - \vec{x}_{i'l'}]\} \cdot$$

$$\cdot\int dt \exp\{i\omega t\}\Sigma_{\alpha\beta} \langle {}^{\alpha}v_{\gamma\gamma'}(t)\, {}^{\beta}v^{i'l'}_{\gamma\gamma'}(o)\rangle\langle {}^{\alpha}u_{il}(t)\, {}^{\beta}u_{i'l'}(o)\rangle \tag{1}$$

where $\vec{q}$ is the exchanged momentum of the light and ${}^{\alpha}v^{il}_{\gamma\gamma'}=\partial P^{il}_{\gamma\gamma'}/\partial u^{\alpha}_{il}(\vec{u}_{il}=0)$ are the first-order polarizability derivatives. We define i as atom index, l the cell index and α,β,γ Cartesian axes. In the normal coordinate space, with the approximation $\vec{q}\cdot[\vec{x}_{il}-\vec{x}_{i'l'}]\simeq \vec{q}\cdot[\vec{x}_{l}-\vec{x}_{l'}]$, (1) becomes:

$$I_{\gamma\gamma'}(\omega) \propto \sum_{j,\vec{k}} \omega_j(\vec{k}) \Sigma_{ll'} \exp\{-i[\vec{q}-\vec{k}]\cdot[\vec{x}_l-\vec{x}_{l'}]\} \Sigma_{\alpha\beta} e^{\alpha}(\vec{k},j|i)\, e^{*\beta}(\vec{k},j|i')\cdot$$

$$\int d\omega' F^{\gamma\gamma'}_{\alpha\beta}[ii';ll';(\omega-\omega')]\cdot[n(\omega)+1]\cdot \mathrm{Im}\, D_{\vec{k},j}(\omega') \qquad (2)$$

where:

$$F^{\gamma\gamma'}_{\alpha\beta}(ii';ll';\omega) = \int dt\, e^{i\omega t} \langle v^{\alpha\, il}_{\gamma\gamma'}(t)\; v^{\beta\, i'l'}_{\gamma\gamma'}(o)\rangle =$$

$$= dt \int e^{i\omega t}\, \tilde{F}^{\gamma\gamma'}_{\alpha\beta}(ii',ll',t). \qquad (3)$$

In (2) j indicates the branch index, $\vec{k}$ is the phonon wavevector, $\omega_j(\vec{k})$ and $\vec{e}(\vec{k},j|i)$ are the eigenfrequency and the eigenvector respectively; $D_{\vec{k},j}(\omega)$ is the phonon propagator and finally $n(\omega)+1$ is the thermal population factor. Introducing a Gaussian correlation function between the local polarizability derivatives, (2) reduces to:

$$I_{\gamma\gamma'}(\omega) \propto J_{\gamma\gamma'}\cdot[n(\omega)+1]\Sigma_{\vec{k},j} \exp\{-|\vec{k}-\vec{q}|^2/2\overline{K}^2\}/\overline{K}^3\, \mathrm{Im}\, D_{\vec{k},j}(\omega)/\omega_j(\vec{k}). \qquad (4)$$

The wavevector $\overline{K}$ can be considered as the inverse "correlation length" for the polarizability derivatives and $J_{\gamma\gamma'} = \Sigma_{ili'l'} \Sigma_{\alpha\beta} F^{\gamma\gamma'}_{\alpha\beta}(ii',ll',\underset{\to}{t})$. (5) Taking into account the partial disorder of α-AgI we assume $0<\overline{K}<k_B$ (size of the Brillouin zone); a first rough evaluation [4] limits $\overline{K}$ in the range 0.2-0.3 A^{-1}. The dispersion curves [5] show that below 40 cm^{-1} two acoustical phonon branches can contribute to the light scattering: a longitudinal acoustical (LA) and a transverse acoustical (TA). So we can separate the spectral intensity:

$$I^{AC}_{\gamma\gamma'}(\omega) = I^{LA}_{\gamma\gamma'}(\omega) + I^{TA}_{\gamma\gamma'}(\omega) \qquad \text{with} \quad \omega \lesssim 40\ cm^{-1}. \qquad (6)$$

Because of the quite linear dispersion relation for the LA branch [5], $\omega_{LA}(\vec{k}) \simeq c_L|\vec{k}|$, for $|\vec{k}|<\overline{K}$, so we can choose for that branch an harmonic propagator:

$$D_{\vec{k},LA}(\omega) = \lim_{\varepsilon\to o^+} 2\,\omega_{LA}(\vec{k})/[\omega^2-\omega^2_{LA}(\vec{k})+2\, i\omega\varepsilon]. \qquad (7)$$

The associated intensity becomes:

$$I^{LA}_{\gamma\gamma'}(\omega) \propto G^{LA} J_{\gamma\gamma'}\cdot[n(\omega)+1]\omega/\overline{K}^3 \exp\{-\omega^2/2c_L^2\overline{K}^2\} \qquad (8)$$

where $G^{LA} = \Omega/\pi^2 c_L^3$ and Ω is the volume of the sample. For the TA branch, which appears strongly anharmonic [5] we can choose an overdamped propagator:

$$D_{\vec{k},TA} = 2\omega_{TA}(\vec{k})\, D_{TA}/[\omega^2-\omega^2_{TA}(\vec{k}) + i\omega\Gamma_{TA}(\vec{k})] \qquad (9)$$

with D_{TA} representing an adimensional oscillator strength. The intensity related to TA can be explicitly calculated:

$$I^{TA}_{\gamma\gamma'}(\omega) \propto G^{TA} J_{\gamma\gamma'}\cdot[n(\omega)+1]\ \omega\, F(\omega,\overline{K}) \qquad (10)$$

where $G^{TA} = 2\omega D_{TA}/\pi^2 c_T^3$ and:

$$F(\omega,\overline{K}) = c_T^3 \int_o^{k_B} k^2\, dk\, \exp\{-k^2/2\overline{K}^2\}\Gamma_{TA}(\vec{k})/\{[\omega^2-\omega^2_{TA}(\vec{k})]^2+\omega^2\Gamma^2_{TA}(\vec{k})\}. \qquad (11)$$

The whole acoustical spectral density becomes:

$$I^{AC}_{\gamma\gamma'}(\omega)/[n(\omega)+1] \propto J_{\gamma\gamma'}\,\omega\{\exp[-\omega^2/2C_L^2\overline{K}^2] + G\cdot F(\omega,\overline{K})\}\ . \qquad (12)$$

This expression can be fitted to the experimental reduced intensity for $\omega \leq 40\ cm^{-1}$, using only two parameters: $G = 2(C_L/C_T)^3 D_{TA}$, completely free, and $\overline{K}$ limited in the range 0.2 - 0.3 A^{-1}. In Fig. 1 we compare the calculated and experimental spectral shapes. The agreement is excellent in the range 2-40 cm^{-1}, using the same value of G for the spectra at different temperatures, while $\overline{K}$ slightly increases versus T (see Fig. 2). This behaviour is quite reasonable, considering the physical meaning of $\overline{K}$ as inverse correlation length.

The equations we have used to fit the intensities predict also a linear correspondence between the parameter $\overline{K}(T)$ and the ratio $\chi(T)$ of the integrated "optical" intensity over the acoustical one. As "optical" intensity we assume the difference between the experimental integrated intensity up to 250 cm^{-1} and the acoustical one, calculated by means of (12). Such optical spectral density is found substantially temperature independent, confirming the realizability of its definition. If we compare the relative increase of $\chi(T)$ in Fig. 3 with that of $\overline{K}(T)$ in Fig. 2 we can verify a very striking correspondence between the plots.

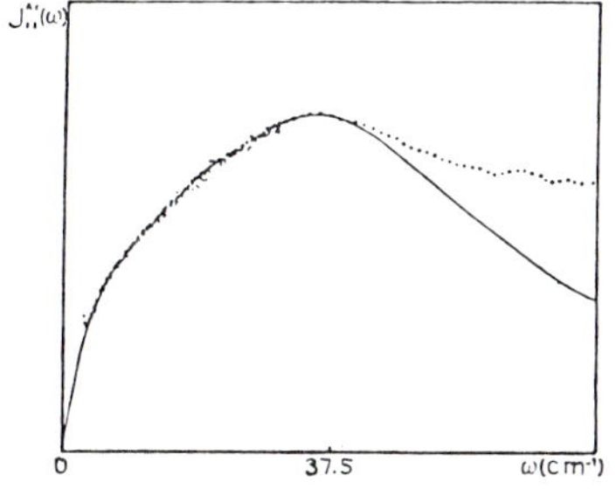

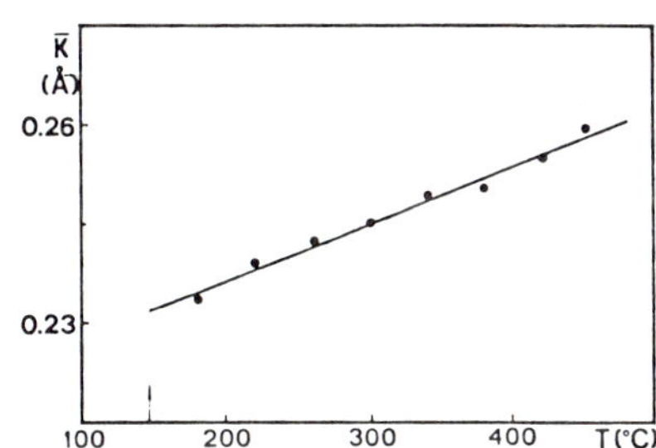

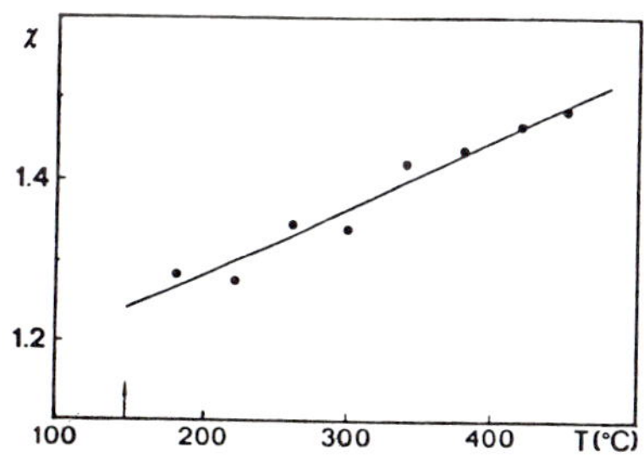

Fig.1. Acoustical spetral density of α-AgI obtained by (12) (full line) together with reduced experimental data (dots)

Fig.2. Plot of K: dots are values obtained by fit of experimental data, full line represents the linear best fit

Fig.3. Plot of ratio between optical intensity over the acoustical one in α-AgI vs T(^{o}C) (dots) together with linear best fit

[1] A. Fontana, G. Mariotto and M.P. Fontana: Phys. Rev. B 21, 1102 (1980)
[2] G. Mariotto, A. Fontana, E. Cazzanelli, F. Rocca, M.P. Fontana, V. Mazzacurati and G. Signorelli: Phys. Rev. B 23, 4782 (1981)
[3] V. Mazzacurati, G. Ruocco, G. Signorelli, E. Cazzanelli, A. Fontana and G. Mariotto: Phys. Rev. B 26, 2216 (1982)
[4] E. Cazzanelli, A. Fontana, G. Mariotto, V. Mazzacurati, G. Ruocco and G. Signorelli: Phys. Rev. B (in press)
[5] P. Brüesch, W. Bührer and H.J.M. Smeets: Phys. Rev. B 22, 970 (1980)

Discussions

LOW-ENERGY EXCITATIONS IN DISORDERED SOLIDS: NEW ASPECTS page 378
S. Hunklinger

A. C. Anderson: You are correct in saying that neutron irradiation of crystalline SiO_2 appears to produce small islands of amorphous material containing TLS. However, a more interesting result is that, following an anneal which returns the melt density to that of a crystal, the TLS appear to remain.

H. M. Rosenberg: Is there any connection between your model of fast and slow processes and the excess T^3 specific heat term?

S. Hunklinger: No, the original tunneling model does not take into account the excess T^3 term in the specific heat. According to our low frequency experiments, the density of states of the two-level system is constant up to at least 10 K. In my opinion the origin of this term is not understood at all.

ANOMALOUS LOW TEMPERATURE ULTRASONIC BEHAVIOUR IN A FLUORIDE GLASS CONTAINING Mn page 386
P. Doussineau, A. Levelut, M. Matecki, W. D. Wallace

B. Golding: I observed that your glass contains relatively high concentrations of Th. This undergoes radioactive decay and may be dangerous.

W. K. Arnold: Is the logarithmic temperature dependence of the sound velocity due to intrinsic TLS or possibly due to magnetic impurities?

A. Levelut: It is due to TLS as shown by experiments done in an applied magnetic field.

R. Pohl: What was the reason for your choice of this glass,in particular its thorium content?

A. Levelut: This glass was the only disordered material containing a high concentration of magnetic atoms and available to us.

LOW-ENERGY EXCITATIONS IN $(KBr)_{1-x}(KCN)_x$ IN THE ORIENTATIONAL GLASS STATE page 392
M Meissner, J. J. De Yoreo, R. O. Pohl, S. Susman

B. Golding: Do you have any idea whether elastic or electric interactions are dominant? If electric dipolar or quadrupolar interactions are dominant the thermal conductivity, which is sensitive only to elastic interactions, may not be the most suitable.

M. Meissner: I agree that elastic interaction is the more attractive process in understanding our thermal conductivity data. On the other hand, if dipolar or quadrupolar electric interaction contributes to larger relaxation times, they would show up in the short-time specific heat measurements, which might be a way to test importance of electric CN- interaction.

H. Jex: The CN- dumbell is known to have a remarkable quadrupole moment which may be important for the glass state. You did not mention this point. Is it not important in your experiment?

M. Meissner: There is much discussion of the question whether the dipole or quadrupole moment is more important for the formation process of the orientational disorder. Recent measurements with dielectric relaxation, ultrasonic techniques and neutron scattering show the existence of relaxation times over many orders of magnitude between ~10K and ~100K. But at least below 5K there is no indication for relaxation processes in the 25% sample, so that a frozen in orientational disorder is rather likely. Thus from our temperature thermal data we can make no conclusive contribution to the question whether the quadrupole moment is important for the glass formation.

S. Hunklinger: Can you estimate the coupling coefficient from your experimental data?

M. Meissner: From the κ and c_p data for the $(KBr)_{0.75}(KCN)_{0.25}$ sample a value for the coupling coefficient $\gamma \approx 0.5$ eV could be estimated. This value is similar to that for PMMA.

A. C. Anderson: From similar measurements we also concluded that (i) individual CN were not the source of TLS and (ii) there exists a broad spectrum of relaxation time extending to long times.

M. Meissner: We were interested in short-time c_p measurements on μsec time scales and we used sample-to-bath time constants of about 50 msec. Thus any time dependence beyond 50 msec was not observable in this experiment. But as we did c_p measurements on 50 sec time scales, too, we found agreement between the 50 msec and 50 sec data within the overall experimental error of both methods. This uncertainty of about 20% may allow for a time dependence on longer time scales. I agree that more careful studies are needed on sec time scale.

G. P. Singh: In the rather concentrated glassy samples could one not imagine that there are small ferroelastically correlated regions? In such cases TLS could arise as the molecules on the boundaries between such regions can move from one correlated region to the other by small changes in energy or can you completely rule out any such mechanism?

M. Meissner: There has been no evidence for any such ferroelectrically correlated regions.

SPECTRAL HOLE BURNING OF DYES, PROBING PHONON PROCESSES AT SURFACES AND IN AMORPHOUS SYSTEMS page 395
U. Bogner, G. Röska

B. Golding: Is it possible that optical excitation may modify the structure in the vicinity of the absorbing site and create a new double-well potential?

U. Bogner: Only in the case of optical exitation by a high power laser (e.g. Q-switched Nd-YAG Laser) with a power density near to the order of the damage threshold of the surface, we obtained results indicating a modification of the structure in the vicinity of the absorbing site.

W. K. Arnold: Do the tunneling sites in this type of material extend to an energy as high as 50 K (≃ 1 THz)?

U. Bogner: Our results, in particular the results concerning the phonon memory, suggest that at higher temperatures phonon processes are dominant in which the barriers of the double-well potentials are crossed. Obviously the largest barrier heights in the Langmuir-Blodgett films correspond to the energy of phonons with frequencies up to 1 THz ; in the case of dye molecules adsorbed at the surface of sapphire single crystals there must be even substantially higher barriers.

L. J. Challis: What is the spatial resolution of the system if used as a phonon detector? Could it be used for phonon imaging?

U. Bogner: The spatial resolution of the system is of the order of 1μ. The limitation is given by the smallest diameter of the focus of the He-Cd laser beam. It could be used for phonon imaging in particular in the range of very high frequencies and with a time resolution better than 20 ns.

EVIDENCE OF TWO-LEVEL SYSTEMS IN ELECTROLYTE GLASS BY BRILLOUIN SCATTERING page 398
J. Pelous, R. Vacher, A. Essabouri, U. Reichert, M. Schmidt

A. C. Anderson: The Cohen model suggests a density of TLS $\bar{P} \propto V_F/T_g$ where V_F is the free volume. Thus, to test the model, one must control both V_F and T_g.

J. Pelous: We have planned to study lower and higher concentration, so we will confirm our results.

T. Ishiguro: You measured the Brillouin scattering down to 1K. How did you reduce the heating effect by laser light? Or how did you evaluate the sample temperature?

J. Pelous: In this temperature range (below 5K) the measurements of the ratio of the Stokes/anti-Stokes intensity give an indication on the actual temperature inside the scattering volume.

W. Rehwald: How is the material prepared?

J. Pelous: The liquid mixture is prepared at room temperature in a silicon cell. The transparent glass is obtained simply by cooling down the cell into the cryostat. Crystallization does not take place for the two compositions studied.

ACOUSTIC ABSORPTION DUE TO HYDROGEN TUNNELING IN $NbN_{0.0015}H_{0.0025}$ page 401
J.L. Wang, G. Weiss, H. Wipf, A. Magerl

S. Hunklinger: Is there additional evidence for the presence of the Orbach level?

G. Weiss: No. Not to my knowlege.

M. Wagner: I have a simple question which refers to the geometry of your tunneling system. Do you assume that each defect (e.g.0) bears more than one or only one single potential well for the tunneling particle?

G. Weiss: In fact, one should try to take into account the geometry of the interstitial sites near the trapping atom. This means the potential has more than two wells, producing more than two levels. But, first of all, you have to make some assumptions about the sites, accessible for the hydrogen atom. In this first approach, I assumed only a double-well potential, i.e., a simple two-level system per OH center.

W. K. Arnold: The H impurities have apparently one energy splitting E. Is there also a distribution of relaxation times associated with it (because it seems to me that the absorption curves versus temperature are rather broad)?

G. Weiss: Assuming only one energy splitting E = 4K or 5K without distribution in relaxation times, I find theoretical absorption peaks which are probably 30% to 40% narrower than the experimental ones. The shift of the peak and its change in height with frequency, however, is very nicely reproduced by this simple assumption.

W. K. Arnold: Is there any connection between the result presented here on hydrogen tunneling and the tunneling systems in amorphous metals?

G. Weiss: From our experiment we can evaluate a deformation potential of 0.1 eV for the hydrogen tunneling system in NB, which is about 10 times weaker than for the tunneling systems present in amorphous metals. On the other hand, we had a concentration of OH centers of about 10^{20} cm^{-3}. This means that even if you have a few ppm hydrogen in your amorphous metal sample (which you probably can never avoid), this could never account for the strong scattering of acoustic and thermal phonons.

Part X

Phonon Echoes

Chairman: **J.K. Wigmore**

Rotary Phonon Echoes in Silica Glass

B. Golding, D.L. Fox, and W.H. Haemmerle

Bell Laboratories
Murray Hill, N.J. 07974, USA

Acoustic resonance experiments have played a central role in developing a quantitative picture of the low-temperature properties of glasses [1,2]. The strong coupling between phonons and two-level tunneling systems, the two intrinsic low energy excitations of disordered insulators, has stimulated the development of novel acoustic propagation methods. Among these, coherent phonon echo studies [2,3] have provided an abundance of detailed information on phonon - two-level system couplings and on relaxation times. At temperatures below 0.1 K, where coherence persists for relatively long times ($\gtrsim 10^{-6}$ sec), two and three pulse phonon echoes, the acoustic analogs of magnetic spin echoes, have been used to study tunneling systems in silica glass [2,3], and neutron-irradiated quartz [4].

We discuss here the observation of a different, yet equally useful, type of coherent echo: the rotary phonon echo. It also has analogs in spin [5] and optical resonance [6]. The rotary phonon echo allows the direct visualization of the nutation rate, or Rabi frequency, for acoustic resonance in glass. Its usefulness in studying disordered systems arises from its insensitivity to inhomogeneities in driving fields or distributions of "spin-phonon" matrix elements. It is the latter intrinsic property of glass which leads, for example, to a time-dependent specific heat. The present study in pure silica glass (Suprasil W) has yielded: (1) direct measurement of the Rabi frequency for acoustic two-level system resonance, (2) determination of the maximum spin-phonon coupling, and (3) dephasing times T_2 for tunneling systems to temperatures as high as 300 mK.

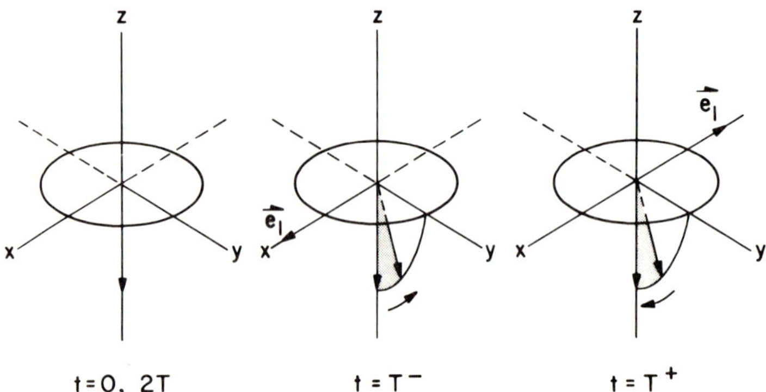

Fig. 1. Schematic of state vector rotations during rotary echo sequence. See text for description. Note that all vectors rephase at t = 2T regardless of their coupling to the resonant field

The principles of rotary echo formation are shown in Fig. 1. Consider initially a narrow energy distribution of two-level systems of energy splitting $E \simeq 1$ GHz at 5 mK in zero external field so that they are condensed into the ground state. The Hamiltonian is $H = \frac{1}{2} E\sigma_z + \overleftrightarrow{M} \cdot \vec{e}\sigma_x$, where $\overleftrightarrow{M}$ is the deformation potential coupling tensor, $\vec{e} = \vec{e}_0 \cos(\Omega t + Kz - \phi(z))$ is the longitudinal acoustic strain and σ_i are Pauli operators for $s = \frac{1}{2}$. At $t = 0$ the states at $z = 0$ are exposed to the coherent acoustic field and at $t = T$ have nutated through an angle $\theta = (M/\hbar) e_0 (z = 0)T$. At this time a $\phi = \pi$ phase shift is imposed causing the states to nutate in the opposite sense with rephasing at $t = 2T$. Note that the rephasing at 2T is completely unaffected by a distribution in M. At 2T the system is in its ground state; maximum radiation is detected at $2T \pm \dfrac{\pi}{2\chi_0}$ when the state vector is directed along $\pm\hat{y}$, where χ_0 is the Rabi frequency, $\chi_0 = (M/\hbar)e_0$. The influence of inhomogeneous broadening is not eliminated, however, and its role will be discussed below.

Fig. 2 shows a typical experimental result at a temperature of 5 mK. A coherent acoustic wave at 0.94 GHz is introduced into the SiO_2 sample at $t = 0$. After 0.5 μs, the phase is shifted by π radians for 1.0 μs. After time $\tau = 2L/v$ where $L = 0.635$ cm, the sample length, and v is the speed of longitudinal sound (5.8×10^5 cm/sec), a reflection is detected at the transducer face. The π phase shift at $\tau + T$ is seen as a change in sign of the detected signal, since a phase-sensitive receiver is used. The echo appears at $\tau + 2T$ as a modulation in the transmitted signal with frequency $\approx\chi_0$. At 2τ a second acoustic reflection is observed and a second rotary echo at $2\tau + 2T$. The second echo is somewhat reduced and its period lengthened as a result of attenuation and not coherency decay. At 5 mK $\alpha_0(2L) = 7.7$, where α_0 is the small-amplitude absorption coefficient of SiO_2, so that $e_0(z)$ is non-uniform and the echo period results from a spatial average over 2L.

For a thin sample the transient strain is given by Bloch-like equations coupled to the elastic wave equation. Decay of the rotary echo is governed by $\Gamma = \frac{1}{2}(1/T_1 + 1/T_2)$. In order to calculate the echo response for a glass the effect of the tunneling distribution

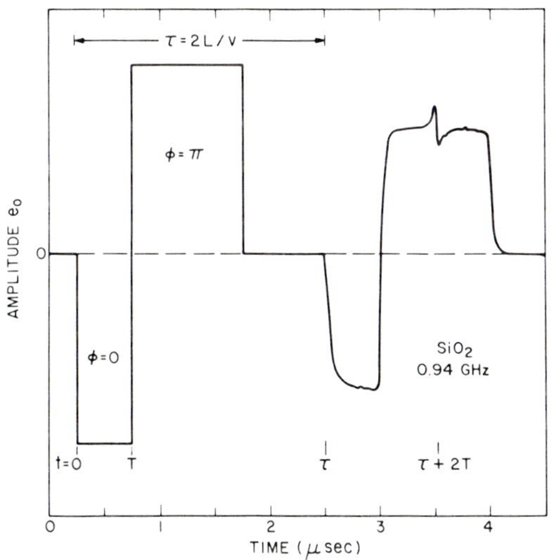

Fig. 2. Experimental observation of the rotary phonon echo in SiO_2 glass at 0.94 GHz and 5 mK

must be introduced. This leads to a forward propagating rotary echo strain amplitude

$$\tilde{e}(L,t) \sim e^{i\phi} w_o \overline{P} M_o e^{-\Gamma t} \sin^2(\phi/2) \chi_0 \int_{r_{min}}^{r} dr \int_{1}^{\infty} dy \frac{r \sin[\chi_0 y(t-2T)]}{(1-r)^{1/2} y^2 (y^2-r)^{1/2}}$$

where: ϕ is the phase shift
w_o is the inversion
$y = \beta/\chi_0$, where $\beta = (\chi^2 + \Delta^2)^{1/2}$
$P(r,E) = \overline{P}(E)/2r(1-r)^{1/2}$ with $r^2 = \Delta_o/E$
Δ_o is the tunneling matrix element
M_o is the maximum tunneling system-phonon coupling
P is the tunneling system density of states, and
Γ is the damping factor.

The integral has been evaluated numerically [7] and shows a damped sinusoid with a period between half-cycles at 2T which is close to 0.8 π/χ_o. The tunneling distribution has negligible effect on the nutational signal at times close to 2T.

The maximum coupling M_o can be determined from the period of the nutation signal shown in Fig. 2 if the absolute value of the strain amplitude is known. Since there is considerable uncertainty in measuring this quantity, it is more reasonable to compare the matrix element obtained by rotary echoes with the matrix element obtained from two-pulse (free-induction decay) phonon echoes. This can be done in situ and at the same frequency by changing the excitation sequence.

It is easily shown [7] that the matrix element ratio is

$$\frac{M_r}{M_{2p}} = \frac{\pi}{\theta^{max}} \frac{\Delta t}{P_{1/2}} \frac{e_{2p}^{max}}{e_r}$$

where subscripts r and 2p refer to rotary and two-pulse respectively, θ^{max} is the pulse area for maximum two-pulse echo amplitude, Δt is the two-pulse duration, and the Rabi period is $P_{1/2} \cong (2.2/\pi)P_{1/2}$ (meas) using the results of the calculation described above. Of course this ratio should be unity. It is found that $M_r/M_{2p} = 1.3$, which is in good agreement and within experimental uncertainties.

Above 100 mK, $T_2'/T_1 << 1$ so the rotary echo decay yields a direct determination of T_2'. Comparison of T_2' obtained from rotary echoes and earlier phonon echoes [3] indicates a continuation of the $T_2' \sim 1/T^2$ temperature dependence found earlier [3]. This reinforces the suggestion that the dephasing process occurs as a result of short-time spectral diffusion arising from interacting pairs of tunneling systems [8].

References

1. S. Hunklinger and M. v. Schickfus: *Amorphous Solids: Low Temperature Properties* edited by W. A. Phillips (Springer-Verlag, Berlin, 1981) p. 81.
2. B. Golding and J. E. Graebner: ibid., p. 107.
3. B. Golding and J. E. Graebner: Phys. Rev. Lett. *37*, 852 (1976), and J. E. Graebner and B. Golding: Phys. Rev. *B*19, 964 (1979).
4. B. Golding and J. E. Graebner: *Phonon Scattering in Condensed Matter*, edited by H. J. Maris (Plenum, NY, 1980), p. 11.
5. I. Solomon: Phys. Rev. Lett. *2*, 301 (1959).
6. N. C. Wong, S. S. Kano, and R. G. Brewer: Phys. Rev. *A*21, 260 (1980).
7. B. Golding and D. L. Fox: (to be published).
8. J. L. Black and B. I. Halperin: Phys. Rev. *B* 16, 2879 (1977).

Pseudospin Echoes in Borate Glasses

M. Devaud, J.-Y. Prieur, and W.D. Wallace[1]

Laboratoire d'Ultrasons[2], Université Pierre et Marie Curie, Tour 13
4 place Jussieu, F-75230 Paris Cedex 05, France

1 INTRODUCTION

The amorphous properties of a family of ionic conducting lithium borate glasses (B_2O_3 - x Li_2O - y LiCl) have been measured recently [1]. From desaturation experiments an evaluation of the transverse relaxation time T_2 (10^{-7} sec) of the pseudospins was possible when their minimum longitudinal relaxation time T_{1m} was in the range of 10^{-5} sec. Thus it seemed worthwhile to try to observe what we will call "pseudospin echoes". They are of the same type as the so-called phonon echoes observed by GRAEBNER and GOLDING [2] in fused silica glass or as the "dipole echoes" observed in γ-irradiated quartz by SAINT-PAUL and JOFFRIN [3].

2 EXPERIMENTS

The experiments were done in the temperature range 0.07 K < T < 0.6 K and at a frequency of about 500 MHz. Several borate glasses have been studied with various values of x. The samples had two opposite flat and parallel faces, on one of which a tranducer was glued. The sample was mounted inside a coaxial type cavity. A hole in the bottom of the cavity permitted the excitation of the transducer. With this arrangement we could excite a sequence of either two acoustic pulses or two electromagnetic pulses or one acoustic pulse followed by one electromagnetic pulse.

The well-known wave vector selection rule for spontaneous echoes is $2\vec{k}_2 - \vec{k}_1 = \vec{k}_e$, where $\vec{k}_1$ and $\vec{k}_2$ are the wave vectors of the first and second exciting waves and $\vec{k}_e$ is that of the echo. With two acoustic waves for $\vec{k}_1$ and $\vec{k}_2$ the echo is a forward acoustic wave. With two electromagnetic waves, the echo is an electromagnetic wave. With one acoustic and one electromagnetic wave, the echo is a backward acoustic wave. We observed both the forward and the backward acoustic echoes. This indicates that the pseudospins have both an elastic and electric dipole moment. Nevertheless we failed to observe the electromagnetic echo.

The results presented here are from measurements with transverse waves on the forward acoustic echo in the glass B_2O_3 - *0.5* Li_2O - *0.7* LiCl. The power level (P_ℓ) of this echo is given by the formula [2] :

$$P_\ell = \frac{P_0[nh\tanh(\hbar\omega/2kT)V]^2 \sin^2\theta_1 \sin^4(\theta_2/2)\exp(-4\tau/T_2)}{[\mathrm{Max}(\Delta t_1,\Delta t_2)]^2} \qquad (1)$$

with $\theta_i = (\gamma_T e_0 \Delta t_i)/\hbar$; $e_0 = (2P_a/S\rho v_s^3)^{1/2}$, where n is the density of states, V the volume sampled by the ultrasonic beam, ω the frequency, T the temperature, Δt_1, Δt_2 the duration of pulse 1 or 2, P_0 the power radiated by a single

[1] Permanent address: Department of Physics, Oakland University Rochester, MI 48063, USA

[2] Associated with the Centre National de la Recherche Scientifique

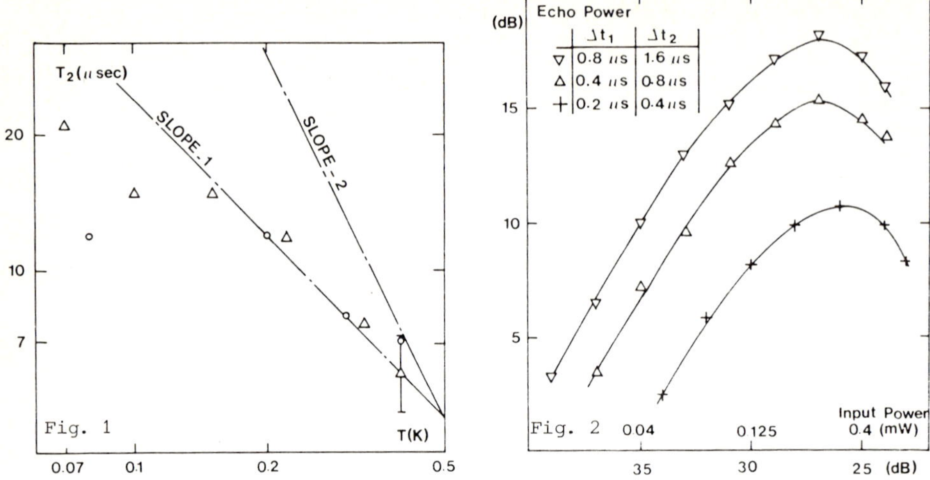

Figure 1. - T_2 pseudospin relaxation time in B_2O_3 - *0.5* Li_2O - *0.7* LiCl versus temperature. The data show a T^{-1} dependence. The open circles and open triangles indicate measurements on two different samples coming from the same boule

Figure 2. - Echo power in arbitrary units versus acoustic input power with different pulse widths. Note that Δt_2 is always twice Δt_1. There is almost no variation of the input power which gives the maximum echo power, in contrast to (1) which implies that for increasing Δt_1, this input power should decrease

pseudospin, P_a the acoustic power, S the area of the acoustic beam (1 mm^2), ρ the density (2.05 g/cm^3), γ_T the transverse deformation potential, v_s the sound velocity (3.75 km/sec) and τ the time delay between the two pulses.

We determined T_2 by measuring the decrease in echo amplitude versus the time delay at different temperatures. The results are shown in Fig. 1.

We verified that for low constant acoustic input power the echo power changes as indicated by the small signal approximation of the above law, i.e.

$$P_\ell \simeq [\Delta t_1^2 \Delta t_2^4] / [\mathrm{Max}(\Delta t_1, \Delta t_2)]^2. \tag{2}$$

$\mathrm{Max}(\Delta t_1, \Delta t_2)$ means use the value of the largest of Δt_1 or Δt_2.

The variation of the echo power versus the input power for different pulse widths, but always keeping $\Delta t_2 = 2\,\Delta t_1$, is shown in Fig. 2 for three different widths. We will comment on them in the discussion part.

Finally we tried to observe stimulated echoes with a sequence of three pulses. Seven different types of sequences can be looked for. We observed echoes in three of them. This shows that T_1 is of the order of T_2.

3 DISCUSSION

From these echo experiments we find values for T_2 and γ_T (see below) which differ significantly from those found in the previous acoustic measurements [1]. This is puzzling since the echo experiments of GRAEBNER and GOLDING (GG) [2] gave results consistent with previous acoustic measurements.

We note, however, that there are several differencies between our echo experiments and those of GG. First of all, the temperature range of our experiments is about ten times higher. Secondly, the magnitude of our pseudospin echo was always much lower (- 33 dB) than our first ultrasonic echo. Third, our echoes were observed in a saturated acoustic regime (strong excitation) when GG found the maximum of the echo appeared for an excitation of the order of the critical flux (non-saturated regime). Fourth, the T_2 time measured here is much longer (100 ×) than that deduced from desaturation experiments [1]. Fifth, from the maximum of the curve P_ℓ versus input power, we deduce a deformation potential $\gamma_T \simeq 10^{-3}$ eV which is a thousand times smaller than our measured value [1].

However these echoes must be due to localized excitations since in our opinion it is not possible to explain the observation of a forward echo, which obeys (2), without them. Thus, we think that the echoes we observed are due to a set of pseudospins which are weakly coupled to the phonon system and do not contribute significantly in a normal acoustic experiment. This would explain why : *(i)* we observed echoes at high temperatures ; *(ii)* we need strong acoustic power to excite them ; *(iii)* the echo is weak ; and *(iv)* that the T_2 measured in our experiment is so much different from the T_2 measured by desaturation experiments.

The fact that the input power needed to obtain the maximum echo amplitude does not change when we change the width of the excitation pulse (Fig. 2) can probably be explained by a saturation effect due to the high power involved. It follows from (1) and the results in Fig. 2 that the γ_T factor deduced from this experiment must be considered as a maximum value of the deformation potential. More experiments are in progress to determine the nature of these new pseudospins.

REFERENCES

1. M. Devaud, J.-Y. Prieur and W.D. Wallace : 4th International Conference on Solid State Ionics, Grenoble, France, July 1983 ;
   M. Devaud and J.-Y. Prieur : J. Physique Coll. 43, C9 - 497 (1982)
2. J.E. Graebner and B. Golding : Phys. Rev. B 19, 964 (1979)
3. M. Saint-Paul and J. Joffrin : J. Low Temp. Phys. 49, 195 (1982)

The Enhancement of Phonon Echo Generation by Defects in Crystals

D.J. Meredith, H. Mkhwanazi, and J.K. Wigmore

Physics Department, University of Lancaster
Lancaster LA1 4YB, England

T. Miyasato

Institute of Scientific and Industrial Research, University of Osaka
Suita, Osaka 565, Japan

The amplitudes of backward wave phonon echoes generated in single crystals of piezoelectric and pyroelectric materials are notoriously unreproducible. In a very few experiments it has been possible to correlate the observed anomalies with the presence of clearly identifiable defects. In CdS, for example, a variety of effects are observed which it is believed are due to the presence of shallow donors together with deep double acceptors [1]. The non-linearity responsible for the parametric coupling between the primary acoustic wave and the electric field of the microwave pump is an electric field-induced tunnelling of electrons from donor states into the conduction band. The tunnelling probability is proportional to $\exp\{-1/|E|\}$, where E is the total electric field at the donor, so that the echo amplitude is a very rapid function of the pump power, P_M. A second defect mechanism for backward wave echo generation arises through the saturated absorption of two-level systems. This mechanism has been identified as being responsible for backward wave echo generation by OH^- defects in glasses [2], and by indium acceptors in silicon [3]. The exact power dependence on P_M depends on the detailed coupling parameters of the particular transition. Clearly, however, it must be more complicated than a simple linear or quadratic dependence, since the echo power, P_E, must tend to zero both at low powers, where the transition is not saturated at all, and at high powers where the populations are equalised.

We focus attention on the dependence of P_E on P_M because in phonon echo experiments, the signature of an intrinsic mechanism of echo formation, that is, a property of the pure crystal, has invariably been taken to be a simple power dependence, either $P_E \propto P_M$ for ω-2ω experiments, or $P_E \propto P_M^2$ for ω-ω echoes. Conversely, a more complicated power dependence has been taken as evidence for the presence of defects of some sort or another. In contradiction to this idea, we wish to propose a further mechanism by which defects may enhance the generation of backward wave echoes. In contrast to the two mechanisms outlined above, it does not require the defects to possess internal structure, and as a consequence the pump power dependence remains the same linear relationship as that of the intrinsic process. Expressed most simply, we believe that the intense lattice distortion that surrounds many defects can modify the echo generation process within the framework of the intrinsic lattice mechanism. The relevant terms in the Taylor expansion of the crystal free energy V in S and E are [4]

$$V = \ldots.\ eSE + \beta S^2E + \eta SE + \gamma S^2E^2 + \chi S^3E + \ldots.$$

Considering as an example the generation of ω-2ω echoes, it is easy to show that the echo amplitude is proportional to $\beta + 2\eta e/d$. If the strain S is written explicitly as the sum of S_A, the acoustic strain, and S_0, a static strain, then the coefficients β, e, and η should be replaced by $\beta + \chi S_0$, $e + \beta S_0$, and $\eta + \gamma S_0$. Since generation still takes place through the S^2E term, the power dependence remains $P_E \propto P_M$. Previous results by GAGNEPAIN and BESSON [5] on electrostriction, and by GRAHAM [6] on non-linear piezoelectricity enable us to estimate values for β and η of 2.6 C m^{-2} and 5×10^{-2} F m^{-1}. We take figures of 0.20 C m^{-2} and 4×10^{-11} F m^{-1} for e and d. Substituting these values into the expressions above shows that the strain-dependent contribution is equal to the truly intrinsic term for $S_0 \sim 8 \times 10^{-2}$. This seems a very large strain, but Gagnepain and Besson did actually observe considerable variation in electrostriction between samples containing different concentrations of visible defects.

We report observations of phonon echoes in single crystals of alpha quartz that provide plausible evidence for such a mechanism. In addition, some rather more circumstantial evidence is reported for $Bi_{12}GeO_{20}$. What we found was that the magnitude of ω-2ω backward waves in quartz could be varied as desired by quenching or annealing the samples. The samples were polished cylindrical rods, both X and AC cut and we obtained similar results for both longitudinal and transverse modes. Experiments, with $\omega/2\pi$ = 17.27 GHz, were carried out at temperatures between 1 and 4 K. Quenching a crystal in liquid nitrogen could change the subsequently observed P_E by more than 10 dB, whilst annealing for 24 hours could reduce it by the same amount. That such treatment of quartz has the effect of creating or removing structural defects is well established, for example, by thermoluminescence. Confirmation in our own experiments was obtained by observing concurrently the reflection echo pattern. We found that quenching drastically lowered the number of reflection echoes, suggesting that the amount of strain in the samples had increased. Furthermore, the decay time of the echo T_2 also became shorter. Only irreversible scattering processes can contribute to T_2, in contrast to straightforward ultrasonic attenuation measurements, so that this observation indicates an actual increase in the number of defects in the sample.

Table I

Crystal	Run	Echo dB	T_2 10^{-6} sec	Number of Reflection Echoes	Treatment
AC 2	18/5	24	54	15	as received
	19/5	26	non-exp.	10	quenched
	24/5	3	100	> 30	annealed
	25/5	11	54	10	quenched

Table I summarises the results obtained on one of the crystals; similar data were obtained on other samples. Initially, and after each treatment the dependence of P_E on P_M was measured. On every run, a result close to $P_E \propto P_M$ was observed. The quality of a straightline fit is illustrated in Fig.1 showing typical data. Thus, although the echo amplitude depended on defect content, the power dependence implied an interaction term of the form S^2E.

Similar indications were also found in $Bi_{12}GeO_{20}$. The backward wave echoes this time were of the ω-ω variety with $\omega/2\pi$ = 17.27 GHz. For this type of echo the intrinsic term is γS^2E^2, so that the expected intrinsic power dependence would be $P_E \propto P_M^2$. Firstly, no echo at all was observed in pure $Bi_{12}GeO_{20}$, or in several lightly doped samples. However, in a crystal con-

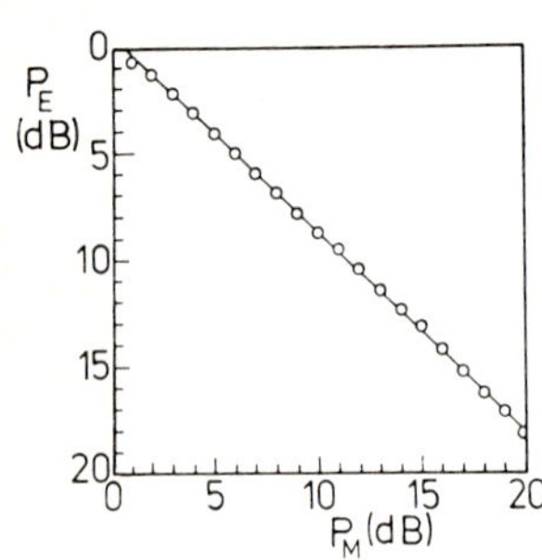

Figure 1: P_M power dependence for ω-2ω echoes in quartz

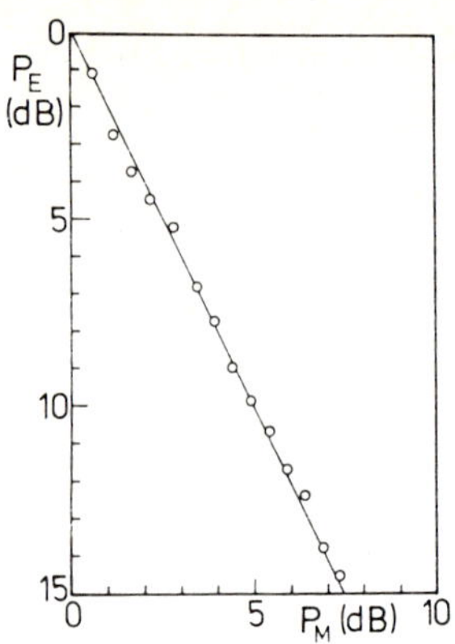

Figure 2: P_M power dependence for ω-ω echoes in $Bi_{12}GeO_{20}$

taining 5% Cr a very large echo was observed, with a power dependence close to the expected intrinsic slope (see Fig.2). The average of several runs gave $P_E \propto P_M^{2\cdot0\pm\cdot05}$. In pursuing this idea, it will clearly be important to characterise the defects as carefully as possible. From evidence concerning T_2, we believe that in quartz the active centre is the same structural defect created by other workers using particle irradiation, and associated with a two-level system [7]. The defect in $Bi_{12}GeO_{20}$ has also been studied using ultrasonic attenuation [8] and tentatively identified as an excess Bi atom substituting for a Ge atom on a trigonal site.

We are grateful to the Science and Engineering Research Council for supporting this work.

References

1. N.S.Shiren and R.L.Melcher: J.Elect.Materials 4 1143 (1975)
2. N.S.Shiren, W.Arnold and T.G.Kazyaka: Phys.Rev.Letters 39 239 (1977)
3. R.L.Melcher: Phys.Rev.Letters 43 939 (1979)
4. D.J.Meredith, J.A.Pritchard and J.K.Wigmore: Phys.Letts. 86A 376 (1981)
5. J.J.Gagnepain and R.Besson: Physical Acoustics vol.11 (Academic Press, New York, 1975) p.245
6. R.A.Graham: J.Appl.Phys.48 2153 (1977)
7. C.Laermans: Phys.Rev.Letters 42 250 (1979)
8. P.K.Grewal and M.J.Lea; P.Phys.C (Solid State) 16 247 (1982)

On Theory of Echo Phenomena in Amorphous Materials

V.S. Kuz'min and A.P. Sayko

Institute of Solid State Physics, Byelorussian Academy of Sciences,
P. Browki 17, SU-Minsk, USSR

It is known that most of the low-temperature properties of glasses can be explained in terms of localized two-level defects strongly coupled with long-wavelength strain fields (phonons). These two-level systems (TLS) obey the same escription as that of the spin-$\frac{1}{2}$ particles. Amongst the various methods of probing such defects, the technique of spontaneous and stimulated echoes has proved to be most efficient.

The purpose of this note is to describe spontaneous (two-pulse) electric echo decay in glasses without using spectral diffusion ideas.

A two-level defect strongly interacted with quantum phonon field and classical electric field $E(t)=E_0 \cos \omega t$ is described by the model Hamiltonian

$$H = \varepsilon S^z + \sum_K \hbar \omega_K a_K^+ a_K + \sum_K V_K (D S^z + M S^x)(a_K - a_K^+) - (\tilde{\mu} S^z + \mu S^x) E(t) \quad (1)$$

where the first and the second terms are free Hamiltonians of the defect and Debye phonons, respectively, the third and the fourth terms represent the diagonal (S^z) and off-diagonal (S^x) interactions of the TLS with phonon and electric fields. The quantities D and M are, respectively, the diagonal and off-diagonal deformation potentials, ε is the energy splitting in a TLS , a_K is the phonon destruction operator, $\omega_K = CK$, K and C are the wave vector and the average velocity of a long-wavelength phonon , $V_K = i(\hbar K/2\rho C\Omega)^{1/2}$, ρ and Ω are the density and the volume of glass, μ and $\tilde{\mu}$ are, respectively, the induced and static electric dipole moments S's are the Pauli spin operators.

In an echo experiment not all excited systems will contribute to the signal since various processes may cause the phases of some systems to have become randomized. This may occur from pure phase-disrupting processes (due to the diagonal interaction of the elastic dipole moment of TLS with phonons) with characteristic time T_ϕ , or by longitudinal relaxation with time T_1 . The measurement of the echo decay as a function of time delay τ between exciting pulses yields T_2' , the homogeneous relaxation time, where $T_2' = T_\phi^{-1} + \frac{1}{2} T_1^{-1}$. Since in glasses $T_1 \gg T_2'$ [1] , i.e. $T_2' \approx T_\phi$ and $\tau \ll T_1$, $\Delta t \ll T_2', T_1$ (Δt is the duration of the exciting pulse) then in (1) it can be assumed that $M = 0$, while the diagonal interaction (T_ϕ processes) will be taken into account only after switching off the exciting pulses.

In order to eliminate the term $-\tilde{\mu} S^z E(t)$ in (1), we perform the following unitary transformation:

$$\tilde{H} = U^+ H U,$$

$$U = \exp\{ i S^z [(\tilde{\mu} E_0/\hbar\omega) \sin\omega t + (m+1)\omega t]\}.$$

As a result of this transformation, taking into account the abovementioned remarks, the Hamiltonian (1) during exciting pulse action is written as follows:

$$\tilde{H}' = (\Delta\varepsilon - m\hbar\omega) S^z + (\mu E_0/2) J_{-m}(\tilde{\mu} E_0/\hbar\omega) S^x$$

and after switching off the pulses as:

$$\tilde{H}'' = \varepsilon S^z + \sum_K \hbar\omega_K a_K^+ a_K + \sum_K V_K D S^z (a_K - a_K^+).$$

Here we have assumed that $(\mu E_0/\hbar\omega) J_n(\tilde{\mu} E_0/\hbar\omega) \ll 1$ and that $\Delta\varepsilon - m\hbar\omega$ $(m = 1,2,3,\ldots)$ is near to zero, where $J_n(x)$ is Bessel function of order n and $\Delta\varepsilon \equiv \varepsilon - \hbar\omega$.

Using the technique of the evolution operator [2] and assuming that the defect is initially in the ground state and that phonon system is in the thermal equilibrium, we obtain the following expression for the resulting echo amplitude in the case of a two-pulse sequence

$$A(2\tau) \sim \sin\theta_1 \sin^2(\theta_2/2) \exp\{-\alpha[4 I(\vartheta, x) - I(2\vartheta, x)]\} \qquad (2)$$

where $\theta_{1,2} = (E_0 \mu/2\hbar) J_{-m}(\tilde{\mu} E_0/\hbar\omega) \Delta t_{1,2}$, $\alpha \equiv 3D^2 \varepsilon_m^2/4\pi^2\hbar^3 \rho c^5$, ε_m is the extreme energy for Debye phonons,

$$I(\vartheta, x) = \int_0^1 dw\, w \coth(wx)[1 - \cos(w\vartheta)], \qquad (3)$$

$x = \varepsilon_m/2k_B T$, $\vartheta = \varepsilon_m \tau/\hbar$, $w = \hbar\omega_K/\varepsilon_m$, T is temperature. It is clearly seen from (2) that the dependence of the echo intensity on the pulse separation τ shows nonexponential decay and that the rate of decay depends on temperature. It is easy to find the approximate expression for the echo decay from (2) and (3) when $w\vartheta \ll 1$:

$$A(2\tau) \sim \exp\{-(\alpha\varepsilon_m^4/2\hbar^4) I(\pi/2w, x)\tau^4\}.$$

References

1. J.L. Black and B.I. Halperin: Phys. Rev. B16, 2879 (1977)
2. M. Aihara: Phys. Rev. B21, 2051 (1980)

Discussions

ROTARY PHONON ECHOES IN SILICA GLASS page 446
B. Golding, D.L. Fox, W.H. Haemmerle

M. v.Schickfus: 1.) How do you determine the amplitude of the echo signal in view of the rapid change of phase and amplitude shown on your figure? 2.) Can you give a value for the matrix element derived from the Rabi frequency?

B. Golding: 1.) The echo amplitude is determined by simply subtracting background from a smoothed positive echo maximum. 2.) The matrix element has been obtained by comparison with two-pulse free-induction decay echoes as described in the text of the conference proceedings. The agreement is within 20% to 30%,i. e. about 1.6 to 2.0 eV.

W. K. Arnold: Are the positive part and the negative part of the echo of the same size?

B. Golding: In situations in which the in- and out-of-phase components are slightly different in amplitude or shape, we believe that attenuation of the propagation signal plays some role. This could be examined in more detail or possibly eliminated by studying systems with low attenuation.

D. R. Taylor: Would it be possible, with this technique, to learn anything about the distribution of the nutation matrix elements?

B. Golding: The nutation echo is dominated by the strongly coupled tunnelling level since the interaction is resonant. Very little information is thus available on the matrix element distribution function.

PSEUDOSPIN ECHOES IN BORATE GLASSES page 449
M. Devaud, J.-Y. Prieur, W. D. Wallace

J. K. Wigmore: How can you estimate the coupling constant γ if the maximum of the power dependence does not depend on pulse width?

J.-Y. Prieur: The fact that the maximum of the amplitude of the echo versus the input acoustic power does not depend on the width of the echo can be explained by invoking some nonlinearity. Therefore, the true maximum of the sine function should occur at a higher power and then the γ factor, determined from the present maximum, must be considered as a maximum value for γ. This reinforces the main result of this experiment which is that γ (from the echo) is much lower than the γ factor found from a normal acoustic experiment.

W. K. Arnold: How large is the deformation potential determined from saturation threshold in comparison with the value of 10^{-3} eV?

J.-Y. Prieur: The γ value from normal acoustic experiment is for this glass 0.5 eV. This discrepancy between the two values makes me think that I must have two different kinds of two-level systems. One kind is strongly coupled and gives rise to the ultrasonic attenuation. The second kind which is weakly coupled does not induce any measurable attenuation but can be seen in the present echo experiment.

S. Hunklinger: What is the physical process determining T2?

J.-Y- Prieur: According to a paper by Joffrin and Levelut, the T_2 time is determined by the spin-spin interaction. The fact that sometimes, as Golding just showed, the transverse relaxation time (T_2) changes with the temperature as T^{-2} and sometimes, as I just showed, it changes as T^{-1}, is not explained. However, there might exist two different temperature dependences, one for the short time decay and one for the long time decay.

THE ENHANCEMENT OF PHONON ECHO GENERATION BY DEFECTS IN CRYSTALS page 452
D. J. Meredith, H. Mkhwanazi, T. Miyasato, J. K. Wigmore

W. K. Arnold: 1.) The phonon echo should depend on the external stress? Is that true? 2.) It is relatively easy to measure internal stress by the change of sound velocity due to the higher-order elastic constant.

D. J. Meredith: The echo amplitude should indeed depend on the size of an applied stress. We have applied a compressional stress to one of the samples but the experiment was inconclusive. It is possible that we cannot generate a sufficiently large stress to reproduce the intense local strain associated with the defects.
Yes, sound velocity measurements should be made.

S. Hunklinger: Did you look for the internal strain optically?

D. J. Meredith: No, we have not done so. In extreme cases the samples were sufficiently damaged for internal cracks to be visible.

Part XI

Spin-Phonon Interaction

Chairman: **V. W. Rampton**

Evidence for Phonon Scattering by Magnetic Two-Level Systems in Crystalline Spin-Glass $Eu_xSr_{1-x}S$

C. Arzoumanian, B. Salce, and A.M. de Goër

Centre d'Etudes Nucléaires de Grenoble, SBT/LCP-85X
F-38041 Grenoble Cedex, France

F. Holtzberg

IBM, P.O. Box 218, Yorktown Heights
Yorktown, NY 10598, USA

During the last years, different experiments have given evidence for the spin-glass properties of the insulating system $Eu_xSr_{1-x}S$ and the phase diagram is well known [1]. Recently, we have measured the low-temperature thermal conductivity K(T) of a x = 0.44 single crystal, and studied the effect of a magnetic field up to 7 T in the temperature range below 1 K [2]. For this concentration, spin-glass behaviour is observed below the low-frequency freezing temperature $T_g^\circ = 1.8$ K [3]. From this study, we have suggested that the thermal phonons are resonantly scattered by low-lying magnetic excitations or "magnetic TLS" by analogy with the properties of structural glasses [2]. We report here additional magneto thermal conductivity measurements carried out to test this hypothesis.

1. EXPERIMENTS and RESULTS

Sample 1 is the same crystal (x = 0.44) as used before [2]. The concentration of sample 2 (x = 0.017) is well outside the range where spin-glass behaviour is observed [1]. Samples with similar concentration display only superparamagnetic behaviour above $T_p = 100$ mK [4]. The experimental results (H = 0 and H = 7 T) are plotted in fig.1. An important result is the fact that there is no detectable magnetic field effect (that is less than ≈ 5 %) up to 7 T on the thermal conductivity of sample 2. In the case of sample 1, the effect of H = 7T is present up to about 15 K, well above T_g°. Measurements made at intermediate fields have shown that the magnetic field effect becomes undetectable at a temperature which decreases as H decreases (about 12 K at 5 T, 9 K at 1 T...). The field dependence of K at constant temperature is illustrated in fig.2. For T < 1.2 K (fig. 2a) the thermal conductivity has been directly measured as a function of H. For T > 2 K (fig.2b) the values of K have been obtained from the set of K(T) curves measured at constant fields. It is seen that the saturation of the effect, which is observed below 1 K, is not achieved at 7 T above 2 K. Magnetization measurements carried out on the same sample between 1.7 and 4.2K seem to indicate that saturation is achieved at lower fields. We note that very recently, experiments have shown similar behaviour in a x = 0.54 sample (a "frustated ferromagnet") [5].

2. DISCUSSION

It is clear that the phonon scattering which is magnetic-field dependent is actually related to the spin-glass disorder, as it is completely absent in the lightly doped paramagnetic sample. The results below 1 K have been previously analyzed with the simple relaxation time $\tau^{-1} = v/L + G\omega$ where the second term describes the phonon scattering by "magnetic" TLS [2]. We have now extended the quantitative analysis to the whole temperature range, by including the phonon scattering from point defects and phonon-phonon interactions, so that

$$\tau^{-1} = v/L + G\omega + A\omega^4 + B\omega^2T^2 .$$

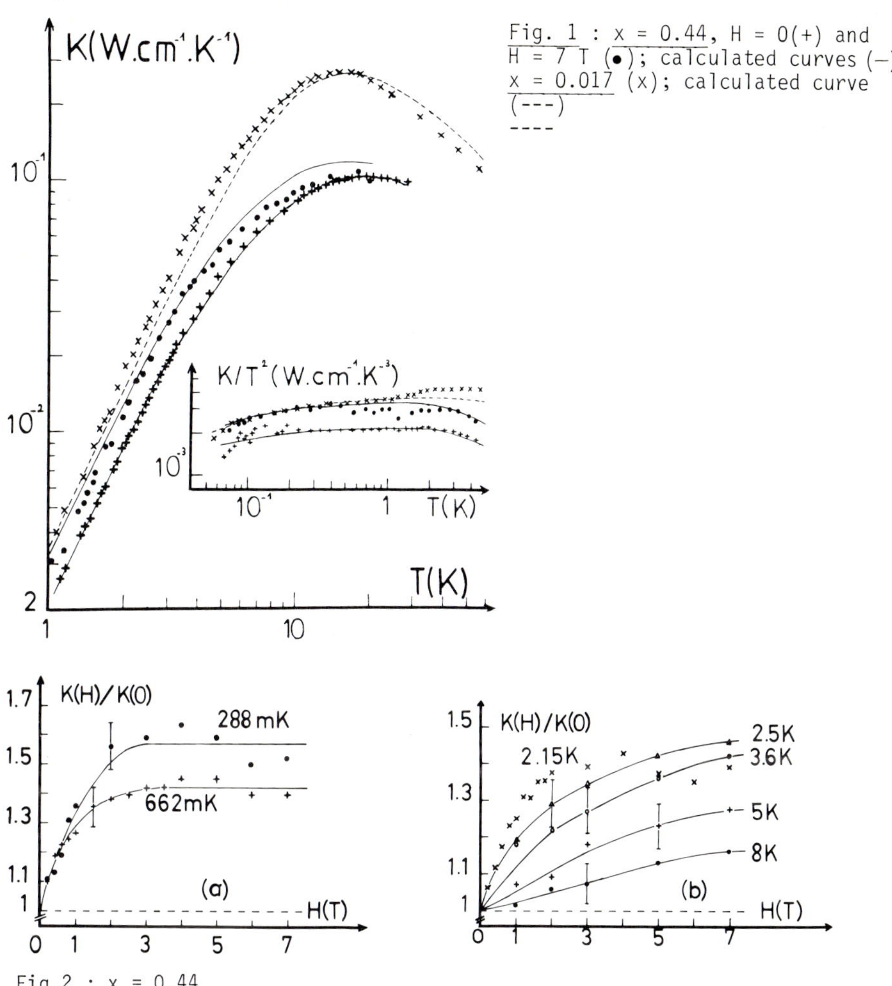

Fig. 1 : x = 0.44, H = 0(+) and H = 7 T (•); calculated curves (—) x = 0.017 (x); calculated curve (---)

Fig.2 : x = 0.44

Relatively satisfactory fits have been obtained with the parameters given in table 1 (and B = 1.3 10^{19} s K^{-2}) and are illustrated in figure 1

Table 1

sample	x	section (mm^2)	length (mm)	(v/L) calc. (s^{-1})	(v/L) exp. (s^{-1})	G	$A_{calc.}$ (s^3)	$A_{exp.}$ (s^3)
1	0.44(H=0) 0.44(H=7T)	1	5.5	2.9 10^6	1.1 10^6	13.4 10^{-5} 8.6 10^{-5}	9.10^{-41}	3.10^{-42}
2	0.017	2.5	4.8	1.6 10^6	1.5 10^6	8. 10^{-5}	3.5 10^{-42}	4.10^{-44}

For sample 1, only the G coefficient is changed between H = 0 and H = 7 T, and the difference $G_0 = G_H = 4.8\ 10^{-5}$ is the same as determined before [2]. The fit is very satisfactory for the H = 0 curve, but for H = 7 T and T > 5 K, some phonon scattering is not taken into account by the calculation (cf. fig 1). This is not too surprising as the saturation is not achieved at these temperatures (cf. fig 2b), so that some magnetic scattering is still present. Therefore it appears that the simplified model of a constant density of states n_0 of the TLS used so far is not appropriate, and that not only n_0 but also the high energy cut-off (and perhaps the shape of the distribution) could be magnetic-field dependent. The G value for sample 2 is very comparable to that obtained for the H = 7 T curve of sample 1. Therefore this phonon scattering is probably due to imperfections of the crystal, though the corresponding density of dislocations would be very high, of the order of 10"/cm^2. In fact a preliminary X-ray experiment performed on sample 2 has shown that it is not a good single crystal. The fact that the experimental values of A are smaller than the calculated ones could indicate again that the contribution $G\omega$ in τ^{-1} is not valid in the whole frequency range, though other sets of experimental parameters would perhaps give fits of equivalent quality.

3. CONCLUSION

The presence of phonon scattering by low-lying magnetic excitations has been confirmed in the spin-glass $Eu_{0.44}Sr_{0.56}S$. The simple description of this scattering by a constant density of states of "magnetic TLS" which decreases with increasing magnetic field is too crude to explain quantitatively the results in the whole temperature range. A possible contribution to K(T) from "spin waves" seems unlikely as such a contribution must decrease with increased field. Specific heat measurements under magnetic field are planned in the near future as only results at $H \lesssim 1$ T and $T \lesssim 4$ K on a x = 0.4 sample are available [6].

1 H.MALETTA, W.FELSCH : Z.Phys. B37 (1980) 55
2 C.ARZOUMANIAN, A.M.DE GOER, B.SALCE, F.HOLTZBERG : J.Phys. Lettres 44, (1983) L-39
3 F.HOLTZBERG, T.L.FRANCAVILLA, C.Y.HUANG, J.L.THOLENCE : J. Appl. Phys. 53, (3) (1982) 2229
4 F.HOLTZBERG, J.L.THOLENCE, R.TOURNIER : in Amorphous Magnetism II, eds. R.A. Levy, R.J. Hasegawa (1977) p.155
5 G.V.LECOMTE, H.V.LOHNEYSEN, M.ZINN : to be published
6 D.MESCHEDE, F.STEGLICH, W.FELSCH, H.MALETTA, W.ZINN : Phys. Rev. Lett. 44 (1980) 102

Phonon Spectroscopy of MnF_2/ZnF_2 Mixed Crystals

P.J. King, D.T. Murphy*, and V.W. Rampton

Department of Physics, University of Nottingham, University Park
Nottingham NG7 2RD, England

Superconducting tunnel junction phonon spectroscopy has shown resonant phonon scattering at magnon-phonon cross-overs in mixed crystals of MnF_2/ ZnF_2. A resonance at 440 GHz and another between 208 GHz and 185 GHz are unexplained.

Manganese fluoride is a uniaxial antiferromagnetic crystal. In such a material a strong magnon-phonon interaction is expected when the frequency and wavelength of magnons is equal to those of the phonons at the cross-over of the dispersion curves. Strong resonant phonon scattering has been reported by Mattes et al. [1] at this cross-over in pure manganese fluoride. Manganese fluoride forms mixed crystals with zinc fluoride, a non-magnetic material. The added zinc perturbs the spin-waves and moves the dispersion curves to lower frequencies as the interaction between the manganese ions is reduced. The shift in the magnon dispersion curves when zinc is added has been studied by Svensson et al. [2] using neutron scattering.

We have used superconducting tunnel junction phonon spectroscopy to examine a range of samples containing various concentrations of zinc in manganese fluoride. Pb/Bi tunnel junctions were used as phonon generators and Al tunnel junctions as detectors. For most experiments sine-wave modulation was used though some experiments employed pulse modulation when phonon modes could be distinguished. The experiments were made at 1.06K. When the phonon propagation direction was the c axis of the crystals up to four resonant phonon scattering lines were found. Figure 1 shows the result for the sample with 0.4% Zn. Two samples were also examined with phonons propagating along the a axis. In this case a very broad resonance stretching over a wide frequency range was obtained. The results are summarised in Table 1.

Two of the resonances from the c-axis samples and the broad resonance from the a-axis samples can be accounted for by magnon-phonon cross-overs but the lowest frequency and highest frequency resonances from the c-axis samples are so far unexplained.

We have used the work of Melcher [3] to calculate the phonon dispersion curves making the assumption that dilution with zinc does not much change them. This seems reasonable in view of the similarity between the ultrasonic velocities in pure MnF_2 and pure ZnF_2 at room temperature [4].

---

*Present address: VG Gas Analysis, Nat Lane, Winsford
Cheshire CW7 3QH, England

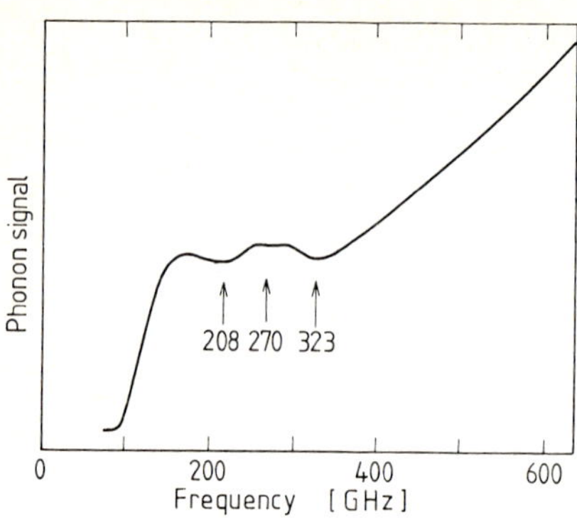

Fig. 1. Phonon signal from MnF_2 with 0.4% Zn

Table 1 Resonant phonon scattering from MnF_2/ZnF_2 with phonons along the c axis

Concentration of Zinc/At. %	Frequencies/GHz			
0.4	208	270	323	
3.0	195	261	298	429
8.0	191	243 to 278		440
10.0	180	236 to 272		
14.0	185	226		428

The magnon dispersion curves were calculated using the parameters of Okazaki et al. [5] and assuming that the perturbation due to zinc could be treated by the "average" model [2]. The spin S was replaced by S(1-C) and the dipole-dipole anisotropy field H_A replaced by $H_A\ (1-C)^2$ where C is the concentration of ZnF_2. The average model appears to work fairly well in the long-wavelength limit, [2], the region we are most concerned with. Figure 2 shows the predicted cross-overs as a function of Zn concentration for

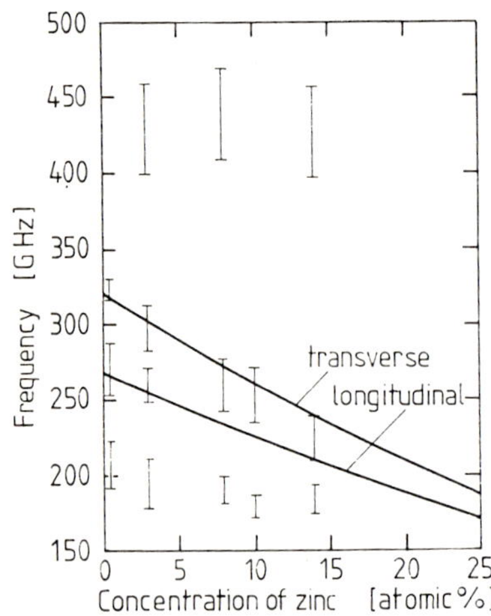

Fig. 2. Predicted magnon-phonon cross-over frequencies along c axis against concentration of Zn. Experimental points are also shown

transverse and longitudinal phonons along the c axis. The experimental resonances clearly fit the predicted cross-overs for low Zn concentration. At higher concentrations we observe a broad resonance probably including both. The cross-overs become closer at higher concentration and the magnon dispersion curves become much wider [2]. Longitudinal phonons are not expected to couple strongly with magnons in either the c or a directions [6]. The low-frequency resonances from 208 GHz to 185 GHz do not fit the cross-overs but appear to depend on the zinc concentration. Mattes et al. [1] observed a resonance in pure MnF_2 at 184 GHz and suggested it might be due to OH^- impurities which are known to cause a resonance in thermal conductivity. Our resonances are at a higher frequency and the dependence on zinc concentration should rule out OH^- impurities. The resonance at about 430 GHz only appears in some samples, it is very broad and weak and was only observed with sine-wave modulation. It may be due to phonons scattered from the side walls of the sample. In some directions the slow transverse phonon-magnon cross-over can occur at frequencies up to 430 GHz.

Phonon propagation along the a axis gives a very broad resonance. This can be explained as there are two transverse phonon modes in this direction and the slow transverse phonon dispersion curve crosses the magnon dispersion curve at a very shallow angle, while doping with zinc increases the frequency width of the magnon dispersion curve [2]. Thus, the resonance scattering would be expected to extend over a considerable range of frequencies and merges with the scattering due to the fast transverse phonons and also the longitudinal phonons.

References

[1]. J. Mattes, P. Berberich and H. Kinder : J. de phys. C6 suppl. 8, 39 988 (1978).

[2] E. C. Svensson, T. M. Holden, W.J.L. Buyers, R. A. Cowley and R.W.H. Stevenson : Solid St. Comm. 7, 1693 (1969).

[3] R. L. Melcher : Phys. Rev. B2, 733 (1970).

[4] D. S. Rimai : Phys. Rev. B16, 2200 (1977).

[5] A. Okazaki, K. C. Turberfield and R.W.H. Stevenson : Phys. Lett. 8, 9 (1964).

[6] F. Keffer : "Spin Waves" in Handbuch der Physik XVIII / 2 Ferromagnetismus (Springer Berlin 1966) pp 1 - 273.

Heat Transport by Phonons and Magnons in Ferromagnetic EuS

G.V. Lecomte and H. v. Löhneysen
2. Physikalisches Institut der RWTH, Templergraben 55
D-5100 Aachen, Fed. Rep. of Germany

W. Zinn
Institut für Festkörperforschung der KFA
D-5170 Jülich, Fed. Rep. of Germany

1 Introduction

The thermal conductivity κ of ferromagnetic substances in the vicinity of and below their Curie temperature T_c can be strongly affected by the spin waves present in the magnetic system. In particular, it has been shown that in some solids the magnons themselves can act as heat carriers, constituting an additional transport channel parallel to the phonons, with $\kappa = \kappa^m + \kappa^{ph}$. In the present study, we have measured κ in the Heisenberg ferromagnet EuS over a temperature range spanning 0.1 T_c to 2 T_c in various applied magnetic fields, in order to elucidate the role played by the magnons in the heat transport occurring in this solid.

EuS is a semiconductor, and has NaCl structure. The isotropic ferromagnetic interaction between nearest neighbors J_1 dominates the weaker antiferromagnetic coupling J_2 existing between next-nearest neighbors, with $J_1/J_2 = -2$ /1/. This leads to ferromagnetic order below $T_c = 16.5$ K. In earlier investigations /2,3/, the thermal conductivity of sintered powder samples has been measured at low temperatures, showing no particular features at T_c, a roughly quadratic variation of κ with T below 4 K, and a marked reduction of κ within an applied magnetic field. Because of the uncertainty linked to the use of powdered specimens, and of the limited range of these early experiments, we have measured the thermal conductivity of a single crystal in fields up to 6.5 T.

2 Experimental Procedure

The sample was a prismatic rod of approximate dimensions 6x1x1 mm^3, with its long axis oriented along the (100) crystal direction. This axis was also parallel to the applied field. Details concerning the cryostat and the calibration method have been given elsewhere /4/. Due to the uncertainty in the specimen geometry, the absolute accuracy was of the order of 15 %. The relative accuracy was however better than 0.5 %.

3 Results

The thermal conductivity as a function of T with and without a large applied field is plotted in double logarithmic representation in Fig. 1a. For B = 0, $\kappa(T)$ grows monotonically between 1.5 and 40 K, with the rounding off at the high-temperature end indicating presumably the existence of a maximum in the vicinity of 40 K. No particular feature can be observed at the Curie temperature, but a marked change in the slope occurs at 4.5 K. Below this point, κ varies approximately as $T^{1.6}$, while directly above it presents a variation roughly quadratic. While the value of κ at 4 K is some 20 % higher than that observed by McCOLLUM /2/, its variation with T below 4 K is less rapid in our crystal than in the pressed powder specimen.

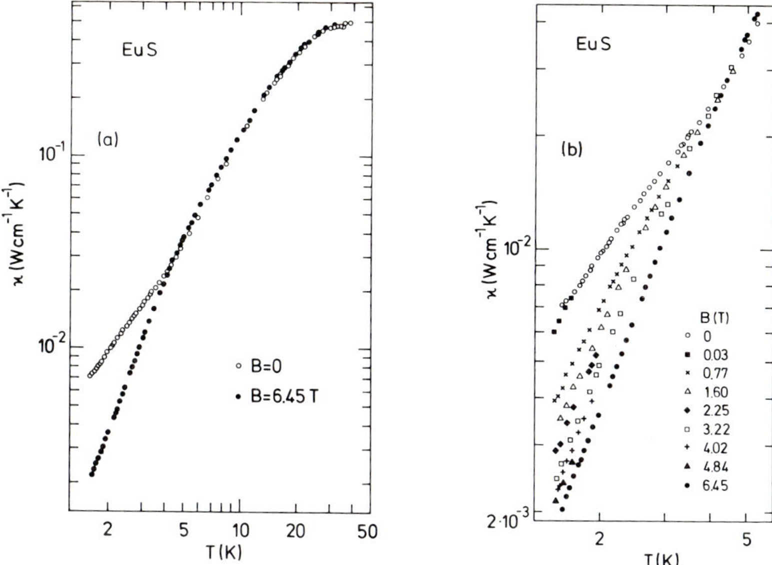

Fig. 1 (a) Thermal conductivity κ of EuS as a function of temperature T in a magnetic field B = 0 and B = 6.45 T. (b) Thermal conductivity at low temperature in various magnetic fields

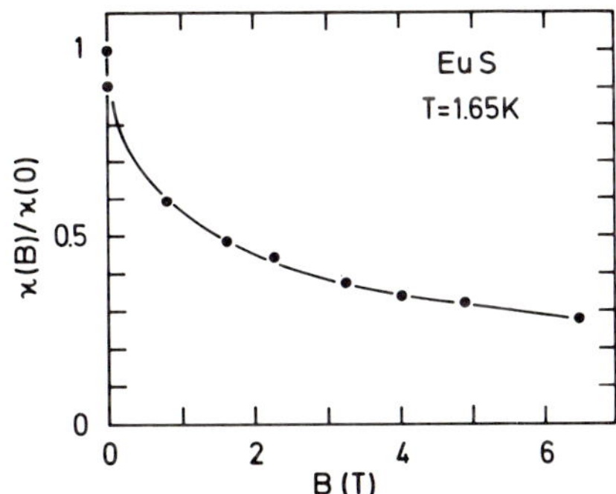

Fig. 2 Magnetic field dependence of the thermal conductivity $\kappa(B)/\kappa(0)$ for T = 1.65 K

In a large applied field (6.45 T), κ remains essentially unaffected above 10 K. Between 5 and 9 K, one can observe a small (5 %) but systematic increase. Below 4.5 K, the temperature dependence becomes markedly stronger, to such extent that by 1.6 K the corresponding reduction of κ reaches 73 %. Figure 1b shows a magnification of the low-temperature region, with intermediate field values. Finally, the field dependence of κ at the fixed temperature of 1.65 K is pictured in Fig. 2, showing that the field effect has not quite reached saturation at 6.45 T.

4 Discussion

The behavior of κ can be qualitatively understood with the hypothesis that the magnons contribute significantly to the heat transport below 4.5 K, as has been suggested earlier /2/. If both magnon and phonon mean free paths are only limited by boundary scattering, κ^{ph} should be proportional to T^3, while κ^m should vary as T^2 /5/. The weaker T dependence observed in B = 0 indicates that other scattering mechanisms are still active in this temperature range. These magnon-magnon or magnon-phonon processes become more important as T increases, and the faster κ^{ph} variation leads to a complete overcome of the magnon conductivity above 4.5 K. The slight increase in κ between 5 and 9 K in a high field could then be due to a reduction of the phonon-magnon scattering, i.e., an increase of κ^{ph}, because in a large magnetic field the magnon energies become so large that they can no longer be thermally excited.

On the other hand, a large magnetic field also suppresses κ^m, which is dominant below 3 K, by freezing the spin-wave excitations. At the complete saturation, the conductivity is reduced to κ^{ph}. In the present case, full saturation was not attained, so that the relative magnitude of κ^m and κ^{ph} could not be established. It is however interesting to observe that in B = 6.45 T, the low-temperature variation of κ is approximately $T^{2.6}$, and the measured κ curve lies appreciably below the expected limit of T^3 for boundary scattered phonons calculated from the specimen dimensions. This suggests additional phonon scattering by dislocations and other spatially extended defects. Detailed X-ray analysis of our sample revealed a few small-angle (< 1^0) grain boundaries. The observed difference between the temperature variation of κ in the single crystal and in the sintered sample can be explained the following way: the dislocations present in our sample limits the phonon conductivity to a value barely higher than in the sintered sample. These dislocations however do not affect strongly the magnon propagation, while these are scattered at the grain boundaries. κ^m will hence reach its T^2 behavior at higher temperatures in the sintered sample than in a single crystal specimen.

Further investigations of the field dependence at lower temperatures are presently in progress, to make possible a more quantitative analysis of this transport phenomenon.

References

1 H.G. Bohn, W. Zinn, B. Dorner, A. Kollmar: Phys. Rev. B 22, 5447 (1980)
2 D.C. McCollum, R.L. Wild, and J. Callaway: Phys. Rev. 136 A, 426 (1964)
3 W.D. Sewart, D.J. Taylor: J. Am. Ceram. Soc. 56, 599 (1973)
4 G.V. Lecomte, H.v.Löhneysen, and W. Zinn: to appear in J. Magn. Magn. Mat.
5 H. Sato: Progr. Theor. Phys. 13, 119 (1955)

Discussions

EVIDENCE FOR PHONON SCATTERING BY MAGNETIC TWO-LEVEL SYSTEMS IN CRYSTALLINE SPIN-GLASS $Eu_xSr_{1-x}S$ page 460
C. Arzoumanian, B. Salce, A.M. de Goër, F. Holtzberg

L.J. Challis: Could I ask if there is evidence yet that the magnetic defects are definitely two-level systems or could the data possibly fit a system with more than two levels ?

C. Arzoumanian: As we have no microscopic model for the magnetic TLS we can only say that the data could be explained taking a constant density of TLS.

T. Nakayama: Can you compare your estimated coupling strength between magnetic TLS and phonons with that between TLS in glasses and phonons? If your result is weaker compared with the case of TLS in glasses, can you interpret qualitatively why it is so weak?

C. Arzoumanian: The coupling strength between magnetic TLS and phonons is smaller than between TLS in structural glasses and phonons. This can be due to the ground state of the isolated Kramers ion Eu^{2+} which is an orbital singulet ground state $^8S_{7/2}$- so that its coupling is probably quite small.

G.V. Lecomte: One cannot dismiss so easily any possible conduction by induced magnons in moderate fields, since the field dependence of this contribution would be first to increase with field, as indeed observed. It would ultimately decrease for very large fields, but possibly not before 10 T or more.

C. Arzoumanian: We think that such a contribution is unlikely as we have no indication of a decrease of the thermal conductivity at 7 T, especially at the lowest temperature,but measurements with higher fields are planned to test this point.

HEAT TRANSPORT BY PHONONS AND MAGNONS IN FERROMAGNETIC EuS page 466
G.V. Lecomte, H. v.Löhneysen

V. Narayanamurti: Is the sample single domain? If not, sizes of the domains will affect the conductivity.

G.V. Lecomte: The sample cooled in zero field was probably not a single domain. We would indeed expect some magnon scattering on domain walls, depending on the relative orientation of walls and heat flux.

B. Salce: What is the size of the sample? Did you apply the field before to cool down the sample?

G.V. Lecomte: The sample was a 6x1x1 mm^3 rod, oriented along the 100 direction. The results were found independent of magnetic history, i.e. of whether the field was applied to a cooled sample or the sample cooled in a field.

Index of Contributors

Abhvani, A.S. 355
Ambrosy, A. 61,361
Anderson, A.C. 348
Armbruster, D. 75
Arzoumanian, C. 460
Ayache, C. 284

Baird, M.J. 206
Basso, H.C. 212
Bates, C.A. 355
Batey, G.J. 197
Baumgartner, R. 37
Beamish, J.R. 269
Berberich, P. 18,40
Bēthoux, O. 410
Bodensohn, J. 266
Bogner, U. 395
Bonfait, G. 241
Bray, R. 121,133
Bron, W.E. 34,88,106
Brown, I.J. 313
Brown, M.A. 313
Bury, P. 355

Castaing, B. 241
Cazzanelli, E. 437
Challis, L.J. 2,358,368
Chin, M.A. 91,249,328
Chopra, P.N. 182
Culbertson, J.C. 338

Dangelmayr, G. 75
de Goër, A.M. 361,460
de Groot, K. 316
de la Cruz, F. 434
de Oliveira, J.E. 422
de Wijn, H.W. 46,112
Devaud, M. 449
De Yoreo, J.J. 392
Dietsche, W. 40,212,419
Dijkhuis, J.I. 46,112, 118,130

Dittrich, E. 209
Dixon, G.S. 115
Dolindo, I. 136
Doussineau, P. 386,431
Dynes, R.C. 163

Eichele, R. 72
Eisenmenger, W. 209
Engelhardt, M. 124
Esquinazi, P. 434
Essabouri, A. 398
Esteves, V. 407
Every, A.G. 194

Farrell, D.E. 422
Feuillet, G. 404
Fieseler, M. 97
Fontana, A. 389,437
Fortier, D. 304
Fox, D.L. 446
Franck, J.P. 269
Fuchs, P. 185
Fujii, I. 257
Fukuhara, M. 257
Fulton, R.C. 133

Geiser, V. 316
Gmelin, E. 425
Golding, B. 446
Goossens, R.J.G. 46,112
Gornik, E. 331
Graham, J.T. 278
Grill, W. 55
Grimshaw, M. 404
Grondey, S. 413
Guillon, F. 200
Güntherodt, H.-J. 316, 425
Güttinger, W. 75

Haavasoja, T. 91,249
Haemmerle, W.H. 446
Halperin, B.I. 163
Happek, U. 37

Harrison, J.P. 200,203
Heil, J. 341
Heim, U. 209
Hensel, J.C. 163
Heraud, A.P. 358,368
Hiki, Y. 254,285
Hirschbiegel, L. 55
Höpfel, R.A. 331
Holdik, K. 61
Holtzberg, F. 460
Holzwarth, N. 103
Hope, F.R. 206
Huebener, R.P. 72
Hunklinger, S. 378
Hwang, J.C.M. 328

Ikushima, A.J. 257
Ipatova, I.P. 319
Ishiguro, T. 292
Iwasa, I. 260

Jandl, S. 284
Jansen, A.G.M. 46
Jay-Gerin, J.P. 284
Joffrin, J. 252
Jongerden, G.J. 130

Kaneko, K. 257
Kim, J.J. 288
Kinder, H. 18,40,212, 419
King, P.J. 355,463
Klein, P.B. 338
Klitsner, T. 188
Knaak, W. 416
Koblinger, O. 209
Kogure, Y. 254
Koo, J.Y. 288
Koos, G.L. 194
Kouroudis, I. 341
Kuhl, J. 106
Kuz'min, V.S. 455

Laermans, C. 407
Lakhani, A.A. 284
Lasjaunias, J.C. 410
Laßmann, K. 361
Lax, M. 103,133
Lecomte, G.V. 466
Leiderer, P. 212,266
Leotin, J. 310
Levelut, A. 386,431
Levinson, Y.B. 88
Lingner, C. 341
Löhneysen, H.v. 413,466
Lüthi, B. 341
Lurie, F.M. 34

Magerl, A. 401
Maier, J. 364
Main, P.C. 197
Maneval, J.P. 43
Mariotto, G. 389,437
Maris, H.J. 234
Matecki, M. 386
Maurer, D. 316
Mazzacurati, V. 437
Meissner, G. 263
Meissner, M. 392,416
Meltzer, R.S. 115
Meredith, D.J. 452
Metzger, W. 72
Miyasato, T. 302,307, 452
Mkhwanazi, H. 452
Montagna, M. 389
Müller, V. 316,322
Murphy, D.T. 463
Mutscheller, W. 298

Nakayama, T. 155
Narayanamurti, V. 91, 103,133,249,328
Nava, R. 252
Neumann, W. 322
Northrop, G.A. 100,191, 194

O'Connor, J.M. 88
Odoni, W. 185
Okubo, K. 109
Ott, H.R. 185
Overton Jr., W.C. 325

Page, J.H. 278,281
Pannetier, B. 43
Parshin, A.Ya. 226
Pelous, J. 398,431
Perrin, B. 371
Perrin, N. 52
Pohl, R.O. 182,188,392
Pooler, D.R. 355
Prieur, J.-Y. 449
Puech, L. 241

Rampton, V.W. 355,358, 463
Ravex, A. 410
Rehwald, W. 295
Reichert, U. 398
Renk, K.F. 10,37,124
Rhee, B.K. 106
Rives, J.E. 115
Röska, G. 395
Rosenberg, H.M. 422
Rothenfusser, M. 419
Rugar, D. 26
Ruocco, G. 437
Rutherford, A.R. 203

Sachrajda, A. 200
Saint-Paul, M. 252
Saker, M.K. 358
Sakurai, H. 302
Salce, B. 58,361,460
Samwer, K. 413
Savignac, D. 266
Sayko, A.P. 455
Schaich, W.L. 34
Scheitler, W. 61
Schick, A. 40
Schickfus, M.v. 425
Schink, H.J. 413
Schmidt, M. 398,431
Schulze, H. 61
Seifert, H. 72
Shchukin, V.A. 319
Sheard, F.W. 81
Siemon, M. 55
Sigmund, E. 364
Signorelli, G. 437
Sildos, I. 136
Singh, M. 310
Slater, G.W. 49
Smith, S.R.P. 281
Sota, T. 304
Sox, D.J. 115
Srivastava, V.P. 94
Stock, B. 97
Stormer, H.L. 328
Stott, M.J. 203
Strom, U. 338
Subashiev, A.V. 319
Susman, S. 392
Suzuki, H. 260
Suzuki, K. 302,304,307

Taborek, P. 148
Tamura, J. 285
Tamura, S. 78,109
Thomas, N. 428
Tietje, H. 425
Tokumoto, H. 292
Tokumura, M. 307
Toombs, G.A. 81
Tremblay, A.-M.S. 49
Tsen, K.T. 121,133
Tsui, D.C. 163
Tsuruoka, F. 254

Ulbrich, R.G. 97,103
Ullersma, P. 127
Unoki, H. 292
Unterhorst, E.-J. 322

Vacher, R. 398,431
van Miltenburg, J.G.M. 118,130
Vandersande, J.W. 182
Vanneste, C. 52
Vass, E. 331

Wagner, M. 298
Wallace, W.D. 386,449
Wan, K. 121,133
Wang, J.L. 401
Weis, O. 179
Weiss, G. 401
Welte, M. 209
Wigmore, J.K. 452
Williams, F.I.B. 171
Williams, S.R. 81
Wilson, T.E. 34
Wipf, H. 401
Wiscombe, P.C. 355
Wolf, S.A. 338
Wolfe, J.P. 100,194
Wu, C.-C. 335
Wyatt, A.F.G. 206
Wybourne, M.N. 358
Wyder, P. 46

Yoo, T.W. 288

Zavt, G. 136
Zeile, H. 361
Zinn, W. 466